# Artificial Intelligence and Machine Learning Techniques for Civil Engineering


Vagelis Plevris
*Qatar University, Qatar*

Afaq Ahmad
*University of Engineering and Technology, Taxila, Pakistan*

Nikos D. Lagaros
*National Technical University of Athens, Greece*



A volume in the Advances in Civil and Industrial Engineering (ACIE) Book Series

IGI Global
PUBLISHER of TIMELY KNOWLEDGE

Published in the United States of America by
IGI Global
Engineering Science Reference (an imprint of IGI Global)
701 E. Chocolate Avenue
Hershey PA, USA 17033
Tel: 717-533-8845
Fax: 717-533-8661
E-mail: cust@igi-global.com
Web site: http://www.igi-global.com



Library of Congress Cataloging-in-Publication Data

Names: Plevris, Vagelis, 1976- editor.
Title: Artificial Intelligence and machine learning techniques for civil engineering / Vagelis Plevris, Afaq Ahmad, and Nikos Lagaros, editors.
Description: Hershey, PA : Engineering Science Reference, [2023] | Includes bibliographical references and index. | Summary: "This reference book offers state-of-the-art contributions in the area of AI and its applications in the field of civil engineering presenting methods and implementation of AI and machine learning in multiple facets of civil engineering"-- Provided by publisher.
Identifiers: LCCN 2022033195 | ISBN 9781668456439 (h/c) | ISBN 9781668456446 (eISBN)
Subjects: LCSH: Civil engineering--Data processing. | Artificial intelligence. | Machine learning.
Classification: LCC TA347.A78 A7935 2023 | DDC 624.0285--dc23/eng/20220907
LC record available at https://lccn.loc.gov/2022033195

This book is published in the IGI Global book series Advances in Civil and Industrial Engineering (ACIE) (ISSN: 2326-6139; eISSN: 2326-6155)

British Cataloguing in Publication Data
A Cataloguing in Publication record for this book is available from the British Library.

All work contributed to this book is new, previously-unpublished material. The views expressed in this book are those of the authors, but not necessarily of the publisher.

For electronic access to this publication, please contact: eresources@igi-global.com.

# Advances in Civil and Industrial Engineering (ACIE) Book Series

Ioan Constantin Dima
University Valahia of Târgovişte, Romania

ISSN:2326-6139
EISSN:2326-6155

## Mission

Private and public sector infrastructures begin to age, or require change in the face of developing technologies, the fields of civil and industrial engineering have become increasingly important as a method to mitigate and manage these changes. As governments and the public at large begin to grapple with climate change and growing populations, civil engineering has become more interdisciplinary and the need for publications that discuss the rapid changes and advancements in the field have become more in-demand. Additionally, private corporations and companies are facing similar changes and challenges, with the pressure for new and innovative methods being placed on those involved in industrial engineering.

The **Advances in Civil and Industrial Engineering (ACIE) Book Series** aims to present research and methodology that will provide solutions and discussions to meet such needs. The latest methodologies, applications, tools, and analysis will be published through the books included in **ACIE** in order to keep the available research in civil and industrial engineering as current and timely as possible.

## Coverage

- Coastal Engineering
- Production Planning and Control
- Ergonomics
- Earthquake engineering
- Transportation Engineering
- Urban Engineering
- Engineering Economics
- Hydraulic Engineering
- Quality Engineering
- Operations Research

IGI Global is currently accepting manuscripts for publication within this series. To submit a proposal for a volume in this series, please contact our Acquisition Editors at Acquisitions@igi-global.com or visit: http://www.igi-global.com/publish/.

The Advances in Civil and Industrial Engineering (ACIE) Book Series (ISSN 2326-6139) is published by IGI Global, 701 E. Chocolate Avenue, Hershey, PA 17033-1240, USA, www.igi-global.com. This series is composed of titles available for purchase individually; each title is edited to be contextually exclusive from any other title within the series. For pricing and ordering information please visit http://www.igi-global.com/book-series/advances-civil-industrial-engineering/73673. Postmaster: Send all address changes to above address. 

## Titles in this Series

*For a list of additional titles in this series, please visit: www.igi-global.com/book-series*

***Implications of Digital Systems on Mobility as a Service***
John M. Easton (University of Birmingha, UK)
Engineering Science Reference • © 2023 • 335pp • H/C (ISBN: 9781799841388) • US $250.00

***Geoinformatics in Support of Urban Politics and the Development of Civil Engineering***
Sérgio António Neves Lousada (University of Madeira, Porugal)
Engineering Science Reference • © 2023 • 315pp • H/C (ISBN: 9781668464496) • US $250.00

***Impact of Digital Twins in Smart Cities Development***
Ingrid Vasiliu-Feltes (University of Miami, USA)
Engineering Science Reference • © 2023 • 314pp • H/C (ISBN: 9781668438336) • US $260.00

***AI Techniques for Renewable Source Integration and Battery Charging Methods in Electric Vehicle Applications***
S. Angalaeswari (Vellore Institute of Technology, India) T. Deepa (Vellore Institute of Technology, India) and L. Ashok Kumar (PSG College of Technology, India)
Engineering Science Reference • © 2023 • 288pp • H/C (ISBN: 9781668488164) • US $260.00

***Urban Life and the Ambient in Smart Cities, Learning Cities, and Future Cities***
H. Patricia McKenna (AmbientEase, Canada)
Engineering Science Reference • © 2023 • 261pp • H/C (ISBN: 9781668440964) • US $215.00

***Remapping Urban Heat Island Atlases in Regenerative Cities***
Hisham Abusaada (Housing and Building National Research Center, Egypt) Abeer Elshater (Ain Shams University, Egypt) and Marwa Khalifa (Ain Shams University, Egypt)
Engineering Science Reference • © 2022 • 267pp • H/C (ISBN: 9781668424629) • US $250.00

***Clean Technologies and Sustainable Development in Civil Engineering***
Raluca- Andreea Felseghi ("Ştefan cel Mare" University of Suceava, Romania) Nicoleta Cobîrzan (Technical University of Cluj-Napoca, Romania) and Maria Simona Raboaca (National Research and Development Institute for Cryogenic and Isotopic Technologies, Râmnicu Vâlcea, Romania)
Engineering Science Reference • © 2022 • 290pp • H/C (ISBN: 9781799898108) • US $250.00

***Health and Well-Being Considerations in the Design of Indoor Environments***
Roberto A. González-Lezcano (Universidad CEU San Pablo, Spain)
Engineering Science Reference • © 2021 • 286pp • H/C (ISBN: 9781799872795) • US $240.00

IGI Global
PUBLISHER of TIMELY KNOWLEDGE

701 East Chocolate Avenue, Hershey, PA 17033, USA
Tel: 717-533-8845 x100 • Fax: 717-533-8661
E-Mail: cust@igi-global.com • www.igi-global.com

# Editorial Advisory Board

# Table of Contents

# Detailed Table of Contents

The recent advances in sensing technology and in the field of artificial intelligence (AI) have boosted the development of techniques based on machine learning (ML) for structural health monitoring (SHM). The chapter provides a discussion on the application of machine learning (ML) methods for vibration-based SHM. The fundamental principles of machine learning are presented and the steps of vibration-based SHM for damage detection are outlined. The application of ML with respect to every step of SHM process is subsequently discussed. Among the vast existing literature, the milestones are identified and the endless possibilities for the successful application of ML are presented.



The design and analysis of structures is performed with the use of national and international design codes that usually suggest the use of semi-empirical formulae. Often the formulae are oversimplified, in some cases are not available to engineers, or are time-consuming and challenging to implement. The objective of this chapter is to demonstrate the use of artificial intelligence and machine learning to develop more accurate formulae for different types of applications related to structural design. The applications that are discussed in this work include predicting the shear capacity of reinforced concrete slender and deep beams without stirrups, calculating the fundamental period of reinforced concrete and steel structures, and predicting the deflection of horizontally curved steel I-beams.



Reverse cyclic loading is applied to the RC knee joint, which causes the knee joint to experience either opening or closing moments. To calculate the joint's shear strength, an effective nonlinear model is created herein. The shear strength of the knee joint either in opening or closing is predicted by two distinct and practical equations developed utilizing a regression-based methodology. The effects of material and geometric factors are taken into account, including the compressive strength of concrete, the joint aspect ratio, and reinforcement strength. The proposed equations are developed using a statistical analysis of a database containing 61 experiments. The proposed equations are statistically evaluated and examined with the models that were earlier put out. Thus, the suggested models can fairly accurately forecast the shear capacity, which makes them ideal for usage in real applications.

The study utilizes an artificial neural network model for determining the fundamental period of Light-Frame Wooden Buildings, employing the Bat algorithm on a data set of 71 measured periods of wooden buildings. The number of stories, floor area, storey height, maximum length, and maximum width are selected as input parameters to estimate the fundamental period of light-frame wooden buildings. The accuracy and the competitiveness of the developed model were evaluated by comparing it with a similar particle swarm optimization (PSO)- ANN scheme, the formulas provided in the National Building Code of Canada, an equation obtained from the Eureqa software, and a non-linear regression (NLR) model. The results of the research show that the bat-ANN model exhibited the best overall performance with the lowest RMSE and MAE error values and the highest values of the Coefficient of determination, R2, in comparison to the other examined models. Therefore, the proposed Bat-ANN model can be considered as a reliable, robust, and accurate tool for predicting the fundamental period of wooden buildings.



This chapter addresses the shear capacity prediction of reinforced concrete elements with transverse reinforcement through a hybrid approach in which a mechanical model (i.e., the variable-angle truss model incorporated in the Eurocode 2) is enhanced with two machine-learning-calibrated corrective coefficients aimed at improving the accuracy of the numerical predictions. Genetic programming is adopted to derive closed-form expressions of the corrective coefficients, thus making the proposed formulation suitable for design purposes and usable by practitioners. The model predictive performance and the improvements over alternative code-based formulations is demonstrated through a wide database of experimental results of reinforced concrete beams and columns with plain and hollow sections failing in shear under both monotonic and cyclic loading conditions. The proposed approach leads to numerical-to-experimental shear capacity ratios having mean value close to one and coefficients of variation equal to 32%, 28% and 24% for beams, columns with plain and hollow sections, respectively.

Water resources are directly related to the economic conditions of a region. Precise estimation of groundwater is an important step toward better planning and management. This book chapter is dedicated to modelling groundwater in terms of both quantity and quality utilizing ANN (artificial neural networks), ANFIS (adaptive neuro-fuzzy inference system), and the numerical-hydraulic modeling by MODFLOW (modular three-dimensional finite-difference groundwater flow model). The model performance was determined using mean of square error and Nash-Sutcliffe efficiency of model. The pumping data of the area was used to determine the parameters of Saq Aquifer (Qassim, Saudi Arabia) including specific storage and transmissivity. It has been found that the ANFIS model is the most effective for qualitative and quantitative modelling of the aquifer. Following sensitivity analysis, different future scenarios for sustainable groundwater pumping were examined. This book chapter presents research findings that will be useful for engineers, planners, and managers of water systems in arid areas.



Over the past couple of decades, a significant rise in utilization of artificial neural network (ANN) in the field of civil engineering has been observed. ANNs have been proven to be very helpful for researchers working in concrete technology. Reinforced cement concrete (RCC) shear walls play an important role in the stability of high-rise reinforced concrete structures. Current study is focused on using ANN-based design technique as an alternative to conventional design codes and physical models to estimate the ultimate load carrying capacity of RCC shear walls. In this study, database of 95 RCC wall samples has been collected from previously published literature. Various critical parameters considered for current research are; length of web portion of the wall (Lw), thickness of wall boundary member (bw), effective depth of wall (d), height of wall (H), shear span ratio (av/d), vertical steel ratio (ρv), horizontal steel ratio (ρh), yield strength of vertical and horizontal steel (fy), compressive strength of concrete (fc), and the ultimate load carrying capacity (Vexp).

In the context of the fourth industrial revolution, machine –learning based applications, such as 3D Printing, computer vision and robotics, are envisaged to digitize construction. This chapter introduces the reader to the basics of AI and focuses on the role of machine learning in the development of technological applications aimed at the improvement of the construction industry's efficiency and sustainability. It reviews the potential benefits of these applications and identifies factors like e.g. high costs, skills gap, dynamic and risky working environment that typically keep them at a very low practical usage in construction. Furthermore, a bibliometric search reveals that, despite its key role in the digitization vision, machine learning is rarely acknowledged in the relevant 'Construction 4.0' literature. The chapter also identifies the cybersecurity issues emerging from machine learning applications' use and discusses mitigation strategies and technologies aimed at minimizing the impact of threats.



The success of Machine Learning (ML) approaches as promising solutions has encouraged their widespread implementation across different fields. Owing to the high accident rate, the construction industry embraced ML in the risk assessment procedure. What if the machine produces knowledge of the relationship between the risk features and accident outcomes contained in the safety dataset? What if machines can explain an accident dataset without human intervention? Unsupervised ML techniques offer several advantages over supervised approaches, including their explainability to analyze and understand complex datasets. This chapter demonstrates the practical implementation of the unsupervised learning method, clustering, and dimensionality reduction to explain the similarities, differences, variances, and patterns that exist between the feature spaces of an occupational safety risk dataset. Principal component analysis (PCA) and K-means clustering with silhouette analysis were selected as two unsupervised ML approaches to demonstrate their implementation in enhancing data-centric decision-making during the construction risk assessment procedure.



Near-fault ground motion effects have been of great importance in recent decades due to the importance of these ground motions in vibration serviceability of structural systems. Fuzzy logic controller as one of the well-known control systems which is formulated based on the human expertise and knowledge is mostly incapable of providing proper control responses for structures under strong ground motions, and therefore the optimization of these controllers is concerned in recent years. In this chapter, optimal design of fuzzy controllers while considering the nonlinear characteristics of the structure is investigated. To

# Preface

In recent years, artificial intelligence (AI) has drawn significant attention with respect to its applications in several scientific fields, varying from big data handling to medical diagnosis and more. A tremendous transformation has taken place with the emerging applications of AI and machine learning (ML) techniques. The use of these technologies is already present in our everyday lives with several applications, such as personalized ads, personal virtual assistants, autonomous driving, etc. AI and ML work in principle by combining large amounts of data with fast, iterative processing, and intelligent algorithms (e.g., neural networks, process mining, convolutional neural networks, deep learning, and others), allowing the machine to learn automatically from patterns or features in the data. This way, AI can provide a wide range of solutions to address many challenging problems.

As expected, AI and ML techniques have been increasingly adopted in various engineering fields, including civil engineering, which is undergoing a major transformation, driven by recent advances in these technologies. AI and ML have started revolutionizing the way we design, construct, and maintain structures and infrastructures, and are enabling engineers to make more informed decisions, leading to more efficient and sustainable practices. These techniques have proven to be beneficial in predicting and optimizing various aspects of civil engineering projects such as structural design, construction, maintenance, and management. For instance, AI and ML techniques can help predict the performance of structural systems under various loads and environmental conditions, thus aiding in the design of more efficient and resilient structures. Additionally, AI and ML techniques can be used for monitoring construction progress and identifying potential defects, thus enabling timely corrective actions. Overall, the use of AI and ML techniques in civil engineering has the potential to improve efficiency, safety, and sustainability of civil engineering projects.

This book is designed to provide an overview of the present thinking and state-of-the-art developments in the application of AI and ML techniques for civil engineering, and to explore their potential applications in various domains. It brings together leading researchers and practitioners from academia and industry to share their experiences, insights, and best practices. It is intended for a wide audience, including professionals, researchers, practitioners, and students in the civil engineering field who are interested in incorporating AI and machine learning techniques into their work or research. It can serve as a valuable reference for those seeking to keep up to date with the latest developments in this exciting field. We hope that it will inspire readers to explore new ideas and opportunities, and to contribute to the ongoing evolution of civil engineering.

Being a collection of innovative research on the methods and implementation of AI and ML in multiple facets of civil engineering, the book is divided into 13 chapters, each covering a different aspect, as described in detail in the following paragraphs.

**Chapter 1** is about the application of AI techniques in Building Information Modeling (BIM). AI in BIM can be used to streamline operations, advancements and innovations that were previously based on human ability, saving costs, delays, and hazards while improving project output quality. The authors examine the technical advancements based on ML algorithms that propelled AI to the top of the BIM list.

**Chapter 2** discusses the importance of detecting and assessing damage in concrete structures for safety reasons and investigates the use of Deep Learning (DL) techniques for the detection and assessment of the damaged condition of these structures. Some indicators of damage are difficult to detect automatically, but deep learning can effectively extract features from multi-source data and perform related tasks such as classification and detection. The authors highlight the potential of DL in addressing crucial problems related to damage inspection for concrete structures, including component recognition, local damage recognition, and structural anomaly analysis.

**Chapter 3** discusses the use of ML applications in vibration-based Structural Health Monitoring (SHM). The chapter examines how recent advances in sensing technology and AI have led to the development of techniques based on ML for SHM. The chapter focuses on the application of ML for vibration-based SHM and outlines the fundamental principles of ML and the steps involved in vibration-based SHM for damage detection. It includes the application of ML at each step of the SHM process, identifying key milestones from existing literature and highlighting the many possibilities for successful ML applications in SHM.

**Chapter 4** discusses the limitations of national and international design codes in providing oversimplified and inaccessible formulas for designing and analyzing structures. It aims to address this issue by demonstrating the use of AI and ML in developing more accurate formulas for different types of applications related to structural design. The applications discussed include predicting shear capacity, calculating the fundamental period of structures, and predicting deflection for steel I-beams with a curved design.

**Chapter 5** investigates a predictive model for the shear strength of reinforced concrete (RC) roof corner joint subjected to dynamic load. Reverse cyclic loading is applied to a RC roof corner joint to determine its shear strength. The study develops a nonlinear model to calculate the joint's shear strength in opening or closing moments. Two equations are developed using a genetic algorithm to predict the joint's shear strength, taking into account material and geometric factors such as compressive strength, joint aspect ratio, and reinforcement strength. The equations are developed using data from 61 experiments and are statistically evaluated and compared to existing models. The proposed equations are shown to be accurate and suitable for real-world applications.

**Chapter 6** is about a research study that aims to develop a reliable and suitable method for calculating the fundamental period of wooden buildings, which is critical for the seismic analysis and design. An artificial neural network (ANN) model is used with the Bat algorithm to determine the fundamental period based on input parameters such as the number of stories, floor area, story height, maximum length, and maximum width. The model is trained and tested on a data set of 71 measured periods of wooden buildings. The model is evaluated against other models, including a Particle Swarm Optimization (PSO)-ANN scheme, formulas of the National Building Code of Canada, an equation from Eureqa software, and a Non-linear regression (NLR) model. The results show that the proposed model is an accurate and reliable tool for predicting the fundamental period of wooden buildings.

**Chapter 7** is about the shear capacity prediction of RC elements with transverse reinforcement through a hybrid approach in which a mechanical model (the variable-angle truss model of Eurocode 2) is enhanced with two ML-calibrated corrective coefficients aimed at improving the accuracy of the numerical predictions. Genetic programming is adopted to derive closed-form expressions of the corrective coefficients, thus making the proposed formulation suitable for design purposes and usable by practitioners. The model predictive performance and the improvements over alternative code-based formulations is demonstrated through a wide database of experimental results of reinforced concrete beams and columns with plain and hollow sections failing in shear under both monotonic and cyclic loading conditions. The proposed formulation is demonstrated to be effective and accurate for practical design purposes.

**Chapter 8** discusses the modelling of groundwater quantity and quality using ANNs, adaptive neuro-fuzzy inference system (ANFIS), and MODFLOW numerical-hydraulic modelling. The study focuses on the Saq Aquifer in Qassim, Saudi Arabia, and uses pumping data to determine specific storage and transmissivity parameters. The performance of the models is evaluated using Mean Square Error and Nash-Sutcliffe Efficiency of Model. The ANFIS model is found to be the most effective for modelling the aquifer, and different scenarios for sustainable groundwater pumping are examined. The research findings presented in the chapter are expected to be most helpful for water system engineers, planners, and managers in arid areas.

**Chapter 9** is about the use of ANN in the reliability analysis of RC code for predicting the load carrying capacity of RC walls. The ANN-based design technique is used as an alternative to conventional design codes and physical models to estimate the ultimate load carrying capacity of RCC shear walls. A database of 95 RC wall samples has been collected from the literature. Various critical parameters have been considered, such as the length of web portion of the wall, thickness of wall boundary member, effective depth of wall, height of wall, shear span ratio, vertical steel ratio, horizontal steel ratio, yield strength of vertical and horizontal steel, compressive strength of concrete, and the ultimate load carrying capacity. The study shows that once the ANN model has been trained, it can accurately predict the response of RC members even with different geometric and material properties.

**Chapter 10** examines the value proposition of ML in construction management and explores the trends in Construction 4.0. According to the study, the use of AI in construction management has long been studied as a way to improve effectiveness and efficiency while addressing common industry issues. However, the unpredictable and harsh nature of construction, combined with the high cost of technology adoption and skills gaps, have presented challenges. ML is seen as a key enabling factor for AI in construction, but it is often overlooked in discussions of the Construction 4.0 paradigm. Despite their potential, ML models are vulnerable to attacks that can decrease performance and reliability, so new models using adversarial and federated learning and the integration of blockchain are being explored for improved security and performance.

**Chapter 11** is about using unsupervised learning techniques for explainable safety risk management in the construction industry. Specifically, the focus is on unsupervised ML techniques, such as clustering and dimensionality reduction, which offer advantages in analyzing and understanding complex datasets without human intervention. The chapter demonstrates the practical implementation of unsupervised ML approaches, including Principal Component Analysis (PCA) and K-means clustering, in enhancing data-centric decision-making during construction risk assessment.

**Chapter 12** investigates the optimization of fuzzy logic controllers for vibration suppression of a 20-story steel structure with nonlinear behavior subjected to strong ground motions. The study considers the limitations of assuming linear behavior for structural elements and presents an enhanced optimization method using the Stochastic Paint Optimizer (SPO) with an improved method based on the Levy flight method. The fuzzy parameter optimization is conducted by the SPO and ESPO to determine the optimal rule base and membership function for fuzzy rules. The study shows that the ESPO method is capable of determining better-controlled schemes for vibration suppression of structural systems under severe seismic conditions.

To conclude, the book covers a wide range of AI- and ML-related topics which are relevant to the civil engineering industry. It also discusses the challenges and opportunities associated with the implementation of AI and ML in civil engineering, such as the need for reliable and high-quality data, the importance of interpretability and transparency, and some ethical considerations surrounding AI and ML. By exploring the latest developments and applications of AI and ML techniques in civil engineering, it is expected to have a significant impact on the field, to enhance the understanding of the subject matter, promote interdisciplinary collaboration, and inspire further research and development. The book will contribute to the advancement of the field and will help engineers to apply AI and ML techniques to solve complex problems related to infrastructure planning, design, construction, and maintenance.

*Vagelis Plevris*
*Qatar University, Qatar*

*Afaq Ahmad*
*University of Engineering and Technology Taxila, Pakistan*

*Nikos D. Lagaros*
*National Technical University of Athens, Greece*

# Acknowledgment

The editors of the book would like to express their deep gratitude to all the contributors for their most valuable support during the preparation of this volume, for their time and effort devoted to the completion of their contributions, and for their great expert help in the review process. In addition, we are most appreciative to the members of the Editorial Advisory Board (EAB) of the book for their constructive comments and suggestions offered during the preparation process and for their contribution to reviewing some of the book chapters. We are also grateful to all the colleagues who, although did not contribute chapters to the book, were kind enough to offer their expert help during the review process. Last but not least, we would also like to thank the personnel of IGI Global, especially Ms. Jocelynn Hessler, Ms. Rowan Rumbaugh, Ms. Jan Travers and Ms. Mikaela Felty for their most valuable continuous support for the publication of this book.

*Vagelis Plevris*
*Qatar University, Qatar*

*Afaq Ahmad*
*University of Engineering and Technology Taxila, Pakistan*

*Nikos D. Lagaros*
*National Technical University of Athens, Greece*

# Chapter 1
# Artificial Intelligence-Assisted Building Information Modelling

**Muhammad Abubakar Tariq**
*International Islamic University, Islamabad, Pakistan*

**Shiraz Ahmed**
*GIK Institute of Engineering, Sciences, and Technology, Pakistan*

**Erum Aamir**
*National University of Sciences and Technology, Pakistan*

**Fahad Iqbal**
*GIK Institute of Engineering, Sciences, and Technology, Pakistan*

**Muhammad Noman**
*International Islamic University, Islamabad, Pakistan*

## ABSTRACT

*Building information modeling gained popularity and recognition in the 2000s, making the building process productive, saving time and money, and improving collaboration. Though the building sector took some time to adopt new technology, when utilized correctly, technology can significantly boost productivity in the conventional construction industry. While BIM has significantly advanced building activities, artificial intelligence contributed by applying machine learning to increase efficiency. Since the advent of artificial intelligence for computer-aided design in the 1970s, technology has improved substantially. Numerous technical advancements are based on machine learning algorithms, propelling AI to the top of the BIM list. Presently different techniques rely primarily on human ability; these advancements and innovations are critical in the construction business. AI can be used to streamline these operations, saving costs, delays, and hazards while improving project output quality.*

DOI: 10.4018/978-1-6684-5643-9.ch001

## INTRODUCTION: ARTIFICIAL INTELLIGENCE (AI)

The term 'Artificial Intelligence (AI)' can be simply defined as a concept in which an object mimics human intelligence to do certain tasks, that is why it is called 'artificial' intelligence. The objects may include computers, machines and systems to impersonate human's intelligence. The idea is to perform tasks through a thinking process backed-up by large datasets. The tasks might be of iterative nature for problem solving and decision making, for instance: chat bots for customer service, Apple's Siri/Android Alexa and virtual assistant to manage work schedule efficiently.

## MACHINE LEARNING AND DEEP MACHINE LEARNING

Machine Learning and Deep Machine Learning are at times used interchangeably with AI. The concept of 'learning' is to develop systems that can learn and consequently improve its accuracy based on the amount of data delivered to the system. Nevertheless, it is worth nothing that both machine learning and deep machine learning are sub-sets of AI. Furthermore, deep machine learning is even a sub-category of machine learning. So it can be stated that all machine learning and deep machine learning is AI but not all AI is machine learning and/or deep machine learning.

Additionally, machine learning and deep machine learning are not exactly the same phenomena. It is important to differentiate between machine learning and deep machine learning. The subtle differences between the two are: human intervention and data structure. In the case of machine learning, structured data with labels are fed to the algorithm and it requires human intervention to produce an output. On the other hand, deep machine learning does not necessarily require labelled data in the form of categories and/or human intervention; it can take raw data and improve its algorithm backed by large data sets. So, deep machine learning is further specialized subset of machine learning.

## HISTORICAL DEVELOPMENT OF AI

The concept of AI emerged afterwards World War II when some independent individuals started working on intelligent machines. Alan Turing, an English mathematician is considered as a first person to delve into this domain through his lecture in 1947 and his published work in 1950, titled "Computing Machinery and Intelligence". He is also referred to as "father of computer science". This article is considered as a groundbreaking and pioneering work in this domain. Turing's work shifted the focus from building machines to programming computers and the idea was taken up by researchers in late 1950s to produce computer programs. (IBM Cloud Education, 2020)

In his work he considered a question: "Can machines think?" and went on to examine this question by deducting counter reasoning and objections based on theology, mathematics, heads in the sand, consciousness, disabilities and so on. In this research work, Turing laid out the conditions to consider a machine, intelligent. The argument to consider a machine to be intelligent is based on a premise that if a machine can imitate being a human to an observer than it should be regarded as intelligent. Turing gave a test, known as, "Turing Test" in which a human interrogator has to differentiate between a machine and a human through type-writer written replies to avoid human voice, hand writing or any sort of clue that distinguish between a machine and a human. The test besides scrutinized since its development remains an important part of AI development. The test might satisfy most of the people nonetheless not all the philosophers. (Turing, 1950)

The critics, such as, John McCarthy in his research work "What is artificial intelligence?" in 2004 argued that the test by Turing is one-sided. The one-sided label is based on the premise that Turing considered a machine intelligent if it can imitate a human but a machine can also be regarded as intelligent without imitating a human or even knowing adequate about humans. Furthermore, John McCarthy referred to Daniel Dennett's book called Brainchildren; he stated about Turning test that observers having limited knowledge about AI asked under restricted conditions on subject would yield results in believing dumb programs as intelligent. Hence, John McCarthy gave his definition of AI: *It is the science and engineering of making intelligent machines, especially intelligent computer programs. It is related to the similar task of using computers to understand human intelligence, but AI does not have to confine itself to methods that are biologically observable.* (McCarthy, 2007)

There is a lot of hype today related to AI development, which is a usual phenomenon when it comes to emerging technologies. According to Gartner's hype cycle this hype or expectation is divided in five stages: innovation trigger, peak of inflated expectations, trough of disillusionment, slope of enlightenment and plateau of productivity. Thus, as per typical pattern of progression in the realm of latest technologies, AI development is at the peak of inflated expectations and approaching towards the trough of disillusionment before finally getting on slope of enlightenment and plateau of productivity.

*Figure 1. The hype cycles*
*(Blosch & Fenn, 2018)*

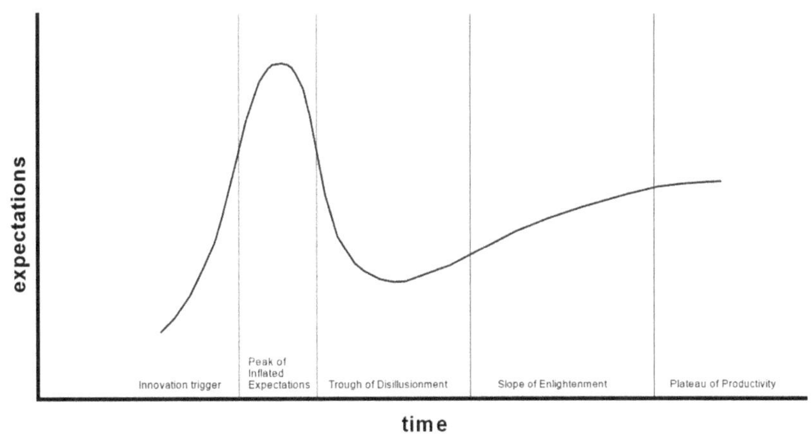


## TYPES OF AI (IBM CLOUD EDUCATION, 2020)

There are two types of AI:

1. Weak AI
2. Strong AI

Weak AI, also known as Narrow AI/Artificial Narrow Intelligence (ANI) is narrow and specific to do a certain task for which it is trained for, thus called weak AI. Most of the prevailing application of AI around us today are basically ANI. For instance, Amazon's Alexa, Apple's Siri, self-driving cars and IBM Watson.

Strong AI, also known as, Artificial General Intelligence (AGI) or Artificial Super Intelligence (ASI) is a conception in which artificial intelligence would be equated to human's intelligence. ASI would be able to learn, solve problems and do future planning self-consciously. The extreme form of ASI might surpass human's intelligence. Scientists are pursing this domain through research. Nevertheless, it is yet not in existence in any shape or form and no application is being utilized based on ASI other than in science fiction.

## APPLICATIONS OF AI (IBM CLOUD EDUCATION, 2020)

Here are some of the applications of AI that are used at present:

### Automatic Speech Recognition (ASR)

Automatic Speech Recognition (ASR), which is also known as, speech recognition or computer speech recognition is a process that utilizes natural language processing (NLP) to convert human's speech into text. Mobile devices have this ability now to recognize speech. ASR is used by Siri, Alexa and in texting to provide ease while operating devices.

### Customer Care

Human agents are replaced with virtual agents for customer service. For instance, to facilitate about different sizes of customers, suggesting products, and engaging with customers on e-commerce websites. Chat bots on platforms like Facebook and Slack are also common.

### Computer Vision

AI has the ability to take visual inputs and give out meaningful outputs. The visual inputs can be in the form of photographs and videos etc. The ability to give output in the form of a recommendation differentiates it from the image recognition technology. Based on convolutional neural networks, this is used in social media for tagging, healthcare sector for radiology and autonomous cars in automobile industry.

### Recommendation Engines

E-commerce industry uses this ability of AI to help their selling operation by recognizing a customer's consumption pattern. Based on the customer's consumption pattern through algorithms enable the website to offer customized recommendations at the checkout for improving customer's engagement and sales.

## Automatic Stock Trading

Trading platforms build using AI technology allow undertaking high-frequency millions of traders without human intervention.

## EVOLUTION/HISTORICAL DEVELOPMENT OF AI

The concept of thinking machines is in existence since Greece. The advent of computers paves the way for this development. Some of the key developments in the evolution of AI is given below:

1950- Alan Turing, famous English mathematician who broke ENIGMA code during World War-II published his seminal work *Computing Machinery and Intelligence* in 1950 followed by his lecture on the subject in 1947. The basic questions addressed in this article was 'Can machines think?". Turing introduces a test called Turing Test to consider a machine intelligent if it can act as a human being. Nonetheless, the test is debatable since then by critics to be confined to be set of conditions. Furthermore, Turing's work brought up the attention to the development of algorithms rather than machines.

1956-The term "artificial intelligence" was coined for the first time at Dartmouth College in a conference by John McCarthy. Afterwards, John McCarthy developed Lisp language in 1958 at MIT. Moreover, in 1956, Allen Newell, Herbert Simon and Cliff Shaw created Logic Theorist software to proof mathematical theorems. So, it is also known as first-ever AI software.

1967- Frank Rosenblatt build first computer on the concept of neural network known as Mark 1 Perceptron. The computer used to learn on trial and error mechanism. Afterwards, Seymour Papert and Marvin Minsky wrote a book called Perceptrons that became an important work on neural networks at that point.

1980-Wide spread utilization of neural networks in AI that use back-propagation algorithm for training.

1997-Garry Kasparov who was world champion in Chess, got beaten twice by IBM Deep Blue in chess matches.

2011-Ken Jennings and Brad Rutter, then champions were beaten by IBM Watson at Jeopardy.

2015-Baidu's supercomputer called Minwa that used convolutional neural network, a special type of deep neural network for identifying and categorizing images at a higher accuracy as compared to an average human.

2016-A program called AlphaGo developed by Deep Mind based on deep neural network competes with Lee Sodol who was world champion of Go game in a five round match. The program beats Lee Sodol. This is a significant development because there can be 14.5 trillion possibilities in the game after four moves. Reportedly Google purchased DeepMind's AlphaGo program for about USD 400 million. (IBM Cloud Education, 2020)

## AI AND ENTERPRISES (RAMASWAMY, 2017)

It is interesting to know how enterprises are using AI for their business. According to a Harvard Business Review study, enterprises are primarily using AI for: information technology, marketing, finance and accounting, and customer service. Evidently, business around the world are using AI predominantly for IT related activities at present. The top four results are as follows:

- Security issues 44%
- Troubleshooting technology 41%
- Automating production 34%
- Quality checks and compliance of technology products 34%

*Figure 2. How companies around the world are using artificial intelligence*
*(Ramaswamy, 2017)*

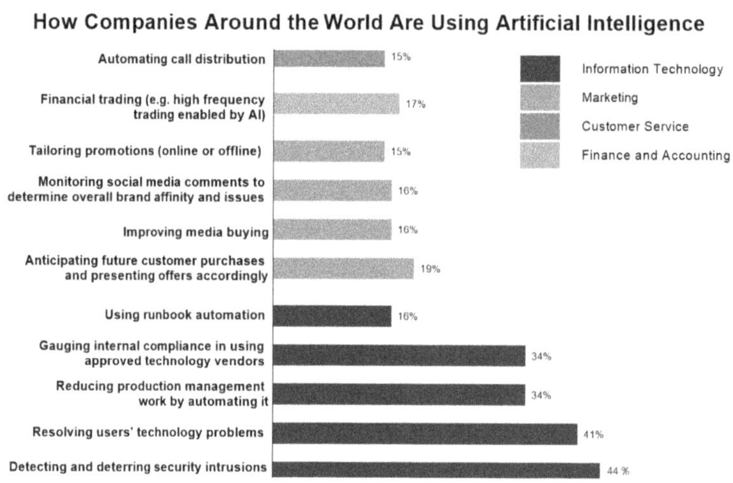


## AI ADOPTION FACTORS

There are many factors for the adoption of AI. Here are three main factors that are currently driving the utilization of AI.

Computing Capability: Readily available cloud-based and non-cloud based computing power enables ease of operations.

### Voluminous Data

As AI requires voluminous data to train its algorithms; the more the data is the better the predictions. Additionally, availability of tools to label data, and readiness to keep and process structured, as well as, unstructured data are making it possible for organization to make AI algorithms better.

## Competitive Advantage

AI is providing competitive advantage by processing large sets of data to give valuable insights. These insights can lead to reduction in cost and risks, better and faster approach to market.

## AI Myths

One of the barriers in the adoption of AI is myths that is surrounding this technology. Some of the the myths encompassing AI are given below:

## Build It Yourself

One of the difficulties in the adoption of AI is the development of algorithms and data to train them. Data scientist handcraft this data to feed the algorithms. Moreover, as objective of using AI can be different in each case so it might require specific algorithms to address those needs. Nevertheless, there are available algorithms for common tasks that can be used and for more specialized needs specific algorithms can be developed, thus a balance of both approaches would be better.

## Magical Results

A common misconception is that AI will yield magical results immediately. For instance, like it will double or triple revenues for the businesses. It is worth noting that comprehensive strategic planning to get clear objectives is required to begin with. This process will take time to produce positive benefits.

## Lack of Human Intervention

A fallacy about AI is that robots taking over the world. It is assumed that AI doesn't need human intervention. Conversely, AI will better serve when it augments human beings by taking up common tasks and freeing human resource for strategic tasks. AI needs humans to feed the data in the right way and work with it right way.

## The More, The Better

The more the data there is to be feed the algorithms the better it is. Nonetheless, quality of data also matters. Erroneous, redundant, and irrelevant data won't do anything. Thus, updated, relevant and rich data is necessary for AI.

## Data and Model Are Sufficient

Only data and model are required for successful implementation of AI. The problem is that these are just the tools to begin with. After initiation, fine tuning by data scientists is required to make necessary adjustments. Additionally, AI solutions have to be scalable to be deployed successfully. There are certain things that are manually adjusted in these solutions that can be regarded as maintenance. A solution to this manual maintenance could be machine learning to automate that process, otherwise, it effectiveness would be reduced. (Oracle Cloud, 2022)

## BUILDING INFORMATION MODELING (BIM)

Building Information Modeling (BIM) is a progressive technology and a process that has unexpectedly changed the way construction is perceived, designed, built and functioned. It is a new and revolutionary technology, which has emerged in the latest years and makes possible the efficient success of greater sustainable designs. (Azhar & Khalfan, 2012)

BIM is an effective tool to manage construction process in a timely manner, and its information-oriented interface is an effective tool for every step of construction (Shin, Lee and Kim, 2018). Moreover, it is an intelligent parametric virtual illustration of a building that offers all the quintessential data required through different number of users for evaluation and to generate information that can be utilized in decision making techniques (AGC, 2005). (Shin et al., 2018) (AGC Research Foundation, 2005)

Building Information Modelling (BIM) is a system that consists of developing and managing of a buildings' physical and functional characteristics in a digitalized 3D model. It is one of the biggest advancements in the Architecture, Engineering and Construction (AEC) industry. It replicates the construction project in a virtual environment. (Eastman et al., 2011)

The US National BIM Standard (2007) defines BIM in three dimensions:

- The Building Information Model is a structured dataset describing a building (a product).
- Building Information Modeling is the act of creating a Building Information Model (a process).
- Building Information Management comprises the commercial enterprise work and verbal exchange structure that extends the of (a system).

Aranda Mena et. al., (2008) described BIM as "for some BIM is just a software, for others, it is a whole new method to practice and evolving the profession which requires the implementation of the new laws, contracts, and relationship amongst project stakeholders". (Aranda-Mena et al., 2009)

BIM is a process that involves advent and usage of digital models for designing, construction and/or operational purposes. (Hill, 2009)

## BENEFITS OF BUILDING INFORMATION MODELING (BIM)

BIM helps make better decisions, reduces abstractions and incorporates several disciplines (Walter, 2006). BIM includes facts about the whole building and a massive set of project documents saved in a combined/build-in database. Whole data is parametric and accordingly linked. Slight change in an item inside the prototype automatically which effects the related services and constructions because the model incorporates the compulsory interpersonal data. This is completely different from the 2D construction illustration of traditional CAD-based sketches. Therefore, BIM has now commenced to change the way designers work with consultants and builders and has the possibility to guide the industry to the production of structures that meet workable progress requirements. (Walter, 2006)

In conventional 3D CAD, a construction block is introduced by using impartial 3-dimensional views in the shape of plans, sections, and elevations, etc. whereas in BIM objects are described as elements/ space such as walls, beams, and columns etc. it includes each physical and practical trait of a building and tasks life-cycle data.

According to (Azhar et al., 2008) a building Information Model can be used to:

1. Visualize the building.
2. Generate or fabrication of shop drawings.
3. Review codes.
4. Facilities Management,
5. Estimation of cost.
6. Finding a sequence of activities.
7. Detection of collisions conflicts and interferences.

Using BIM technology in construction projects, all group members can collaborate at some stage in the design and execution cycle to enhance the process of construction. (Zollinger III et al., 2010)

Utilizing BIM in 32 exceptional construction projects, the following benefits were observed: (CIFE, 2007).

- 40% elimination of unbudgeted change.
- Cost estimation accuracy less than 3%.
- 80% discount in the time taken to generate a value estimate.
- A saving-of up to 10% of the contract value through clash detection.
- Up to 7% reduction in task completion time.

During the design phase, BIM contributes to the success of a project by means of enabling practitioners to effectively manipulate schedule, price range, and quality, and to reduce dangers (Ku & Mills, 2010). BIM improves the quality of design by increasing efficiency and precision. Improvement of design evaluation and communication is a major benefit of BIM (Cheng & Song, 2010). BIM not only saves time but also the cost of designing. It allows the design to be accessed earlier/faster (Yan & Demian, 2008). (Sacks & Barak, 2006) document that the complete quantity of hours spent on three initiatives was reduced via 21 percent, fifty-five percent & 61 percent respectively owing to the use of 3D modeling alternatively of 2D.

BIM enhances the quality of the project by ensuring the information is consistent throughout the execution phase. BIM provides tabular views of components and interaction with their elements such as name, type, attributes and relationship. All this information can be used to assess the significance of variation between design and in-built conditions instead of manual evaluation of drawings. (Zollinger III et al., 2010)

Building Information Modeling (BIM) has the potential to reduce cost and delivery time of project while enhancing its quality and productivity if used properly (Azhar et al., 2008). The cost cycle of maintenance is reduced by BIM as it provides timely & relevant information as early as the design stage. (Wang et al., 2013)

According to the results of a questionnaire survey conducted by McGraw Hill construction (2008), 82% showed that adopting BIM has had a positive effect on company's productivity. 79% of BIM users say that the using BIM has improved results of their projects i.e. reduction in problems due to the lack of coordination on the field and reduced requests for information (RFI). 66% believed that adopting BIM has increased their chances of winning projects. (Hill, 2009)

The performance and operation of the building become more predictable with the use of BIM (Azhar, 2008). The information provided by BIM is helpful in any way by combining BIM with other modern technologies that join digital processes to physical processes (Wang et.al, 2013).

## AI-ASSISTED BIM

According to a McKinsey report *The Next Normal in Construction: How Disruption is Shaping the World's Largest Ecosystem* $10 trillion per year are spent regarding construction activities and that is projected to grow 4.2% till 2023 (McKinsey & Company, 2020). Construction industry plays a major role in the development of the world by providing shelter, commercial spaces, mobility, energy and infrastructure. But the construction sector is regarded as an inefficient, unproductive and unsafe due to its primitive working practices. Any technology that can address to the aforementioned problems would be in great demand. (McKinsey & Company, 2020)

BIM is one such technology that has changed the design creation process and the design itself. The design process is driven through 3D virtual models having geometry, as well as, associated data of the design. The data rich 3D model offers much more detail during different design processes to the stakeholders. The availability of the information in 3D model that is available to project members supports collaborative working environment. Hence, it can be stated that BIM reduces inefficient working practices and positively impacts productivity of the construction sector.

AI is the ability of computers, machines and software to mimic human's like ability to get something done; it requires data as an input to perform a given task. Generally speaking, the more the input data there is, the better the accuracy of the required output. BIM generates a lot of data in the design creation process that can be utilized for different phases of construction process, i.e., design, planning and scheduling, construction, and post-construction. AI can use the data generated by the BIM model and process it for optimum results to surpass the current improvements in the construction sector by resolving issues like health and safety, shortage of labor, planning and scheduling. For instance, Request for Information (ROI) and Change Order/Variation Order is a common practice in the sector. AI can help project managers to efficiently utilize their time and attend to the most critical requests on priority so that project schedule doesn't get affected by the delay in the process of responding to the requested information.

The integration of BIM and AI can couple the strengths of the each other. Emergence of these technologies and its combination give birth to AI-Assisted BIM. It is the advent of combining the strength of BIM to generate data and use that data by AI to automate the process and analyze to get an output in terms of planning, health and safety measures, reduce cost overruns and so on. AI-Assisted BIM can automate different tasks during the life-cycle of a construction project, removing redundancies, performing with greater speed and fewer errors, thus, bringing efficiency and productivity in the construction sector.

## Cost Overrun

Mega construction projects and infrastructure projects are very expensive and can cost a fortune to be build and operate. Example of some of the mega projects that can be referred to as white elephants are:

- Russky Bridge ($1 billion)
- Ciudad Real Central Airport (€1 billion)
- Ambedkar Memorial Park ($1.3 billion)
- Montreal Olympic Stadium (£1 billion)
- The Ryugyong Hotel ($1.3 billion)
- Romanian Palace of the Parliament (€4 billion)
- South China Mall ($1.3 billion)

These projects usually go over-budget despite of the fact that experienced teams are employed to conduct such projects. BIM along with its 3D model of the design, also offers 4D cost information. Change in the design would reflect change in the cost as well. AI can be a step forward to create predictive models regarding cost overruns using artificial neural networks. Project related information such as: planned start and end time, type of contract, project manager's experience, historical data etc. can be given as an input to envision cost overruns predictions and realistic timeline to complete a project. Furthermore, AI can remotely support workers to upgrade their skills when needed rather to bring new resources to complete a task so as to expedite the project delivery timeline.

## Generative Design

BIM is based on a 3D model having geometry of the building and infrastructure. It helps the project team to participate in design and planning of the construction facility. Different design consultants can work on their model and combine it hereafter for a coordinated BIM model. The challenge here is to create clash free designs or even remove them in the virtual design before construction on-site to reduce rework and ROI, thus improving efficiency and productivity. Here AI can support BIM with the help of generative design. The idea is to generate clash free models based on numerous permutations. For instance, add on like GenMEP developed by Building Smart Planning focuses MEP design and automatically design routing of the MEP components, like wires, based on building's geometry. It creates different solutions and learns from each permutation iteratively. Thus, using machine learning of AI, it provides routing of MEP components so that there are no clashes in the model.

## Health and Safety Risk Mitigation

The construction sector is hazardous industry as there are many people working in confined spaces along with high voltage wires, heavy equipment, and objects that can fall. The visualization power of BIM model enables project participants to put in place safety measures before the commencement of the construction on-site. AI can take it further with the help of machine learning that can be trained to detect health and safety risks through images. The algorithm can then detect risks pertaining to falling objects, height, slips and so on.

## MACHINE LEARNING APPLICATIONS TO BIM DATA IN DESIGN PHASE

Within the AECO sector, design has always been a subject of great attention (Zimina et al., 2012) (Sharma et al., 2011) (Zabin et al., 2022). Multiple research initiatives have been taken between 2013 and 2021, to expose the utility of Machine Learning on designers to analyze and automate the data generated by BIM models with the aim to improve the design, well understand the end user needs, and make well-informed decisions (Lomio et al., 2018) (Karan & Asadi, 2019) (Shao & Zhang, 2018). Although BIM has enhanced the AECO industry's design phase but still the design process is long and time-consuming with largely dependent on the designer's expertise and judgment, which might be enhanced by utilizing various data sets made available by BIM models. A 3D BIM model is created from data from many sources and is used by various project stakeholders. After that, the data is transferred into a new form with distinct features. In past researchers applied Machine Learning techniques to get various outcomes related to design automation and better prediction of clashes, waste and energy analysis. Few of the Machine learning applications to BIM data during design phase will be covered in the subsections below.

To evaluate numerous alternatives for building potential locations, facilities, and transportation, created (Ustinovichius et al., 2017) a territorial spatial planning BIM model based on Artificial Neural Network (ANN) and territorial documents. This model learns during its development and chooses the best solutions. As per set criteria, the model efficiency was tested and found out to be 0.8796 which is excellent. Further the ANN's based BIM model accuracy was evaluated against erroneous solutions, and the results were 0.0053, which is quite close to 0 (Ustinovichius et al., 2017)

The potential benefits for the AECO sector are highlighted by the capacity related to design automation and modeling process utilizing digital technology and related algorithms and the power to manage information from many sources in order to make these processes optimal. (Paoletti, 2018). In this relation, Genetic Algorithm (GA) technique was used by (Bianconi et al., 2019) that allowed design experts to visualize hundreds of created design possibilities for a mass timber based BIM home model. The model is meant to be used for the early selection of design options and is easily adaptable by an average user to adjust and test various dimensions and associated parameters. With this novel method, the designers can easily and rapidly evaluate various options and produce a better fit for homes that would satisfy all necessary requirements. (Bianconi et al., 2019)

To categorize photos of building designs, (Lomio et al., 2018) employed both traditional ML methods like Support Vector Machine (SVM) and contemporary ANN techniques. Pictures of the structures were taken from multiple BIM models of randomly selected structures and divided into three groups: commercial buildings, residential buildings, and other building types. This will enable the designer to scale these models to classify the numerous BIM structures that have not yet been labeled under any category. The findings indicate that ANN image recognition approaches performed more accurately than the SVM methodology. This form of Artificial Neural Network (ANN) classification demonstrates that accurate results may be achieved both with rigorously trained ANN on realistic images and with ANN with randomly generated structures and initialized loadings. The platform aims to serve as an intelligent decision support tool that unifies the applications and data collection sources.

BIM also offers advantages in spatial processing and visualization, identifying clashes by integrating several BIM models to automatically identifying all clashes. The integrated use of BIM software has dramatically improved the design stage of different projects. Numerous studies have employed supervised learning algorithms, such as Decision Trees, JRip-based rule methods, BL regression, Random Forests and Bayesian network, to automatically distinguish between related and unrelated clashes to enhance the clash detection process (Hu & Castro-Lacouture, 2019) (Lin & Huang, 2019). An ANN model was trained by (Hsu et al., 2020) to predict the sorts of potential disputes in the construction stage using a sample of BIM models for construction. Construction conflicts can be resolved early in the design process, saving the project money and time.

Additionally, Reinforced Learning (RL) is beginning to take off in BIM-based projects. (Liu et al., 2020) developed a computational framework to automatically minimize spatial collisions of rebar in precast concrete pieces and create designs (MARL). In the study, 2D rebar patterns were produced using (GAN). Then, digital surroundings for (MARL) Multi-Agent Reinforcement Learning were created from 2D rebar designs. The RL agent then chooses appropriate actions based on the MARL results and moves toward the demarcated targets without colliding with obstacles. The efficacy and effectiveness of the developed framework with MARL and BIM is evident from the result of paths of rebar designs, success rates, and average time.

Also, most of BIM software generates log files for each specific session when a designer works on a particular file. All modeling actions during a design session are recorded in log files that have been time-stamped. These files are exportable and have a wide range of information, allowing DM techniques to analyze various patterns. The findings can be used to study and track a design team's productivity, find and examine social networks of collaboration within a design team and apply artificial neural networks (ANNs) to forecast the next potential design directive. (Pan et al., 2020) (Pan & Zhang, 2020b) (Kouhestani & Nik-Bakht, 2020) (Pan & Zhang, 2020a) For tasks that facilitate more straightforward information retrieval, collaboration, and visualization, ML based approaches were applied to BIM models during the design phase. BIMbot is an artificial cognitive assistant that (Mutis & Ramachandran, 2021) developed. It was created to improve engagement in design BIM meetings by carrying out routine tasks, including loading, filtering, and viewing section(s) of interest and automating object placement, among others. The architecture of BIMbot is based on a rule-based model powered with Natural Learning engine and feed by text data gathered from many sources. This model eases the team mates in a project to reach a collaborative understanding of knowledge when they utilize BIM based technologies.

Rule checking is another domain of design phase that grasped attention for automation (Sacks et al., 2019). The procedure of confirming design restrictions as per regional by laws is intended to be speed up by automated rule checking. BIM has enhanced automated rule checking during the past ten years due to various uses, such as standardizing standards and developing tools to facilitate the process. However, ML approaches have also shown to be highly effective in this area. The rule testing of a design code has been accelerated and automated with the use of Natural Language Processing, Artificial Neural Networks and semantic logic-based information representations. (Song et al., 2018)

## MACHINE LEARNING APPLICATIONS TO BIM DATA IN CONSTRUCTION PHASE

One of the benefits that BIM delivers to the stage of construction management is the ability to plan and visualize construction operations, generate shop drawings, decrease the number of Requests for Information (RFI), increase risk management, and track progress (Sacks et al., 2018). Between 2010 and 2021, researchers investigated several applications of ML and BIM-generated data to automate and enhance the management and monitoring of construction projects. The information can be acquired from various sources, including site photos, design and construction BIM models, examples, historical material, labor cost data, and construction timetables. After the data has been combined and integrated into a Building

Information Model, it can be processed to get multiple formats that machine learning techniques will use to achieve results like automating the scheduling process, improving construction tracking, improving price approximation, and lowering energy costs during the execution of the project.

The expertise of planning and site engineers is crucial for scheduling construction activities. Linking the construction schedule to BIM models to depict construction simulation, often known as 4D BIM models (3D models + time), is one of the common uses for construction BIM (Charef et al., 2018). A novel schedule was produced by Wang et al. using a Revit-based BIM model, GA, and a matrix of constructability constraints (MoCC). Results indicate that the suggested method can produce building schedule models in two levels of complexity, from simple to sophisticated. The experiment was run on a fundamental BIM model; therefore, the resulting progress schedule only approximately reflects the real world. To generate project schedules, connected BIM models made in Autodesk Revit and Primavera using a knowledge-based expert system technique. To eliminate some tedious and time-consuming procedures in developing a building timetable. A semi-automatic technique was adopted to identify and connect BIM elements with scheduled activities. Torres-Calderon et al. automated the creation of a construction schedule with ANN, creating a detailed plan of work locations and areas. (Torres-Calderon et al., 2019)

Researchers have looked into how the BIM process might improve the project progress tracking on-site in addition to 4D BIM planning (Hallberg & Tarandi, 2011). It has historically been challenging to use photologs to track the progress on a building site because the images are frequently disorganized, uncalibrated, and taken under lighting circumstances that are largely uncontrolled. (Bosche et al., 2009). The first ML-based BIM research to address this issue was put up in 2010. (Golparvar-Fard et al., 2010). They automated the identification of progress using Bayesian probabilistic models, IFC BIM, and un-ordered daily photologs. An image-based 3D viewer that uses BIM objects to display building progress and color-coded deviations was developed in 2011 and 2015 to improve this method (Golparvar-Fard et al., 2011) (Golparvar-Fard et al., 2015). To measure the built-in progress information of a construction site automatically and transmit a real-time display of progress information, Deng et al. devised an automated monitoring and reporting method of indoor tiles using SVM and BIM (Deng et al., 2020). Project managers may make precise and fast judgments using the method's efficient real-time automatic quantity computations. By combining cloud BIM, IoT, and DM, Pan et al. created a digital twin-based framework that is a realistic way to manage and optimize the complicated building process while establishing a high level of automation and intelligence (Pan & Zhang, 2021b) (Pan & Zhang, 2021a). Instead of relying on expert knowledge, subjective judgment, or intuition, cloud BIM and IoT integration can regularly store and update data to provide a reliable basis for data-driven decision-making. This advancement makes construction management a more efficient, automated, continuous, and intelligent process.

## REFERENCES

AGC Research Foundation. (2005). The Contractor's Guide to BIM (1st ed.). AGC Research Foundation.

Aranda-Mena, G., Crawford, J., Chevez, A., & Froese, T. (2009). Building information modelling demystified: Does it make business sense to adopt BIM? *International Journal of Managing Projects in Business*, *2*(3), 419–434. doi:10.1108/17538370910971063

Azhar, S., Khalfan, M., & Maqsood, T. (2012). Building information modelling (BIM): Now and beyond. *Construction Economics and Building*, *4*(12), 15–28. doi:10.5130/AJCEB.v12i4.3032

Azhar, S., Nadeem, A., Mok, J. Y. N., & Leung, B. H. Y. (2008). Building Information Modeling (BIM): A new paradigm for visual interactive modeling and simulation for construction projects. *Proc., First International Conference on Construction in Developing Countries,* (*vol. 1*, 435–446). Scientific Research Publishing.

Bianconi, F., Filippucci, M., & Buffi, A. (2019). Automated design and modeling for mass-customized housing. A web-based design space catalog for timber structures. *Automation in Construction, 103*, 13–25. doi:10.1016/j.autcon.2019.03.002

Blosch, M., & Fenn, J. (2018). *Understanding Gartner's hype cycles.* Gartner Research.

Bosche, F., Haas, C. T., & Akinci, B. (2009). Automated recognition of 3D CAD objects in site laser scans for project 3D status visualization and performance control. *Journal of Computing in Civil Engineering, 23*(6), 311–318. doi:10.1061/(ASCE)0887-3801(2009)23:6(311)

Charef, R., Alaka, H., & Emmitt, S. (2018). Beyond the third dimension of BIM: A systematic review of literature and assessment of professional views. *Journal of Building Engineering, 19*, 242–257. doi:10.1016/j.jobe.2018.04.028

Cheng, B., & Song, Y. (2010). Using Parametric Modeling to Improve the Equipment Systems Coordination Process for Buildings. *2010 International Conference on Management and Service Science*, (pp. 1–4). IEEE. 10.1109/ICMSS.2010.5577740

Deng, H., Hong, H., Luo, D., Deng, Y., & Su, C. (2020). Automatic indoor construction process monitoring for tiles based on BIM and computer vision. *Journal of Construction Engineering and Management, 146*(1), 04019095. doi:10.1061/(ASCE)CO.1943-7862.0001744

Eastman, C. M., Eastman, C., Teicholz, P., Sacks, R., & Liston, K. (2011). *BIM handbook: A guide to building information modeling for owners, managers, designers, engineers and contractors.* John Wiley & Sons.

Golparvar-Fard, M., Pena-Mora, F., & Savarese, S. (2011). Monitoring changes of 3D building elements from unordered photo collections. *2011 IEEE International Conference on Computer Vision Workshops (ICCV Workshops)*, (pp. 249–256). IEEE. 10.1109/ICCVW.2011.6130250

Golparvar-Fard, M., Pena-Mora, F., & Savarese, S. (2015). Automated progress monitoring using unordered daily construction photographs and IFC-based building information models. *Journal of Computing in Civil Engineering, 29*(1), 04014025. doi:10.1061/(ASCE)CP.1943-5487.0000205

Golparvar-Fard, M., Savarese, S., & Peña-Mora, F. (2010). Automated model-based recognition of progress using daily construction photographs and IFC-based 4D models. *Construction Research Congress 2010: Innovation for Reshaping Construction Practice*, (pp. 51–60). ASCE Library. 10.1061/41109(373)6

Hallberg, D., & Tarandi, V. (2011). On the use of open bim and 4d visualisation in a predictive life cycle management system for construction works. [ITcon]. *Journal of Information Technology in Construction, 16*, 445–466.

Hill, M. (2009). *The business value of BIM: getting to the bottom line.* Opgehaald van BIM Construction. http://www. bim. construction. com/research

Hsu, H.-C., Chang, S., Chen, C.-C., & Wu, I.-C. (2020). Knowledge-based system for resolving design clashes in building information models. *Automation in Construction, 110*, 103001. doi:10.1016/j.autcon.2019.103001

Hu, Y., & Castro-Lacouture, D. (2019). Clash relevance prediction based on machine learning. *Journal of Computing in Civil Engineering, 33*(2), 4018060. doi:10.1061/(ASCE)CP.1943-5487.0000810

IBM Cloud Education. (2020). *What is Artificial Intelligence (AI)?* IBM Cloud Learn Hub. https://www.ibm.com/cloud/learn/what-is-artificial-intelligence

Karan, E., & Asadi, S. (2019). Intelligent designer: A computational approach to automating design of windows in buildings. *Automation in Construction, 102*, 160–169. doi:10.1016/j.autcon.2019.02.019

Kouhestani, S., & Nik-Bakht, M. (2020). IFC-based process mining for design authoring. *Automation in Construction, 112*, 103069. doi:10.1016/j.autcon.2019.103069

Ku, K., & Mills, T. (2010). Research needs for building information modeling for construction safety. *International Proceedings of Associated Schools of Construction 45nd Annual Conference, Boston, MA*. Associated Schools of Construction.

Lin, W. Y., & Huang, Y.-H. (2019). Filtering of irrelevant clashes detected by BIM software using a hybrid method of rule-based reasoning and supervised machine learning. *Applied Sciences (Basel, Switzerland), 9*(24), 5324. doi:10.3390/app9245324

Liu, J., Liu, P., Feng, L., Wu, W., Li, D., & Chen, F. (2020). Towards automated clash resolution of reinforcing steel design in reinforced concrete frames via Q-learning and building information modeling. *Automation in Construction, 112*, 103062. doi:10.1016/j.autcon.2019.103062

Lomio, F., Farinha, R., Laasonen, M., & Huttunen, H. (2018). Classification of building information model (BIM) structures with deep learning. *2018 7th European Workshop on Visual Information Processing (EUVIP)*, (pp. 1–6). Semantic Scholar.

McCarthy, J. (2007). *What is artificial intelligence?* McKinsey & Company.

Mutis, I., & Ramachandran, A. (2021). The Bimbot: Mediating technology for enacting coordination in teamwork collaboration. *J. Inf. Technol. Constr., 26*, 144–158. doi:10.36680/j.itcon.2021.009

Oracle Cloud. (2022). *What is AI?* Oracle Cloud. https://www.oracle.com/artificial-intelligence/what-is-ai/

Pan, Y., & Zhang, L. (2020a). BIM log mining: Exploring design productivity characteristics. *Automation in Construction, 109*, 102997. doi:10.1016/j.autcon.2019.102997

Pan, Y., & Zhang, L. (2020b). BIM log mining: Learning and predicting design commands. *Automation in Construction, 112*, 103107. doi:10.1016/j.autcon.2020.103107

Pan, Y., & Zhang, L. (2021a). A BIM-data mining integrated digital twin framework for advanced project management. *Automation in Construction, 124*, 103564. doi:10.1016/j.autcon.2021.103564

Pan, Y., & Zhang, L. (2021b). Automated process discovery from event logs in BIM construction projects. *Automation in Construction, 127*, 103713. doi:10.1016/j.autcon.2021.103713

Pan, Y., Zhang, L., & Li, Z. (2020). Mining event logs for knowledge discovery based on adaptive efficient fuzzy Kohonen clustering network. *Knowledge-Based Systems*, *209*, 106482. doi:10.1016/j.knosys.2020.106482

Paoletti, I. (2018). Mass customization in the era of industry 4.0: towards immaterial building technology. In *Informed Architecture* (pp. 77–87). Springer. doi:10.1007/978-3-319-53135-9_8

Ramaswamy, S. (2017). How companies are already using AI. *Harvard Business Review*, *14*(April), 2017.

Sacks, R., & Barak, R. (2006). Quantitative assessment of the impact of 3D modelling of building structures on engineering productivity. *Joint International Conference on Computing and Decision Making in Civil and Building Engineering*, (pp. 14–16). IEEE.

Sacks, R., Bloch, T., Katz, M., & Yosef, R. (2019). Automating design review with artificial intelligence and BIM: State of the art and research framework. *Computing in Civil Engineering 2019: Visualization, Information Modeling, and Simulation*, (pp. 353–360). Research Gate.

Sacks, R., Eastman, C., Lee, G., & Teicholz, P. (2018). *BIM handbook: A guide to building information modeling for owners, designers, engineers, contractors, and facility managers.* John Wiley & Sons. doi:10.1002/9781119287568

Shao, T., & Zhang, C. (2018). Architectural Design Model based on BIM Management System Model and Data Mining. *International Journal of Performability Engineering*, *14*(11), 2574. doi:10.23940/ijpe.18.11.p3.25742580

Sharma, A., Saxena, A., Sethi, M., Shree, V., & Varun. (2011). Life cycle assessment of buildings: A review. *Renewable & Sustainable Energy Reviews*, *15*(1), 871–875. doi:10.1016/j.rser.2010.09.008

Shin, M. H., Lee, H. K., & Kim, H. Y. (2018). Benefit-Cost analysis of Building Information Modeling (BIM) in a Railway Site. *Sustainability (Basel)*, *10*(11), 4303. Advance online publication. doi:10.3390u10114303

Song, J., Kim, J., & Lee, J.-K. (2018). NLP and deep learning-based analysis of building regulations to support automated rule checking system. *ISARC. Proceedings of the International Symposium on Automation and Robotics in Construction,* (*vol. 35*, pp. 1–7). IAARC.

Torres-Calderon, W., Chi, Y., Amer, F., & Golparvar-Fard, M. (2019). Automated mining of construction schedules for easy and quick assembly of 4D BIM simulations. *ASCE International Conference on Computing in Civil Engineering 2019: Visualization, Information Modeling, and Simulation, I3CE 2019*, (pp. 432–438). ASCE Library. 10.1061/9780784482421.055

Turing, A. M. (1950). Computing Machinery and Intelligence. *Mind, New Series, 59*(236). Oxford University Press.

Ustinovichius, L., Peckienė, A., & Popov, V. (2017). A model for spatial planning of site and building using BIM methodology. *Journal of Civil Engineering and Management*, *23*(2), 173–182. doi:10.3846/13923730.2016.1247748

Walter, M. (2006). Return on Interoperability: the new ROL. UK: John Jageurs. *CAD User, March/April.*

Wang, Y., Wang, X., Wang, J., Yung, P., & Jun, G. (2013). Engagement of facilities management in design stage through BIM: Framework and a case study. *Advances in Civil Engineering*, *2013*, 2013. doi:10.1155/2013/189105

Yan, H., & Demian, P. (2008). *Benefits and barriers of building information modelling*. Loughborough University.

Zabin, A., González, V. A., Zou, Y., & Amor, R. (2022). Applications of machine learning to BIM: A systematic literature review. *Advanced Engineering Informatics*, *51*, 101474. Advance online publication. doi:10.1016/j.aei.2021.101474

Zimina, D., Ballard, G., & Pasquire, C. (2012). Target value design: Using collaboration and a lean approach to reduce construction cost. *Construction Management and Economics*, *30*(5), 383–398. doi:10.1080/01446193.2012.676658

Zollinger, W. R. III, Sutton, D. L., Montler, G., & Seifried, M. M. (2010). *BIM: Sharing project data reduces conflict, AACE International Transactions*. BIM.

# Chapter 2
# Deep Learning-Based Damage Inspection for Concrete Structures

**Jiangpeng Shu**
*Center for Balance Architecture, College of Civil Engineering and Architecture, Zhejiang University, China*

**Zhenfen Jin**
*The Architectural Design and Research Institute, Zhejiang University, China*

## ABSTRACT

*It is important to detect and assess the damaged condition of concrete structures to ensure their safety. Several common indicators of the damaged condition such as dimension deviation of precast concrete (PC) components, propagation of the local crack, anomaly of vibration response, and decrease in structural bearing capacity are difficult to be detected automatically and efficiently. The rapid development of deep learning (DL) provides new ideas to solve the above problems. DL can effectively extract features from a large amount of multi-source data such as images, point clouds, and acceleration data, and perform related classification and detection tasks. The authors demonstrate the potential of DL in engineering practice by employing DL to address the crucial problems of damage inspection for concrete structures.*

## INTRODUCTION

Most concrete structures have a limited design life, during which they suffer multiple types of damage such as crack and component deformation, and are accompanied by abnormal vibration response and reduced bearing capacity (Ye et al., 2019). Therefore, the assessment of concrete structures is very important. However, existing various inspection methods are accompanied by the shortcomings of low efficiency, low accuracy, and greater subjective influence by the inspector. Take crack detection as an example, manual visual inspection is often conducted to examine crack width, length, and distribution

DOI: 10.4018/978-1-6684-5643-9.ch002

to ensure the safety and durability of structures. As for precast concrete (PC) dimension measurement, traditional methods such as measuring tapes are time-consuming and unsuitable for large-scale components, which exhibit certain limitations.

Therefore, several advanced transducers and equipment are developed for structural inspection in engineering practice. For the most common crack damage, the camera mounted on unmanned aerial vehicles can easily capture images for recognition, and image processing techniques (IPTs) have been developed over the years to enhance the effectiveness of damage recognition. These IPTs, including automatic thresholding(Ng, 2006), edge detection Hutchinson & Chen, 2006), and percolation-based modeling(Qu et al., 2015), extract cracks from concrete surfaces. However, unfavorable illumination conditions and complex backgrounds in the images affect the performance of IPTs.

As for the PC dimension measurement, several studies have investigated the application of three-dimensional (3D) laser scanning and images. Wang et al. (2017) used colored laser scan data for automated estimation of the reinforced PC rebar position. Herein, rebars were extracted from the laser scan data based on geometric and color features of points using a one-class classifier. Furthermore, Wang et al. (Wang et al., 2019) compared four different edge line estimation algorithms for edge length estimation. However, the dimensional information and positions between two adjacent PC components during the construction process have not been compared thus far. Therefore, an automatic and efficient approach that inspects the rebar, sleeve, and concrete of PC components is necessary.

Additionally, the deployment of advanced structural health monitoring (SHM) systems in large-scale civil structures collects large amounts of data. Note that these data may contain multiple types of anomalies (e.g. missing, minor, outlier, etc.) caused by harsh environment, sensor faults, transfer omission and other factors. These anomalies seriously affect the evaluation of structural performance. Therefore, the effective analysis and mining of SHM data is an extremely important task.

In recent years, due to the end-to-end ability and strong robustness, advanced deep learning (DL) techniques have been applied to overcome the limitation of feature extraction from a large amount of multi-source data such as images, point clouds, and acceleration data, and perform related classification and detection tasks(Avci et al., 2021; Baduge et al., 2022; Gordan et al., 2022). This chapter focus on four crucial problems of damage inspection for concrete structures to demonstrate the potential of DL in engineering practice, including (1) PC component recognition and measurement based on point clouds, (2) local damage recognition based on images, (3) structural anomaly analysis based on vibration response, and (4) finite element analysis based on images and point clouds.

## MAIN FOCUS OF THE CHAPTER

### PC Component Recognition and Measurement Based on Point Clouds

Precast concrete (PC) components are widely used in prefabricated buildings and infrastructure. In comparison with cast-in-situ structures, prefabricated structures are environment-friendly, exhibit modular design and production, consume less energy, and require short construction periods . The performance of prefabricated structures significantly relies on the quality of PC components and their connections (Ghayeb et al., 2017). Therefore, dimensional mismatches of rebars and sleeves of PC components during assembly could cause connection failures, resulting in severe structural destruction (Belleri et al.,

2015;). Typically, adjacent PC components are connected using rebars and sleeves, and grout is poured into the sleeves to ensure the integrity of the connection (Xu et al., 2019). Therefore, it is essential to locate the positions of rebars and sleeves precisely according to blueprints.

At present, quality assessment is primarily conducted using traditional manual inspection methods, which exhibit certain limitations. For instance, traditional methods are time-consuming and unsuitable for large-scale components. Moreover, manual inspection is highly prone to error. In recent years, researchers have explored several non-contact inspection methods for quality control of PC components. Tommaselli and Reiss (2005) reported an image-based approach for determining the dimensions of flat surfaces using a camera and a laser meter. Jáuregui et al. (2003) proposed a close-range noncontact photogrammetric method to measure vertical deflection in bridges. Shi et al. (2009) developed a prototype of an augmented reality system for steel column inspection, which considers column anchor bolt positions and steel column plumbness. Although these dimension inspection methods are economical, their significant reliance on human interactions and lighting conditions decreases the accuracy.

To obtain a faster and more accurate approach for the quality control of connecting PC components, a 3D laser-scanning technique is used to obtain the visual information of PC components, such as spatial coordinates and color information. Subsequently, the visual information is transferred to the point clouds of PC components for further processing. A 3D laser scanner emits laser beams and receives the reflected signals from the scanned objects to determine the distance. Several researchers adopted 3D laser scanners to inspect different types of geometric quality. Kim et al. (2020) proposed a laser-scanning-based approach that automatically evaluates the key dimensional quality assessment checklists of reinforced concrete components. Thus, most existing studies have focused on the geometric quality inspection of components. However, the inspection of subcomponents with multiple geometric or color characteristics requires further investigation. The subcomponents comprise rebars, sleeves, and concrete, and it is important to verify the accuracy of the dimensions and positions of rebars and sleeves before connecting the PC components.

To this end, an approach using algorithms for automated recognition and measurement of rebars, concrete, and sleeves of PC components is developed (Zhao, Jiang, et al., 2022). The proposed technique is based on the 3D point cloud acquired by a 3D laser scanner. The contributions of this study can be summarized as follows: (1) novel approaches are presented for recognizing rebars, concrete, and sleeves based on the 3D point cloud, and (2) the dimensions and positions of subcomponents are automatically estimated using multiple algorithms.

Figure 1 depicts the workflow of the proposed recognition and measurement approach. Initially, a 3D laser scanner acquires the raw data of the PC components, which are preprocessed by denoising and downsampling based on the point density. After the completion of data preprocessing, the data is used for recognizing the rebar, concrete, and sleeve. Subsequently, boundary estimation is performed to determine the boundaries of the sleeve ports, concrete planes, and rebar cross-sections. The final step estimates the position and dimension of the subcomponents. The details of each step are explained in the subsequent sections.

*Figure 1. Overall procedure of the proposed recognition and measurement approach*

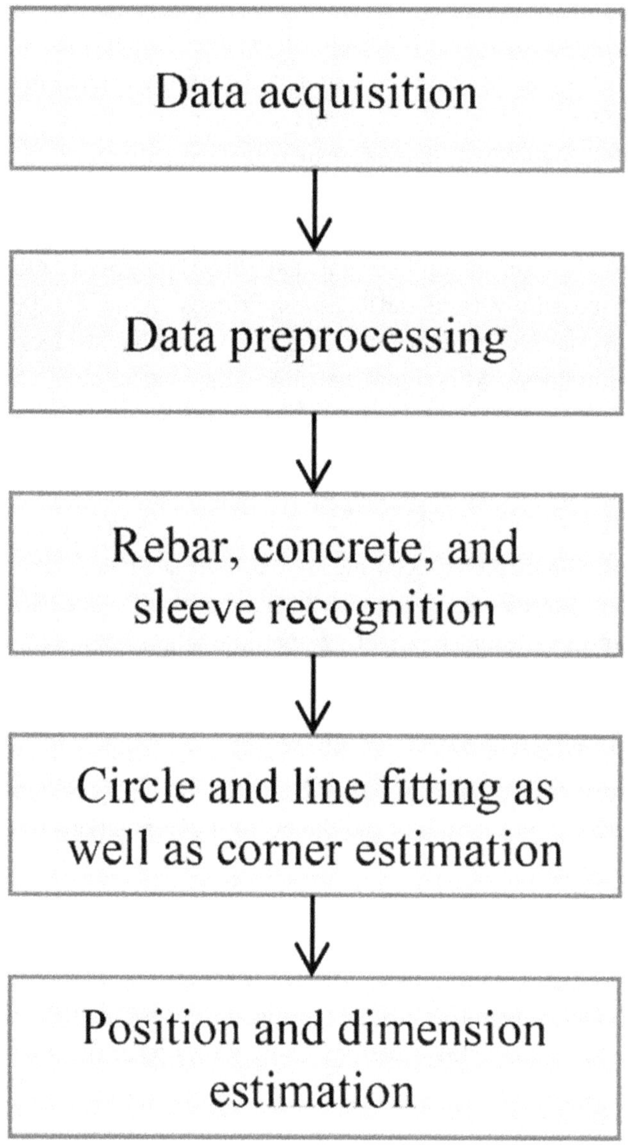


## Data Preprocessing

Considering the influence of the scanning angle, scanning distance, surface material of the inspected object, and environmental factors, the acquired 3D point cloud model may include noise (Hebert & Krotkov, 1992). Therefore, it is necessary to eliminate as much noise as possible during the data preprocessing step. The *k*-nearest neighbors (KNN) algorithm (Peterson, 2009) was used to reduce the noise in the point cloud. The KNN algorithm is a non-parametric method used for classification or regression. The input comprises *k* closest training examples in a dataset. During classification, an object is assigned to the class that is most common among its *k* nearest neighbors. When the KNN algorithm is applied to denoise a point cloud, a point is considered as noise if the average distance from a point to its *k* nearest neighbors is larger than a self-defined threshold value. Additionally, a K-dimensional tree (KD-tree) was used to accelerate the process of searching for the nearest neighbors. Several types of 3D laser scanners contain spatial coordinates and color information, which, if required, can also be used to improve the denoising effect caused by the KNN algorithm (Wang et al., 2017).

Additionally, a density-based downsampling algorithm was developed to reduce the point density of the overall data while simultaneously retaining the relative data integrity of low-density subcomponents. Initially, the point cloud was divided into multiple grids, and the number of grids in the *x*-, *y*-, and *z*-directions was customized. Each grid was assigned an index, wherein all the points in a grid were designated the same index. The point density for grids with the number of points is greater than a self-defined value was calculated.

## Rebar, Concrete, and Sleeve Recognition

As mentioned above, the performance of prefabricated structures predominantly relies on the quality of PC components and their connections. Therefore, it is important to recognize rebars, concrete, and sleeves, which are used to connect the PC components, and verify the accuracy of their dimensions and positions to avoid dimensional mismatches that may lead to construction failures.

As the rebars do not exist in the concrete plane and a significant amount of external interference may appear in the point cloud, the features of rebars that are different from other parts are extracted to distinguish the rebars. The one-class support vector machine (OC-SVM) algorithm (Mazhelis, 2006; Wang et al., 2017) was used to identify the points of the rebar class based on the geometric and color features of a training set of rebars. Typically, one-class classifiers are used to extract data of an expected class among all the points by learning from a training set containing only the points of the expected class. The OC-SVM algorithm maps the points of the expected class to a region where the value of the function is non-zero, and the remaining points are mapped to the origin (Schölkopf et al., 2001).

The density-based spatial clustering of applications with noise (DBSCAN) algorithm (Ester et al., 1996) was used to separate different concrete planes. The DBSCAN algorithm is a data clustering method that identifies arbitrary shape clusters in the presence of noise. Herein, a set of points in dense regions are grouped, whereas those in low-density regions are marked as outliers.

The sleeves, in the point cloud of the component with sleeves, are generally buried inside the concrete with their ports on the surface of the concrete. Therefore, the primary objective was to recognize sleeve ports. The alpha-shape algorithm (Edelsbrunner et al., 1983; Widyaningrum et al., 2020a) was employed to extract the edge points of the sleeve ports. In computational geometry, the alpha-shape is a concept associated with the convex hull, and the algorithm is based on the Delaunay triangulation of the point set. Delaunay triangulation aims to generate disjoint triangular grids for points in a point cloud. The vertices of each triangular grid indicate the three closest points, and no other point exists in the grid.

## Boundary and Corner Estimation

Figure 2(a) illustrates the estimation of the edge points of the hollow square point set using the alpha-shape algorithm. The edge points evaluate the performance of the algorithms by first fitting the inner circle boundary, estimating the corner points, and then fitting the outer square boundary. As the detected edge points of circle-shaped boundaries are generally not on a circle and are unevenly distributed, circle centers cannot be estimated by calculating the mean value of the edge points. Therefore, it is necessary to fit a circle that passes closest to all the edge points on a circle-shaped boundary. A least-squares circle fitting algorithm (Maisonobe, 2007) was used to fit the inner circle boundary of the point set based on its edge points. The algorithm is a regression analysis approach to identify the best function matching the data by minimizing the sum of the squared errors. Figure 2(b) depicts the circle fitting result.

*Figure 2. Edge point detection result of a hollow square point set at $\alpha = 2$ and (b) location of the inner circle boundary and outer square boundaries*

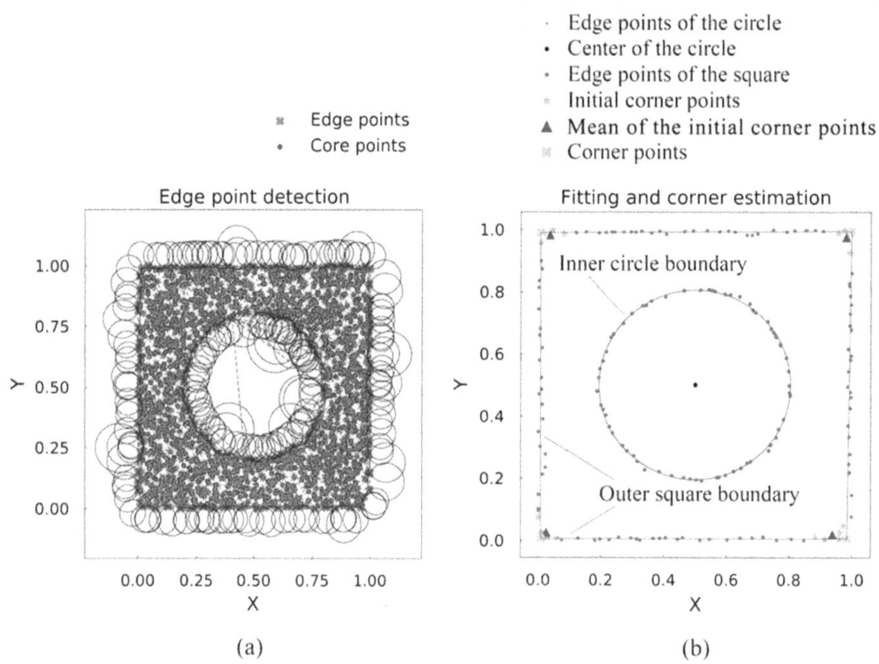


## Position and Dimension Estimation

In the point cloud model of the component with rebars, rebars were first recognized and segmented from the point cloud. The concrete segment was then processed using the concrete part recognition approach to separate different concrete planes, which were divided into the concrete plane with rebars and other planes. Subsequently, the former concrete plane was projected onto the 2D plane placed parallel to the concrete plane. The edge points of the outer square boundary of the rebar-located concrete plane were extracted using the alpha-shape algorithm. Furthermore, rebar-associated points were projected onto the 2D plane. Eventually, the corner points of the concrete plane were estimated, and the sidelines of the concrete plane were fitted using the least-squares line fitting algorithm. Additionally, the circle boundaries of the rebar cross-sections were fitted using the least-squares circle fitting algorithm.

## Local Damage Recognition Based on Images

The most common local damage is cracks, which severely affect concrete structures. Traditional methods of crack inspection are inefficient and inconvenient as they rely on manual measurement. Therefore, several IPTs have been developed over the years to enhance the effectiveness of damage inspection in civil engineering. However, IPTs are easily affected by unfavorable illumination conditions and complex backgrounds in the images.

To overcome the limitations of IPTs, deep learning was initiated in the 1990s based on convolutional neural networks (CNNs) that could identify handwritten zip codes(LeCun et al., 1989). In 2015, Shelhamer et al.(2017) proposed a fully convolutional neural network (FCN) that uses deconvolution in up-sampling. Since then, FCN has promoted the use of end-to-end CNNs in semantic segmentation. Based on this concept, U-Net(Ronneberger et al., 2015), which can perform end-to-end training using only a few images, was developed. However, the process of reducing the receptive field before increasing the image size caused information loss in the FCN model. Thereafter, Yu and Koltun (2016) proposed dilated convolutions, wherein multi-scale contextual information is systematically aggregated without compromising the resolution; this increases the accuracy of semantic segmentation systems. Additionally, Lin et al. (2017) proposed feature pyramid networks to fuse multilevel semantic information.

Due to the strong ability and relevance of DL method, it was introduced to enable crack detection in concrete structures, and this method typically is used to classify, locate, and segment cracks. Based on the U-Net method, Huyan et al. (2020) proposed Crack U-Net for pixel-wise crack detection, which outperforms both classical FCN approaches and U-Net. Ni et al. (2019) presented a crack delineation network wherein cracks are delineated accurately and rapidly from images; thus, this method represents a fully autonomous vision-based approach to structural crack detection. Liu et al. (2019) proposed DeepCrack, a deep hierarchical CNN for segmenting cracks based on multi-scale and multi-level features.

However, several deficiencies exist in the research on crack segmentation and measurement The existing crack segmentation network based on DL is not sufficiently effective to perform crack segmentation under complex conditions. Furthermore, owing to the poor image resolution, accurate measurement of crack widths is challenging. Additionally, the core component of any DL-based application of segmentation methods is data (Bao, Chen, et al., 2019); this also applies to the crack segmentation task. Currently, the number of available image databases of structural systems and other infrastructure components is very limited for SHM purposes; this leads to poor performance of the available trained models when new conditions arise, such as texture, joints, light, environment, and pollution (Azimi et al., 2020). Building a new dataset is not easy, which means the manual annotation of crack images relied on experts, making it costly, time-consuming, and laborious. Therefore, the high cost of annotation is an obstacle to increasing limited databases. In particular, the amount of unlabeled crack images is increasing dramatically with the increasing development of diversified image acquisition methods.

To address the above issues about crack segmentation networks and data limitation, an active learning framework is proposed to identify the most informative crack images (Shu et al., 2022). Annotating these selected crack images to build a dataset can support sufficient supervision information and reduce the requirement of labeled crack images dramatically. In addition, a novel crack feature pyramid network (Crack-FPN) is proposed (Zhao, Liu, et al., 2022), wherein the segmentation result is obtained by fusing the high-resolution and high-level semantic information of low- and high-level features, respectively.

## Active Learning Framework With Difficulty Learning Mechanism

An active learning method with a difficulty learning mechanism is introduced to select the most informative crack images. Considering the background diversities and shape disparity of cracks, a semantic difficulty difference exists among different semantic areas. To capture this difference, a two-branch network with difficulty learning mechanism is adopted in the method. One branch acts as a segmentation means, and the other one works for difficulty learning. The former segmentation branch adopts a common segmentation network. Through the segmentation branch, the probability map and prediction result are generated. After comparing the prediction results and ground truth, the wrong prediction is obtained. For the latter difficulty learning branch, the wrong prediction result is leveraged as the supervision. In the second branch, a pixel-wise attention module is adopted first to aggregate similar pixels, and learn the proportion of misclassified pixels as the difficulty score. In the last of the second branch, the semantic difficulty map is obtained. Based on the difficulty map and the uncertainty map from segmentation, an acquisition function is utilized to measure the informativeness of images in the unlabeled dataset. These unlabeled images are ranked by their informativeness. The most informative *m* samples of them are selected. After annotation by experts, the selected *m* samples form a subset. This subset is added to the labeled dataset. The final is then ready for the next active learning stage.

The overview of the active learning method with the learning mechanism is shown in Figure 3. Subsequently, a more detailed active learning framework is introduced. Then, the probability attention module and loss functions are presented.

*Figure 3. Overview of the active learning method with difficulty learning mechanism*

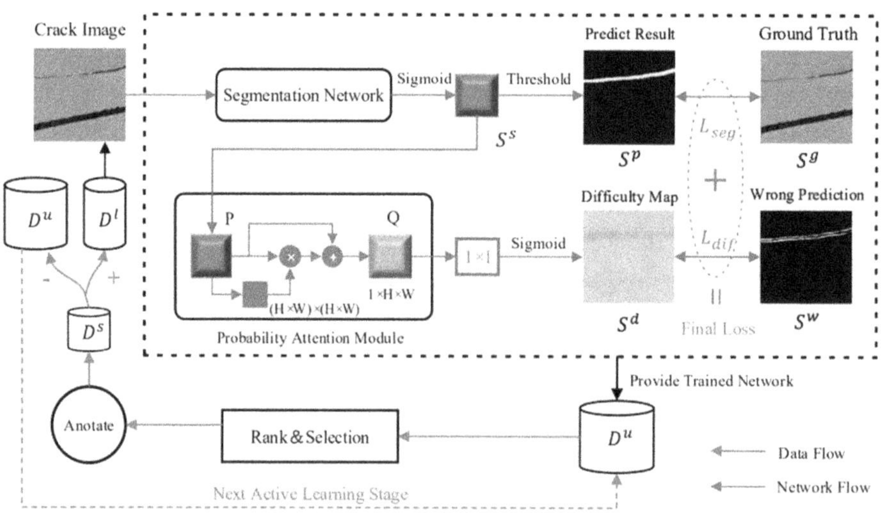


A two-branch network is built to complete a multi-task manner, and generates crack segmentation predictions and difficulty maps. The first branch is a common segmentation network. As illustrated in Figure 3, the output of the sigmoid layer is assumed as $S^s$, and the prediction result $S^p$ is generated at the threshold of 0.5. By comparing $S^p$ and the ground truth $S^g$, the wrong prediction result $S^w$ that will be used at the second branch to learn semantic difficulty, is computed using the following equation:

$$S_i^w = \begin{cases} 1 & if\ S_i^P \neq S_i^g \\ 0 & if\ S_i^P \neq S_i^g \end{cases} \tag{1}$$

where $S_i^w$ is the $i^{th}$ pixel value of the wrong prediction result, and $S_i^P$ and $S_i^g$ represent the $i^{th}$ pixel value of prediction result and ground truth, respectively.

The second branch comprises two parts. The first part is a probability attention module. The sigmoid output of the first branch $S^s$ is directly used as the input of the second branch. For the crack segmentation task, $S^s$ is a one-channel probability map. Thus, the input of the second branch, named P, is the shape of C × H × W (Chanel × Height × Width). There is an advantage by directly using probability maps, because pixels with similar difficulty are likely to have a similar probability vector. If they are combined with a pixel-wise attention module, similar pixels are easily aggregated and similar difficulty scores are attached to them. The second part of the second branch is a simple 1 × 1 convolution layer to process the output of the probability attention module for binary classification.

In conclusion, the whole semantic difficulty learning process can be divided into two steps. First, a learned segmentation network is trained with the wrong prediction results $S^w$. Thereafter, every pixel will learn a semantic difficulty value. Finally, the semantic difficulty map $S^d$ is obtained.

The probability attention module is explained in detail in this section. According to (Fu et al., 2019), pixels with similar probability will be aggregated through this module. The input of the second branch, i.e., the probability map $P$, is the shape of $C \times H \times W$. $P$ is first reshaped into $C \times K$, in which $K$ is calculated by $H \times W$. Then, the attention matrix $A$, which is the shape of K × K, is generated with $P^TP$ and a sigmoid operation as follows:

$$A_{ji} = \frac{exp\left(P_i^T \cdot P_j\right)}{\sum_{i=1}^{K} exp\left(P_i^T \cdot P_j\right)} \tag{2}$$

$$Q_j = \gamma \sum_{i=1}^{K}\left(A_{ji}P_i\right) + P_j \tag{3}$$

where $A_{ij}$ is the $i^{th}$ pixel's impact on the $j^{th}$ pixel, $P_j$ is the probability vector of the $j^{th}$ pixel in the probability map $P$, $Q_j$ is the result after attention $\gamma$ is a learnable weight factor. The final probability map after the attention module (Q) is obtained. Moreover, the probability attention module ((Fu et al., 2019) is designed to capture the position relationship of pixels. The boundaries of objects, like 'pole', are paid more attention and predicted more precisely.

## Architecture of Crack-FPN

Figure 4(a) depicts the structure of Crack-FPN. The quadrilaterals represent the input, output, and intermediate layers, and the arrows represent various operations of the neural network.

*Figure 4. Architecture of Crack-FPN. (a) Crack-FPN structure, (b) Se_ResNeXt module, (c) skip connection, and (d) W operation*

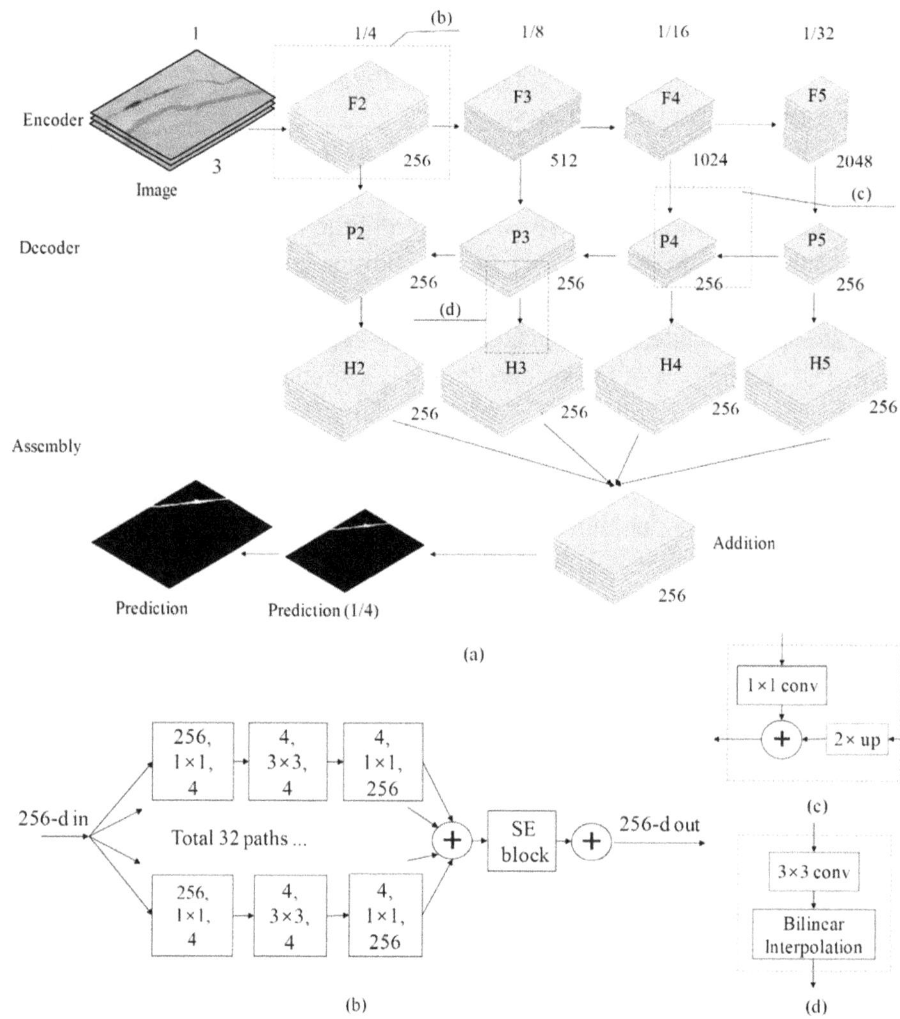


The proposed network structure is divided into three parts—the encoder, decoder, and assembly. The encoder is a feedforward process in the convolutional network, extracting feature vectors at different stages and reducing the feature resolution continuously. The decoder performs bottom-up feature map enlargement, wherein the top and bottom levels of the feature map are merged via up-sampling to enhance the semantic information. After the fusion, a 3 × 3 convolution kernel is used to convolve each fusion result to eliminate the aliasing effect of up-sampling. Finally, in the assembling process, the feature maps of each stage are combined to obtain the information of each layer. Thus, Crack-FPN can utilize valuable semantic information to satisfy the speed and memory requirements for crack detection. The unique characteristic of Crack-FPN is that predictions are performed independently at each feature layer, which facilitates the detection of different-sized cracks.

The encoder is a feature extraction network that generally uses VGGNet (Simonyan & Zisserman, 2015) and ResNet (He et al., 2016) as a backbone. However, the Se_ResNeXt50_32x4d model (Hu et al., 2020) (Figure 4 (b)), which includes ResNet, an SE block, and ResNeXt (X), is adopted to ensure enhanced feature extraction. The residual block in ResNet enables a deeper network, faster convergence, and easier optimization. Additionally, it has fewer parameters and is less complex than the previous model like VGGNet, which renders it suitable for various computer vision tasks. The SE block is a computational unit adopted from SENet (Hu et al., 2020). The squeeze function executes global average-pooling to generate feature maps with global receptive fields, and the excitation function uses a fully connected neural network to perform a nonlinear transformation on the squeeze operation result. The resultant transformation is used as a weight to multiply the input feature. The SENet primarily learns the correlation between channels, filters the channel attention, and slightly increases the calculation cost; however, an effective result is obtained. The core innovation of ResNeXt (Xie et al., 2017), an upgraded version of ResNet, lies in the aggregated transformations that it performs. The original three-layer convolutional block in ResNet is replaced with a parallel stacked block of identical topology, which improves the model accuracy without significantly increasing the parameter magnitude. Owing to the identical topology, the hyperparameters are reduced, which renders the network convenient for model transplantation. ResNeXt50_32x4d is an enhanced version of ResNet50 with 50 layers, wherein 32x4d represents 32 paths, with each path containing four channels. The SE block is embedded in ResNeXt50_32x4d to generate the final network, Se_ResNeXt50_32x4d.

Therefore, the encoder is a bottom-up crack feature extraction network with se_resnext50_32x4d as the backbone. A stage corresponds to a level of the feature pyramid in subsequent operations. The features extracted from the conv2, conv3, conv4, and conv5 layers are denoted as $\{F2, F3, F4, F5\}$, respectively, which are the four levels in Crack-FPN. Here, the information of the bottom feature, $F1$, is unused, which significantly reduces the calculation cost for network training without affecting prediction results. The feature vectors are $F2 = (N, 256, 80, 80)$, batch size is denoted as $N$, $F3 = (N, 512, 40, 40)$, $F4 = (N, 1024, 20, 20)$, and $F5 = (N, 2048, 10, 10)$. As the feature map of $F5$ is 1/32 that of the original image, the length and width of the input crack image must be multiples of 32.

The decoder performs top-down enlargement of the crack feature map. $P5 = (N, 256, 10, 10)$ is directly obtained from $F5$ using a $1 \times 1$ convolutional layer. In the subsequent operations, $P5$ is magnified by a factor of 2 using an interpolation algorithm to generate $(N, 256, 20, 20)$. $F4$ is passed through a $1 \times 1$ convolutional layer and becomes $P4 = (N, 256, 20, 20)$. The two aforementioned feature vectors are combined to form $(N, 256, 20, 20)$. Figure 4(c) depicts these operations, called skip connections, which are further used to generate $P3$ and $P2$. The skip connection utilizes both the high-level semantic features of the top layer (helpful for crack feature classification) and the high-resolution information of the bottom layer (helpful for crack feature location).

In the assembling operation, the output feature vectors at each level of the pyramid must have identical resolutions. To ensure this, an operation $W$ involving a $3 \times 3$ convolution layer and double linear interpolation amplification is performed (Figure 4(d)). $P5$ changes to $H5 = (N, 256, 80, 80)$ after three $W$ operations; $H4$ and $H3$ undergo two and one $W$ operations respectively, changing to $(N, 256, 80, 80)$. However, $H2$ need not be amplified because it already has the same size as other $H_i(i{=}3\text{–}5)$. Subsequently, the values of $H_i(i{=}2\text{–}5)$ are directly added to obtain the vector $(N, 256, 80, 80)$, which is passed through a $3 \times 3$ convolution layer and undergoes a bilinear interpolation to attain the original crack image size, $(N, 1, 320, 320)$. To facilitate the prediction, a mask = $(N, 1, 320, 320)$ is determined using a sigmoid activation function whose values are 0–1. If the value of a point exceeds a fixed threshold, the point is predicted to be a crack. In our experiment, the threshold value was 0.5, and the IoU was calculated accordingly.

In general, the structure of Crack-FPN has its own unique advantages. The encoder as the feature extraction network generally uses VGG or ResNet as the backbone. To improve the feature extraction efficiency of cracks here, Se_ResNeXt50_32x4d, a combination of ResNet, SE module, and X module of ResNeXt network, is used as the encoder of the feature pyramid network. The simple addition of feature maps with different resolutions at different levels in the sampling affects the information integrity and accuracy, called aliasing effect. To avoid this, the authors use W operations including 3×3 convolutions to unify the four different resolution outputs of the feature pyramid network into the same size for addition. Unlike most classical networks such as VGG and U-Net, which only consider the feature maps of the last convolutional layer, Crack-FPN extracts four different levels of feature maps during the encoding-decoding process and assembles these feature maps for final prediction. It uses high-level features to supplement low-level semantics, thus obtaining high-resolution and strong semantic features.

## Structural Anomaly Analysis Based Vibration Response

With the rapid development of advanced sensing techniques, a substantial number of large-scale civil infrastructures are equipped with sophisticated structural health monitoring (SHM) systems(Gatti, 2019; Heinlein et al., 2019; Li & Ou, 2016; Y. L. Xu, 2018), aiming to ensure the safe and stability of the structure. After long-term monitoring of structures, these SHM systems have collected a large volume of data including wind, temperature, displacement, strain, acceleration and so on. Such information can be employed to evaluate the structural performance, judge whether the structure is damaged, and predict the future behavior of the structure. However, it is noteworthy that massive data in SHM systems generally present various types of anomalies caused by a series of factors like environment interference, sensor faults and system malfunctions (Huang et al., 2016; J. Park et al., 2017; Yuen & Mu, 2012). These anomalies have a great impact on the structural performance assessment, because it is difficult to distinguish whether these anomalies are caused by structural damage or by incorrect data. Therefore, to overcome this problem, it is essential to identify and remove anomalies from SHM data, which requires data processing methods to have high recognition accuracy and high calculation efficiency.

To date, many development traditional anomaly detection methods have been utilized in the field of SHM (Kerschen et al., 2005; S. Park et al., 2018; Yang & Nagarajaiah, 2014). For instance, Yi et al. (Yi et al., 2016) employed a statistical process control approach, called as the cumulative sum (CUSUM) chart, for detecting persistent shifts in global positioning system (GPS) monitoring. Ding et al. (Ding et al., 2016) proposed a latent correlation-based anomaly detection approach that effectively identifies potential anomalies from a large number of correlative isomerous monitoring data. Note that the traditional anomaly detection approaches are inefficient. Meanwhile, the uniqueness and the robustness of the features extracted by the traditional approaches are not good enough.

Nowadays, AI-based method is capable of detecting multiple anomaly data fast and automatically with a higher accuracy. Therefore, it has become popular research topics in SHM. For example, Bao et al. (Bao, Tang, et al., 2019) put forward a novel data anomaly detection approach based on computer vision and deep neural network. In this approach, the original time series signals are first transformed into image vectors, and these image vectors were then input into a deep neural network for training and classifying data pieces. Subsequently, Ni et al. (Ni et al., 2020) designed a one-dimensional CNN for extracting the essential features of input signals and detect the anomaly of SHM data. CNN is difficult to obtain high classification accuracies for class-imbalanced problems, because it is based on a class-balanced hypothesis (Yin & Gai, 2015). The development of generative adversarial network (GAN) offers a new perspective for overcoming this problem (Jiang et al., 2019). The basic principle of GAN is that it generates prototypical samples via a generator with random samples that satisfy a definite distribution (e.g. Gaussian distribution) (Jiang et al., 2019). However, in the field of SHM, GAN-based approaches for anomaly detection are rarely seen.

To address these problems and obtain accurate detection results, a novel data anomaly detection method is developed (Liu et al., 2022). This method mainly includes three modules (as presented in Figure 5. Framework of the proposed data anomaly detection method): (1) A three-channel input including time response image, frequency response image and Gramian Angular Field (GAF) image is built; (2) A GANomaly network is introduced to solve the class-imbalanced issue; (3) Different types of anomalies are distinguished by a designed CNN network.

In the first module, to better distinguish multiple anomaly types, extra frequency domain features are introduced via using fast Fourier transform (FFT) method. Meanwhile, the original time series are encoded into two-dimensional vector images by using GAF method. In addition, a two-stage of hybrid sampling strategy called MUS1 and MUS2 is adopted. In the second module, only normal samples are considered in the training stage. All different types of anomalies fall into one class. Such a class of samples can be discriminated by higher anomaly scores compared to normal samples in the testing stage. In the third module, a training set consisting of labeled samples is fed into a CNN network, and then multiple types of anomalies can be detected via the trained CNN.

## Three-Channel Input Establishment and Multistage Sampling

Generally, abnormal samples include six types, i.e. missing, minor, outlier, square, trend and drift. Considering that it may be difficult to distinguish multiple anomaly types only by using time response information. A three-channel input including time response image, frequency response image and GAF image is established. Particularly, GAF encodes time series into image by polar coordinates based matrix and it can preserve absolute temporal correlation (Barra et al. 2020).

FFT method is employed to obtain frequency response information. Then, the frequency response signal is converted into grayscale image and input to the network. Thus far, a three-channel input is derived. In addition, to addressing class-imbalance issue in SHM data, a multistage sampling method is introduced. In anomaly detection, normal data usually plays a dominant role in the whole dataset. The increasing ratio of examples between majority and minority classes as well as the number of minority classes had a negative effect on performance of the resulting classifiers. Hybrid sampling method can simply rebalance the datasets by taking constant numbers of samples from different classes. Nevertheless, the number of samples from different classes being equal did not always correspond to the optimal results. Therefore, a two-stage of hybrid sampling called MUS1 and MUS2, is adopted to effectively relieve severe imbalance of training samples and achieve the optimal sampling ratios.

*Figure 5. Framework of the proposed data anomaly detection method*

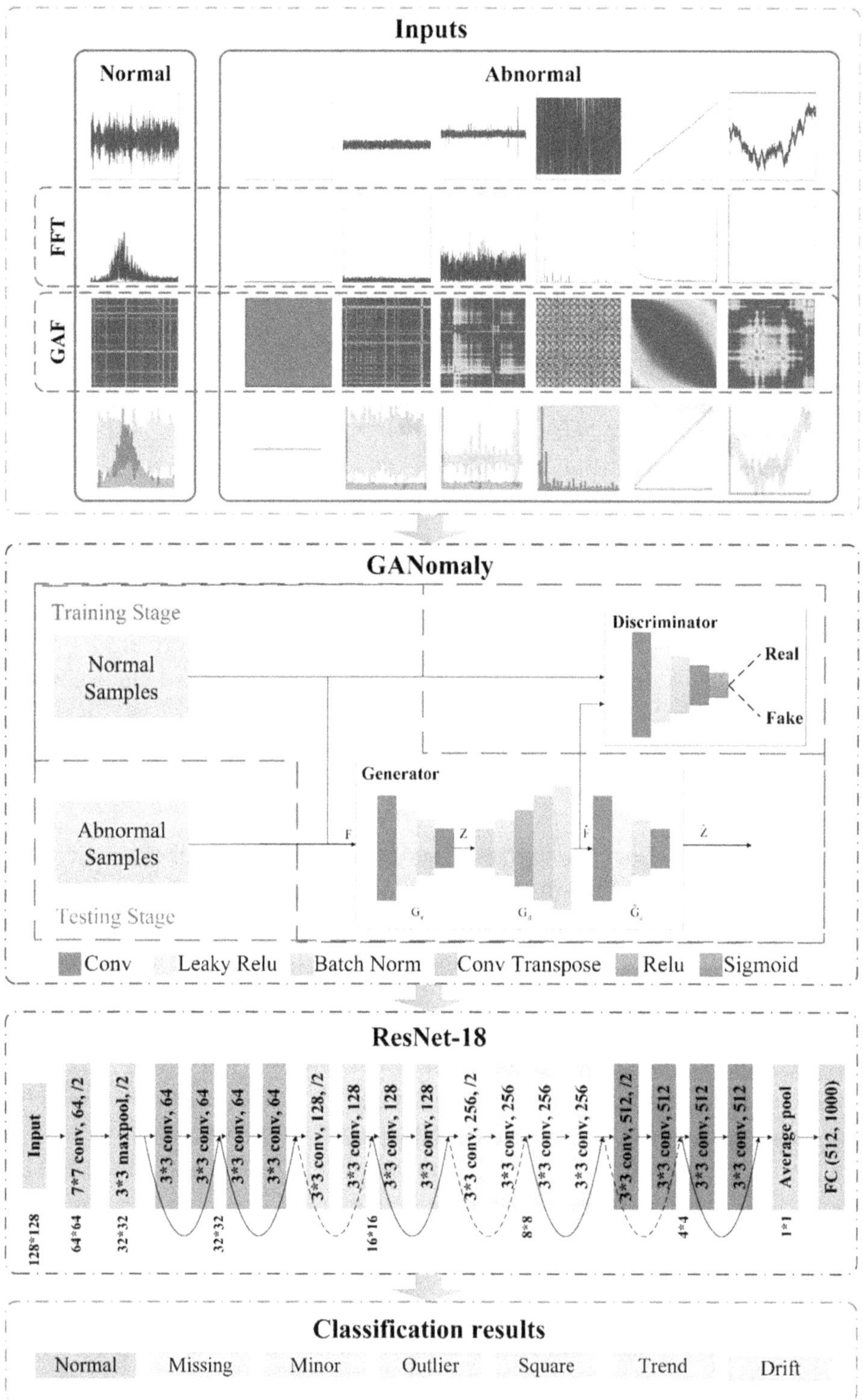

## GANomaly

For class-imbalance problem, the dataset-oriented method (i.e. multistage sampling) is an available scheme. Beyond this, GANomaly has also been proved to be an effective method for processing class-imbalance problem (Jiang et al., 2019). Therefore, it is introduced to detect abnormal samples from normal samples involving imbalanced dataset, which consists of a generator and a discriminator. An encoder-decoder-encoder three-sub-network is employed in the design of the generator. Two encoders $G_e$ and $\tilde{G}_e$ learn to get the input representations of original images $X$ and regenerated images $\widehat{X}$ respectively. The decoder $G_d$ tries to reconstruct $\widehat{X}$ at the same time. The corresponding process of the generator is the following:

a) With the use of convolutional layers followed by batch-norm and leaky ReLU activation, respectively, the encoder $G_e$ downscales $x$ into latent vector $z$ via $z=G_e(x)$;
b) The decoder $G_d$ adopts convolutional transpose layers, ReLU activation layers and batch-norm layers. This method upscales the latent vector $z$ to reconstructs the output, defined as $\hat{x}$, where $\hat{x} = G_d(z)$;
c) The last encoder $\tilde{G}_e$ has the same architecture as $G_e$ but with different parametrization. $\tilde{G}_e$ downscales $\hat{x}$ to obtain its feature representation $\hat{z}$ via $z = \tilde{G}_e(\hat{x})$;
d) The discriminator $D$ is based on the standard discriminator introduced in DCGAN (Radford et al., 2016). It is applied to distinguish whether input data is real or fake.

The objective function $L$ combines three loss functions:

$$L= w_{adv}L_{adv} + w_{con}L_{con} + w_{enc}L_{enc} \tag{4}$$

$$\begin{cases} L_{adv} = \left\| f(x) - f(\hat{x}) \right\|_2 \\ L_{con} = \left\| x - \hat{x} \right\|_1 \\ L_{enc} = \left\| z - \hat{z} \right\|_2 \end{cases} \tag{5}$$

where $L_{adv}$, $L_{con}$, $L_{enc}$ are adversarial loss, contextual loss and encoder loss, respectively, and $f$ is a function that outputs an intermediate layer of the discriminator $D$.

During the testing stage, the encoder loss $L_{enc}$ is used for scoring the abnormality of a given image. The anomaly score $A(x)$ for a test sample $x$ is defined as:

$$A(x) = \left\| G_E(x) - E(G(x)) \right\|_2 \tag{6}$$

For an anomaly sample $A(x)$ will be much higher compared to a normal sample, which help us to identify the anomalies. The procedure above should be enough to solve anomaly detection in the scenario of unsupervised learning, which aims at distinguishing between normal and abnormal samples. For a supervised learning task, the probability of a normal sample $p(x)$ is given by a sigmoid function (Han & Moraga, 1995).

$$p(x)=1-\frac{1}{1+e^{-2\gamma A(x)}} \tag{7}$$

where $\gamma$ is a parameter for the slope of the function.

### CNN

A residual neural network, i.e. ResNet-18 (He et al. 2016), is established and employed for the automated classification of different types of anomalies. This network consists of five convolutional layers (see Figure 1 for further details). The classification layers consist of the full connection layer and the softmax layer, and each classification label corresponds to one output type.

## Finite Element Analysis Based on Images and Point Clouds

Natural aging and increasing service demands on existing reinforced concrete (RC) structures have produced a significant number of structures experiencing damage and at risk of failure. As a critical stage in the lifecycle of structures, inspection and assessment have become an important emergent topic, and researchers have investigated numerous methods for automating this process and enhancing structural assessment, including automatic modeling and finite element (FE) analysis. One interesting prospective structural assessment method is the use of digital twin (DT) models obtained from geometric modeling and damage inspection results (Zandi et al., 2019). However, despite significant advances in point cloud-based automated modeling and the automated inspection of images, DT models remain ineffective in incorporating damages.

New and innovative 3D scanning technology allows for the reliable and rapid collection of points on structural surfaces based on the measured distances between scanner and objects. For instance, Yan et al. (2021) and Lu et al. (2019) detected structural components in the point clouds of steel girder and RC beam bridges, respectively. Lu et al. (2019) generated geometric DTs of existing RC beam bridges using swept solid representations based on industry foundation classes. Overall, geometry modeling based on point clouds significantly automate the process of modeling existing conditions, which is a basis for structural assessment.

However, as noted by Blomfors et al. (Blomfors et al., 2020), cracks uncovered during the inspection of existing structures must be considered in their structural evaluation. To address this issue, a large number of studies have focused on the detection of damages such as cracks and concrete spalling using point clouds. Guldur et al. (Guldur & Hajjar, 2014) developed a point cloud-based approach to detect and quantify concrete cracks and spalling involving the comparison of normal variations. Liu et al. (2019) studied convolutional neural networks and U-net structures as tools for segmenting cracks in concrete surfaces. Overall, there is a data gap between geometry modeling and damage detection approaches, which says that different source data is used to obtain geometries and damages of an existing structure, and the results are unable to be integrated directly. To overcome the challenges described above, some attempts at data fusion between point clouds and images have been made. The key problem in such approaches is to determine the position and posture of the camera used to obtain the image data. Because point clouds and images are obtained from different devices and perform different roles (for example, imaging focuses on collecting color information whereas point cloud approaches are primarily used to record 3D coordinates and intensities), registration of images into point clouds remains a challenge.

To automatically model the geometries of RC structures that incorporate existing damages and then generate FE models for structural assessment, a method based on the alignment of point clouds and images is developed (Yu et al., 2021; Zhang et al., 2022). The workflow for generating an FE model of a cracked RC beam is shown in Figure 6. The process comprises three components: (1) damage detection, (2) an Iterative Closest Point-based Direct Linear Transformation (ICP-based DLT), and (3) model generation. The required inputs for the proposed process include preprocessed point clouds and images, with the point clouds used to model the geometry of the RC beam using a slice-based method, and the images used to extract and quantify cracks. Feature points such as corners and lines are extracted from both images and 3D models. Subsequently, the proposed ICP-based DLT algorithm is used to find the relationship between the 2D image and 3D object spaces. Based on the obtained relation, cracks in the images are back-projected to the 3D model. To aid in structural assessment, a corresponding FE model is finally generated based on the extracted geometric data and back-projected crack points.

*Figure 6. Overall workflow of the proposed method*

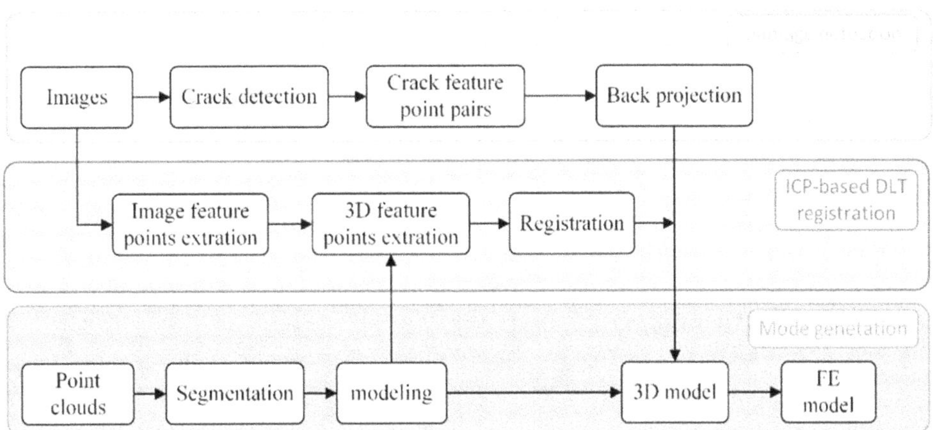


## Damage Detection

Damage processing involves two primary steps: (1) crack segmentation and followed by the extraction of crack edges and (2) the calculation of crack parameters, including crack length and width. In the proposed method, image thresholding and edge detection algorithm are employed instead of using deep learning methods, because the background of the image is clear enough to perform image processing algorithms.

Two approaches for extracting cracks are investigated in this method: edge detection and image thresholding technology. The canny edge detector is a commonly used tool for crack detection (Abdel-Qader et al., 2003), and a non-maximum suppression and double thresholding strategy are implemented to extract true edges. The threshold approach determines cracks using global and local thresholding techniques such as adaptive and global thresholding. In applying this method, a further step is required to extract crack edges, which can be realized by implementing the alpha-shape algorithm (Widyaningrum et al., 2020b).

After obtaining the crack edges, skeleton points are computed by applying the medial axis transform (MAT) algorithm, a powerful tool for generating skeleton points that represent the topology of an object. To implement the shrinking circle principles, as explained by Widyaningrum et al. (2020b), to generate 2D MAT crack points, the normal vectors of each edge point are first estimated using $k$ neighbors searched by a k-nearest neighbor algorithm. Under the proposed approach, only internal MAT points are required, and the calculation of external MAT points can be avoided by orienting the normal vectors of the crack edge points toward the outside of the crack area.

To obtain crack width feature pairs as shown in Figure 7(a), for each MAT point, four edge points that define two edge lines on two respective sides are searched, enabling the determination of the intersections of the two edge lines and the normal vector of the skeleton point. In the implementation, a B-Spline curve is first fitted to the skeleton points and then a number of points on the fitted B-Spline curve are redefined with the goal of filling the missing MAT points and reducing the influence of noise (see Figure 7(b)). In the following discussion, the redefined points are still referred to as skeleton points for convenience. The normal vectors of the B-Spline curve at each point are calculated by deviating the B-Spline curve (see Figure 7(c)). The result of calculated crack feature pairsis shown in Figure 7(d).

*Figure 7. Method for calculating crack width feature pairs: (a) Schema of crack width feature pair calculation. (b) B-Spline curve fitting. (c) Normal estimation. (d) Results of crack width feature pairs*

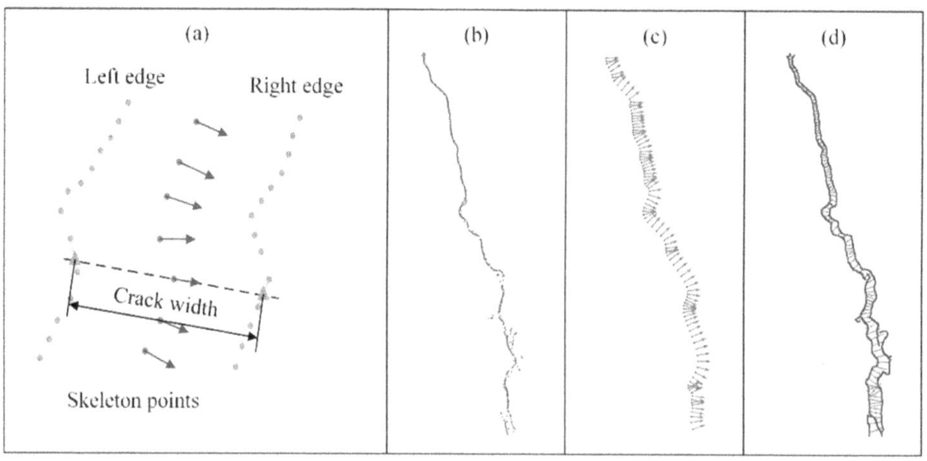


## ICP-Based DLT Registration

The goal of recovering the camera pose of an image is to minimize the total distance between the projected feature points and the 2D image features based on the application of DLT algorithm. Corners and lines are common features of structures that can be used for registration. However, the distance between image points and projected lines is often difficult to calculate. Therefore, the proposed method discretizes 3D lines into scattered points to convert optimal objects to distances between pairs of point sets. The overall ICP-based DLT algorithm is described in Algorithm 1.

*Table 1. Algorithm 1*

| **Algorithm 1: ICP-based DLT registration** |
|---|
| **Input**: the feature point sets in 2D image space ($p_m$) and 3D object space ($p_w$) |
| **Output**: the projective matrix ($P_{ult}$) between 2D image space and 3D object space |
| Compute initial $P_{initial}$ using at least 6 corners |
| $N \leftarrow$ the number of $p_m$ |
| $i \leftarrow 0$ |
| change $\leftarrow 0$ |
| $P_{iter} = P_{initial}$ |
| **While** $i <$ maximum iteration and change < change threshold |
| Projecting 3D points $p_w$ into 2D image space denoted as $p_w^p$ using $P_{iter}$ |
| $t \leftarrow 1$ |
| **For** $j = 1$ to $N$ by step 1 |
| Search the nearest projected point ( $p_{wk}^p$ ) in $p_w^p$ for $p_m[j]$ |
| $d \leftarrow$ Distance ( $p_m[j]$, $p_{wk}^p$ ) |
| **If** $d_{min} < d < d_{max}$ |
| $p_w^s[t] \leftarrow p_w[k]$ |
| $p_m^s[t] \leftarrow p_m[j]$ |
| $D[t] \leftarrow d$ |
| $t \leftarrow t+1$ |
| **end If** |
| **End For** |
| $P_{iter} \leftarrow$ Compute projective matrix using $p_w^s$, $p_m^s$ |
| $D_{mean} \leftarrow$ the mean value of $D$ |
| Change $\leftarrow (D_{mean} - D_{pre})/D_{mean}$ |
| $D_{pre} \leftarrow D_{mean}$ |
| $i \leftarrow i+1$ |
| **End While** |
| $P_{ult} = P_{iter}$ |

## FE Model Generation

The process of back-projection of crack width feature pairs onto the 3D geometric model is developed to generate a damaged FE model. The crack width feature pairs, camera parameters, and geometric model described in the preceding section provide a sufficient number of parameters to finish the back-projection process. In the thin lens camera model, camera rays start from the camera center through image points and finally intersect with the geometric model at the corresponding 3D points. The schema of the back-projection process is demonstrated in Figure 8.

Based on the obtained geometry of the RC beam and crack topology, the automatic generation of an FE model of a cracked RC beam is achieved. Note that the reinforcement positions are predetermined rather than obtained from the point cloud because the point cloud only records surface information. The Python (version 3.6.1) API of DIANA (DIANA FEA BV & TNO DIANA BV, 2019) is used to build the FE model. Particular attention must be paid to the integration of cracks into the FE model. Blomfors et al. (2021) discussed two methods of incorporating existing cracks into FE models; for this study, the method involving the use of reduced element properties was selected owing to its convenience and applicability to the FE model. It assumes that cracks are localized within one element and assigns reduced material properties to the elements at the position of the crack according to the relationship between residual tensile strength and crack width. As a result, selecting the cracked elements and setting their material properties become the primary tasks in the incorporation of existing cracks. Because the skeleton points of the cracks have been back-projected into 3D space, the elements that coincide with the back-projected skeleton points can be selected as cracked elements. The mean value of crack width associated with the skeleton points in each element is calculated and used to set the material properties of the element. Selected cracked elements and their corresponding crack widths are illustrated in Figure 9.

*Figure 8. Schema of back-projection of cracks*

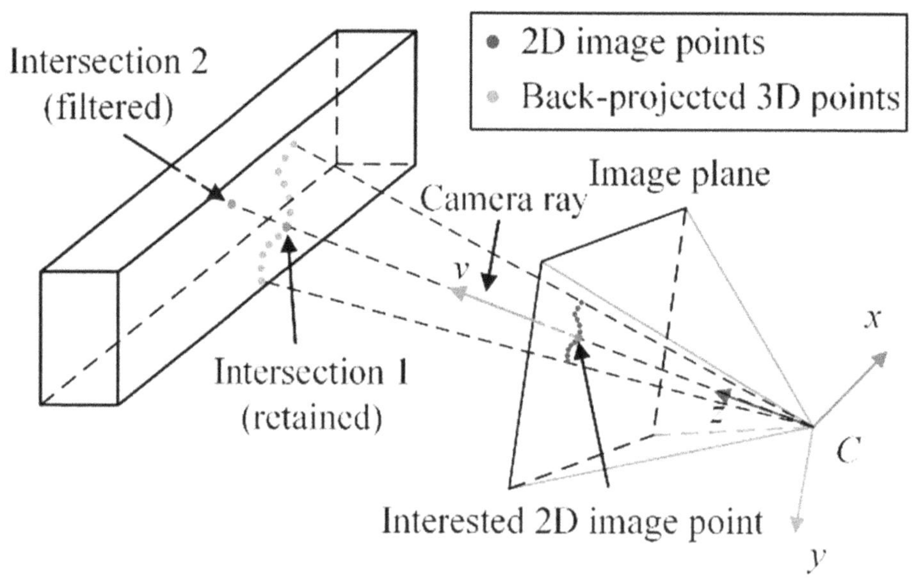


*Figure 9. Schema for incorporating cracks into FE model*

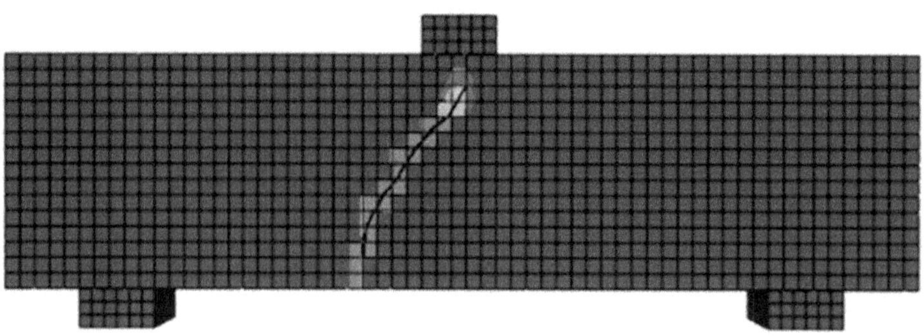

## CONCLUSION

This chapter demonstrates the potential of DL in engineering practice by employing DL to address the crucial problems of damage inspection for concrete structures. The primary conclusions of this chapter are as follows:

(1) A new method is developed for automated recognition and measurement of rebars and sleeves of PC components based on the 3D point cloud. The point cloud model of a prefabricated column joint obtained from a 3D laser scanner can be processed using the proposed algorithms. PC components with rebars and sleeves can be recognized and the dimensions and positions of the subcomponents, such as rebars, are subsequently obtained from the proposed method.

(2) To address the issues of the insufficient feature of crack segmentation networks and labeled data limitation, an active learning framework is proposed to identify the most informative crack images. Annotating these selected crack images to build a dataset can support sufficient supervision information and reduce the requirement of labeled crack images dramatically. Additionally, a novel crack feature pyramid network (Crack-FPN) is proposed that uses high-level features to supplement low-level semantics, thus obtaining high-resolution and strong semantic features.

(3) To detect multiple anomaly features for SHM data, a novel GANomaly-Convolutional Neural Network (CNN) method is developed. In particular, the raw time series are encoded as two-dimensional image vectors by using Gramian Angular Field (GAF) method. Moreover, in order to better distinguish abnormal samples, extra frequency domain information is also introduced.

(4) To automatically model the geometries of RC structures that incorporate existing damages and then generate finite element (FE) models for structural assessment, a method based on the alignment of point clouds and images is developed. The generated FE model is practically useful in the rapid assessment of existing structures as it provides detailed information on features such as cracks and deformation. The proposed method provides a new solution to the automated generation of FE models of existing structures and analysis of structure bearing capacity subject to deformation and cracks.

## REFERENCES

Abdel-Qader, I., Abudayyeh, O., & Kelly, M. E. (2003). Analysis of Edge-Detection Techniques for Crack Identification in Bridges. *Journal of Computing in Civil Engineering*, *17*(4), 255–263. doi:10.1061/(ASCE)0887-3801(2003)17:4(255)

Avci, O., Abdeljaber, O., Kiranyaz, S., Hussein, M., Gabbouj, M., & Inman, D. J. (2021). A review of vibration-based damage detection in civil structures: From traditional methods to Machine Learning and Deep Learning applications. *Mechanical Systems and Signal Processing*, *147*, 107077. doi:10.1016/j.ymssp.2020.107077

Azimi, M., Eslamlou, A. D., & Pekcan, G. (2020). Data-driven structural health monitoring and damage detection through deep learning: State-of-the-art review. In Sensors, 20(10). doi:10.339020102778

Baduge, S. K., Thilakarathna, S., Perera, J. S., Arashpour, M., Sharafi, P., Teodosio, B., Shringi, A., & Mendis, P. (2022). Artificial intelligence and smart vision for building and construction 4.0: Machine and deep learning methods and applications. *Automation in Construction*, *141*, 104440. doi:10.1016/j.autcon.2022.104440

Bao, Y., Chen, Z., Wei, S., Xu, Y., Tang, Z., & Li, H. (2019). The State of the Art of Data Science and Engineering in Structural Health Monitoring. *Engineering (Beijing)*, *5*(2), 234–242. doi:10.1016/j.eng.2018.11.027

Bao, Y., Tang, Z., Li, H., & Zhang, Y. (2019). Computer vision and deep learning–based data anomaly detection method for structural health monitoring. *Structural Health Monitoring*, *18*(2), 401–421. doi:10.1177/1475921718757405

Beck, A., & Teboulle, M. (2009). Fast gradient-based algorithms for constrained total variation image denoising and deblurring problems. *IEEE Transactions on Image Processing*, *18*(11), 2419–2434. doi:10.1109/TIP.2009.2028250 PMID:19635705

Belleri, A., Brunesi, E., Nascimbene, R., Pagani, M., & Riva, P. (2015). Seismic Performance of Precast Industrial Facilities Following Major Earthquakes in the Italian Territory. *Journal of Performance of Constructed Facilities*, *29*(5), 04014135. doi:10.1061/(ASCE)CF.1943-5509.0000617

Blomfors, M., Lundgren, K., & Zandi, K. (2020). Incorporation of pre-existing longitudinal cracks in finite element analyses of corroded reinforced concrete beams failing in anchorage. *Structure and Infrastructure Engineering*, *0*(0), 1–17. doi:10.1080/15732479.2020.1782444

Blomfors, M., G. Berrocal, C., Lundgren, K., & Zandi, K. (2021). Incorporation of pre-existing cracks in finite element analyses of reinforced concrete beams without transverse reinforcement. *Engineering Structures, 229*, 111601. doi:10.1016/j.engstruct.2020.111601

. Ferreira, D. & Manie, J. (eds). (2019). *DIANA finite element analysis; Documentation release 10.3.* DIANA FEA BV, & TNO DIANA BV.

Ding, J., Liu, Y., Zhang, L., Wang, J., & Liu, Y. (2016). An anomaly detection approach for multiple monitoring data series based on latent correlation probabilistic model. *Applied Intelligence*, *44*(2), 340–361. doi:10.100710489-015-0713-7

Edelsbrunner, H., Kirkpatrick, D., & Seidel, R. (1983). On the shape of a set of points in the plane. *IEEE Transactions on Information Theory*, *29*(4), 551–559. doi:10.1109/TIT.1983.1056714

Ester, M., Kriegel, H.-P., Sander, J., & Xu, X. (1996). A density-based algorithm for discovering clusters in large spatial databases with noise. *KDD : Proceedings / International Conference on Knowledge Discovery & Data Mining. International Conference on Knowledge Discovery & Data Mining*, *96*(34), 226–231. doi:10.5120/739-1038

Fu, J., Liu, J., Tian, H., Li, Y., Bao, Y., Fang, Z., & Lu, H. (2019). Dual attention network for scene segmentation. *Proceedings of the IEEE Computer Society Conference on Computer Vision and Pattern Recognition, 2019-June*, (pp. 3141–3149). IEEE. 10.1109/CVPR.2019.00326

Gatti, M. (2019). Structural health monitoring of an operational bridge: A case study. *Engineering Structures, 195*(November 2018), 200–209. doi:10.1016/j.engstruct.2019.05.102

Ghayeb, H. H., Razak, H. A., & Sulong, N. H. R. (2017). Development and testing of hybrid precast concrete beam-to-column connections under cyclic loading. *Construction & Building Materials*, *151*, 258–278. doi:10.1016/j.conbuildmat.2017.06.073

Gordan, M., Sabbagh-Yazdi, S.-R., Ismail, Z., Ghaedi, K., Carroll, P., McCrum, D., & Samali, B. (2022). State-of-the-art review on advancements of data mining in structural health monitoring. *Measurement*, *193*, 110939. doi:10.1016/j.measurement.2022.110939

Guldur, B., & Hajjar, J. (2014). *Laser-based structural sensing and surface damage detection*. Citeseer. doi:10.17760/d20009283

Han, J., & Moraga, C. (1995). The influence of the sigmoid function parameters on the speed of back-propagation learning. Lecture Notes in Computer Science (Including Subseries Lecture Notes in Artificial Intelligence and Lecture Notes in Bioinformatics), 930, 195–201. doi:10.1007/3-540-59497-3_175

He, K., Zhang, X., Ren, S., & Sun, J. (2016). Deep residual learning for image recognition. *Proceedings of the IEEE Computer Society Conference on Computer Vision and Pattern Recognition, 2016-Decem*, (pp. 770–778). IEEE. 10.1109/CVPR.2016.90

Hebert, M., & Krotkov, E. (1992). 3D measurements from imaging laser radars: How good are they? *Image and Vision Computing*, *10*(3), 170–178. doi:10.1016/0262-8856(92)90068-E

Heinlein, S., Cawley, P., & Vogt, T. (2019). Validation of a procedure for the evaluation of the performance of an installed structural health monitoring system. *Structural Health Monitoring*, *18*(5–6), 1557–1568. doi:10.1177/1475921718798567

Hu, J., Shen, L., Albanie, S., Sun, G., & Wu, E. (2020). Squeeze-and-Excitation Networks. *IEEE Transactions on Pattern Analysis and Machine Intelligence*, *42*(8), 2011–2023. doi:10.1109/TPAMI.2019.2913372 PMID:31034408

Huang, L., Duan, S. W., Zhu, K. M., & Chen, W. G. (2016). Study on Deformation Monitoring of Subway Station Deep Foundation Construction. *Applied Mechanics and Materials*, *847*, 425–430. . doi:10.4028/www.scientific.net/AMM.847.425

Hutchinson, T. C., & Chen, Z. (2006). Improved Image Analysis for Evaluating Concrete Damage. *Journal of Computing in Civil Engineering*, *20*(3), 210–216. doi:10.1061/(ASCE)0887-3801(2006)20:3(210)

Huyan, J., Li, W., Tighe, S., Xu, Z., & Zhai, J. (2020). CrackU-net: A novel deep convolutional neural network for pixelwise pavement crack detection. *Structural Control and Health Monitoring*, *27*(8), 1–19. doi:10.1002tc.2551

Jacqueline, G. (2000). *The future for precast concrete in low-rise housing*. British Precast Concrete Federation.

Jaillon, L., Poon, C. S., & Chiang, Y. H. (2009). Quantifying the waste reduction potential of using prefabrication in building construction in Hong Kong. *Waste Management (New York, N.Y.)*, *29*(1), 309–320. doi:10.1016/j.wasman.2008.02.015 PMID:18434128

Jáuregui, D. V., White, K. R., Woodward, C. B., & Leitch, K. R. (2003). Noncontact Photogrammetric Measurement of Vertical Bridge Deflection. *Journal of Bridge Engineering*, *8*(4), 212–222. doi:10.1061/(ASCE)1084-0702(2003)8:4(212)

Jiang, W., Hong, Y., Zhou, B., He, X., & Cheng, C. (2019). A GAN-Based Anomaly Detection Approach for Imbalanced Industrial Time Series. *IEEE Access : Practical Innovations, Open Solutions*, *7*, 143608–143619. doi:10.1109/ACCESS.2019.2944689

Kerschen, G., De Boe, P., Golinval, J. C., & Worden, K. (2005). Sensor validation using principal component analysis. *Smart Materials and Structures*, *14*(1), 36–42. doi:10.1088/0964-1726/14/1/004

Kim, M. K., Thedja, J. P. P., & Wang, Q. (2020). Automated dimensional quality assessment for formwork and rebar of reinforced concrete components using 3D point cloud data. *Automation in Construction*, *112*(August 2019). doi:10.1016/j.autcon.2020.103077

LeCun, Y., Boser, B., Denker, J. S., Henderson, D., Howard, R. E., Hubbard, W., & Jackel, L. D. (1989). Backpropagation applied to digit recognition. *Neural Computation*, *1*(4), 541–551. doi:10.1162/neco.1989.1.4.541

Li, H., & Ou, J. (2016). The state of the art in structural health monitoring of cable-stayed bridges. *Journal of Civil Structural Health Monitoring*, *6*(1), 43–67. doi:10.100713349-015-0115-x

Lin, T.-Y., Dollar, P., Girshick, R., He, K., Hariharan, B., & Belongie, S. (2017). Feature pyramid networks for object detection. *Proc. IEEE Conf. Comput. Vis. Pattern Recognit*, (pp. 936–944). IEEE. 10.1109/ISPA-BDCloud-SustainCom-SocialCom48970.2019.00217

Liu, G., Niu, Y., Zhao, W., Duan, Y., & Shu, J. (2022). Data anomaly detection for structural health monitoring using a combination network of GANomaly and CNN. *Smart Structures and Systems*, *39*, 195–206. doi:10.12989ss.2022.29.1.053

Liu, Y., Yao, J., Lu, X., Xie, R., & Li, L. (2019). DeepCrack: A deep hierarchical feature learning architecture for crack segmentation. *Neurocomputing*, *338*, 139–153. doi:10.1016/j.neucom.2019.01.036

Liu, Z., Cao, Y., Wang, Y., & Wang, W. (2019). Computer vision-based concrete crack detection using U-net fully convolutional networks. *Automation in Construction*, *104*(January), 129–139. doi:10.1016/j.autcon.2019.04.005

Lu, R., & Brilakis, I. (2019). Digital twinning of existing reinforced concrete bridges from labelled point clusters. *Automation in Construction*, *105*(May), 102837. doi:10.1016/j.autcon.2019.102837

Lu, R., Brilakis, I., & Middleton, C. R. (2019). Detection of Structural Components in Point Clouds of Existing RC Bridges. *Computer-Aided Civil and Infrastructure Engineering*, *34*(3), 191–212. doi:10.1111/mice.12407

Maisonobe, L. (2007). *Finding the circle that best fits a set of points*. SpaceRoots. https://www.spaceroots.org/documents/circle/circle-fitting.pdf

Mazhelis, O. (2006). One-class classifiers: A review and analysis of suitability in the context of mobile-masquerader detection. *South African Computer Journal = Suid-Afrikaanse Rekenaartydskrif*, *2006*(36), 29–48. doi:10.46298/arima.1877

Ng, H. F. (2006). Automatic thresholding for defect detection. *Pattern Recognition Letters*, *27*(14), 1644–1649. doi:10.1016/j.patrec.2006.03.009

Ni, F. T., Zhang, J., & Chen, Z. Q. (2019). Pixel-level crack delineation in images with convolutional feature fusion. *Structural Control and Health Monitoring*, *26*(1), 1–18. doi:10.1002tc.2286

Ni, F. T., Zhang, J., & Noori, M. N. (2020). Deep learning for data anomaly detection and data compression of a long-span suspension bridge. *Computer-Aided Civil and Infrastructure Engineering*, *35*(7), 685–700. doi:10.1111/mice.12528

Park, J., Kim, K., & Cho, Y. K. (2017). Framework of Automated Construction-Safety Monitoring Using Cloud-Enabled BIM and BLE Mobile Tracking Sensors. *Journal of Construction Engineering and Management*, *143*(2), 05016019. doi:10.1061/(ASCE)CO.1943-7862.0001223

Park, S., Kim, S., & Choi, J. H. (2018). Gear fault diagnosis using transmission error and ensemble empirical mode decomposition. *Mechanical Systems and Signal Processing*, *108*, 58–72. doi:10.1016/j.ymssp.2018.02.028

Peterson, L. E. (2009). K-nearest neighbor. *Scholarpedia Journal*, *4*(2), 1883. doi:10.4249cholarpedia.1883

Qu, Z., Lin, L. D., Guo, Y., & Wang, N. (2015). An improved algorithm for image crack detection based on percolation model. *IEEJ Transactions on Electrical and Electronic Engineering*, *10*(2), 214–221. doi:10.1002/tee.22056

Radford, A., Metz, L., & Chintala, S. (2016). Unsupervised representation learning with deep convolutional generative adversarial networks. *4th International Conference on Learning Representations, ICLR 2016 - Conference Track Proceedings*, (pp. 1–16).

Ronneberger, O., Fischer, P., & Brox, T. (2015). U-net: Convolutional networks for biomedical image segmentation. Lecture Notes in Computer Science (Including Subseries Lecture Notes in Artificial Intelligence and Lecture Notes in Bioinformatics), 9351, 234–241. doi:10.1007/978-3-319-24574-4_28

Saatcioglu, M., Mitchell, D., Tinawi, R., Gardner, N. J., Gillies, A. G., Ghobarah, A., Anderson, D. L., & Lau, D. (2001). The August 17, 1999, Kocaeli (Turkey) earthquake—Damage to structures. *Canadian Journal of Civil Engineering*, *28*(4), 715–737. doi:10.1139/cjce-28-4-715

Schölkopf, B., Platt, J. C., Shawe-Taylor, J., Smola, A. J., & Williamson, R. C. (2001). Estimating the support of a high-dimensional distribution. *Neural Computation*, *13*(7), 1443–1471. doi:10.1162/089976601750264965 PMID:11440593

Shelhamer, E., Long, J., & Darrell, T. (2017). Fully Convolutional Networks for Semantic Segmentation. *IEEE Transactions on Pattern Analysis and Machine Intelligence*, *39*(4), 640–651. doi:10.1109/TPAMI.2016.2572683 PMID:27244717

Shin, D. H., & Dunston, P. S. (2009). Evaluation of Augmented Reality in steel column inspection. *Automation in Construction*, *18*(2), 118–129. doi:10.1016/j.autcon.2008.05.007

Shu, J., Li, J., Zhang, J., Zhao, W., Duan, Y., & Zhang, Z. (2022). An active learning method with difficulty learning mechanism for crack detection. *Smart Structures and Systems*, *29*(1), 195–206. doi:10.12989/SSS.2022.29.1.195

Simonyan, K., & Zisserman, A. (2015). Very deep convolutional networks for large-scale image recognition. *3rd International Conference on Learning Representations, ICLR 2015 - Conference Track Proceedings*, (pp. 1–14).

Tommaselli, A. M. G., & Reiss, M. L. L. (2005). A photogrammetric method for single image orientation and measurement. *Photogrammetric Engineering and Remote Sensing*, *71*(6), 727–732. doi:10.14358/PERS.71.6.727

Wang, Q., Cheng, J. C. P., & Sohn, H. (2017). Automated Estimation of Reinforced Precast Concrete Rebar Positions Using Colored Laser Scan Data. *Computer-Aided Civil and Infrastructure Engineering*, *32*(9), 787–802. doi:10.1111/mice.12293

Wang, Q., Sohn, H., & Cheng, J. C. P. (2019). Development of high-accuracy edge line estimation algorithms using terrestrial laser scanning. *Automation in Construction*, *101*, 59–71. doi:10.1016/j.autcon.2019.01.009

Widyaningrum, E., Peters, R. Y., & Lindenbergh, R. C. (2020a). Building outline extraction from als point clouds using medial axis transform descriptors. *Pattern Recognition*, *106*, 107447. doi:10.1016/j.patcog.2020.107447

Widyaningrum, E., Peters, R. Y., & Lindenbergh, R. C. (2020b). Building outline extraction from ALS point clouds using medial axis transform descriptors. *Pattern Recognition*, *106*, 107447. doi:10.1016/j.patcog.2020.107447

Xie, S., Girshick, R., Dollár, P., Tu, Z., & He, K. (2017). Aggregated residual transformations for deep neural networks. *Proceedings - 30th IEEE Conference on Computer Vision and Pattern Recognition, CVPR 2017, 2017-Janua*, (pp. 5987–5995). IEEE. 10.1109/CVPR.2017.634

Xu, L., Pan, J., & Cai, J. (2019). Seismic performance of precast RC and RC/ECC composite columns with grouted sleeve connections. *Engineering Structures*, *188*(March), 104–110. doi:10.1016/j.engstruct.2019.03.022

Xu, Y. L. (2018). Making good use of structural health monitoring systems of long-span cable-supported bridges. *Journal of Civil Structural Health Monitoring*, *8*(3), 477–497. doi:10.100713349-018-0279-2

Yan, Y., & Hajjar, J. F. (2021). Automated extraction of structural elements in steel girder bridges from laser point clouds. *Automation in Construction*, *125*, 103582. doi:10.1016/j.autcon.2021.103582

Yang, Y., & Nagarajaiah, S. (2014). Data Compression of Structural Seismic Responses via Principled Independent Component Analysis. *Journal of Structural Engineering*, *140*(7), 04014032. doi:10.1061/(ASCE)ST.1943-541X.0000946

Ye, X. W., Jin, T., & Yun, C. B. (2019). A review on deep learning-based structural health monitoring of civil infrastructures. *Smart Structures and Systems*, *24*(5), 567–585. doi:10.12989ss.2019.24.5.567

Yi, T. H., Li, H. N., Song, G., & Guo, Q. (2016). Detection of Shifts in GPS Measurements for a Long-Span Bridge Using CUSUM Chart. *International Journal of Structural Stability and Dynamics*, *16*(4), 1–21. doi:10.1142/S0219455416400241

Yin, H., & Gai, K. (2015). An empirical study on preprocessing high-dimensional class-imbalanced data for classification. *Proceedings - 2015 IEEE 17th International Conference on High Performance Computing and Communications, 2015 IEEE 7th International Symposium on Cyberspace Safety and Security and 2015 IEEE 12th International Conference on Embedded Software and Systems, H*, (pp. 1314–1319). IEEE. 10.1109/HPCC-CSS-ICESS.2015.205

Yu, F., & Koltun, V. (2016). Multi-scale context aggregation by dilated convolutions. *4th International Conference on Learning Representations, ICLR 2016 - Conference Track Proceedings*. Research Gate.

Yu, K., Zhang, C., Shooshtarian, M., Zhao, W., & Shu, J. (2021). Automated finite element modeling and analysis of cracked reinforced concrete beams from three dimensional point cloud. *Structural Concrete*, *22*(6), 3213–3227. doi:10.1002uco.202100194

Yuen, K. V., & Mu, H. Q. (2012). A novel probabilistic method for robust parametric identification and outlier detection. *Probabilistic Engineering Mechanics*, *30*, 48–59. doi:10.1016/j.probengmech.2012.06.002

Zandi, K., Ransom, E. H., Topac, T., Chen, R., Beniwal, S., Blomfors, M., Shu, J., & Chang, F. (2019, November 15). A Framework for Digital Twin of Civil Infrastructure—Challenges & Opportunities. *Structural Health Monitoring*, *2019*. Advance online publication. doi:10.12783hm2019/32288

Zhang, C., Shu, J., Shao, Y., & Zhao, W. (2022). Automated generation of FE models of cracked RC beams based on 3D point clouds and 2D images. *Journal of Civil Structural Health Monitoring*, *12*(1), 29–46. doi:10.100713349-021-00525-5

Zhao, W., Jiang, Y., Liu, Y., & Shu, J. (2022). Automated recognition and measurement based on three-dimensional point clouds to connect precast concrete components. *Automation in Construction*, *133*, 104000. doi:10.1016/j.autcon.2021.104000

Zhao, W., Liu, Y., Zhang, J., Shao, Y., & Shu, J. (2022). Automatic pixel-level crack detection and evaluation of concrete structures using deep learning. *Structural Control and Health Monitoring*, *e2981*(8). doi:10.1002tc.2981

# Chapter 3
# Machine Learning Applications for Vibration-Based Structural Health Monitoring

**Zeinep Achmet**
*National Technical University of Athens, Greece*

**Michalis Fragiadakis**
*National Technical University of Athens, Greece*

## ABSTRACT

*The recent advances in sensing technology and in the field of artificial intelligence (AI) have boosted the development of techniques based on machine learning (ML) for structural health monitoring (SHM). The chapter provides a discussion on the application of machine learning (ML) methods for vibration-based SHM. The fundamental principles of machine learning are presented and the steps of vibration-based SHM for damage detection are outlined. The application of ML with respect to every step of SHM process is subsequently discussed. Among the vast existing literature, the milestones are identified and the endless possibilities for the successful application of ML are presented.*

## INTRODUCTION

Structural systems, such as transportation, power and energy infrastructure, as well as critical buildings (e.g., schools, hospitals, government buildings) are essential for the functioning of modern societies and their possible failure may have significant consequences. For example, Figure 1 shows some notable examples of bridge collapses around the world. Among the reasons for bridge collapse is their ageing and the increased traffic loads compared to the loads considered at time that the bridge was designed. Environmental reasons, such as flood and scour are also common reasons that cause the collapse of structures which usually were already deteriorated. When the anticipated service life of these structures comes to an end, or when their capacity is reduced, replacing them is quite challenging, or just not

DOI: 10.4018/978-1-6684-5643-9.ch003

possible. Therefore, it is often necessary to monitor the response of an important structure in order to evaluate its health condition and to detect early-stage damage for ensuring operational safety.

Structural Health Monitoring (SHM) involves the observation and the analysis of a system over time using periodically sampled response measurements, or/and digital images in order to monitor changes in the material, or the geometry of the system. The major families of SHM algorithms are *vibration-based*, and *vision-based*. Vibration-based SHM can be seen as the traditional approach for monitoring a structural system where sensors (usually accelerometers) are used to measure the structural response under various loading conditions. The collected measurements are then processed in order to extract the dynamic characteristics, or other damage-sensitive features after applying signal processing techniques. Post-processing also aims on extracting meaningful damage indication parameters (*damage indices* or *damage-sensitive features*) that will be used to determine structural health. Thanks to the recent technological advances in sensors, drones, high-speed internet, cloud computing, as well as the advances in Machine Learning (ML) techniques, the combination of ML techniques with both SHM families is gaining more and more interest. This Chapter focuses on ML applications for vibration-based SHM offering a detailed and critical examination of the problems and the potential of ML methods.

*Figure 1. Examples of recent bridge collapses: (a) Pittsburg bridge, USA (collapsed 28/01/2022), (b) Mirepoix-sur-Tarn, France (collapse 18/11/2019), (c) Kalambaka, Trikala bridge (collapsed 2016), (d) Plaka bridge, Greece (collapsed 1/02/2015)*

Vibration-based SHM can be either *model-based* (known as *physics-based*), or *data-based* (known as *data-driven*). Model-based SHM usually includes the development of a *digital twin* of the monitored structure. A *digital twin* might be a numerical model, an analytical equation, an ML algorithm or a surrogate model, while in model-based SHM, Finite Element (FE) models are commonly used. An example of a structure and its digital twin can be seen in Figure 2. The model properties can be updated through the solution of an inverse problem using the measured data. On the other hand, data-based approaches require a statistical model instead of a digital twin. The development of the statistical model depends on the use of ML or pattern recognition algorithms, and the extracted damage-sensitive features, or on the raw sensor data, since this approach is entirely based upon data. Some researchers have also proposed hybrid SHM approaches which combine the best features of model-based and data-based methods. The major principles of hybrid methods can be found in Barthorpe (2010). It is worth mentioning that in the context of ML applications for SHM, numerical models are used not only for damage detection through model updating but also as a tool for designing an optimal SHM sensor network, and as a proxy for damage induction in order to enable the supervised training of data-based damage detection algorithms (Figueiredo et al., 2019).

The strengths and the weaknesses of model-based, data-based and hybrid approaches to SHM according to Wang et al., (2020) are compared in Table 1. Model-based methods generally are quite efficient and descriptive. The advantage of these approaches is that they are relatively accurate and precise in terms of model fidelity and ease of validation. The disadvantages are that creating a high-fidelity model is time-consuming, computationally intensive, and it carries the risk of not achieving the desired result since it has limited reusability. The quick implementation and the non-requirement of knowledge about the physics of the system are some of the main advantages of data-driven methods for vibration-based SHM. Their major shortcoming is that they require large amounts of data and they have a short prediction horizon. On the other hand, hybrid approaches are often the optimal choice since they attempt to combine advantages of both model-based and data-based methods. However, hybrid approaches also carry the shortcomings of both approaches. The Chapter first discusses the vibration-based SHM and the fundamentals of ML methods aiming to provide the necessary background for understanding the vast possibilities of using AI techniques for SHM. The application of ML methods providing examples of vibration-based SHM applications from the literature will be subsequently presented.

## VIBRATION-BASED STRUCTURAL HEALTH MONITORING

### Fundamentals of Damage Detection

Prior to further discussing the steps of the vibration-based SHM it is useful to define what damage detection means in the context of structural engineering. Farrar & Worden (2013), define damage as the intentional, or the unintentional, changes to the material and/or the geometric properties of structural systems, including changes to the boundary conditions and system connectivity. These changes adversely affect the current, or the future, performance of the structure. The damage identification problem can be considered as a hierarchical process (Rytter, 1993) that attempts to answer the following questions:

*Figure 2. Example of a three-storey reinforced concrete building: (a) actual structure, (b) corresponding digital twin and building first mode*
*(Fragiadakis et al., 2009)*

(a)

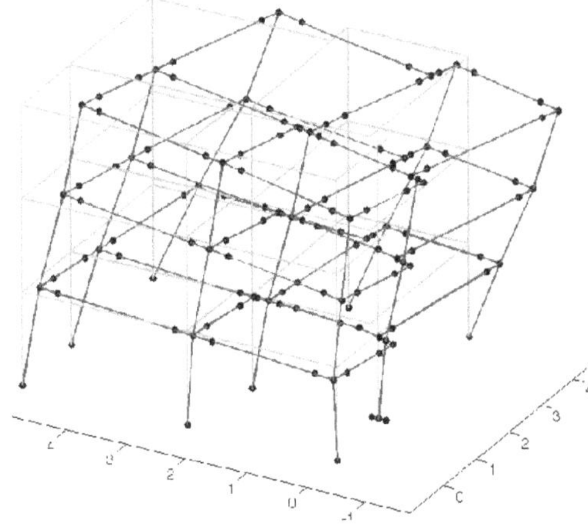

(b)

*Table 1. Strengths and weaknesses of SHM approaches*

| **Method** | *Strengths* | *Weaknesses* |
|---|---|---|
| Model-based | • *Efficient and descriptive*<br>• *High accuracy and precision*<br>• *Easy to validate*<br>• *The dynamics of the states of the structure can be estimated at every time interval* | • *High fidelity models are costly, time-consuming and computationally intensive*<br>• *Limited reusability*<br>• *Difficult in understanding the physics of damage for complex systems*<br>• *Too stochastic and complex to model the damage*<br>• *Require the examination of simplifying assumptions* |
| Data-based | • *Quick implementation and deployment*<br>• *Low cost of algorithm development and require little knowledge about system physics*<br>• *Algorithms can be tuned to be used for other systems*<br>• *No requirement on the assumptions and empirical estimations of physics parameters*<br>• *Able to transform high-dimensional noisy data into lower dimensional information for prognostic decisions* | • *Requirement of a robust algorithm*<br>• *Requirement of uncertainty management*<br>• *Accuracy affected by the overgeneralization and overfitting during the training of the algorithm*<br>• *Inadequate data for newly developed systems*<br>• *Counter-intuitive results due to the absence of physical knowledge of the system*<br>• *Require a large amount of data*<br>• *Have a short prediction horizon*<br>• *Low accuracy with high dimensional data* |
| Hybrid | • *Eliminate the weaknesses of model-based approaches and gain their benefits*<br>• *Do not require highly accurate models or enormous amount of data* | • *Carry the weaknesses of both methods but in minor extent*<br>• *Increased computational resources required*<br>• *Require both a model and sufficient data*<br>• *Inaccuracy caused by incorrect model of noisy data* |

(Wang et al., 2020)

1. *Detection*: is the structural system damaged?
2. *Location*: where is the damage located?
3. *Classification*: what is the type of damage?
4. *Assessment:* how severe is the damage?
5. *Prognosis:* what is the residual life of the structural system following this damage?

Each of the above questions, in the order given, provides an increased knowledge of the structural damage state. Therefore, the above questions offer the basis of a holistic approach for the health monitoring of a system. For a structure, for example a bridge, damage can be identified from changes of its mass, stiffness, and energy dissipation characteristics, which, in turn, modify the measured dynamic response properties of the system. A very interesting discussion on this topic can be found in the work of Friswell (2008). It is worth mentioning that damage detection can be either global or local, while the majority of the proposed damage detection methods in civil engineering operate at the global level. Global damage detection is performed observing the changes in the dynamic characteristics of the structural system, mainly related to the principal modes, while local damage is identified through traditional methods, such as visual inspection or/and through more sophisticated methods, e.g., processing images from drones. Commonly used damage indicators, also known as "damage-sensitive features" are: (i) natural frequencies, (ii) mode shapes, (iii) mode shape curvature/modal strain energy, (iv) operational deflection shapes, (v) frequency response functions and transmissibility, (vi) dynamic flexibility, (vii) damping coefficients (Barthorpe, 2010).

## Steps of Vibration-Based Structural Health Monitoring

Monitoring of engineering structures is traditionally based on inspections, i.e. on visual checks if there is cracking, spalling, or other visual signs of structural damage. Even at this simplest form, the monitoring process involves many of the steps of modern SHM processes, such as data collection and processing followed by diagnosis. Moreover, damage detection through observing changes of the dynamic properties has been applied in a qualitative manner, using acoustic or vibration-based techniques since many years ago (Farrar & Worden, 2013). During the last years, a significant increase in the number of studies that attempt to evolve SHM from these early qualitative approaches to heuristic, sensitive, and quantifiable damage detection methods has been recorded.

According to Farrar et al. (2001), the steps of vibration-based SHM are: (i) operational evaluation, (ii) data acquisition, (iii) feature selection, and (iv) updating of the FE model (model-based), or the development of a statistical model (data-based). The operational evaluation step is common for both model and data-based approaches, and it determines what will be monitored and how the monitoring will be accomplished. Regarding the *data acquisition* step, the requirements for the two approaches may differ. However, usually the limited resources hamper the data acquisition process, regardless of the approach adopted. ML methods can facilitate the decision-making process regarding the optimum location of a limited number of sensors in order to obtain the best possible measurements. *Feature selection* in SHM is used to identify the features, or the damage indices, which are highly correlated with damage. It also concerns which parameters to discard since they contain redundant information and hence successful feature selection leads also to a reduction of the problem dimensions. Commonly used features in SHM, are the natural frequencies, mode shapes, frequency response functions, among others. Moreover, depending on the availability of data, supervised or unsupervised ML algorithms can be used to discover the relationship of such features with the damage state in order to facilitate the feature selection step. The last step of vibration-based SHM differs depending on the approach adopted. If a model-based approach is chosen, the updating of the numerical model of the monitored structure is performed through the solution of an inverse problem. In the case of the data-based approaches, a statistical model is developed instead. The steps of a modern SHM procedure are shown in Figure 3 and they are further discussed in the lines that follow.

*Figure 3. Application steps of vibration-based SHM*

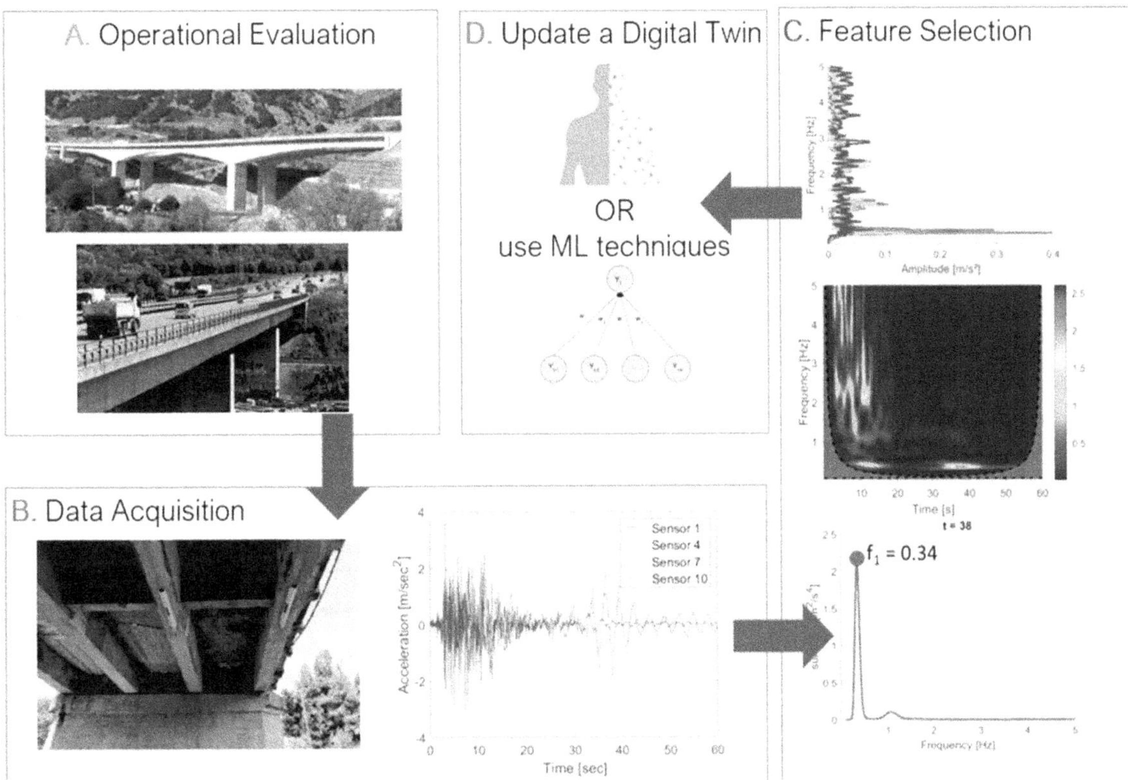


## Operational Evaluation

Operational evaluation concerns the decisions regarding what exactly will be monitored and how the monitoring will be accomplished. This is always the first step, regardless of the approach adopted, i.e. model-based, data-based, or hybrid. Therefore, during this step the damage identification process is associated with the features that are unique for the structural system under study. The successful operational evaluation should determine what is the economic and/or safety needs for performing structural health monitoring, how is damage defined for the investigated structural system and if there are multiple possible damage patterns. Moreover, it has to be defined what are the operational and the environmental conditions under which the structural system will be monitored and what are the limitations of data acquisition in the given operational environment (Farrar & Worden, 2013).

## Data Acquisition

The selection and the validation of the type of the sensors is one of the most important elements of the data acquisition step. Depending on the structural system and the available resources, it is necessary to choose the type of sensors that will be used, e.g. piezoelectric, fiber optics, microelectromechanical or nanoelectromechanical. Apart from the selection of the sensor type, the number of the sensors, their locations and the data acquisition/storage/transmittal hardware requirements have to be decided. Furthermore, the frequency of the data acquisition must be determined. For example, if earthquake applications are of interest, it would be meaningful to collect data immediately before and at periodic intervals after the severe event. If the critical damage type under study is crack growth, the data should be collected continuously.

*Data normalization* is another challenging problem that has to be considered for the successful collection of information. Data normalization is the process of separating changes in the measured system response caused by benign operational and environmental variability from the changes due to structural damage (Farrar & Worden, 2013). A common practice is to normalize the measured responses by the measured inputs, or depending on the problem, the measured data are temporally normalized and compared against those of different measuring periods. Furthermore, the sources of variability in the data acquisition process must be identified and reduced to the extent possible. The final step of data acquisition is the *cleansing* of the collected data. Cleansing includes the acceptance, or the rejection, of the collected data that will be later used for the feature selection process. Several signal processing techniques, such as filtering and resampling are included within data cleansing.

## Feature Selection

Feature selection includes the identification of the features of the measured structural response data which will indicate the existence of damage and will also quantify it. A damage-sensitive feature, or a damage indicator, is extracted from the signal of the measured structural response, while the complexity of the extraction process can be simple or not, depending on the feature at hand. Well-known signal processing techniques, such as Fast Fourier Transform (FFT), Hilbert Transform (HT) and Wavelet Transform (WT), among others, can be used to process the data and to extract meaningful damage indicators. According to Farrar & Worden (2013), another useful source of damage indicators is the development of a data-based, parametric model of the structural system. Common examples of such models are the Autoregressive Models (AR), simple NNs, etc. The parameters of these models, or the predictive errors associated with them, can be used as damage indicators. Moreover, the fusion of data arising from multiple sensors is often necessary for the successful feature selection. Methods of data normalization are also necessary for the feature extraction step in order to distinguish changes in the response caused by the variability of the operational and environmental conditions from changes produced due to structural damage (Sohn et al., 2002).

## Development of Statistical Model and Model Updating

The final step of vibration-based SHM differs for model-based and data-based SHM approaches. In the model-based approach, the existing numerical model is updated using the measured data and possible changes of the structural properties are subsequently observed. In the data-based approach, a statistical model is used to detect anomalies in the measured features which are identified as "damage". Even though ML techniques can be applied for model-based or hybrid approaches, their use for the data-based approach is undeniably more common and of vital importance. Depending on the structure of interest and the availability of data from both the undamaged and the damaged structure, three categories of ML can be used for the development of the statistical model (Farrar et al., 2001). The first category is "group classification" and requires data from both damaged and undamaged structure. It includes the assignment of the selected features into undamaged, or damaged states. The second category of algorithms is the "outlier analysis" or "novelty detection". This family of algorithms is used when data from the damaged state of the structure is not available for comparison and subsequently they observe if there is a significant change from the previously observed features which cannot be explained by their previous distribution. Finally, the third category is "regression analysis". This approach is used to relate the selected features with types, locations or extends of damage. All three algorithm families analyze statistical distributions of the measured data, or of the derived features, in order to enhance the SHM process.

## Principles of Structural Health Monitoring

Some general principles of SHM that have been based on the literature (e.g. Farrar & Worden (2013), Farrar et al. (2010) and Worden et al. (2007)) are briefly discussed in this section. First, all structural materials have inherent flaws or defects. At the microstructure level, every material has flaws including impurities, inclusions, and empty spaces. However, engineers have learned to work and to design systems under the presence of material flaws. Another fundamental principle is that damage assessment requires a comparison between two system states. A baseline (typically undamaged) state of the structure is necessary for every SHM strategy. The training data for damage identification can be only undamaged states, or a combination of damaged and undamaged states, depending on whether the purpose is only to detect the existence of damage or more information such as location, type, severity, etc. is sought. Moreover, damage increases the complexity of the structure. Structural damage will lead an initially linear structure to demonstrate nonlinear behavior. Applying statistical and signal processing methods to the data from the damaged and undamaged states is a way to create quantifiable measures of complexity. For instance, the probability density functions of the damaged and the undamaged structural responses may be used in order to compare the two states. Identifying the existence and the location of damage can be done in an unsupervised learning mode but identifying the type of damage present and its severity, in general, is possible only in a supervised learning mode. Unsupervised learning algorithms are appropriate for damage detection when the training dataset is limited to undamaged states of the structure. However, getting more information about damage also requires response data from damaged states and hence the application of supervised learning methods.

Sensors cannot measure damage. For this purpose, feature extraction through signal processing and statistical classification are necessary to convert sensor data into damage information. A major challenge of SHM is relating a damage condition "$D$" to a measurable quantity "$x$", in the simplest form $x=f(D)$. The data acquired by the sensors must be used to discover this function "$f$", which is obviously unknown. Without intelligent feature extraction, the more sensitive a measurement is to damage, the more sensitive it is also to operational and environmental conditions. The main idea of the axiom is that the features deduced from measured data will depend on the damage to be diagnosed as well as on operational and environmental factors. Therefore, there is need for algorithms where the measured response is primarily damage-dependent and its sensitivity to other factors can be reduced, if not completely removed. The length and time scales associated with damage initiation and its evolution dictate the required properties of the SHM sensing systems. Structural damage may develop over an extended period (years or longer), but it may also happen suddenly. The a priori quantification of these time scales enables the sensor system to operate more effectively by selecting the appropriate hardware components. There is a trade-off between the sensitivity to damage of an algorithm and its noise rejection capability. Similar to environmental and operational effects, the response measurements of a structure may include both damage and noise effects and it is challenging to distinguish the influence of each source.

## MACHINE LEARNING

Machine learning is undeniably one of the most influential and powerful technology in the modern world. It is certain that it will continue to make headlines for the foreseeable future as a tool for turning information into knowledge. In the past fifty years, there has been an explosion in information generation related to all aspects of life including all engineering disciplines. This mass of data is useless unless it is analyzed and its hidden patterns are revealed. Machine learning techniques are used to automatically, or heuristically, find the valuable underlying patterns within complex data that otherwise would be challenging to discover. The hidden patterns can provide knowledge about a problem that can be used to predict future events and to perform all kind of complex decision-making.

Figure 4 shows the relationship between artificial intelligence (AI), machine learning (ML) and deep learning (DL). AI is the general field that uses computers in order to solve tasks that require human intelligence, while ML can be seen as a subset of AI that solves specific tasks through learning from data and making predictions. More specifically, *Machine Learning* is the scientific study of algorithms and statistical models used by computer systems in order to effectively perform a specific task without using explicit instructions, relying on models and inference instead. Machine Learning algorithms first build a mathematical model using sample data, known as training data, in order to make predictions or decisions without being explicitly programmed to perform the task (Bishop, 2006). Deep Learning is a subset of ML, that in essence is a neural network (NN) technology with many hidden layers. DL is supposed to imitate the complex way that humans gain certain kinds of knowledge.

Machine learning algorithms are categorized as *supervised*, *unsupervised*, and *reinforcement learning*, as explained in Figure 5. In *supervised learning*, every training instance is a pair that consists of an input and the corresponding output value. The algorithm analyzes the training data and produces an inferred function, which can be used for mapping new instances (Russell & Norvig, 2010). Supervised learning algorithms are commonly applied for classification and regression tasks. *Classification* categorizes a set of data into classes, while *regression* algorithms are used to predict continuous outcomes depending on the input features. Examples of supervised ML algorithms used for both classification and regression are $k$-nearest neighbors (kNN), decision tree (DT), random forest (RF), support vector machine (SVM), Gaussian process (GP), Neural Networks and others. An empirical comparison of supervised learning algorithms can be found in the study by Caruana & Niculescu-Mizil (2006). In the field of deep learning, commonly used algorithms are the Convolutional Neural Networks (CNN), Recurrent Neural Networks (RNN), long short term memory networks (LSTM), generative adversarial networks (GAN), and others. A review of Deep Learning algorithms can be found in the study by Shyu et al. (2020).

*Figure 4. The relationship between artificial intelligence (AI), machine learning (ML) and deep learning (DL)*

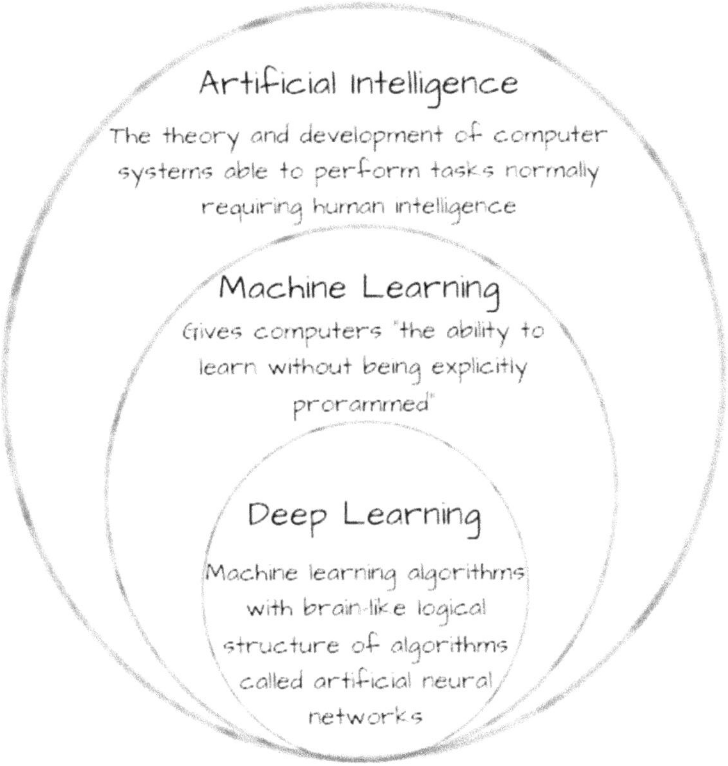


In *unsupervised learning*, the training data consists only of input features. The algorithm identifies commonalities in the data and reacts based on the presence, or on the absence, of such commonalities for every new piece of data. *Clustering* algorithms are used to group unlabeled data and they are also used for anomaly/outlier detection tasks. Commonly used algorithms for clustering are *k*-Means, and Gaussian Mixture Model (GMM). A comparative study on the clustering algorithms is conducted by Rodriguez et al. (2019). A schematic representation of the *k*-Means clustering algorithm is presented in Figure 6, where $X_1$ and $X_2$ are labelled depending on their distance from the mean of the corresponding set. *Dimensionality reduction* refers to techniques for reducing the number of input features in training data to obtain a principal set of features. Principal component analysis (PCA), linear discriminant analysis (LDA), and truncated singular value decomposition (SVD) are some of the popular dimensionality reduction algorithms. A review of the dimensionality reduction methods is offered in Huang et al. (2019).

Finally, *reinforcement learning* (*RL*) is an area of Machine Learning that focuses on how intelligent agents should take actions in an environment in order to maximize the notion of cumulative reward. It differs from supervised learning in the sense that it does not need labelled input/output pairs and it also does not need sub-optimal actions to be explicitly corrected. Instead, the focus of reinforcement learning is on finding a balance between exploration (of an uncharted territory) and exploitation (of current knowledge).The reader can refer to the article of Oh et al. (2020) for a further study on Reinforcement Learning algorithms.

*Figure 5. Categories of machine learning algorithms*

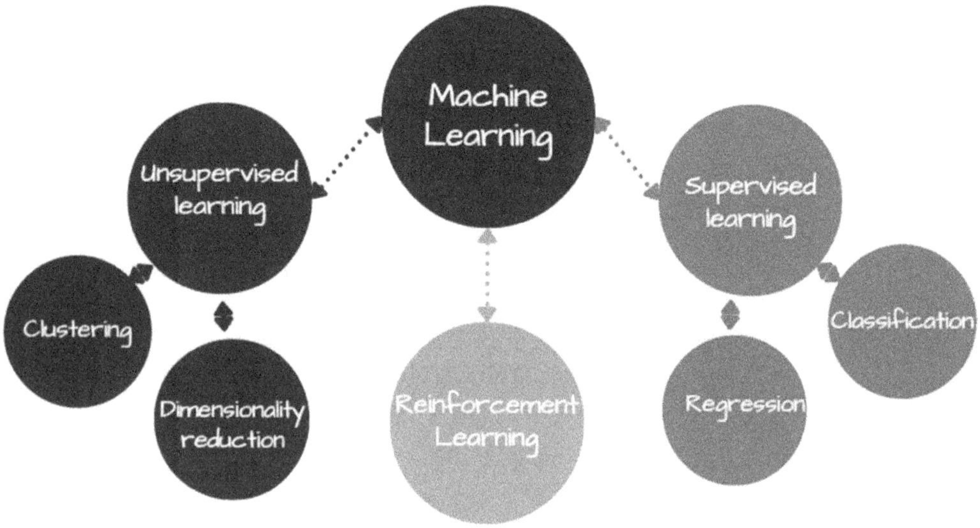


*Figure 6. Schematic explanation of the k-Means clustering algorithm used for Unsupervised Learning. The two points X1 and X2 are classificed, or labelled as red or blue based on their distance from the mean of corresponding set*

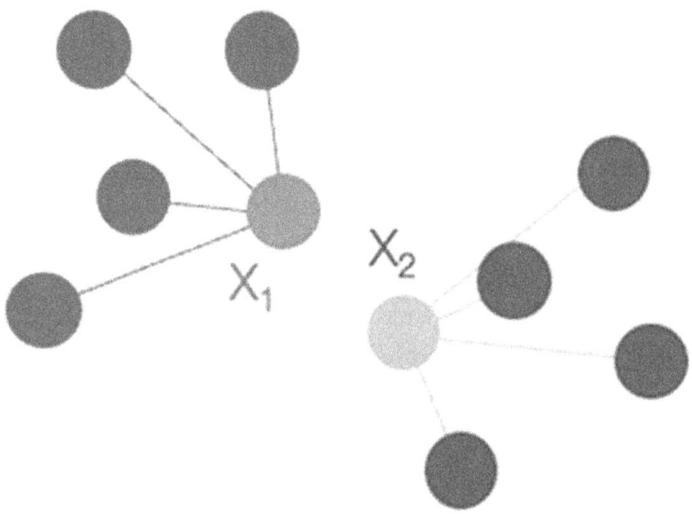


## APPLICATIONS OF MACHINE LEARNING METHODS FOR SHM

In the SHM literature, one can find numerous applications of ML and DL techniques for both vibration-based and vision-based approaches, as well as a considerable amount of review articles. For instance, a comprehensive review of ML applications for vibration and vision-based SHM is provided in Flah et al. (2021). Interesting literature review articles have been conducted by Ye et al. (2019) and Azimi et al.

(2020) which focus on the use of DL algorithms for both SHM approaches. Furthermore, an extensive literature review of vibration-based damage detection which includes both traditional as well as ML and DL applications is conducted by Avci et al. (2021), while a state-of-the-art review of ML applications for SHM of heritage buildings is provided by Mishra (2021). In recent years, the application of DL methods for vibration-based SHM, such as the Convolutional Neural Networks (CNN) among others, gained more attention since these multi-layer algorithms are able to eliminate the data pre-processing and feature extraction process. The application of one-dimensional CNN algorithms for vibration-based SHM is discussed in the study of Kiranyaz et al. (2021). A systematic review of CNN-based condition assessment techniques is conducted by Sony et al.(2021). Applications of CNN algorithms for SHM of railway bridges are discussed by Wang et al. (2022). A review of DL-based applications for safety management in architecture, engineering and construction industry is conducted by Hou et al. (2021). Finally, a state-of-the-art review of data science methods in SHM has been provided in Bao et al. (2019) and Gordan et al. (2022), while interesting application examples can be found in Cury et al. (2022). This section focuses on the ML applications for vibration-based SHM and critically discusses the possibility of using ML methods with respect to each step of vibration-based SHM. The section is divided into four subsections discussing each SHM step and providing interesting application examples from the literature.

## Data Acquisition

Data acquisition concerns the selection of the type, the location and the number of sensors that will be used for the monitoring process. It also concerns the choice of the hardware for the acquisition, storage and data transmittion. Even though these decisions are usually empirical, ML techniques can be quite useful. Finding the optimal sensor layout is perhaps the most critical issue; the goal is to ensure maximum damage sensing capabilities or "maximizing damage observability" using as few sensors as possible (Farrar et al. (2010)). ML models are typically adopted in order to revise the existing sensor locations and increase the overall system performance in terms of optimum damage detection. Commonly used ML algorithms for this purpose are Genetic Algorithms (GA) (Zhou et al., (2017), Gomes et al., (2019), Banik & Das, (2020)) and also Neural Networks (Mallardo & Aliabadi, (2014), Banik & Das, (2020)). There are also some studies that have used Bayesian models in order to optimally re-define the sensor locations, e.g. Flynn & Todd (2010) and Capellari et al. (2016). Other contributions, e.g. Bigoni et al.(2020), use sparse Gaussian processes and one-class SVM (Support Vector Machine) clustering in order to tackle the issue of the limited sensor availability.

## Data Normalization

Distinguishing whether the changes in the damage-sensitive features extracted from the structural responses are due to structural damage or environmental/operational conditions is a major challenge in SHM. Everyday changes in temperature, humidity, operational loads, etc. may also cause changes in the damage-sensitive features and may give false alarms regarding the condition of the structure. The standard first stage approach to signal normalization is to bring the data acquired from different sensors to a common scale. One of the commonly applied methods is the Z-score normalization which subtracts the mean of the response signal from each point of the signal and divides it with the signal standard deviation. But in the SHM framework, the term signal normalization is much wider and includes the detection of the effects of changing environmental and operational conditions to damage-sensitive features.

For this purpose, the damage-sensitive features are parameterized with respect to environmental and operational effects and are treated as known variables. Subsequently, the extracted features are compared to those of a new set of extracted features, or to features obtained from data that have been indirectly developed using ML. The distinction of environmental and operational effects can be achieved with a total data-based approach which requires the analysis of these effects in a data-based approach as well. Most of the times, the lack of data regarding these effects results in the adaptation of model-based, or hybrid approaches. In this case, a numerical model of the monitored structure is used to generate damage scenarios caused by environmental and operational effects which are used as training data for ML damage classification algorithms.

Several studies that try to address the issue of data normalization, using data-based, model-based, or hybrid approaches can be found in the SHM literature. A full data-based approach is adopted in the paper of Figueiredo et al., (2011) where four ML algorithms, namely the "Auto-associative NN", "Factor Analysis", "Mahalanobis Squared Distance" and "Singular Value Decomposition" are used to detect structural damage in the presence of operational and environmental effects. The algorithms were trained using data collected from experiments to classify the damaged and undamaged conditions of the structure. The structural damage is simulated through nonlinear effects introduced by a bumper mechanism that induces a repetitive, impact-type nonlinearity. This mechanism intends to simulate the cracks that open and close under dynamic loads or loose connections that rattle. Their study concluded that the "Mahalanobis Squared Distance" algorithm had the best performance in terms of classification accuracy. The same experimental results were used in Santos et al., (2016) in order to compare the performance of four kernel-based ML algorithms for damage detection under varying operational and environmental conditions. The ML algorithms examined are one-class SVM, "Support Vector Data Description", "kernel PCA" (Principal Component Analysis) and "greedy kernel PCA". The authors concluded that kernel-based approaches have better classification performance than the ones studied in Figueiredo et al., (2011). This finding is justified by the fact that the algorithms studied in Figueiredo et al., (2011) were essentially linear in their formulation, while the kernel-based algorithms operate essentially in the high-dimensional space, hence they are capable to model nonlinear patterns.

Damage detection in bridges for long-term SHM under environmental and operational conditions is studied in Figueiredo & Cross (2013) using the benchmark data sets collected from the Z-24 bridge in Switzerland. It is worth mentioning that most of the commonly applied ML algorithms assume linear correlation between the measured variables or features. However, the environmental or operational variations often cause nonlinear effects. The use of Gaussian mixture models was proposed by Figueiredo & Cross (2013) in an effort to tackle the problem of detection and removal of only linear patterns from the damage-sensitive features. The performance of the proposed method was compared with linear methods, such PCA and the "Mahalanobis Squared Distance", as well as with nonlinear PCA. It has been showed that linear algorithms might not be appropriate for long-term monitoring, since they are not able to remove nonlinear patterns from the baseline condition data during the training process, which leads to a relatively high number of false alarms. On the other hand, the Gaussian mixture model-based algorithm might be suitable for long-term monitoring because it has also the ability to reduce the baseline/normal condition into a limited number of Gaussian component densities. This stands as an advantage in the presence of limited amounts of training data to characterize the normal condition. Comparative studies on the performance of linear and nonlinear algorithms for the removal of the operational and environmental effects conducted by Rainieri, (2022) and Cava et al., (2022). The reader can refer to these studies for further advice on the use of purely data-based methods which deal with variability due to operational and environmental conditions.

In an effort to overcome issues, such as the lack of data, operational and environmental effects, and the long-term behavior of these effects, some researchers have proposed to use of FE models as digital twins, for the generation of data in order to enhance the ML-based damage classifiers. Figueiredo et al., (2019) proposed a hybrid technique that integrates model and data-based approaches into SHM. Data recorded in situ under normal conditions were combined with data obtained from FE simulations of more extreme environmental and operational scenarios and were used as input for the training of ML algorithms for damage detection. The addition of simulated data enabled a sharper classification of damage by avoiding false positives induced by wide environmental and operational variability. The procedure was applied to the Z-24 Bridge, for which one year of continuous monitoring data were available.

Liu & Zhang, (2017) proposed a different approach to model-based SHM for damage detection under varying operational and environmental conditions. The basic principle is that a digital-twin FE model would not be appropriate as the baseline model for localizing structural damage. To address this issue, they proposed the concept of probabilistic baseline of FE models under variable environmental temperature in order to establish reasonable probability distributions of the physical parameters of bridges that are suitable for damage localization with varying environmental temperature. Their method imports cluster analysis into stochastic finite element model updating. Unlike conventional methods, the measured natural frequencies are first classified into different clusters using the Gaussian mixture method, with each cluster consisting of measured data that satisfy the same Gaussian distribution. Then, the conventional methods of stochastic FE model updating are implemented to obtain the probabilistic baseline of FE model for each cluster.

## Feature Selection

The term "Feature Selection" in the context of vibration-based SHM includes the extraction of damage-sensitive features by processing the measured structural responses of interest and the selection of the features that are most sensitive to structural damage. The selection of features which can minimize or eliminate the effects of measurement noise, and/or changing operational and environmental conditions is also necessary for accurate damage detection. Regarding the ML applications for feature selection in SHM mainly two approaches are proposed in the literature: the application of shallow ML algorithms, and deep learning (DL) applications where deep neural networks (DNN) are used. The first approach, can be considered as the conventional approach for feature selection. First, the collected data is transformed to the frequency, or the time-frequency, domain using standard signal processing techniques. Subsequently, several features are extracted, where many of the known ML algorithms can be used to assist the process of selection of the optimal features. A most recent tool for feature selection in vibration-based SHM is the use of DL algorithms. Thanks to their deep, human brain-like architecture, DL algorithms are capable to extract features that are hidden in the measured data without any previous processing. The method can be applied on the frequency, the time-frequency, or directly on the raw data in the time domain skipping the step of feature extraction. In the next two subsections, some application examples of ML and DL algorithms for feature selection are discussed.

## Machine Learning Methods

Principal component analysis (PCA) is an unsupervised ML method used for dimensionality reduction that converts the observations of correlated features into a set of linearly uncorrelated features with the help of orthogonal transformation. When combined with other ML algorithms, it can be quite useful for the identification of the optimum damage-sensitive features in SHM. For instance, in Hoell & Omenzetter (2016) several damage extent scenarios were non-destructively introduced to a wind turbine plate and the response data were collected. Partial autocorrelation coefficients were extracted as the damage-sensitive features from the response measurements and the original features were also transformed to PCA scores. Next, the statistical distances between the feature subsets from the healthy and a reference damage state were estimated with respect to a statistical threshold that plays the role of a measure of optimality. Finally, fast forward selection method and a genetic algorithm were applied to optimize the damage sensitive feature selection. ML classification algorithms based on the optimal features were tested on data from unseen healthy and damaged cases which indicated that the transformed features had superior performance compared to the original ones and they even allowed the identification of slight damage.

Recursive Feature Elimination (RFE) is a simple and popular feature selection algorithm. One application example combined with the Random Forest (RF) algorithm can be found in the study by Zhou et al. (2014). In this study, the response signals were transformed to the time-frequency domain using wavelet transform in order to identify the damage-sensitive features, namely the energy of each frequency band. Next, a two-stage feature selection method using Random Forest Recursive Feature Elimination (RF -RFE) algorithm were applied. First, the damage-sensitive features were sorted using the RF variable importance analysis. The RF-RFE algorithm was afterwards applied to eliminate the least important features and the feature list was reordered to get the new feature importance sequence. Finally, the *k*-Nearest Neighbor (KNN) algorithm was used as a benchmark classifier to evaluate the extracted feature subset. The verification of the proposed method on the experimental results of a four-story steel shear building showed that using a smaller number of features can achieve higher identification accuracy and reduce the detection time cost.

A very comprehensive study for feature selection and extraction methods for structural damage detection was conducted by Bisheh et al. (2019) using the acceleration responses of the Yonghe bridge in mainland China. Initially, several damage-sensitive features in the time and the frequency domains were extracted using "damaged" and "undamaged" responses of the bridge. Several spectral parameters in the time-frequency domain were also introduced as effective features in order to reduce false alarms. Support Vector Machine (SVM) algorithm is then employed as a classifier to compare the detection ability of the introduced and the initial features. Furthermore, several feature selection algorithms were employed to improve feature extraction accuracy. The results of the study showed that the proposed features could effectively reduce false alarms. However, for a generalized applicability of the method, case studies on other cable-stayed bridges with different sensor types and configurations considering long-term vibration data to capture environmental influences are necessary.

## Deep Learning Methods

A Convolutional Neural Network (CNN) is a class of deep neural networks that specializes in processing data that have a grid-like topology, such as an image, or a map. The human brain processes a huge amount of information as soon as we see an image. Every neuron works in its own receptive field and is connected to other neurons in such a way that they cover the entire visual field. Just as each neuron only responds to stimuli in the limited area of the visual field that is called the receptive field in the biological visual system, each neuron in a CNN also only processes data in its receptive field. The layers are arranged in such a way that they initially recognize the simpler patterns and at later stages the more complex patterns. Having this ability in mind, the CNN algorithm can be a valuable tool for feature extraction without any further pre-processing.

Lin et al. (2017) proposed a method for structural damage detection with automatic feature extraction using CNN algorithm and augmented response data obtained from numerical simulations. CNN was designed to learn features and to identify damage locations by training the network directly with low-level waveform signals without any pre-processing. The ability of CNN to detect multiple damage patterns and the corresponding damage level was also tested. The features learned by hidden layers in the network give a physical insight into the black-box algorithm. It is found that the learned features evolve with the depth from rough filters to the concept of vibration mode, indicating its ability to learn essential characteristics behind the data. Furthermore, another ML classifier was trained using the wavelet packet component energy as the input feature for comparison. Their study concluded that the CNN had an excellent localization accuracy on both noise-free and noisy data set, as opposed to the conventional ML algorithm.

A similar approach based on Deep Learning was adopted in Guo et al. (2020) where a CNN was trained to extract the damage-sensitive features and then to detect damage on beams using noisy and incomplete mode shape data as input feature. The training data was obtained from laboratory experiments and FE simulations of beams with single or multiple damage states. The novelty of their work is the architecture of the CNN which is a multi-scale module. This network helps the damage-sensitive feature extraction at various scales and can reduce the interference of contaminated data. The stacked residual learning modules accelerate the CNN convergence and additionally a global average pooling layer helps in reducing the consumption of computing resources and obtaining a regression performance. The transferring parameter methodology was also introduced for reducing retraining requirements. Furthermore, the feature vectors of each layer of the CNN network were visualized to provide a basis for explaining the working principle of the network. It is worth mentioning that in both Lin et al. (2017) and Guo et al. (2020) there is the limitation of dependence on the data generated with the aid of a numerical model/laboratory experiments.

Finally, Shang et al.(2021) proposed an unsupervised deep neural network for damage detection of bridges in an effort to overcome the problems of the supervised approach. They trained a deep convolutional denoising autoencoder using multi-dimensional cross-correlation functions as inputs. The cross-correlation functions were calculated using raw acceleration data collected from numerical and experimental beam models. Their methodology aims to extract damage-sensitive features from field measurements of undamaged structures under the influence of noise and temperature uncertainties. Thus, initially the cross-correlation functions of vibration data were first calculated as basic features and then the deep convolutional denoising autoencoder was developed to reconstruct the cross-correlation functions from their noise-corrupted versions in order to extract the desired features. Subsequently, exponentially weighted moving average control charts were used for these features to identify minor structural damage.

## Damage Identification

The final and perhaps the most important step of the vibration- based SHM procedure is the statistical model development or the updating of the numerical model of the monitored structure in order to identify damage. The hierarchy of the damage identification procedure includes the detection of damage, its location, type, severity, and finally the estimation of the residual life of the structure. In a pure data-based approach for SHM combined with ML methods, damage detection is performed developing ML algorithms. In the model-based or hybrid approaches, either a digital twin of the monitored structure is updated using the response measurements, or a numerical model is used in order to generate missing data for training the ML algorithms. Regardless of the approach adopted, successful damage detection processes are those that include as many of the steps of damage identification as possible. This section offers a literature survey on the use of ML methods for damage identification and discusses in detail some important applications. There are applications that focus on pure data-based methods and also applications that use model-based, and hybrid methods.

### Data-Based Damage Identification

Supervised and unsupervised ML algorithms are common learning algorithms used for damage identification in data-based SHM. The selection of the appropriate algorithm depends mainly on the availability of training data. In rare cases when data from both undamaged and damaged states of the structure is available, supervised ML algorithms can be a promising option for damage identification. For instance, group classification algorithms can be applied for damage type and severity classification, or regression analysis can be used. However, for common engineering structures, such as bridges and buildings, it is hard to obtain both damaged and undamaged state measurements under daily operational conditions. As mentioned in Figueiredo et al. (2011), in such cases unsupervised learning is required due to the lack of undamaged state data. In this context, unsupervised learning is also referred to as "novelty", or as "outlier" detection (Bishop, 2006). Both ML and DL algorithms have been used in the literature for supervised and unsupervised learning for damage detection.

The *k*-Nearest Neighbor (kNN) algorithm is one of the simplest supervised ML algorithms. It has shown a good performance for vibration-based damage detection in cases where the sensor data was incomplete and data imputation was necessary (Salehi et al. 2018, and Salehi et al. 2019). The performance of the algorithm was compared to other ML algorithms and a DL algorithm in Li et al. (2020). Due to its simplicity it has shown the best performance among other ML methods. Similarly, the Decision Tree (DT) algorithm is a simple yet well-established learning method for regression and classification tasks. Implementation examples of the DT algorithm for detecting the existence of damage can be found in the studies of Li et al.(2018), and Mariniello et al. (2021). However, the DT algorithm is not capable to compete with other ML algorithms, such as the NNs, due to its lack of flexibility and complexity (Gordan et al., 2020). The Random Forest (RF) algorithm is an ensemble of decision trees and can overcome the issues of a single DT algorithm, as it has good performance on high-dimensional sparse data. For instance, in the study of Laory et al.(2014), where the effect of the operational and environmental conditions on the extracted natural frequency of a suspension bridge is investigated, it is shown that the RF and the Support Vector Regression (SVR) algorithms had more satisfactory results compared to NN, and DT. On the other hand, the study of Lu et al.(2020) for fatigue damage assessment of a coastal bridge indicated that the results of the RF algorithm have a high degree of variability compared to other algorithms, such as the Support Vector Machine (SVM) and the Gaussian Process (GP).

SVM is one of the most popular ML algorithms used for both regression (SVR) and classification. In the study of Gordan et al. (2020), SVM and DT methods were used for damage detection, localization, and severity assessment of a slab-on-girder bridge. The study showed that the SVM algorithm had a superior performance compared to DT. Chong et al. (2014) proposed a nonlinear multiclass SVM for damage detection of structures with nonlinear, time-varying responses. First, they applied wavelet transform to responses generated from healthy and damaged structures under ambient excitation and calculated wavelet-based autoregressive coefficients as the damage indicators. Consequently, they trained the proposed algorithm for classification. Finally, Kohiyama et al. (2020), combined Deep Neural Network (DNN) with SVM for damage detection to overcome the misclassification issues of deep learning algorithms.

Neural Networks (NN), either shallow or deep, are commonly applied ML algorithms for SHM. Among the vast literature, some interesting NN and DNN applications to structural damage detection have been proposed. For instance, in the study by Athar et al. (2020), a NN is applied for damage detection on a lab-scale bridge. The proposed NN was trained using data collected from both direct (contact-based) and indirect (contactless) sensors aiming to decrease the cost and potential damage of sensors in direct sensing methods, and to increase the accuracy and the reliability of monitoring through indirect techniques. At two-stage bridge damage detection and severity assessment method based to a NN was proposed by Malekjafarian et al. (2019), where the responses of a bridge under operational conditions were used as training data. Initially, a NN was used to predict the response of the bridge to a passing vehicle. In the second stage, the structural damage was identified through the change in the distribution of the prediction error also using a GP algorithm.

In the field of DL, the Recurrent Neural Networks (RNN) and the Long Short-Term Memory (LSTM) networks are useful for vibration-based SHM as they are designed to solve sequential input data, such as time series. These algorithms can also be very useful for structural damage identification when there is direct relation with external factors. In terms of SHM, this can be useful for considering the operational and the environmental conditions, and handling noisy data. Examples of studies where RNNs were used to detect structural damage under the influence of environmental conditions are works of Mousavi & Gandomi (2021a) and Mousavi & Gandomi (2021b). LSTMs have the ability of learning long-term temporal dependencies of the data making them quite powerful for time series prediction. In the studies by Zhang et al. (2019) and Li et al. (2022), LSTMs have been successfully applied for predicting the nonlinear response of structures subjected to damaging seismic excitations.

All methods discussed above perform damage detection using supervised learning, which requires labelled data from damaged and undamaged states of the structure. In real life, usually it is not possible to collect data from damaged states. For instance, consider monitoring a newly built bridge. The collected data include only the responses of the bridge under operational conditions without any damage. In such cases, the use of unsupervised learning for anomaly detection can be a remedy for the lack of data. K-means clustering has been used in the study of Alamdari et al. (2017) for anomaly/damage detection using acceleration data collected from the Sydney Harbour Bridge. Burrello et al. (2020) used it for vehicle identification and for tracking to enhance the monitoring of bridges. Figueiredo & Cross (2013) used unsupervised Gaussian Mixture Model (GMM) for damage identification based on outlier detection for long-term monitoring of bridges.

## Model-Based and Hybrid Approaches

A pure data-based approach might not be sufficient for accurate and robust damage identification. For example, when there is lack of training data from different damage states. In the case of model-based SHM, damage identification is achieved through linear or nonlinear model updating of the properties of the digital twin according to the response measurements of the structure. Although the lack of data is not a primary issue, in this approach, uncertainties and modelling errors arise during the construction of the digital twin of the monitored structure. Furthermore, the numerical modelling might be computationally expensive for complex structures when FE models are used as digital twins.

Physics-guided ML-based SHM procedures have been proposed in the recent years in order to optimally combine the two methods and to tackle their shortcomings. The study of Zhang & Sun (2021) attempts to combine the merits of data-based and model-based SHM methods via physics-guided ML in order to improve the damage identification process. Specifically, they carried out numerical model updating interactively with a NN where a physics-based loss function determined the discrepancies between the NN model output and that of the updated FE model. Owing to this interaction, the NN was able to learn and to detect damage when validated with unseen data. In Ritto & Rochinha (2021), a digital twin was constructed for a damaged structure, where a discrete physics-based numerical model was employed to investigate several damage scenarios. A ML classifier was used as the digital twin and was trained using data obtained from a stochastic numerical model. The proposed methodology allowed the use of an interpretable model (model-based) in order to build a fast digital twin (ML) that is connected to the numerical model to support real time damage detection of engineering structures. Different classifiers (quadratic discriminant, support vector machines, etc.) were tested, and different model parameters (number of sensors, level of noise, damage intensity, uncertainty, operational parameters, etc.) were considered to construct datasets for the training.

*Figure 7. Summary of ML and DL applications for vibration-based SHM*

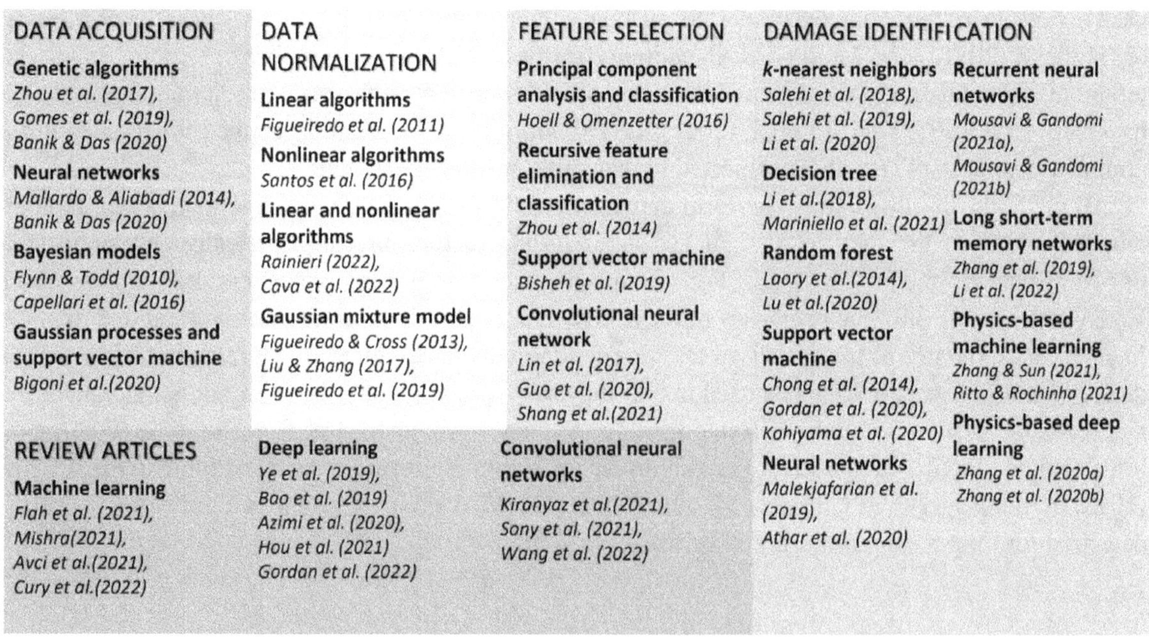

The studies discussed above show that the combination of a numerical model of the structure with an ML classifier in a virtual domain provides interpretability, flexibility and reduced complexity for SHM applications. In Zhang et al. (2020a), a physics-based DL surrogate model for seismic structural response prediction was proposed. A deep physics-based Convolutional Neural Network (PhyCNN) model was trained based on limited seismic input–output datasets and physics constraints in order to establish a surrogate model for predicting structural response. The principles of structural dynamics provide constraints to the network outputs, fix overfitting issues, reduce the need of big training datasets, and improve the robustness of the proposed model. Subsequently, they used the surrogate model for fragility analysis. The performance of PhyCNN was demonstrated through both numerical and experimental examples and it outperformed non-physics-guided Neural Networks. A deep LSTM network with a similar physics-based approach was proposed by Zhang et al. (2020b). The same testing methods were used and significant results were obtained. Shallow NNs like ANN have clear limits when simulating time series of complex nonlinear dynamical systems. A better strategy might be offered by physics-guided DL models like CNN, and RNN, particularly when dealing with constraint data.

The algorithms and the methods discussed in this section are summarized in Figure 7. The ML and DL applications to vibration-based SHM are categorized based on the steps of the SHM procedure in a similar manner with this section. Specifically, ML and DL applications for data acquisition, data normalization, feature selection, damage identification, as well as the literature review articles discussed at the beginning of the section are presented. For every SHM step, the studies are further categorized according to the algorithm/method used in the study. As observed in Figure 7, shallow ML methods seem to be more popular for data acquisition and data normalization than DL methods. On the other hand, CNNs and other DL methods, such as LSTMs and RNNs, are mostly applied for feature selection and damage identification. Furthermore, physics-based ML and DL are promising methods for damage identification since they tackle many of the issues of conventional ML algorithms.

## CONCLUSION

The scope of this Chapter is to briefly provide the theoretical background for vibration-based SHM and damage detection in structural engineering. The approaches of vibration-based SHM, namely model-based, data-based, and hybrid, as well as their strengths and weaknesses were first discussed. Consequently, damage identification and the outline of modern SHM procedures were considered. The steps of SHM and the possible application of ML methods were discussed providing examples from the literature. The Chapter shows that the integration of ML to SHM significantly boosted our damage detection capabilities and offers creative solutions in many engineering practice applications. For example, ML and DL algorithms can reveal the impact of changing operational and environmental conditions. Combined with the data availability in recent years, the SHM community is one step closer to solving the long-standing issue of operational and environmental effects on the response measurements. Additionally, non-stationary and nonlinear sources of variation can be addressed with the use of ML methods. Despite the advancements in machine learning, deep learning, and artificial intelligence, unsupervised SHM frameworks still seem to be lacking. Training opportunities are hindered by unseen states of damage, which can hardly be satisfied by artificial datasets or model-based data generation. To address this issue in SHM methodologies, future developments with unsupervised methodologies, such as population-based SHM, are foreseen.

## ACKNOWLEDGMENT

The research project was supported by the Hellenic Foundation for Research and Innovation (H.F.R.I.) under the "2nd Call for H.F.R.I. Research Projects to support Faculty Members & Researchers" (Project Number:61/5121).

## REFERENCES

Alamdari, M. M., Rakotoarivelo, T., & Khoa, N. L. D. (2017). A spectral-based clustering for structural health monitoring of the Sydney Harbour Bridge. *Mechanical Systems and Signal Processing*, *87*, 384–400. doi:10.1016/j.ymssp.2016.10.033

Athar, S. P. H., Taheri, M., Secrist, J., & Taheri, H. (2020). Neural network for structural health monitoring with combined direct and indirect methods. *Journal of Applied Remote Sensing*, *14*(1), 1–13.

Avci, O., Abdeljaber, O., Kiranyaz, S., Hussein, M., Gabbouj, M., & Inman, D. J. (2021). A review of vibration-based damage detection in civil structures: From traditional methods to Machine Learning and Deep Learning applications. *Mechanical Systems and Signal Processing*, *147*, 107077. doi:10.1016/j.ymssp.2020.107077

Azimi, M., Eslamlou, A. D., & Pekcan, G. (2020). Data-driven structural health monitoring and damage detection through deep learning: State-ofthe- art review. In Sensors (Switzerland), 20(10).

Babajanian Bisheh, H., Ghodrati Amiri, G., Nekooei, M., & Darvishan, E. (2019). Damage detection of a cable-stayed bridge using feature extraction and selection methods. *Structure and Infrastructure Engineering*, *15*(9), 1165–1177. doi:10.1080/15732479.2019.1599964

Banik, M. R., & Das, T. (2020). Application of neuro-GA hybrids in sensor optimization for structural health monitoring. *ACM International Conference Proceeding Series*, (pp. 6–12). ACM. 10.1145/3377049.3377131

Bao, Y., Chen, Z., Wei, S., Xu, Y., Tang, Z., & Li, H. (2019). The State of the Art of Data Science and Engineering in Structural Health Monitoring. *Engineering (Beijing)*, *5*(2), 234–242. doi:10.1016/j.eng.2018.11.027

Barthorpe, R. J. (2010). *On Model and Data Based Approaches to Structural Health Monitoring*. White Rose University Consortium.

Bigoni, C., Zhang, Z., & Hesthaven, J. S. (2020). Systematic sensor placement for structural anomaly detection in the absence of damaged states. In Computer Methods in Applied Mechanics and Engineering, 371. doi:10.1016/j.cma.2020.113315

Bishop, C. M. (2006). *Pattern Recognition and Machine Learning*. Springer.

Burrello, A., Brunelli, D., Malavisi, M., & Benini, L. (2020). Enhancing Structural Health Monitoring with Vehicle Identification and Tracking. *2020 IEEE International Instrumentation and Measurement Technology Conference (I2MTC)*, (pp. 1–6). IEEE. 10.1109/I2MTC43012.2020.9128641

Capellari, G., Chatzi, E., & Mariani, S. (2016). *Optimal Sensor Placement through Bayesian Experimental Design: Effect of Measurement Noise and Number of Sensors*. 41.

Caruana, R., & Niculescu-Mizil, A. (2006). An Empirical Comparison of Supervised Learning Algorithms. *Proceedings of the 23rd International Conference on Machine Learning*, (pp. 161–168). IEEE. 10.1145/1143844.1143865

Chong, J. W., Kim, Y., & Chon, K. H. (2014). Nonlinear multiclass support vector machine-based health monitoring system for buildings employing magnetorheological dampers. *Journal of Intelligent Material Systems and Structures, 25*(12), 1456–1468. doi:10.1177/1045389X13507343

Cury, A., Ribeiro, D., Ubertini, F., & Todd, M. D. (2022). *Structural Health Monitoring Based on Data Science Techniques, 21*. Research Gate.

Farrar, C., Worden, K., & Park, G. (2010). Complexity: A new axiom for Structural Health Monitoring. *Proceedings of the 5th European Workshop: Structural Health Monitoring*, 882–888.

Farrar, C. R., Doebling, S. W., & Nix, D. A. (2001). Vibration-based structural damage identification. *Philosophical Transactions of the Royal Society A: Mathematical, Physical and Engineering Sciences, 359*(1778), 131–149.

Farrar, C. R., & Worden, K. (2013). *Structural health monitoring: a machine learning perspective*. John Wiley & Sons, Ltd.

Figueiredo, E., & Cross, E. (2013). Linear approaches to modeling nonlinearities in long-term monitoring of bridges. *Journal of Civil Structural Health Monitoring, 3*(3), 187–194. doi:10.100713349-013-0038-3

Figueiredo, E., Moldovan, I., Santos, A., Campos, P., & Costa, J. C. W. A. (2019). Finite Element–Based Machine-Learning Approach to Detect Damage in Bridges under Operational and Environmental Variations. *Journal of Bridge Engineering, 24*(7), 04019061. doi:10.1061/(ASCE)BE.1943-5592.0001432

Figueiredo, E., Park, G., Farrar, C. R., Worden, K., & Figueiras, J. (2011). Machine learning algorithms for damage detection under operational and environmental variability. *Structural Health Monitoring, 10*(6), 559–572. doi:10.1177/1475921710388971

Flah, M., Nunez, I., Ben Chaabene, W., & Nehdi, M. L. (2021). Machine Learning Algorithms in Civil Structural Health Monitoring: A Systematic Review. *Archives of Computational Methods in Engineering, 28*(4), 2621–2643. doi:10.100711831-020-09471-9

Flynn, E. B., & Todd, M. D. (2010). A Bayesian approach to optimal sensor placement for structural health monitoring with application to active sensing. *Mechanical Systems and Signal Processing, 24*(4), 891–903. doi:10.1016/j.ymssp.2009.09.003

Fragiadakis, M., Lanese, I. A. P., & Papadrakakis, M. (2009). Experimental And Numerical Investigation of a Reinforced Concrete Building Designed For Gravity Loads Only. *2nd International Conference on Computational Methods in Structural Dynamics and Earthquake Engineering*. COMPDYN.

Friswell, M. I. (2008). Damage identification using inverse methods. *CISM International Centre for Mechanical Sciences, Courses and Lectures, 499*(1993), 13–66.

García Cava, D., Avendaño-Valencia, L. D., Movsessian, A., Roberts, C., & Tcherniak, D. (2022). In A. Cury, D. Ribeiro, F. Ubertini, & M. D. Todd (Eds.), *On Explicit and Implicit Procedures to Mitigate Environmental and Operational Variabilities in Data-Driven Structural Health Monitoring BT - Structural Health Monitoring Based on Data Science Techniques* (pp. 309–330). Springer International Publishing.

Gomes, G. F., de Almeida, F. A., da Silva Lopes Alexandrino, P., da Cunha, S. S. Jr, de Sousa, B. S., & Ancelotti, A. C. Jr. (2019). A multiobjective sensor placement optimization for SHM systems considering Fisher information matrix and mode shape interpolation. *Engineering with Computers*, *35*(2), 519–535. doi:10.100700366-018-0613-7

Gordan, M., Razak, H. A., Ismail, Z., Ghaedi, K., Tan, Z. X., & Ghayeb, H. H. (2020). A hybrid ANN-based imperial competitive algorithm methodology for structural damage identification of slab-on-girder bridge using data mining. *Applied Soft Computing*, *88*, 106013. doi:10.1016/j.asoc.2019.106013

Gordan, M., Sabbagh-Yazdi, S.-R., Ismail, Z., Ghaedi, K., Carroll, P., McCrum, D., & Samali, B. (2022). State-of-the-art review on advancements of data mining in structural health monitoring. *Measurement*, *193*, 110939. doi:10.1016/j.measurement.2022.110939

Guo, T., Wu, L., Wang, C., & Xu, Z. (2020). Damage detection in a novel deep-learning framework: A robust method for feature extraction. *Structural Health Monitoring*, *19*(2), 424–442. doi:10.1177/1475921719846051

Hou, L., Chen, H., Zhang, G., & Wang, X. (2021). Deep Learning-Based Applications for Safety Management in the AEC Industry: A Review. *Applied Sciences (Basel, Switzerland)*, *11*(2), 821. doi:10.3390/app11020821

Huang, X., Wu, L., & Ye, Y. (2019). A Review on Dimensionality Reduction Techniques. *International Journal of Pattern Recognition and Artificial Intelligence*, *33*(10), 1950017. doi:10.1142/S0218001419500174

Kiranyaz, S., Avci, O., Abdeljaber, O., Ince, T., Gabbouj, M., & Inman, D. J. (2021). 1D convolutional neural networks and applications: A survey. *Mechanical Systems and Signal Processing*, *151*, 107398. doi:10.1016/j.ymssp.2020.107398

Kohiyama, M., Oka, K., & Yamashita, T. (2020). Detection method of unlearned pattern using support vector machine in damage classification based on deep neural network. *Structural Control and Health Monitoring*, *27*(8), e2552. doi:10.1002tc.2552

Laory, I., Trinh, T. N., Smith, I. F. C., & Brownjohn, J. M. W. (2014). Methodologies for predicting natural frequency variation of a suspension bridge. *Engineering Structures*, *80*, 211–221. doi:10.1016/j.engstruct.2014.09.001

Li, S., Laima, S., & Li, H. (2018). Data-driven modeling of vortex-induced vibration of a long-span suspension bridge using decision tree learning and support vector regression. *Journal of Wind Engineering and Industrial Aerodynamics*, *172*, 196–211. doi:10.1016/j.jweia.2017.10.022

Li, S., Zuo, X., Li, Z., & Wang, H. (2020). Applying Deep Learning to Continuous Bridge Deflection Detected by Fiber Optic Gyroscope for Damage Detection. *Sensors (Basel)*, *20*(3), 911. doi:10.339020030911 PMID:32046314

Li, T., Pan, Y., Tong, K., Ventura, C. E., & de Silva, C. W. (2022). Attention-Based Sequence-to-Sequence Learning for Online Structural Response Forecasting Under Seismic Excitation. *IEEE Transactions on Systems, Man, and Cybernetics. Systems*, *52*(4), 2184–2200. doi:10.1109/TSMC.2020.3048696

Lin, Y. Z., Nie, Z. H., & Ma, H. W. (2017). Structural Damage Detection with Automatic Feature-Extraction through Deep Learning. *Computer-Aided Civil and Infrastructure Engineering*, *32*(12), 1025–1046. doi:10.1111/mice.12313

Liu, Y., & Zhang, S. (2017). Probabilistic Baseline of Finite Element Model of Bridges under Environmental Temperature Changes. *Computer-Aided Civil and Infrastructure Engineering*, *32*(7), 581–598. doi:10.1111/mice.12268

Lu, Q., Zhu, J., & Zhang, W. (2020). Quantification of Fatigue Damage for Structural Details in Slender Coastal Bridges Using Machine Learning-Based Methods. *Journal of Bridge Engineering*, *25*(7), 04020033. doi:10.1061/(ASCE)BE.1943-5592.0001571

Malekjafarian, A., Golpayegani, F., Moloney, C., & Clarke, S. (2019). A Machine Learning Approach to Bridge-Damage Detection Using Responses Measured on a Passing Vehicle. *Sensors (Basel)*, *19*(18), 4035. doi:10.339019184035 PMID:31546759

Mallardo, V., & Aliabadi, M. H. (2014). Optimal sensor placement for structural, damage and impact identification: A review. *SDHM Structural Durability and Health Monitoring*, *9*(4), 287–323. doi:10.32604dhm.2013.009.287

Mariniello, G., Pastore, T., Menna, C., Festa, P., & Asprone, D. (2021). Structural damage detection and localization using decision tree ensemble and vibration data. *Computer-Aided Civil and Infrastructure Engineering*, *36*(9), 1129–1149. doi:10.1111/mice.12633

Mishra, M. (2021). Machine learning techniques for structural health monitoring of heritage buildings: A state-of-the-art review and case studies. *Journal of Cultural Heritage*, *47*, 227–245. doi:10.1016/j.culher.2020.09.005

Mousavi, M., & Gandomi, A. H. (2021a). Prediction error of Johansen cointegration residuals for structural health monitoring. *Mechanical Systems and Signal Processing*, *160*, 107847. doi:10.1016/j.ymssp.2021.107847

Mousavi, M., & Gandomi, A. H. (2021b). Deep learning for structural health monitoring under environmental and operational variations. *Proceedings of the Society for Photo-Instrumentation Engineers*, 11592. doi:10.1117/12.2582649

Oh, J., Hessel, M., Czarnecki, W. M., Xu, Z., van Hasselt, H., Singh, S., & Silver, D. (2020). Discovering reinforcement learning algorithms. Advances in Neural Information Processing Systems. NeurIPS.

Rainieri, C. (2022). In A. Cury, D. Ribeiro, F. Ubertini, & M. D. Todd (Eds.), *Environmental Influence on Modal Parameters: Linear and Nonlinear Methods for Its Compensation in the Context of Structural Health Monitoring BT - Structural Health Monitoring Based on Data Science Techniques* (pp. 269–288). Springer International Publishing.

Ritto, T. G., & Rochinha, F. A. (2021). Digital twin, physics-based model, and machine learning applied to damage detection in structures. *Mechanical Systems and Signal Processing*, *155*, 107614. doi:10.1016/j.ymssp.2021.107614

Rodriguez, M. Z., Comin, C. H., Casanova, D., Bruno, O. M., Amancio, D. R., Costa, L. da F., & Rodrigues, F. A. (2019). Clustering algorithms: A comparative approach. *PLoS One*, *14*(1), e0210236. doi:10.1371/journal.pone.0210236 PMID:30645617

Russell, S. J., & Norvig, P. (2010). *Artificial Intelligence: A Modern Approach*. Prentice Hall.

Rytter. (1993). *Aalborg Universitet Vibrational Based Inspection of Civil Engineering Structures*. Rytter, Anders.

Salehi, H., Biswas, S., & Burgueño, R. (2019). Data interpretation framework integrating machine learning and pattern recognition for self-powered data-driven damage identification with harvested energy variations. *Engineering Applications of Artificial Intelligence*, *86*, 136–153. doi:10.1016/j.engappai.2019.08.004

Salehi, H., Das, S., Chakrabartty, S., Biswas, S., & Burgueno, R. (2018). A methodology for structural health diagnosis and assessment using machine learning with noisy and incomplete data from self-powered wireless sensors. *Proceedings of the Society for Photo-Instrumentation Engineers*, 10598. doi:10.1117/12.2295990

Santos, A., Figueiredo, E., Silva, M. F. M., Sales, C. S., & Costa, J. C. W. A. (2016). Machine learning algorithms for damage detection: Kernel-based approaches. *Journal of Sound and Vibration*, *363*, 584–599. doi:10.1016/j.jsv.2015.11.008

Shang, Z., Sun, L., Xia, Y., & Zhang, W. (2021). Vibration-based damage detection for bridges by deep convolutional denoising autoencoder. *Structural Health Monitoring*, *20*(4), 1880–1903. doi:10.1177/1475921720942836

Shyu, M., Chen, S., & Iyengar, S. S. (2020). a Survey on Deep Learning Techniques. *Strad Research, 7*(8).

Sony, S., Dunphy, K., Sadhu, A., & Capretz, M. (2021). A systematic review of convolutional neural network-based structural condition assessment techniques. *Engineering Structures*, *226*, 111347. doi:10.1016/j.engstruct.2020.111347

Wang, Y., Zhao, Y., & Addepalli, S. (2020). Remaining useful life prediction using deep learning approaches: A review. *Procedia Manufacturing, 49*(2019), 81–88.

Wang, Y.-W., Ni, Y.-Q., & Wang, S.-M. (2022). Structural health monitoring of railway bridges using innovative sensing technologies and machine learning algorithms: a concise review. *Intelligent Transportation Infrastructure, 1*, liac009.

Worden, K., Farrar, C. R., Manson, G., & Park, G. (2007). The fundamental axioms of structural health monitoring. *Proceedings - Royal Society. Mathematical, Physical and Engineering Sciences*, *463*(2082), 1639–1664. doi:10.1098/rspa.2007.1834

Ye, X. W., Jin, T., & Yun, C. B. (2019). A review on deep learning-based structural health monitoring of civil infrastructures. *Smart Structures and Systems*, *24*(5), 567–585.

Zhang, R., Chen, Z., Chen, S., Zheng, J., Büyüköztürk, O., & Sun, H. (2019). Deep long short-term memory networks for nonlinear structural seismic response prediction. *Computers & Structures*, *220*, 55–68. doi:10.1016/j.compstruc.2019.05.006

Zhang, R., Liu, Y., & Sun, H. (2020a). Physics-guided convolutional neural network (PhyCNN) for data-driven seismic response modeling. *Engineering Structures*, *215*, 110704. doi:10.1016/j.engstruct.2020.110704

Zhang, R., Liu, Y., & Sun, H. (2020b). Physics-informed multi-LSTM networks for metamodeling of nonlinear structures. *Computer Methods in Applied Mechanics and Engineering*, *369*, 113226. doi:10.1016/j.cma.2020.113226

Zhang, Z., & Sun, C. (2021). Structural damage identification via physics-guided machine learning: A methodology integrating pattern recognition with finite element model updating. *Structural Health Monitoring*, *20*(4), 1675–1688. doi:10.1177/1475921720927488

Zhou, K., Wu, Z. Y., Yi, X. H., Zhu, D. P., Narayan, R., & Zhao, J. (2017). Generic Framework of Sensor Placement Optimization for Structural Health Modeling. *Journal of Computing in Civil Engineering*, *31*(4), 04017018. doi:10.1061/(ASCE)CP.1943-5487.0000662

Zhou, Q., Zhou, H., Zhou, Q., Yang, F., & Luo, L. (2014). Structure damage detection based on random forest recursive feature elimination. *Mechanical Systems and Signal Processing*, *46*(1), 82–90. doi:10.1016/j.ymssp.2013.12.013

## KEY TERMS AND DEFINITIONS

**Artificial Intelligence (AI):** Theory and development of computer systems that can carry out tasks that would typically need human intelligence.

**Data-based SHM:** Approach of vibration-based SHM where the assessment of structural health depends on the development of statistical models on the collected response data.

**Deep Learning (DL):** Subset of ML, that in essence is a deep Neural Network (NN) technology with many hidden layers. DL is supposed to imitate the complex way that humans gain certain kinds of knowledge.

**Machine Learning (ML):** Subset of AI, that is the development of algorithms and statistical models used by computer systems in order to effectively perform a specific task without using explicit instructions.

**Model-based SHM:** Approach of vibration-based SHM where the assessment of structural health condition is obtained by updating a numerical model (digital twin) of the structure according to the collected measurements.

**Structural Health Monitoring (SHM):** Observation and analysis of structural systems over time using periodically sampled response measurements in order to monitor changes in the material or the geometry of the systems.

**Supervised Learning:** Machine learning task of learning a function that maps an input to an output based on sample input-output pairs.

**Unsupervised Learning:** Machine learning task where the algorithm learns patterns from data consisting of only input features.

**Vibration-based SHM:** Assessment of structural health by processing structural response measurements (usually accelerations) collected from sensors.

# Chapter 4
# Use of AI and ML Algorithms in Developing Closed-Form Formulae for Structural Engineering Design

**George Markou**
*University of Pretoria, South Africa*

**Nikolaos Bakas**
*National Infrastructures for Research and Technology–GRNET, Greece*

**Ashley Megan van der Westhuizen**
*University of Pretoria, South Africa*

## ABSTRACT

*The design and analysis of structures is performed with the use of national and international design codes that usually suggest the use of semi-empirical formulae. Often the formulae are oversimplified, in some cases are not available to engineers, or are time-consuming and challenging to implement. The objective of this chapter is to demonstrate the use of artificial intelligence and machine learning to develop more accurate formulae for different types of applications related to structural design. The applications that are discussed in this work include predicting the shear capacity of reinforced concrete slender and deep beams without stirrups, calculating the fundamental period of reinforced concrete and steel structures, and predicting the deflection of horizontally curved steel I-beams.*

## INTRODUCTION

The use of artificial intelligence (AI) and machine learning (ML) in structural analysis and design problems has been increasing due to the methods' abilities to handle complex nonlinear structural systems under extreme actions (Markou & Bakas, 2021a, 2021b; Thai, 2022). The purpose of the chapter is to

DOI: 10.4018/978-1-6684-5643-9.ch004

discuss the use of AI and ML in structural engineering design based on current work performed by the authors and analyse the future perspectives related to this technology. The applications that are presented herein include the development of predictive formulae for the calculation of the shear capacity of slender reinforced concrete (RC) beams without stirrups, predicting the shear capacity of deep beams without stirrups reinforced with fibre reinforced polymer (FRP) rebars, determining the fundamental period of RC as well as steel structures, and determining the deflection of curved steel I-beams. In some cases, the formulae currently available for the determination of the above structural problems are oversimplified or are not readily available to practising engineers (Ababu et al., 2022), therefore, this is discussed in the relevant sections of this chapter that follow.

It is the objective of this chapter to show the reader the development of closed mathematical form design formulae that can predict the mechanical behaviour of structures using an out-of-the-box approach (Bakas & Markou, 2019), while artificial neural networks (ANN) and other ML methods were adopted. According to the proposed approach, by using finite element software and nonlinear analysis through the use of ReConAn FEA (2020) software, datasets are developed that are used for the development of predictive models. This approach aims to marry state-of-the-art 3D detailed modelling with AI and ML algorithms for the development of predictive models that exhibit a more accurate calculation of the mechanical response of structural members and structures. Furthermore, given that AI and ML algorithms are an effective and efficient tools used to predict analysis outputs of computationally demanding engineering problems, their use by the civil engineering community has emerged over the last decade and increases exponentially (Bakas et al., 2021).

A dataset is obtained from the results of the analysis of these models and is then used to train AI and ML algorithms that can develop predictive models. One of the main obstacles in training and testing models is the lack of datasets that refer to a specific structural problem. This is attributed to the limitations in relation to performing a large number of experiments for any structural problem, where a sufficient number of output data will be made available for the training of our AI and ML models (van der Westhuizen et al., 2022). For this reason, the proposed approach foresees the substitution of the actual experiment through the use of 3D detailed finite element analysis, hence, to be able to proceed with this approach an extensive validation of the adopted finite element algorithm has to be performed. This is also going to be presented by referencing the research work performed towards validating the research finite element analysis (FEA) software ReConAn FEA (2020) that was used for the development of the datasets presented in this work. Additionally, it is important to note that the proposed approach foresees, where feasible, the validation of the predictive models through the use of additional datasets that derived from physical experiments and were found in the international literature.

For the case where physical experiments are not available, the accuracy of the proposed predictive models is validated using out-of-sample data, while for all the structural problems presented herein the mean absolute percentage error (MAPE) and other error parameters are determined and compared to that of the design formulae currently available in the international literature. As is going to be shown in this chapter, the predictive models derived by using AI or ML algorithms proved to outperform design code formulae currently used to determine the mechanical properties and the strength of structures or their members. All studies presented in the chapter show strong evidence that a combination of FEA and ML methods provides reliable predictions that can be used in improving design code formulae currently used to calculate the shear capacity of slender and deep beams, determining the fundamental period of RC and steel structures as well as determining the deflection of curved steel I-beams.

The ML algorithms used in the studies discussed include ANN, linear regression (LR), non-linear higher-order regression (NLR), random forests (RF) and gradient boosting (GB). It is important to note that the ML algorithms are not a one-size-fits-all solution and different methods are used in each case to find the most accurate predictive model (Markou & Bakas, 2021b).

# BACKGROUND

## Machine Learning

The higher-order NLR ML algorithm used extensively in the research work presented herein can be seen below.

Algorithm: Higher Order Regression (Gravett et al., 2021)

```
Input: XX (matrix of Independent Variables), YY (Vector of Dependent Variable),
nlf (number of nonlinear features to be kept in the model)
Output: Prediction Formulae
1. Create all nonlinear features* (anlf)
2. For i from 1 to nlf do
        3. For j from 1 to anlf do
                4. Add j^th feature to the model
                5. Calculate Prediction Error, MAPE_j**
        6. End
7. Keep in the model the j^th feature which yields the minimum prediction error
8. End
Return: Prediction Formula
```

*with all inter-items combinations up to the 3rd degree, **Mean Absolute Percentage Error (MAPE).

The LR and NLR models are useful because they generate a closed-form formula whereas the other algorithms develop predictive models that are numerical. According to the numerous parametric investigations performed by the authors, it was always found that the LR model is unable to capture the non-linear behaviour of variables so more complex ML algorithms are required in achieving this task. For the needs of the numerical investigations related to the accuracy of the proposed predictive models, the ML algorithms produce the MAPE, which is used as an objective metric for measuring the abilities of the proposed models to accurately describe the different at-hand engineering applications. It should be noted that in each case 85% of the database was used for training purposes and 15% for testing. The formulae obtained are then validated using out-of-sample results that were not used in the training nor in the testing procedure, where in some cases the validation phase included solely datasets that were derived from experimental results.

## Shear Capacity of Reinforced Concrete Beams Without Stirrups

Shear failure in structural beam members occurs in a brittle manner and is usually attributed to insufficient shear resistance in the web. It is important to design members, like beams, for shear resistance to avoid these failures that provide no warning to the occupants (Spijkerman et al., 2021). Various design codes divide beams into two categories, namely, slender beams and deep beams. This difference is performed using the span-to-depth ratio (a/d) which is found to be directly proportional to the shear strength of a concrete beam. In some codes, the a/d classifies a beam as slender when it is larger than 2 (Tureyen & Frosch, 2002; El-Slayed & Benmokrane, 2005; El-Slayed et al., 2006; AlHamaydeh et al., 2022), whereas according to Eurocode 2 the ratio a/d must be larger than 3. Unlike slender beams, deep beams develop the so-called arch action which is a mechanism that foresees the transfer of loads to the supports through diagonal compression struts (Yang et al., 2003). It should also be noted that the arch action experienced by these beams as well as the use of FRP rebars within the concrete domain is not accounted for in most current design formulae. The use of FRP rebars instead of conventional steel reinforcement is proposed as an alternative due to their high immunity to corrosion, increasing the service life of concrete (Abed et al., 2012; Dhahrir, 2017, 2018). It is important to investigate how the behaviour differs when compared to steel-reinforced deep beams and these differences should be considered when developing revised formulae (Razaqpur & Isgor, 2006; AlHamaydeh et al., 2022).

Currently, the main approach used in determining the shear capacity of RC beams without stirrups is using formulae derived from experimental data. The use of this method, however, is limited by the availability of experimental data that covers the vast combination of beam geometries, load cases and boundary conditions (Markou & Bakas, 2021; Spijkerman et al., 2021). An in-depth literature review for determining the shear capacity of RC beams can be found in AlHamaydeh et al. (2022). A summary of the formulae that were used in the comparative study that will be discussed in this chapter can be seen in Equations 1, 2 and 3.

ACI 318-19 $$V_c = \left( 8 \cdot \lambda_s \cdot \lambda \cdot (\rho_w)^{1/3} \cdot \sqrt{f'_c} + \frac{N_u}{6A_g} \right) b_w \cdot d \qquad (1)$$

ACI 440 $$\mathbf{V_c} = \left( \frac{5}{2} \cdot \mathbf{k} \right) 0.17 \sqrt{\mathbf{f'_c}} \cdot \mathbf{b_w} \cdot \mathbf{d} \qquad (2)$$

Eurocode 2 $$V_{Rd,c} = \left( C_{Rd,c} \cdot k \cdot (\rho_1 \cdot f_{ck})^{\frac{1}{3}} + k_1 \cdot \sigma_{cp} \right) b_w \cdot d \qquad (3)$$

The ACI Equations are presented in US units. The design code Eurocode 2 mandates the use of β, an approximate strength reduction, when designing for deep beams, which accounts for the arch effect.

## Fundamental Period of Reinforced Concrete and Steel Structures

In structural design it is important to derive safe structures to avoid the loss of life, especially when structures undergo earthquake excitations, thus the understanding of the dynamic behaviour of structures is crucial during the design process. Calculating the fundamental period of structures is one of the most important parameters that is considered when designing seismic-resistant structures (Young, 2011; Gravett et al., 2021; van der Westhuizen et al., 2022).

The natural frequencies are determined using modal analysis which is based on the eigenvalue problem. Eigen-frequencies are important for engineers to understand the dynamic behaviour of their structures and which modal shapes will contribute the most during a seismic event. There are numerous methods proposed to solve eigenvalue problems represented by Equation 4.

$$K\varphi i = \lambda i M_{\varphi} i \tag{4}$$

Where:

K is the stiffness matrix of the model
M is the mass of the model
$\varphi i$ is the vector that contains the eigenvector of the problem
$\lambda i$ is the corresponding eigenvalue

The method presented herein is the subspace iteration algorithm (Bathe, 1971), which is integrated within ReConAn FEA, (2020). Validation work performed on the ability to predict the fundamental period of structures through the use of ReConAn FEA, (2020) can be found in Markou et al., (2018), Mourlas et al., (2019a, 2019b, 2019c), Gravett et al., (2019), and Papadrakakis and Markou (2022). This solution algorithm finds an orthogonal basis of vectors in $E_{K+1}$, calculating in one step the required eigenvectors when $E_{K+1}$ converges to $E_{\infty}$. For k=1,2,…, iterate from $E_k$ to $E_{k+1}$:

$$KX_{k+1} = Mx_k \tag{5}$$

then, find the projections of matrices K and M onto $E_{k+1}$:

$$K_{k+1} = X_{k+1}^T K X_{k+1} \tag{6}$$

$$M_{k+1} = X_{k+1}^T M X_{k+1} \tag{7}$$

and solve for the eigensystem of the projected matrices:

$$K_{k+1} Q_{k+1} = M_{k+1} Q_{k+1} \Lambda_{k+1} \tag{8}$$

Thereafter, find an improved approximation to the eigenvectors:

$$x_{k+1} = X_{k+1} Q_{k+1} \tag{9}$$

and then provided that the vectors $X_1$ are not orthogonal to one of the equilibrium eigenvectors, $\Lambda_{k+1} \rightarrow \Lambda$ and $X_{k+1} \rightarrow \Phi$ as $k \rightarrow \infty$.

It is important to note that the convergence of this method assumes that within the iteration procedure the vectors in $X_{k+1}$ are ordered so the $i^{th}$ diagonal element in $\Lambda_{k+1}$ will always be larger than the previous i–1 element, i=1,…,p. This approach ensures that the $i^{th}$ column in $X_{k+1}$ converges linearly to $\Phi_i$ (Bathe, 1971). A detailed modal analysis is time-consuming and requires excessive computational effort, therefore by taking into account the soil-structure interaction (SSI) phenomenon, the problem becomes more cumbersome and computationally demanding (Gravett et al., 2021). This further highlights the importance of developing a predictive model that will be able to compute the fundamental period of a structure, including the SSI effect, through a minimal computational cost (van der Westhuizen et al., 2022).

In the event of a seismic excitation, the interaction between the superstructure (building) and substructure (soil) becomes critical as it alters the distribution of stresses and strains within the superstructure, influencing the expected results (Mourlas et al., 2020). Current building codes use empirical equations to predict the fundamental period of structures (Jiang et al., 2020; Taljaard et al., 2021; and Gravett et al., 2021; van der Westhuizen et al., 2022) that do not take into account the SSI effect. Therefore, design formulae that can be found in literature and design codes omit the use of SSI that, in some cases, significantly influences the natural frequency of buildings due to flexibility induced by soft soils. A number of design code formulae that are investigated in the international literature for both steel and concrete structures can be seen in Equations 10 to 13.

NEAK (RC frames) $$\mathbf{T}_{\mathbf{NEAK}} = 0,09 \cdot \frac{\mathbf{H}}{\sqrt{\mathbf{L}}} \cdot \sqrt{\frac{\mathbf{H}}{\mathbf{H} + \mathbf{p} \cdot \mathbf{L}}} \tag{10}$$

Eurocode 8 and SANS10160-4 (RC Frames with concrete or masonry shear walls)

$$C_T = \frac{0,075}{\sqrt{A_c}} \tag{11}$$

Where:

$$A_c = \Sigma\left[A_i\left(0,2 + \left(\frac{L_{wi}}{H}\right)^2\right)\right] \tag{12}$$

Eurocode 8 and SANS10160-4 (RC and steel frames)

$$T_{EC} = C_T \cdot H^{0.75} \tag{13}$$

Where:

$C_t$=0.085 for moment resistant space steel frames
$C_t$=0.075 for RC moment-resisting frames and for eccentrically braced frames
$\mathbf{C}_t$=0.05 for all other buildings 12

ASCE 7-05 (steel frames) $\mathbf{T}_1=0.0724(\mathbf{H})^{0.8}$ for steel moment-resisting frames $\mathbf{T}_1=0.0731(\mathbf{H})^{0.75}$ for eccentrically braced steel frames

As stated previously and as can be seen in the formulae described above, there is no presence of accounting for the SSI effect, while the parameters used to compute the fundamental period practically foresee the use of the height of the structure. The only code that takes into account the width of the structure and the RC shear wall ratio is the new Greek earthquake-resistant code (NEAK).

## Deflection of Horizontally Curved Steel I-Beams

The use of curved beams has been increasing in popularity throughout the years due to the aesthetically pleasing designs they produce in buildings as well as in industrial applications and bridges. They can be found in applications ranging from girders in modern highway bridges, interchange facilities, as well as industrial buildings where they are used as crawl beams. The mechanical behaviour of curved steel I-beams is still not well understood and currently, there are no precise formulae available to determine their deflections nor their overall mechanical response due to different types of loads. This lack of understanding is attributed to the mechanical parameters that affect the beams' response when they undergo gravitational loads. These parameters include a combination of shear, flexure and torsion deformations that generate a complicated stress-strain development at a material level, while the lack of using a 3D material model approach does not allow the accurate prediction of the local and global mechanical response of these structures. The design of light structures is often governed by the serviceability considerations such as deflection (Ababu, et al., 2022), which highlights the importance of understanding the behaviour of curved beams and their deflections when in the design phase.

By researching the international literature, one will realise that the most widely used formula for determining the deflection of this type of beam is Castigliano's second theorem presented in Equation 14. This formula is not applied directly but the deflection is derived using the equation and an integration algorithm (Dahlberg, 2004).

$$\delta = \frac{\partial \mathbf{U}}{\partial \mathbf{P}} \tag{14}$$

Other formulae researchers have attempted to provide, based on Castigliano's second theorem, for the determination of the deflection of curved beams, which are not readily available to professional civil engineers and can only be applied to specific load cases and support conditions. Other studies conducted to determine the deflection do not analyse the parameters that may influence the deflection and those that did use Castigliano's theorem (Ababu, et al., 2022).

A few of the formulae found in Dahlberg (2004) for determining the deflections of beams with different constraints are presented below. The first case refers to a curved cantilever beam, which is curved to form a quarter of an ellipse. The beam is clamped at one end and a force P acts at the free end. Using Castigliano's theorem along with other mathematical methods, such as a numerical integration method, Equation 15 is derived.

$$\delta = \frac{\mathbf{Pa}^3}{\mathbf{EI}}\mathbf{I}_1(\beta) + \frac{\mathbf{Pa}^3}{\mathbf{GK}_t}\mathbf{I}_2(\beta) \tag{15}$$

where the integrals $I_1$ and $I_2$ are functions of the parameter $\beta(=b/a)$ only

$$I_1(\beta)=\int_0^1\left(\beta\sqrt{1-\frac{x^2}{a^2}}\frac{dy/dx}{\sqrt{1+\left(\frac{dy}{dx}\right)^2}}-\left(1-\frac{x}{a}\right)\frac{1}{\sqrt{1+\left(\frac{dy}{dx}\right)^2}}\right)^2\sqrt{1+\left(\frac{dy}{dx}\right)^2}\,d\left(\frac{x}{a}\right) \tag{16}$$

and

$$\mathbf{I}_2(\beta)=\int_0^1\left(\beta\sqrt{1-\frac{\mathbf{x}^2}{\mathbf{a}^2}}\frac{\mathbf{dy/dx}}{\sqrt{1+\left(\frac{\mathbf{dy}}{\mathbf{dx}}\right)^2}}+\left(1-\frac{\mathbf{x}}{\mathbf{a}}\right)\frac{\frac{\mathbf{dy}}{\mathbf{dx}}}{\sqrt{1+\left(\frac{\mathbf{dy}}{\mathbf{dx}}\right)^2}}\right)^2\sqrt{1+\left(\frac{\mathbf{dy}}{\mathbf{dx}}\right)^2}\,\mathbf{d}\left(\frac{\mathbf{x}}{\mathbf{a}}\right) \tag{17}$$

where:

a is the length in the x-direction of one half-axis
b is the length in the y-direction of one half-axis

In Equations 15, 16 and 17 it can be seen how complex the derivation of such formulae can get. These are the "basic" formulae, and more are derived in Dahlberg (2004) where the influence of bending, and torsion are investigated. It is noteworthy to state at this point that the formulae become lengthier when indeterminate beams are investigated.

## PROPOSED PREDICTIVE MODELS DERIVED FROM AI AND ML ALGORITHMS

### Predicting the Shear Capacity of Reinforced Concrete Slender Beams

The studies presented herein focused on the development of a single relationship that can predict the shear capacity of slender RC beams without stirrups without using available experimental data. ML algorithms are used to train and test models on a large database that, for the first time in international literature, was generated using a non-linear FEA. It is important to state here that ReConAn FEA, (2020) and its ability to predict the shear capacity of the RC slender beams was presented in Markou and Papadrakakis, (2013), Markou et al., (2017), Mourlas et al., (2017) and Markou et al., (2021).

The first study presented here (Markou & Bakas, 2021a) comprises approximately 36,000 slender beams without stirrups generated using a newly developed program called ReConAn FEA Multirun that generates multiple input files and then uses ReConAn FEA to analyse them. This software reads the input file that contains different material properties and reinforcement ratios, where it develops the finite element (FE) models' input files automatically (Markou & Bakas, 2021a).

A sensitivity analysis was conducted to investigate the effect of each independent variable while keeping all other input variables constant. Through this approach, one can compare the results of each ML model which should be as close as possible (Markou & Bakas, 2021a). Two variables that controlled the shear prediction are the effective depth of the beam, d (Figure 1) and the width of the section, b (Figure 2) as they were derived through the training procedure.

*Figure 1. Sensitivity curves for d*
*(Markou & Bakas, 2021a)*

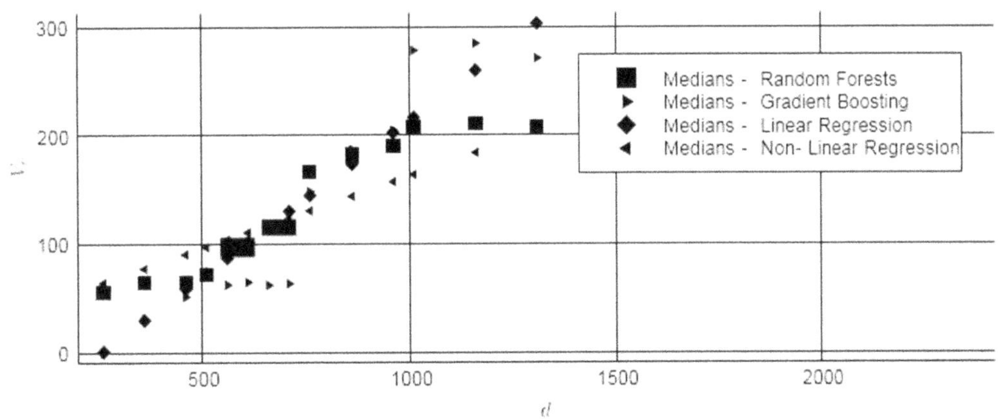


*Figure 2. Sensitivity curves for b*
*(Markou & Bakas, 2021a)*

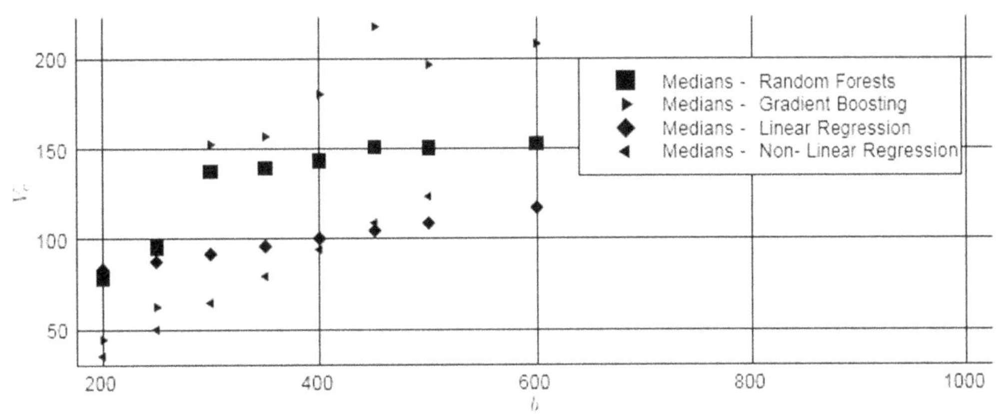


It must be noted here that in order to create a dataset that would have covered a wide range of beam geometries, 95 different beam geometries were constructed, and a summary of the minimum and maximum parameters considered can be seen in Table 1. The modelling of the beams used 20-noded isoparametric hexahedral FE for the concrete, steel plates at the supports and 2-noded rod elements for the embedded reinforcing, where the mesh used for discretizing the beams can be seen in Figure 3. The RC beam depicted in this figure has a net span of 1,500 mm and a 200x300 mm section.

*Table 1. RC slender beam geometry. Minimum and maximum magnitudes of the 95 beams*

| Variables | Minimum | Maximum |
|---|---|---|
| L [mm] | 1 500 | 8 700 |
| b [mm] | 200 | 600 |
| h [mm] | 300 | 1350 |
| L/h | 5 | 11.6 |
| h/b | 1.0 | 4.5 |

(Markou & Bakas, 2021a)

*Figure 3. Hexahedral and embedded rod element meshes*
*(Markou & Bakas, 2021a)*

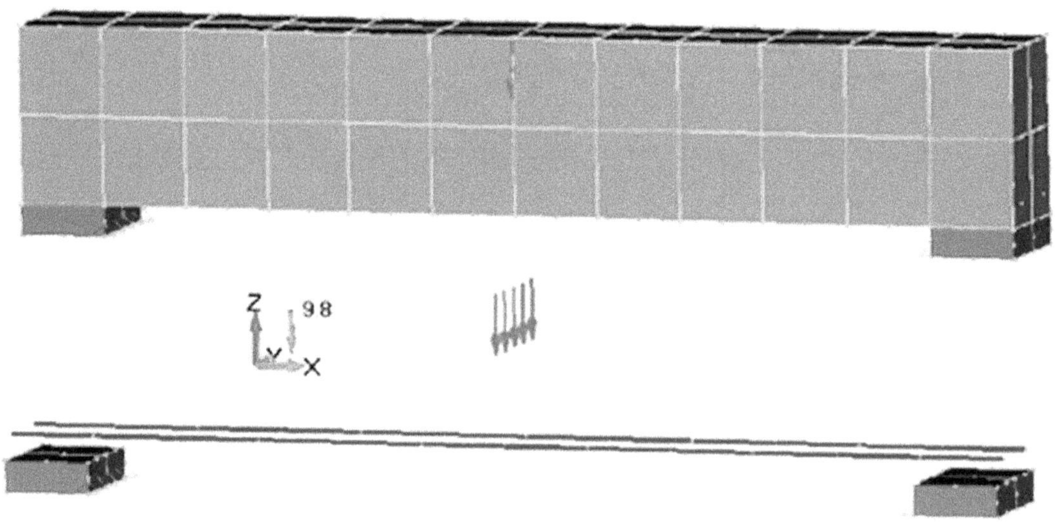


The database is constituted of two groups, with the first group consisting of variables affecting the mesh of the beams, namely the span (*L*), the beam width (*b*) and the effective depth (*d*). The second group consisted of the material properties such as the uniaxial compressive strength in MPa ($f_c$), concrete Young modulus in MPa ($E_c$), concrete tensile strength ratio ($f_t$) as a percentage of the uniaxial compressive strength, the remaining shear capacity strength factor (β) as a percentage of the initial shear resistance of the material point under investigation, the steel Young modulus in MPa ($E_s$), steel yielding stress in MPa ($f_y$) and the tensile longitudinal reinforcement ratio ($\rho$) as a percentage of the concrete area of the beam's section. Four ML methods were utilized in the predictive modelling including LR, higher-order NLR, RF and GB. Multiple error metrics were considered to quantify the ML method response. The metrics are presented in detail in Dimopoulos et al., (2018).

A summary of the metrics obtained can be found in Markou & Bakas (2021a). The predictive formula obtained using the higher-order NLR model can be seen in Equation 18.

$$\begin{aligned} \mathbf{V_c} = & \left(0.168 \cdot \mathbf{d}\right) - \left(0.0196 \cdot \rho \cdot \mathbf{b} \cdot \mathbf{d}\right) - (1.5310^{-10} \cdot \mathbf{L}^3) \\ & + \left(0.0197 \cdot \rho \cdot \mathbf{f_c} \cdot \mathbf{L}\right) - \left(809.8 \cdot \rho^2 \cdot \mathbf{d}\right) + \left(1.55810^{-4} \cdot \rho \cdot \mathbf{E_s} \cdot \mathbf{d}\right) \\ & + (5.7410^{-9} \cdot \mathbf{b} \cdot \mathbf{L}^2) - \left(9.6010^{-12} \cdot \mathbf{E_s} \cdot \mathbf{L}^2\right) - \left(6853.2 \cdot \rho^2 \cdot \mathbf{f_c}\right) \\ & + (6.7810^{-3} \cdot \mathbf{f_t} \cdot \mathbf{f_c} \cdot \mathbf{d}) - 36.172 \end{aligned} \tag{18}$$

For the validation of the proposed formula, 100 RC beams that were experimentally tested were used, where the experimental results were obtained from Reineck et al., (2013). It is important to note that no experimental data were used in the formulation of Equation 18 as opposed to what is usually performed and presented in the international literature, where experimental results are used to train and test the proposed predictive models. The 100 RC beams that were used during the validation procedure were selected according to the geometric minimum and maximum values presented in Table 1. A comparison is made between the RF trained model (RFTM), which was found to be the best predictive model in Markou and Bakas (2021a), and Equation 1, which can be seen in Figure 4.

According to the MAPE, the ML algorithm predictive model produced a MAPE of 11.6%, 36.4% lower than the ACI 2019 formula. From the figures, it can also be seen that for larger sections ($V_c >$ 150kN) the predictions do not favour safety, whereas for the ML model there is a better distribution about the diagonal line. These beams are at higher risk in a failure scenario and the MAPE for these cases for the ML model was 9.4%, 44% lower compared to the ACI formula. It was also found that within the experimental data beams with the same geometry, reinforcement ratio and material properties derived different shear capacities.

A study presented by Spijkerman et al., (2021) that was performed in parallel to the Markou and Bakas (2021a), assumed a smaller number of RC beams for the training and testing procedure. In this study, a total of 10,000 RC beams foresaw a rectangular section with a larger width compared to the height. The database consisted of 20 geometrically different simply supported slender RC beams with minimum and maximum values presented in Table 2.

*Figure 4. Experimental shear strength vs Prediction with (a) ACI (2019) and (b) Random forests (Markou & Bakas, 2021a)*

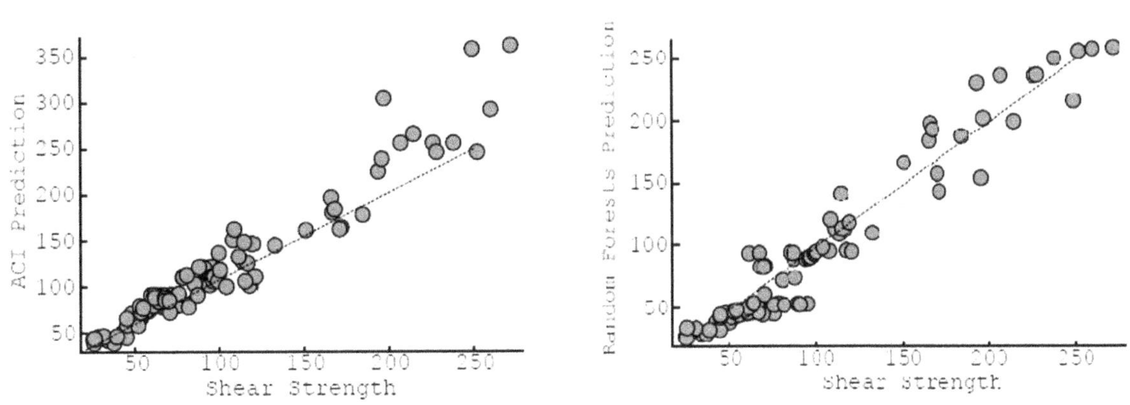


(a) (b)

*Table 2. RC slender beam geometry. Minimum and maximum magnitudes of the 20 beams*

| Variables | Minimum | Maximum |
|---|---|---|
| L [mm] | 1 500 | 2 700 |
| b [mm] | 300 | 750 |
| h [mm] | 150 | 350 |
| L/h | 6 | 10.8 |
| h/b | 0.33 | 0.83 |

(Spijkerman et al., 2022)

Each of the 20 RC geometrically unique beams were analysed 500 times for different material and reinforcement ratio combinations forming the 10 000 results used to train and test the predictive model. The combinations consisted of different magnitude variations of the $f_c$ (kPa), $E_c$ (kPa), $f_t$, $\beta$, $E_s$ (kPa), $f_y$ (kPa) and $A_s$ ($m^2$), similar to the study conducted by Markou & Bakas (2021a) presented above.

The formula developed, was then used to predict the shear strength of experimental data found in the international literature. In total, three formulae were produced to investigate the robustness and accuracy, whereas the formula presented herein in Equation 19 was the one that had the minimum MAPE (5.1%). As it can be seen, the proposed formula has 10-features and was compared to both Eurocode 2 and ACI 318-19.

$$\begin{aligned}\mathbf{Vc} = &\left(0.253939\cdot\rho\cdot\mathbf{d}^2\right)-\left(0.0571682\cdot\mathbf{d}\cdot\mathbf{L}\cdot\mathbf{p}\right)+\left(0.0967954\cdot\rho\cdot\mathbf{b}\cdot\mathbf{d}\right)\\&+\left(19.3334\cdot10^{-9}\cdot\mathbf{E_S}\cdot\mathbf{f_c}\cdot\mathbf{d}\right)-\left(0.234107\cdot10^{-6}\cdot\mathbf{f_c}\cdot\mathbf{L}^2\right)+\left(0.003624\cdot\rho\cdot\mathbf{L}^2\right)\\&-\left(0.00809799\cdot\rho\cdot\mathbf{L}\cdot\mathbf{b}\right)+\left(5.38644\cdot10^{-6}\cdot\mathbf{f_c}\cdot\mathbf{b}\cdot\mathbf{d}\right)\\&+\left(343643\cdot\rho\cdot\mathbf{f_t}\cdot\beta\right)-\left(387.761\cdot\rho^2\cdot\mathbf{d}\right)+0.950862\end{aligned} \tag{19}$$

Figures 5, 6 and 7 contain a comparison between the experimental results and Equations 1, 3 and 19, respectively. The results in all three figures look quite similar, however, for ultimate forces below 50 kN there is a closer and more accurate prediction for the proposed formula. At forces larger than 50 kN the proposed formula becomes less accurate and is more conservative. A validation database was selected from experiments found in the international literature, where 36 beams were selected with properties within the range of the limits presented in Table 2. The proposed formula was found to outperform the existing design code formulae, where the MAPE of the proposed formula was 47.03%, and for the cases of the design codes, it was found that the ACI318-19 derived a respective MAPE of 145.51% and from Eurocode 2 the MAPE was 112.98%. This finding, which states that the proposed formula derives an almost 3 times lower error compared to the two codes, highlights the importance of further investigating the proposed approach.

*Figure 5. Comparison between experimental and ACI prediction*
*(Spijkerman et al., 2022)*

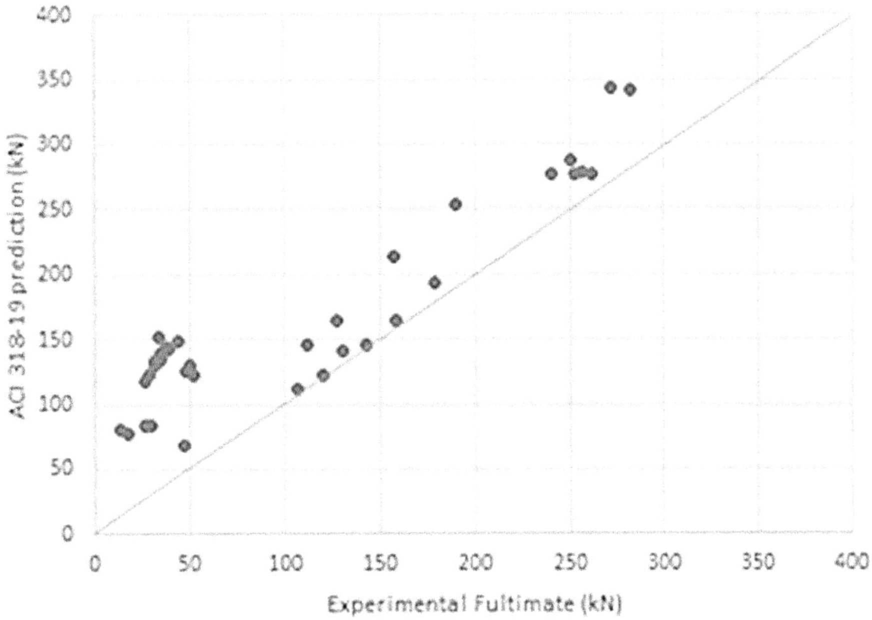


*Figure 6. Comparison between experimental and EC2 prediction*
*(Spijkerman et al., 2022)*

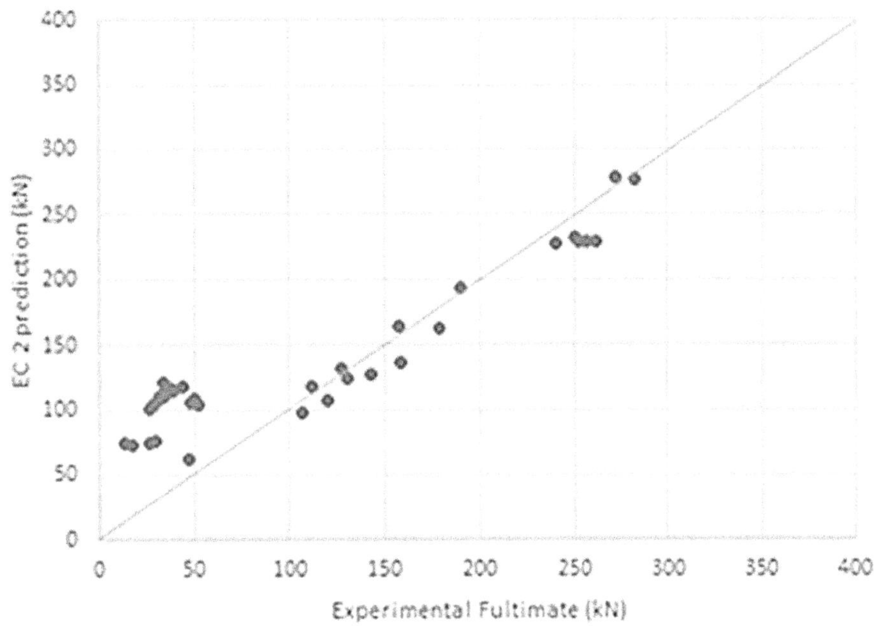

*Figure 7. Comparison between experimental and proposed formula*
*(Spijkerman et al., 2022)*

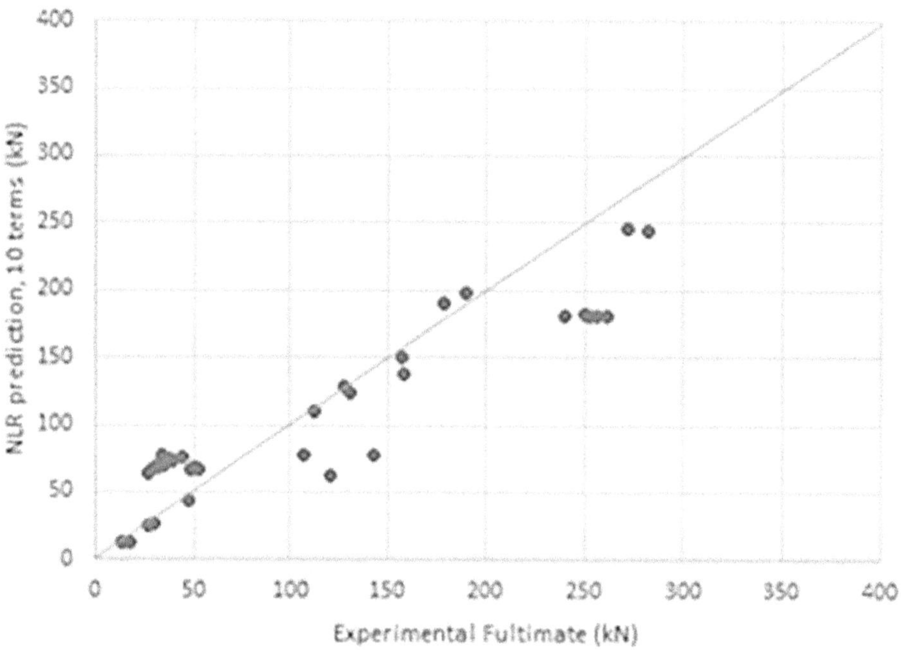


## Predicting the Shear Capacity of FRP Reinforced Concrete Deep Beams

As it was presented in AlHamaydeh et al., (2022), the aim of that research work was to model and interpret the behaviour of FRP-bar RC deep beams, while forming a dataset that was going to allow the development of a predictive model that would be able to overperform the current design code formulae in computing the shear capacity of these structural members.

A sensitivity analysis was conducted, and it was found that the shear span has an inverse relationship with the shear strength, and this is attributed to the arch action phenomenon experienced by deep beams. This relationship can be seen in Figure 8. Similar to the sensitivity analysis performed on slender beams, the geometric characteristics of the beams (b and d) have a significant influence on the shear capacity. A total of 93 models representing various possible FRP-reinforced deep beams were analysed through ReConAn FEA, where a summary of the minimum and maximum parameters used to develop the dataset can be seen in Table 3.

For the needs of this research work, the modelling of the RC deep beams, 20-noded hexahedral FE were used and the FRP reinforcement was modelled through the use of rod elements. The mesh of one of RC deep beams can be seen in Figure 9, where the 20-noded hexahedral elements and the embedded rebar macro-elements can be depicted. It is noteworthy to state here that the beams were tested under four-point bending. The study also included an investigation of the effect of different factors on the shear capacity and the crucial parameters were highlighted. Regarding the validation of the adopted software in capturing the shear capacity of deep RC beams with FRP-bars can refer to Markou and AlHamaydeh, (2018).

*Figure 8. Sensitivity analysis for alpha*
*(AlHamaydeh et al., 2022)*

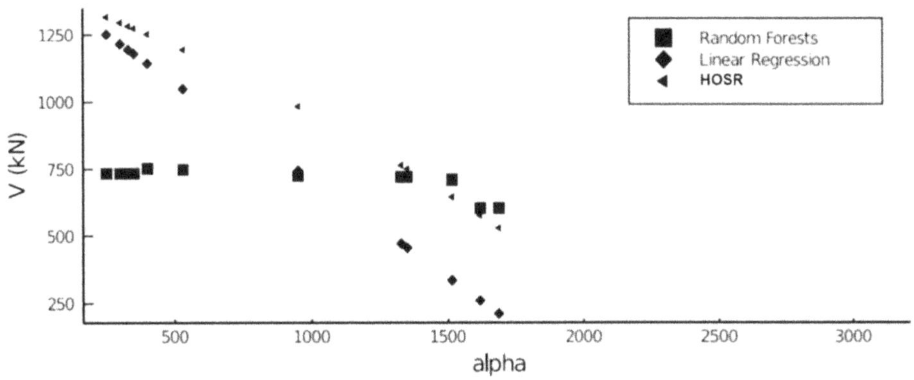


*Table 3. FRP deep beam geometry. minimum and maximum magnitudes of the 93 beams*

| Variables | Minimum | Maximum |
|---|---|---|
| b [mm] | 200 | 400 |
| $E_c/f_c$ (Concrete) | 701 | 940 |
| $E_c/f_c$ (FRP bars) | 71 | 75 |
| a/h | 0.5 | 2 |
| ρ [%] | 0.2991 | 2.595 |

(AlHamaydeh et al., 2022)

*Figure 9. RC beam (a) 20-noded hexahedral mesh and (b) embedded rebar elements*
*(AlHamaydeh et al., 2022)*

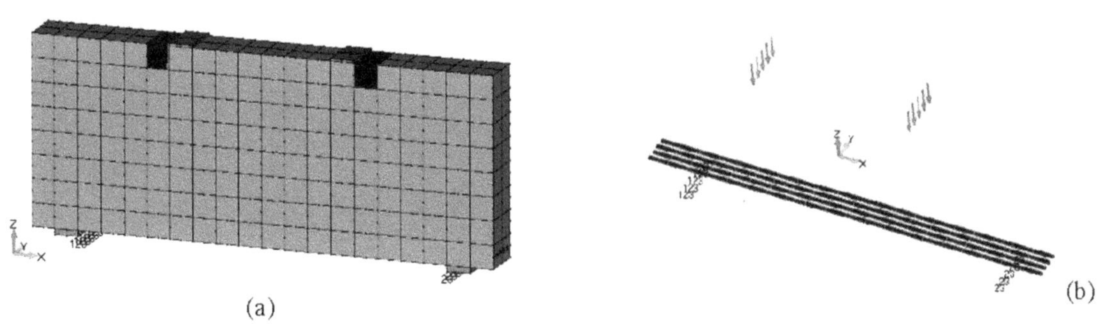


After successfully constructing the database through nonlinear analysis, AlHamaydeh et al., (2022) proposed a 5-feature formula that was generated using a higher-order stepwise regression (HOSR) algorithm, where the resulting relationship can be seen in Equation 20.

$$\begin{aligned}\mathbf{V_{Rk,cp}} &= \left(1.2\cdot10^{-4}\cdot\mathbf{f_c}\cdot\mathbf{d}\cdot\mathbf{b}\right)-\left(1.1\cdot10^{-5}\cdot\mathbf{a}^2\cdot\mathbf{f_c}\right)+\left(6.33\cdot10^{-6}\cdot\mathbf{f_{frp}}.\mathbf{f_c}\cdot\mathbf{b}\right)\\ &-\left(1.47\cdot10^{-6}\cdot\mathbf{a}\cdot\mathbf{b}\cdot\mathbf{d}\right)+\left(9.4210^{-8}\cdot\mathbf{a}^2\cdot\mathbf{L}\right)+115\end{aligned} \tag{20}$$

The correlation between the numerically generated shear strength and those developed from the formula can be seen in Figure 10. The dotted lines in Figure 10 represent a +/-95% confidence intervals and an acceptable range of shear values to check the reliability and validity of Equation 20. So as to investigate the models' predictive capabilities, 43 experimentally tested RC deep beams were used. A comparison of the MAPE of the predictive model as well as the various design code formulae can be seen in Table 4.

It can be seen in Table 4 that Eurocode 2 (using the β parameter) had the minimum MAPE among the design codes presented. The proposed formula was also found to outperform the two design code formulae considered in that research study, where further investigation is deemed necessary in improving the predictive capabilities of the proposed model. It is also important to be noted that the Eurocode equation is originally developed for steel and not FRP reinforcement, which is something that is accounted for in ACI 440.

*Figure 10. Correlation between numerically predicted and formula predicted shear strength*
*(AlHamaydeh et al., 2022)*

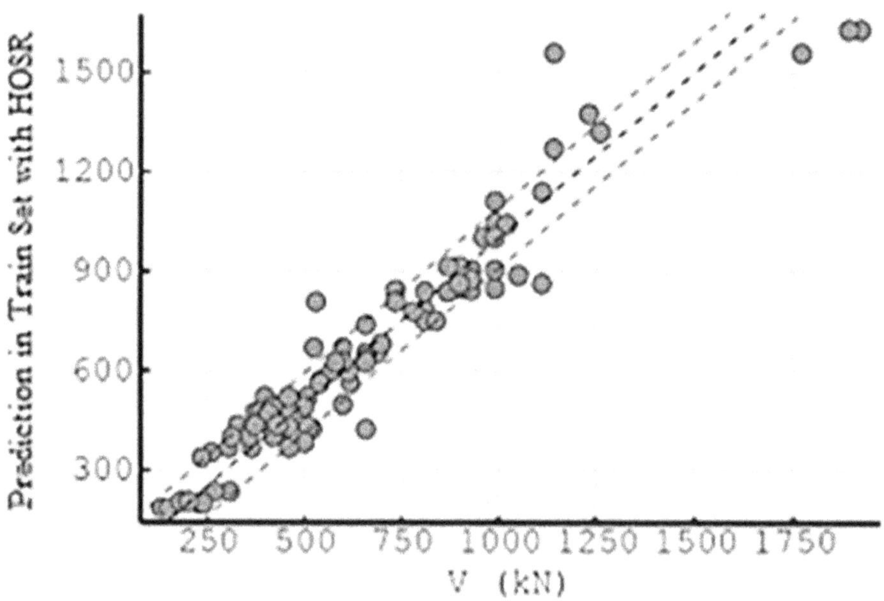


*Table 4. Comparison of MAPE of formulae*

| | AI | ACI 440 | ACI 440 (Modified) | Eurocode (without β) | Eurocode (with β) |
|---|---|---|---|---|---|
| **MAPE** | 57 | 77 | 83 | 91 | 34 |

(AlHamaydeh et al., 2022)

## Determining the Fundamental Period of RC Structures

An initial study was conducted (Gravett et al., 2019) to investigate and then propose an approach that will alleviate numerical constraints regarding performing modal analysis of large-scale models through 3D detailed modeling. The numerical limitations are numerical instabilities that occur when the opening and closing of cracks occur and the excessive computational demand required when dealing with small numerical models. This issue can be solved by performing modal analyses on structures discretized with the hybrid model (HYMOD) approach (Gravett et al., 2019) without compromising the accuracy of the model. The HYMOD approach was adopted for the development of the structures presented herein as described by Markou and Papadrakakis (2015). The HYMOD approach combines hexahedral and beam column finite elements as it is depicted in Figure 11 where it is shown how the coupling between the hexahedral and beam-column elements through kinematic constraints is performed.

Extending the Gravett et al., (2019) research work, a relatively large dataset was developed and presented in Gravett et al., (2021) that foresaw the analysis of multistorey RC buildings initially constructed and investigated in Markou et al. (2018). Initially, 2-, 4-, 6-, 8-, 10 and 20-storey models without and with shear walls were created as shown in Figure 12.

Each storey height was 3 m tall, while each frame consists of 3 bays with an 8.9 m total building width. Furthermore, two different foundation types were considered, namely a raft slab and an isolated footing so as to decide which type of foundation to use according to the soil characteristics, since the SSI effect was also considered in this study (Gravett et al., 2021). In order to account for the SSI effect, two soil types were considered. The isolated footings were used in models with a medium solid (E = 700 MPa), and a raft slab foundation was adopted for the case of soft soil (E = 65 MPa). A total of 475 numerical results were used in this study as well as 60 out-of-sample modal results for the validation of the proposed formula. Figure 13 shows the first 3 modes for a 2-storey as well as a 20-storey RC building.

The numerical models analysed in Gravett et al., (2019) were modified accordingly by Gravett et al., (2021) to include a soil domain discretized with hexahedral finite elements. A total of 475 modal results were generated considering the minimum and maximum values in terms of geometry and soil material properties as seen in Table 5. Figure 14 shows different models that foresee the discretization of the soil domain through the use of hexahedral elements.

*Figure 11. Hybrid model showing the interface between the elements*
*(Markou & Papadrakakis, 2015)*

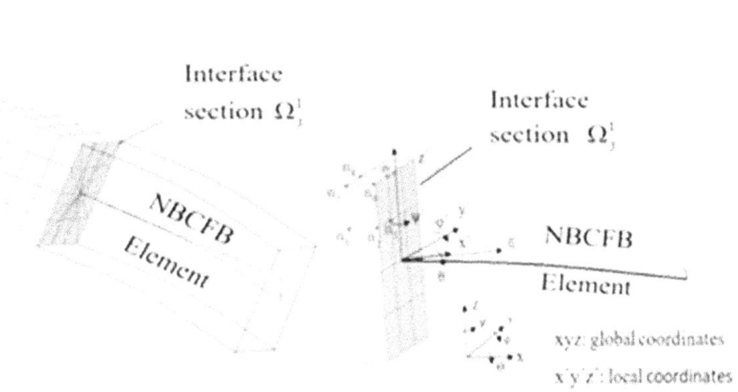

*Figure 12. Initial models (a) without shear walls, (b) with shear walls*
*(Gravette et al., 2019, 2021)*

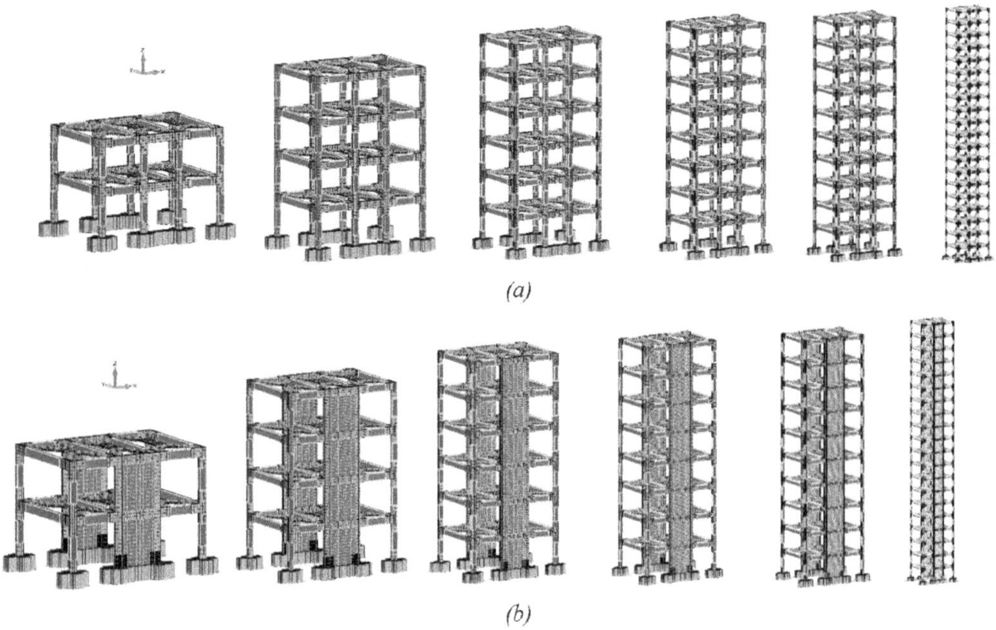

*Figure 13. First 3 modes of (a) 2-storey and (b) 20-storey building without shear walls*
*(Gravette et al., 2019, 2021)*

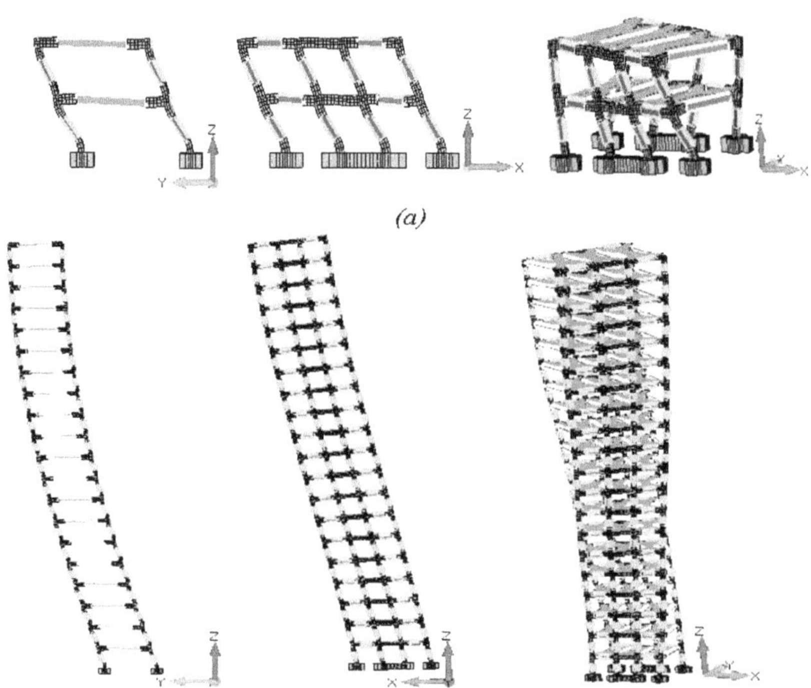

Three design formulae were generated using the NLR ML algorithm (Gravett et al., 2021). It is important to note that the formulae are, at present, only valid for the values as seen in Table 5. The 20-feature formula as seen in Equation 21 is discussed because it derived the minimum MAPE (8.5%) and the highest correlation (91.5%) when used on the 60 out-of-sample validation dataset.

$$\begin{aligned}
\mathbf{T} = &\left(0.032867\,\&\,7\cdot\mathbf{H}\right)-\left(143.475\times10^{-6}\cdot\rho\cdot\mathbf{H}\right)-\left(79.921\times10^{-8}\cdot\mathbf{L}\cdot\mathbf{B}^{2}\right)\\
&-\left(10.458\times10^{-14}\cdot\rho\cdot\mathbf{L}\cdot\mathbf{E_s}\right)-\left(55.061\times10^{-6}\cdot\rho\cdot\mathbf{L}\right)-\left(71.214\times10^{-12}\cdot\mathbf{E_s}\cdot\mathbf{D}^{2}\right)\\
&+\left(5.016\times10^{-6}\cdot\mathbf{L}\cdot\mathbf{H}\cdot\mathbf{D}\right)-\left(2.93\times10^{-18}\cdot\mathbf{B}\cdot\mathbf{E_s^{2}}\right)+\left(7.121\times10^{-10}\cdot\mathbf{B}\cdot\mathbf{E_s}\right)\\
&+\left(8.702\times10^{-6}\cdot\mathbf{L}\cdot\mathbf{H}^{2}\right)+\left(93.621\times10^{-8}\cdot\mathbf{L}\cdot\rho^{2}\right)-\left(6.093\times10^{-6}\cdot\mathbf{L}\cdot\mathbf{H}\cdot\rho\right)\\
&+\left(3.351\times10^{-6}\rho\cdot\mathbf{H}^{2}\right)+\left(60.549\times10^{-12}\cdot\mathbf{B}\cdot\mathbf{E_s}\cdot\mathbf{D}\right)+\left(16.982\times10^{-5}\cdot\mathbf{B}^{2}\right)\\
&-\left(64.89\times10^{-6}\cdot\mathbf{H}^{2}\right)-\left(9.776\times10^{-12}\cdot\mathbf{H}\cdot\mathbf{E_s}\cdot\mathbf{D}\right)+\left(1.477\times10^{-17}\cdot\mathbf{H}\cdot\mathbf{E_s^{2}}\right)\\
&-\left(44.111\times10^{-10}\cdot\mathbf{H}\cdot\mathbf{E_s}\right)-\left(2.891\times10^{-6}\cdot\mathbf{H}\cdot\mathbf{D}\cdot\mathbf{B}\right)+3.8476\times10^{-4}
\end{aligned} \tag{21}$$

*Figure 14. Soil discretized with hexahedral meshes of 10-storey RC models for (a) raft slab foundation and (b) isolated footings*
*(Gravett et al., 2021)*

*Table 5. Minimum and maximum magnitudes of the RC Frames*

| Variables | Minimum | Maximum |
|---|---|---|
| Soil Depth [m] | 1 | 60 |
| Soil E [kPa] | 65 000 | 700 000 |
| H [m] | 6 | 30 |
| L [m] | 6.25 | 18.25 |
| B [m] | 6.25 | 18.25 |
| ρ(%) | 0 | 82.9 |

(Gravette et al., 2019)

The validation dataset consisted of 60 additional modal results from buildings with geometries that differ from those seen in Table 5 and used for the training and testing of the predictive models. The correlation was found to be high for all three formulae, but as stated above only the 20-feature formula is discussed herein due to its superior predictive capabilities illustrated in Figure 15. It is also noteworthy to state here that the corresponding MAPE obtained by Eurocode 8 and other proposed formulae found in the international literature was larger than 30% compared to the 5.68% obtained from the proposed formula.

*Figure 15. Correlation of 20-feature formula on validation data*
*(Taljaard et al., 2021)*

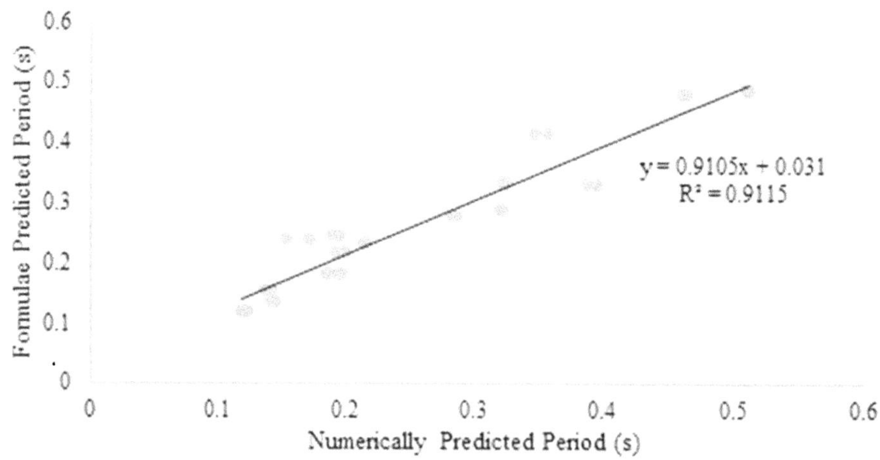


Comparing the performance of the proposed fundamental period formulae, Figure 16 was developed which compares the 20-feature formula with four other formulae proposed in Hatzigeorgiou and Kanapitsas (2013), Chopra and Goel (2000), Amanat and Hoque (2006) and Asteris et al. (2015) . Table 6 shows the MAPE for the different formulae as they resulted when implemented on the validation dataset of the 60 out-of-sample data. It is easy to observe that the proposed fundamental period formulae (Gravett et al., 2021) exhibit superior accuracy when used to predict the fundamental periods of the under-study structures.

*Figure 16. Comparison between the 20-feature formula predictions and other fundamental period formulae proposed in the international literature*
*(adapted from Gravett et al., 2021)*

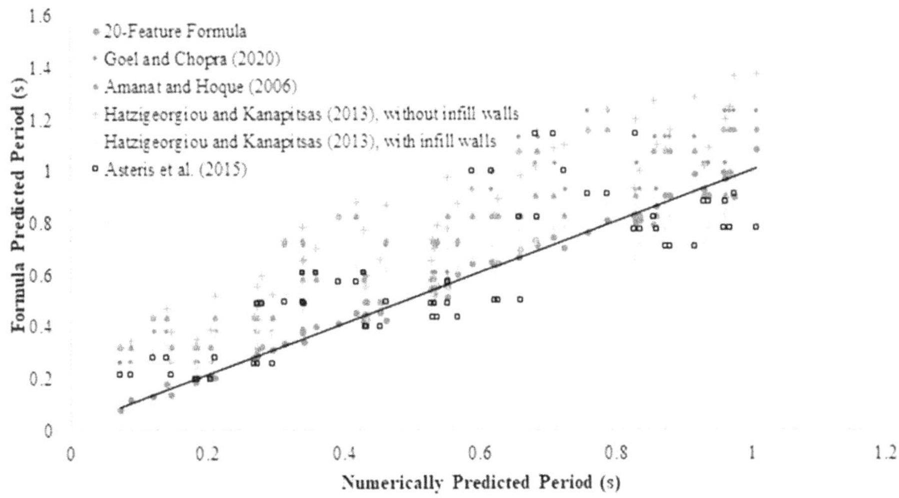


*Table 6. Absolute average errors for different formulae on the validation dataset*

| | Mean Absolute Percentage Error |
|---|---|
| **3-Feature Formula** | 14.4% |
| **5-Feature Formula** | 10.26% |
| **20-Feature Formula** | **5.68%** |
| **Eurocode 8 (2004)** | 45.24% |
| **NEAK (2000)** | 51.67% |
| **Hatzigeorgiou and Kanapitsas, (2013), without infill walls, W = 0** | 75.67% |
| **Hatzigeorgiou and Kanapitsas, (2013), with infill walls, W = 1** | 30.88% |
| **Asteris et al. (2015) without infill walls, $a_w = 1, E_t = 0$** | 34.03% |
| **Amanat and Hoque (2006)** | 56.98% |
| **Goel and Chopra (2000)** | 54.05% |

(Gravett et al., 2021)

Furthermore, in their study, Gravett et al., (2021) developed and implemented shallow and deep neural networks to train, test and validate predictive models by using the developed dataset. Specifically, a 50 neurons network was utilized for the approximation and training of the model with the algorithm defined in Bakas et al. (2020). So as to select 50 neurons, the researchers implemented the recommendation that foresees the use of a 10% on the number of dataset values. Figure 17 shows the structure of the ANN used to train, test, and validate the corresponding predictive model as presented by Gravett et al., (2021).

*Figure 17. Artificial Neural Network Architecture*
*(Gravett et al., 2021)*

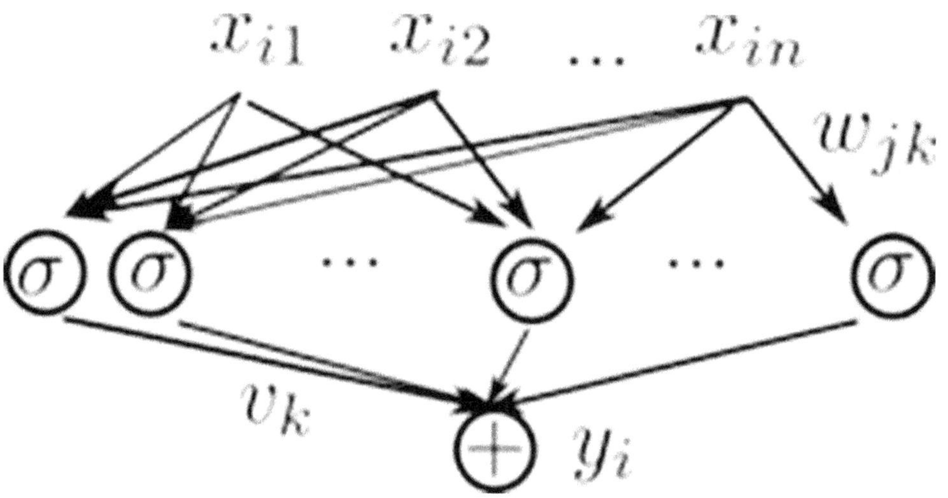


According to the numerical investigation involving the ANN model, it was found that it exhibited an $R^2 = 0.9709$, which was close to the 5-feature formula, ($R^2 = 0.9594$), as well as a corresponding MAPE = 10.71%, which is similar to that obtained from the 5-feature formula (MAPE = 10.26%). Nevertheless, the MAPE on the validation dataset of the 60 out-of-sample modal results was only 3%, deriving a more accurate prediction compared to the 20-feature formula, demonstrating a significantly more accurate prediction compared to all the formulae investigated by Gravett et al., (2021). Similar results were reported for the cased of the deep learning algorithm that was used to train a similar predictive model. As was presented in Gravett et al., (2021), the MAPE of the predictive model obtained from the deep learning algorithm was found to be slightly better than that of the 20 terms nonlinear formula (5.68%). It was also stated that the advantage of using the non-linear regression approach is that of a closed-form solution which is achieved, in contrast with the "black-box" when the ANN is used to develop the predictive model.

The research work performed by Gravett et al., (2021) was extended by Carstens et al., (2022) where the 475 numerical results were increased by constructing and analysing additional buildings with and without accounting for the SSI effect. Table 7 summarises the new minimum and maximum values used in this study.

*Table 7. Minimum and maximum magnitudes of the new RC Frames*

| Variables | Minimum | Maximum |
|---|---|---|
| Soil Depth [m] | 1 | 60 |
| Soil E [kPa] | 65 000 | 700 000 |
| H [m] | 3 | 30 |
| L [m] | 3.4 | 34.4 |
| B [m] | 3.4 | 34.4 |
| ρ(%) | 0 | 85.29 |

(Carstens et al., 2022)

Once more, three formulae were developed (Carstens et al., 2022) and the most accurate proposed formula is presented herein. As it was reported in their study, the most accurate formula was the 20-feature formula which can be seen in Equation 22. The MAPE obtained from this formula was 1.49%, which is a 7% improvement from the Gravett et al., (2021) study. This result was obtained when using the 60 out-of-sample results that were implemented during the validation of the newly proposed formula, whereas on the other hand, the 3-feature formula (Equation 23) resulted in the highest correlation. Figure 18 shows the correlation between the validation dataset and the prediction of the proposed fundamental period formula of Equation 22, as it was presented by Carstens et al., (2022).

$$\begin{aligned} T &= (0.0292939 \cdot \mathrm{H}) - (0.150825 \cdot 10^{-3} \cdot \rho \cdot \mathrm{H}) + (5.82242 \cdot 10^{-6} \cdot \mathrm{H} \cdot \mathrm{B}^2) \\ &+ (3.30369 \cdot 10^{-6} \cdot \rho \cdot H^2) + (0.000215881 \cdot \mathrm{H} \cdot \mathrm{L}) - (1.89375 \times 10^{-15} E_s^2 \cdot \mathrm{D}) \\ &+ (3.23855 \cdot 10^{-6} \cdot \mathrm{L} \cdot \mathrm{H} \cdot \mathrm{D}) - (6.46154 \cdot 10^{-6} \cdot \rho \cdot \mathrm{B} \cdot \mathrm{H}) - (92.5478 \cdot 10^{-12} \cdot H \cdot E_s \cdot D) \\ &- (40.6192 \cdot 10^{-12} \cdot \rho \cdot E_s \cdot D) + (194.394 \cdot 10^{-9} \cdot D \cdot \rho^2) + (0.0037148 \cdot B) \\ &+ (35.8861 \cdot 10^{-9} \cdot \rho \cdot H \cdot D) + (66.2381 \cdot 10^{-12} \cdot E_s \cdot D^2) - (0.278639 \cdot 10^{-6} \cdot D^3) \\ &- (0.113737 \cdot 10^{-9} \cdot L \cdot E_s \cdot D) - (1.6727 \cdot 10^{-6} \cdot B^3) + (30.9934 \cdot 10^{-6} \cdot L \cdot D) \\ &- (0.00178654 \cdot L) + (0.645744 \cdot 10^{-6} \cdot L^3) + 0.00239996 \end{aligned} \tag{22}$$

$$T = (0.0310197 \cdot H) - (0.11254 \cdot 10^{-3} \cdot \rho \cdot \mathrm{H}) + (12.9093 \cdot 10^{-6} \cdot \mathrm{H} \cdot \mathrm{B}^2) + 0.0110165 \tag{23}$$

*Figure 18. Correlation of the 3-feature formula on the validation set*
*(Carstens et al., 2022)*

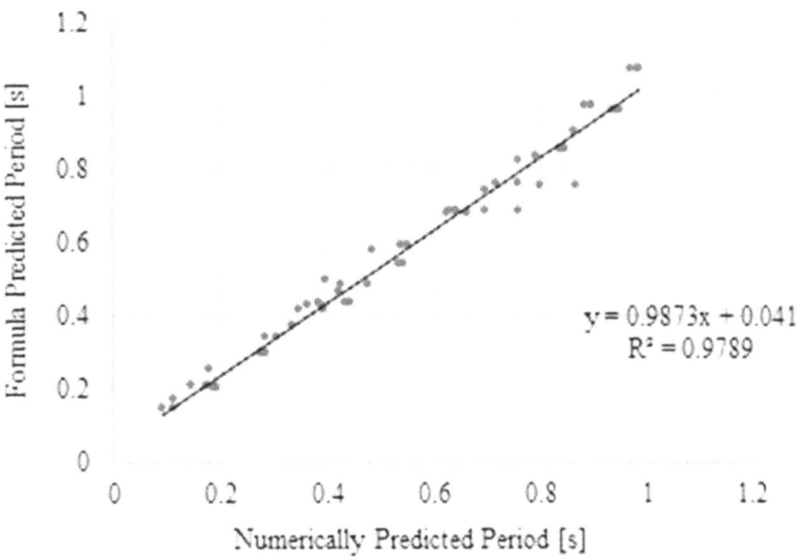

## Determining the Fundamental Period of Steel Structures

In a similar study (van der Westhuizen et al., 2022) the development of fundamental period formulae was conducted for steel structures. The models were created using a varying number of stories, bays, and superstructural geometry conditions. Initially, a model with a single-storey, single bay with a storey height of 3.5 m and a raft foundation assuming a fixed base (see Figure 19) was constructed.

*Figure 19. Initial Model*
*(van der Westhuizen et al., 2022)*

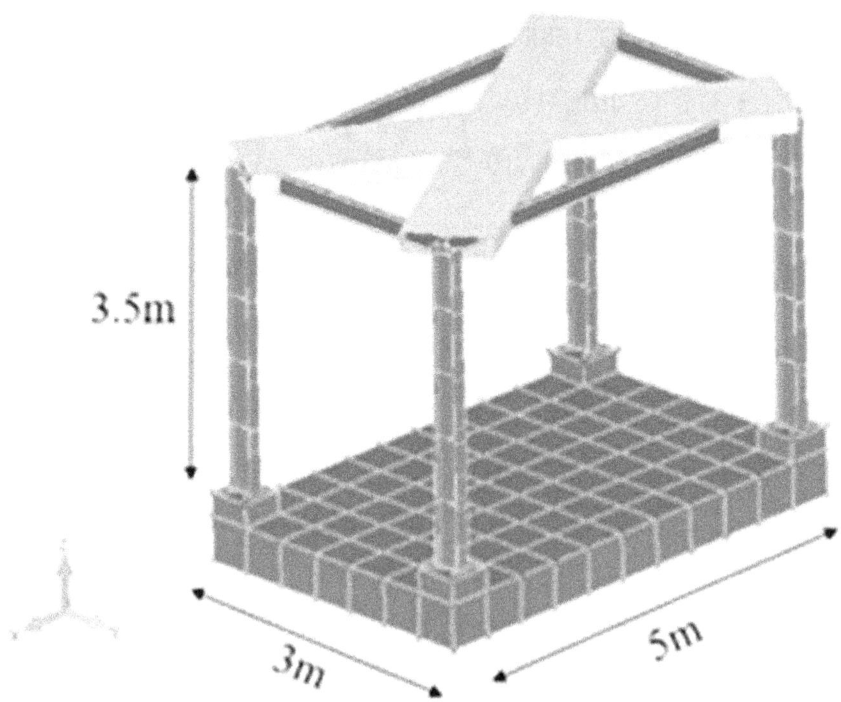


*Figure 20. Initial model modified to have (a) two bays and (b) three bays*

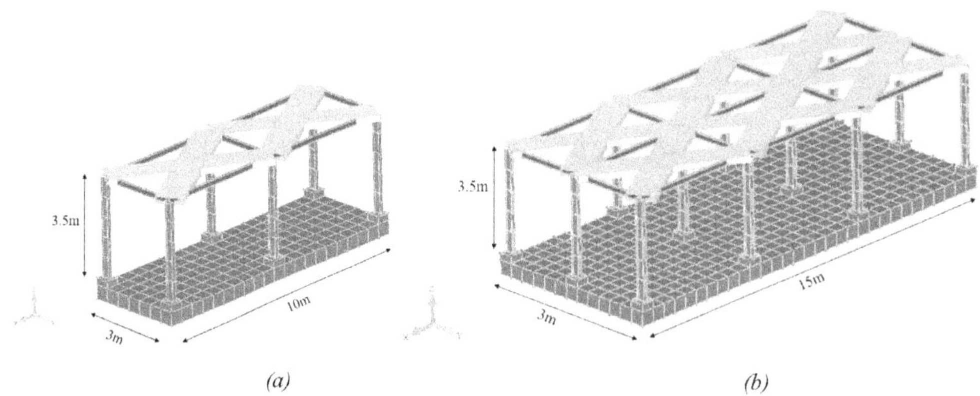

The initial model was then expanded by altering the number of storeys, the geometry of the structure, depth of soil as well as the soil's Young modulus and orientation of columns. The models foresaw 2-, 4-, 6-, 8- and 10 stories, each with a 3.5 m height. The models were further expanded to include double and triple spans along the x-axis as shown in Figure 20. The smallest plan was 5x3 m and the largest foresaw a 15x6 m plan view. The models were further modified to include the SSI effect where a soil domain was discretized with depths of 1, 5, 12.5, 22.5 and 37.5 m. The superstructure was discretized through the use of Natural Beam-Column Flexibility-based (NBCFB) finite elements (Markou & Papadrakakis, 2015), where the raft and soil domains were discretized through 8-noded isoparametric hexahedral finite elements. A total of 1,152 numerical results were produced for the final dataset that was used to train and test the closed-form solutions.

The machine-learning algorithm was used to train and test the database and construct a fundamental period formula for steel structures taking into account the SSI effect. The fundamental period 40-feature formula obtained can be seen in Equation 24.

$$
\begin{aligned}
\mathbf{T} = & \left(0.194630 \cdot \mathbf{lH}^2\right) + \left(0.0580556 \cdot \mathbf{CO}^2 \cdot \mathbf{B}\right) - \left(9.39027 \cdot \mathbf{InvCO} \cdot \mathbf{InvB} \cdot \mathbf{lB}\right) \\
& - \left(8.49213 \cdot \mathbf{InvL} \cdot \mathbf{CO} \cdot \mathbf{H}\right) - \left(41.8498 \cdot \mathbf{InvCO} \cdot \mathbf{lL} \cdot \mathbf{H}\right) - \left(8.14564 \cdot \mathbf{InvE} \cdot \mathbf{E} \cdot \mathbf{H}\right) \\
& - \left(0.800465 \cdot \mathbf{CO} \cdot \mathbf{B} \cdot \mathbf{H}\right) + \left(114.808 \cdot \mathbf{InvCO} \cdot \mathbf{InvB} \cdot \mathbf{H}\right) + \left(46.6778 \cdot \mathbf{InvCO} \cdot \mathbf{InvB}^2\right) \\
& + \left(0.0631499 \cdot \mathbf{B}^2 \cdot \mathbf{H}\right) + \left(4.20803 \cdot \mathbf{lB} \cdot \mathbf{CO} \cdot \mathbf{H}\right) - \left(0.144945 \cdot \mathbf{lL} \cdot \mathbf{H} \cdot \mathbf{L}\right) \\
& + \left(0.847694 \cdot \mathbf{B} \cdot \mathbf{H} \cdot \mathbf{InvL}\right) + \left(9.37930 \cdot \mathbf{InvL}^2 \cdot \mathbf{H}\right) - \left(1.08930 \cdot \mathbf{InvCO}^2 \cdot \mathbf{L}\right) \\
& + \left(4.04342 \cdot \mathbf{InvL}\right) - \left(0.251627 \cdot \mathbf{InvL} \cdot \mathbf{CO} \cdot \mathbf{B}\right) - \left(0.00783561 \cdot \mathbf{InvB} \cdot \mathbf{lCO} \cdot \mathbf{lE}\right) \\
& + \left(0.523388 \cdot \mathbf{lL}^2 \cdot \mathbf{InvCO}\right) + \left(0.0947335 \cdot \mathbf{InvH} \cdot \mathbf{lH} \cdot \mathbf{L}\right) + \left(46.8309 \cdot \mathbf{InvE} \cdot \mathbf{H} \cdot \mathbf{lDs}\right) \\
& + \left(0.00764850 \cdot \mathbf{lH} * \mathbf{B}\right) + \left(0.000161108 \cdot \mathbf{lL} \cdot \mathbf{L} \cdot \mathbf{lE}\right) - \left(20.5554 \cdot \mathbf{InvE} \cdot \mathbf{CO} \cdot \mathbf{Ds}\right) \\
& - \left(0.00474725 \cdot \mathbf{InvL}^2 \cdot \mathbf{InvDs}\right) + \left(2.73101 \cdot \mathbf{InvL} \cdot \mathbf{InvH} \cdot \mathbf{CO}\right) + \left(0.403996 \cdot \mathbf{InvCO} \cdot \mathbf{lB} \cdot \mathbf{L}\right) \\
& - \left(0.0105914 \cdot \mathbf{lL} \cdot \mathbf{L} \cdot \mathbf{B}\right) - \left(0.228100 \cdot \mathbf{lB}^2 \cdot \mathbf{CO}\right) + \left(0.00265642 \cdot \mathbf{InvL} \cdot \mathbf{H}^2\right) \\
& - \left(2.58386 \cdot \mathbf{InvB} \cdot \mathbf{InvH} \cdot \mathbf{CO}\right) + \left(5.84142 \cdot \mathbf{InvCO} \cdot \mathbf{H} \cdot \mathbf{L}\right) + \left(29.5168 \cdot \mathbf{InvCO} \cdot \mathbf{H}\right) \\
& + \left(0.849560 \cdot \mathbf{InvL} \cdot \mathbf{lH} \cdot \mathbf{CO}\right) - \left(2.14776 \cdot \mathbf{InvB} \cdot \mathbf{lH} * \mathbf{lCO}\right) + \left(1.34222 \cdot \mathbf{lB} \cdot \mathbf{lH} \cdot \mathbf{InvH}\right) \\
& - \left(0.00333495 \cdot \mathbf{lE} \cdot \mathbf{L} \cdot \mathbf{InvH}\right) - \left(2.64111 \cdot \mathbf{InvB}^2 \cdot \mathbf{InvH}\right) \\
& + \left(71.1358 \cdot \mathbf{InvH} \cdot \mathbf{Ds} \cdot \mathbf{InvE}\right) - \left(17.9194 \cdot \mathbf{InvE} \cdot \mathbf{lE} \cdot \mathbf{lL}\right) - 1.16636
\end{aligned}
\tag{24}
$$

Where:

$T$ is the fundamental period (s)
$D_s$ is the depth of soil (m)
$E$ is the soils Young's Modulus (kPa)
$H$ is the building height (m)
$L$ is the length of the building parallel to the oscillating direction (m)?
$B$ is the width of the building perpendicular to the oscillating direction (m)
$CO$ is the orientation of the columns (either a 1 or 2)

*l Parameter* is ln(*Parameter*+1) i.e., $lD_s = \ln(D_s+1)$

*Inv Parameter* is $\frac{1}{Parameter+1}$ i.e., $InvD_s = \frac{1}{D_s+1}$

The correlation between the numerically predicted and formula predicted fundamental period can be seen in Figure 21.

*Figure 21. Correlation between numerically predicted and formula predicted fundamental period*
*(van der Westhuizen et al., 2022)*

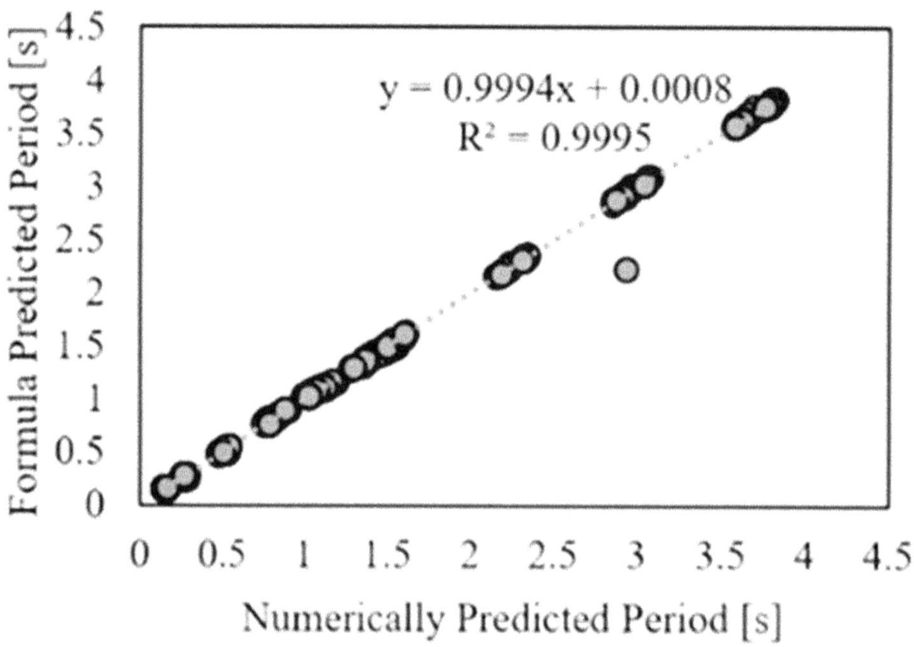

According to the validation phase performed in van der Westhuizen et al., (2022), it was found that the formula had a mean absolute error of 2.8%, compared to Eurocode 8 which had an error of 76%. It can be seen that the proposed formula yields a more accurate result while accounting for the all-important SSI effect.

## Determining the Deflection of Horizontally Curved Steel I-Beams

The aim of this work (Ababu et al., 2022) was to derive an easy-to-apply predictive formula for the determination of the mid-span deflection of curved steel I-beams. It is important to note that Castigliano's' theorem was not used in this case, where machine learning was utilised to determine an explicit closed-form equation.

In order to generate beams using a finite element method, it is important to note what mesh size to adopt to accurately model the beam behaviour as this affects the overall numerical results Models with varying mesh sizes were created and the deflections were compared to a curve obtained from an experimental study conducted by Shanmugam et al. (1995). It was found that the 20-noded isoparametric hexahedral elements best represented the mechanical response of the under study curved steel I-beams. Figure 22 shows an example beam developed using a 20-noded hexahedral mesh.

*Figure 22. (Left) General beam, undeformed mesh and (Right) Deformed shape and von Mises stress contour for a total deflection of 17 mm*
*(Ababu et al., 2022)*

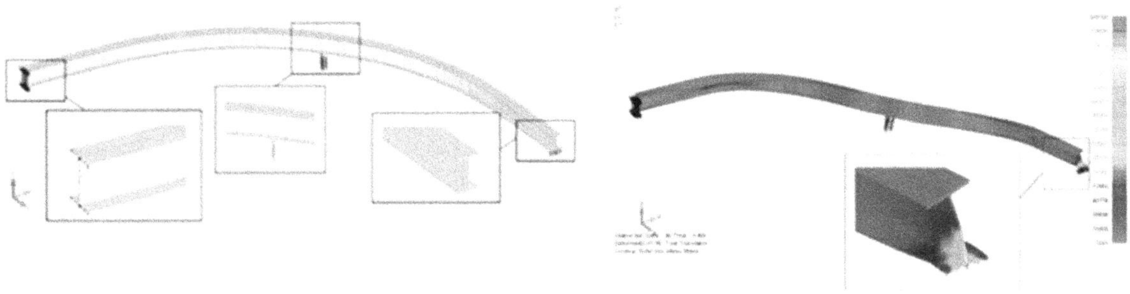

A total of 270 beams were created that were fixed at one end with roller support at the other (also seen in Figure 22) as this represents the boundary conditions of crawl beams. The load was applied at the mid-span and was divided into 10 load increments where the beams were loaded until failure. A load-deflection curve was then developed which led to a total of 1 890 unique data points that were used to train the ML algorithm. It is important to note that the selected points foresaw a maximum of 50% load level of the maximum load in avoiding including deflections that resulted in load levels that might generate significant geometrical nonlinearities.

Specifically, 8 formulae were developed in this case where 5 to 40 features were considered. The correlation remained the same no matter the number of features used but the MAPE was optimum when the 10-feature formula was used (4.11%). The 10-feature formula, which yielded the minimum average error for the validation data, can be seen in Equation 25. The MAPE was derived by using a validation dataset of 15 out-of-sample data points. This pilot project demonstrates the potential of the proposed approach, where further analysis and experimentation were deemed necessary (Ababu et al., 2022).

$$\begin{aligned}\text{Deflection} = &\left(1.08128\cdot10^{-1}\cdot\mathbf{Q}\cdot\mathbf{L}\right)\\ &-\left(8.41124\cdot10^{-6}\cdot\mathbf{Q}\cdot\mathbf{L}\cdot\mathbf{A}\right)-\left(9.81969\cdot10^{-6}\cdot\mathbf{Q}\cdot\mathbf{E}\cdot\mathbf{R}\right)+\left(1.82604\cdot10^{-5}\cdot\mathbf{Q}\cdot\mathbf{R}\cdot\mathbf{R}\right)\\ &-\left(2.71029\cdot10^{-4}\cdot\mathbf{Q}\cdot\mathbf{R}\cdot\mathbf{L}\right)+\left(6.22991\cdot10^{-3}\cdot\mathbf{Q}\cdot\mathbf{L}\cdot\mathbf{L}\right)+\left(2.55963\cdot10^{-6}\cdot\mathbf{Q}\cdot\mathbf{L}\cdot\mathbf{I}_{\mathbf{xx}}\right)\\ &+\left(2.99150\cdot10^{-7}\cdot\mathbf{Q}\cdot\mathbf{f}_{\mathbf{y}}\cdot\mathbf{f}_{\mathbf{y}}\right)-\left(1.07754\cdot10^{-4}\cdot\mathbf{Q}\cdot\mathbf{E}\cdot\mathbf{L}\right)-8.95792\cdot10^{-2}\end{aligned} \tag{25}$$

Where:

$E$ is the Young's modulus (GPa)
$L$ is the curved length of the beam (m)
$A$ is the section area ($mm^2$)
$f_y$ is the yielding stress
$I_{xx}$ is the second moment of area about the strong axis
$f_y$ is the yielding stress ($mm^4$)
$Q$ is the percentage of ultimate load applied to the beam as a number (50%=50)

## FUTURE RESEARCH

This chapter highlighted the use of AI and ML algorithms in developing formulae that successfully determined the mechanical properties and response of structures and structural members. The predictability of these models, however, depends on the datasets that are constructed and used to train in each case. In some instances, it is necessary to expand the datasets to come up with formulae that can predict properties in a larger range. For example, Equation 25 can only be used to predict accurately the fundamental period of a structure that falls within the dataset's limits and requires expansion if it were to be applied to a larger geometrical set of structures. A current research project is being completed in which the dataset of determining the fundamental period of steel structures is being expanded to include an odd number of stories as well as more Young's modulus values. It is expected that the dataset will contain 100,000 numerical results obtained in a similar way as discussed in the chapter through the use of a high-performance computer that facilitates a large number of numerical modelling and ML training.

With the successful use of AI and ML algorithms in the determination of mechanical behaviour of structural elements of buildings, Civil Engineers will be able to incorporate the proposed formulae during their design achieving additional accuracy compared to the current design codes. This method, as stated, will be also implemented in the research of other structural types and configurations, such as the case of predicting of the horizontal displacement of piles embedded in clay material. Furthermore, RC beams with stirrups will be investigated including deep and slender beams which are reinforced with standard and FRP bars. RC structural columns and shear walls will also be analysed thereafter.

In expanding the curved beams dataset, other geometries, boundary conditions and different steel yield strengths are currently being explored, where the rotation of the section of the beams is being investigated. This research work will also be expanded to include curved RC beams that are also found to be of significant interest. Since there are no formulae current available like the one presented herein, it is necessary to compare the results obtained to that of experimental results, which is currently a research project that is being performed at the laboratories of the University of Pretoria at the Civil Engineering department.

Through the presented research work, it has been established that time can be saved in terms of the development and analysis of models using FEA software, however, the ML algorithm may produce equations that are long and difficult to make use of when manual calculations are performed. The development of an automated ML software is currently being developed that will be able to analyse a dataset and perform these calculations for the user, in addition to the ability to use excel sheets or programming calculator apps that incorporate the developed predictive models. This would make the analysis of the results significantly easier. The development of such an application is now being finalised, thus providing with the ability to reduce the required time developing graphs that further help to understand the ML algorithm's output. This will also allow more engineers to be able to use the developed technology, therefore, train their own predictive models.

## CONCLUSION

While there are, in some cases, formulae available to design engineers this does not necessarily mean that they are the most accurate, easily accessible, or easy to apply. The chapter presented through this research work summarized the use of AI and ML in developing predictive models for the design of structural elements and full-scale structures through an innovative approach that combines advanced FEA for the construction of datasets that are then used to train new predictive models. It has been shown that the predictive models are more accurate compared to the current design code formulae, which highlights the prospects of the proposed approach in developing more accurate design codes leading to increased safety, sustainability, and reduced cost. Finally, further numerical investigation and experimentations are needed to establish objective and accurate design formulae that will derive from the proposed methodology.

## REFERENCES

Ababu, E. M., Markou, G., & Bakas, N. P. (2022). *Using Machine Learning and Finite Element Modelling to Develop a Formula to Determine the Deflection of Horizontally Curved Steel I-beams* [Paper presentation]. ICAART. doi:10.5220/0010982400003116

Abed, F., El-Chabib, H., & AlHamaydeh, M. (2012). Shear characteristics of GFRP-reinforced concrete deep beams without web reinforcement. *Journal of Reinforced Plastics and Composites, 31*(16), 1063–1073. doi:10.1177/0731684412450350

AlHamaydeh, M., Markou, G., Bakas, P. N., & Papadrakakis, M. (2022). AI-Based Shear Capacity of FRP-Reinforced Concrete Deep Beams without Stirrups. *Engineering Structures, 264*, 264. doi:10.1016/j.engstruct.2022.114441

Amanat, K. M., & Hoque, E. (2006). A rationale for determining the natural period of RC building frames having infill. *Engineering Structures, 28*(4), 495–502. doi:10.1016/j.engstruct.2005.09.004

Asteris, P. G., Repapis, C. C., Cavaleri, L., Sarhosis, V., & Athanasopoulou, A. (2015). On the fundamental period of infilled RC frame buildings. *Structural Engineering and Mechanics*, *54*(6), 1175–1200. doi:10.12989em.2015.54.6.1175

Committee, A. C. I. (2015). *ACI 440.1 R-15: guide for the design and construction of structural concrete reinforced with FRP bars*. American Concrete Institute.

A.C. Institute. (2019). *Building Code Requirements for Structural Concrete (ACI 318-19): Commentary on Building Code Requirements for Structural Concrete (ACI 318R-19): an ACI Report*. American Concrete Institute.

Bakas, N.P., Langousis, A., Nicolaou, M. & Chatzichristofis, S.A. (2020), A Gradient Free Neural Network Framework Based on Universal Approximation Theorem. arXiv preprint arXiv:1909.13563

Bakas, N. P., Markou, G., Charmpis, D., & Hadjiyiannakou, K. (2021). *Performance and scalability of deep learning models trained on a hybrid supercomputer: Application in the prediction of the shear strength of reinforced concrete slender beams without stirrups* [Paper presentation]. COMPDYN.

Bathe, K.-J. (1971). *Solution Methods for Large Generalized Eigenvalue Problems in Structural Engineering, Report UC SESM 71-20*. Civil Engineering Department, UC Berkeley.

CEN. (2004). *Eurocode 8: Design of structures for earthquake resistance. Part 1: general rules, seismic actions and rules for buildings. European Standard EN 1998-1:2004*. Comite Europeen de Normalisation.

Chopra, A. K., & Goel, R. K. (2000). Building period formulas for estimating seismic displacements. *Earthquake Spectra*, *16*(2), 533–536. doi:10.1193/1.1586125

Dahlberg, T. (2004). Procedure to calculate deflections of curved beams. *International Journal of Engineering Education*, *20*(3), 503–513.

Dhahir, M. K. (2017). Shear strength of FRP reinforced deep beams without web reinforcement. *Composite Structures*, *165*, 223–232. doi:10.1016/j.compstruct.2017.01.039

Dhahir, M. K. (2018). Strut and tie modelling of deep beams shear strengthened with FRP laminates. *Composite Structures*, *193*, 247–259. doi:10.1016/j.compstruct.2018.03.073

Dimopoulos, T., Tyralis, H., Bakas, N. P., & Hadjimitsis, D. (2018). Accuracy measurement of Random Forests and Linear Regression for mass appraisal models that estimate the prices of residential apartments in Nicosia, Cyprus. *Advances in Geosciences*, *45*, 377–382. doi:10.5194/adgeo-45-377-2018

El-Sayed, A. K., El-Salakawy, E. F., & Benmokrane, B. (2005). Shear strength of concrete beams reinforced with FRP bars: design method. *7th International Symposium on Fiber-Reinforced (FRP) Polymer Reinforcement for Concrete Structures, 230*, (pp. 955-974). QuakeWrap.

El-Sayed, A. K., El-Salakawy, E. F., & Benmokrane, B. (2006). Shear capacity of high-strength concrete beams reinforced with FRP bars. *ACI Materials Journal*, *103*(3), 383.

European Committee for Standardization. (2015). *Eurocode 2: Design of concrete structures: Part 1-1: General rules and rules for buildings*. European Committee for Standardization.

Gravett, D. Z., Mourlas, C., Markou, G. & Papadrakakis, M. (2019) *Numerical performance of a new algorithm for performing modal analysis of full-scale reinforced concrete structures that are discretized with the HYMOD approach [Paper presentation].* COMPDYN, Athens.

Gravett, D. Z., Mourlas, C., Taljaard, V.-L., Bakas, N., Markou, G., & Papadrakakis, M. (2021). New fundamental period formulae for soil-reinforced concrete structures interaction using machine learning algorithms and ANNs. *Soil Dynamics and Earthquake Engineering*, *144*, 106656. doi:10.1016/j.soildyn.2021.106656

Hatzigeorgiou, G. D., & Kanapitsas, G. (2013). Evaluation of fundamental period of low-rise and mid-rise reinforced concrete buildings. *Earthquake Engineering & Structural Dynamics*, *42*(11), 1599–1616. doi:10.1002/eqe.2289

Markou, G., & AlHamaydeh, M. (2018). 3D finite element modelling of GFRP-reinforced concrete deep beams without shear reinforcement. *International Journal of Computational Methods*, *15*(02), 1850001. doi:10.1142/S0219876218500019

Markou, G., & Bakas, N. P. (2021a). Prediction of the Shear Capacity of Reinforced Concrete Slender Beams without Stirrups by Applying Artificial Intelligence Algorithms in a Big Database of Beams Generated by 3D Nonlinear Finite Element Analysis. *Computers and Concrete*, *28*(6), 433–447.

Markou, G., & Bakas, P. N. (2019). AI reinforced concrete: Algorithms predict concrete strength without training on experimental data. *Innovate (North Miami Beach, Fla.)*, *14*, 8–10.

Markou, G., & Bakas, P. N. (2021b). Developing reinforced concrete structures through AI algorithms and large-scale modelling. *Innovate (North Miami Beach, Fla.)*, *16*, 38–40.

Markou, G., Garcia, R., Mourlas, C., Guadagnini, M., Pilakoutas, K., & Papadrakakis, M. (2021). A new damage factor for seismic assessment of deficient bare and FRP-retrofitted RC structures. *Engineering Structures*, *248*, 113152. doi:10.1016/j.engstruct.2021.113152

Markou, G., Mourlas, C., Bark, H., & Papadrakakis, M. (2018). Simplified HYMOD non-linear simulations of a full-scale multistory retrofitted RC structure that undergoes multiple cyclic excitations–An infill RC wall retrofitting study. *Engineering Structures*, *176*, 892–916. doi:10.1016/j.engstruct.2018.08.002

Markou, G., Mourlas, C., & Papadrakakis, M. (2017). Cyclic nonlinear analysis of large-scale finite element meshes through the use of hybrid modelling (HYMOD). *International Journal of Mechanics*, *11*, 218–225.

Markou, G., & Papadrakakis, M. (2013). Computationally efficient 3D finite element modelling of RC structures. *Computers and Concrete*, *12*(4), 443–498. doi:10.12989/cac.2013.12.4.443

Markou, G., & Papadrakakis, M. (2015). A simplified and efficient hybrid finite element model (HYMOD) for non-linear 3D simulation of RC structures. *Engineering Computations*, *32*(5), 1477–1524. doi:10.1108/EC-11-2013-0269

Mourlas, C., Gravett, D. Z., Markou, G., & Papadrakakis, M. (2019a). *Investigation of the soil structure interaction effect on the dynamic behavior of multistorey RC buildings. [Paper presentation].* VIII international conference on computational methods for coupled problems in science and engineering (coupled problems), Spain.

Mourlas, C., Khabele, N., Bark, H. A., Karamitros, D., Taddei, F., Markou, G., & Papadrakakis, M. (2020). Effect of Soil–Structure Interaction on Nonlinear Dynamic Response of Reinforced Concrete Structures. *International Journal of Structural Stability and Dynamics, 20*(13), 2041013. doi:10.1142/S0219455420410138

Mourlas, C., Markou, G., & Papadrakakis, M. (2019b). Accurate and computationally efficient nonlinear static and dynamic analysis of reinforced concrete structures considering damage factors. *Engineering Structures, 178*, 258–285. doi:10.1016/j.engstruct.2018.10.034

Mourlas, C., Markou, G., & Papadrakakis, M. (2019c). *3D Detailed Modeling of Reinforced Concrete Frames Considering accumulated damage during static cyclic and dynamic analysis–new validation case studies* [Paper presentation]. COMPDYN. doi:10.7712/120119.7207.19727

Mourlas, C., Papadrakakis, M., & Markou, G. (2017). A computationally efficient model for the cyclic behavior of reinforced concrete structural members. *Engineering Structures, 141*, 97–125. doi:10.1016/j.engstruct.2017.03.012

*New Greek Seismic Code.* (2000). NEAK.

Papadrakakis, M., & Markou, G. (2022). Assessment of the Structural Response of Steel Reinforced and Steel-Fibre Reinforced Concrete Structures with 3D Detailed Modeling: Limitations and Remedies. In *Current Trends and Open Problems in Computational Mechanics* (pp. 395–405). Springer. doi:10.1007/978-3-030-87312-7_38

Razaqpur, A. G., & Isgor, O. B. (2006). Proposed shear design method for FRP-reinforced concrete members without stirrups. *ACI Structural Journal, 103*(1), 93.

Reconan, F. E. A. (2020). *v2.00, User's manual.* Research Gate. https://www.researchgate.net/publication/342361609_ReConAn_v200_Finite_Element_Analysis_Software_User's_Manual

Reineck, K.-H., Bentz, E. C., Fitik, B., Kuchma, D. A., & Bayrak, O. (2013). ACI-DAFSTB Database of shear tests on slender reinforced concrete beams without stirrups. *ACI Structural Journal, 110*(5). doi:10.14359/51686819

SABS Standards Division. (2009). Seismic actions and general requirements for building, South African National Standard (SANS), SABS Standards Division.

Spijkerman, Z., Bakas, P. N., Markou, G., & Papadrakakis, M. (2021). Predicting the shear capacity of reinforced concrete slender beams without stirrups by applying artificial intelligence algorithms [Paper presentation]. COMPDYN 2021. Eccomas Proceedia. doi:10.7712/120121.8749.18602

Taljaard, V-L., Gravett, D. Z., Mourlas, C., Markou, G., Bakas, N., & Papadrakakis, M. (2021) *Development of a new fundamental period formula by considering soil-structure interaction with the use of machine learning algorithms [Paper presentation].* COMPDYN 2021, Athens.

Thai, H-T. (2022). Machine learning for structural engineering: A state-of-the-art review. *Structures, 38*, 448-491.

Tureyen, A. K., & Frosch, R. J. (2002). Shear tests of FRP-reinforced concrete beams without stirrups. *ACI Structural Journal, 99*(4), 427–434.

van der Westhuizen, A. M., Markou, G., & Bakas, N. P. (2022). *Development of a New Fundamental Period Formula for Steel Structures Considering the Soil-structure Interaction with the Use of Machine Learning Algorithms* [Paper presentation]. ICAART. doi:10.5220/0010978400003116

Yang, K.-H., Chung, H.-S., Lee, E.-T., & Eun, H.-C. (2003). Shear characteristics of high-strength concrete deep beams without shear reinforcements. *Engineering Structures, 25*(10), 1343–1352. doi:10.1016/S0141-0296(03)00110-X

Young, K. C. (2011). *An Investigation of the Fundamental Period of Vibration of Irregular Steel Structures* [Unpublished Thesis, The Ohio State University, Columbus].

# Chapter 5
# A Predictive Regression Model for the Shear Strength of RC Knee Joint Subjected to Cyclic Load

**Azam Khan**
*National University of Sciences and Technology, Pakistan*

**Moiz Tariq**
*National University of Science and Technology, Pakistan*

## ABSTRACT

*Reverse cyclic loading is applied to the RC knee joint, which causes the knee joint to experience either opening or closing moments. To calculate the joint's shear strength, an effective nonlinear model is created herein. The shear strength of the knee joint either in opening or closing is predicted by two distinct and practical equations developed utilizing a regression-based methodology. The effects of material and geometric factors are taken into account, including the compressive strength of concrete, the joint aspect ratio, and reinforcement strength. The proposed equations are developed using a statistical analysis of a database containing 61 experiments. The proposed equations are statistically evaluated and examined with the models that were earlier put out. Thus, the suggested models can fairly accurately forecast the shear capacity, which makes them ideal for usage in real applications.*

## INTRODUCTION

Due to the fact that they transmit forces from the member ends, reinforced concrete (RC) beam-column joints must be comprehensively understood. These primary joints of structural components typically experience failure (Akguzel & Pampanin, 2010, 2012; Anderson et al., 2008; Bakir & Boduroğlu, 2002; Lowes & Altoontash, 2003a; Mitra & Lowes, 2007a; M. L. Moretti et al., 2014; Pampanin et al., 2002; Shafaei & Nezami, 2019; Wang et al., 2012). The geometrical and structural arrangement of the joint

DOI: 10.4018/978-1-6684-5643-9.ch005

causes complicated stress condition, particularly when it is subjected to lateral stresses like earthquakes or wind. In contrast to other joints, the beam-column roof corner joints exhibit more complicated behavior. The opening moments, which produce compression outside the joint, or the closing moments, which create compression inside the joint, are both responsible for this complicated behavior of beam-column roof corner joints under cyclic loading. The erratic behavior of the roof corner joints is also related to the small to moderate axial stresses acting on the exterior columns of ordinary RC frames. The difference in cyclic response between the closing and opening actions of the roof corner joint causes this phenomenon. But there hasn't been enough research done on the roof corner joint before. The design of these joints is not even given any attention by the bulk of seismic construction regulations.

When the roof corner joint opens and closes, it travels through many load-resisting processes. Due to its unique behavior, the roof corner joint is more sensitive and sophisticated than other RC joints. An experimental study (Priestley et al., 1996) has determined the nominal joint shear strength to be $0.29\sqrt{f_c^{'}}$ MPa and $0.3f_c^{'}$ MPa, respectively, is based on the principal tensile and compressive stresses. This study served as the basis for the recommendation (Meggeta, 2003) of reducing the roof corner joint strength from 0.2f c' to 0.1f c', which was also approved in New Zealand code of practice (NZS 3101, 2006). However, experimental tests have shown that the roof corner joint's shear strength for opening behavior is lower than for closing (Cote & Wallace, 1994; Mazzoni et al., 1991; McConnell & Wallace, 1995; Megget, 1998). This is due to a weaker diagonal concrete strut exposed to excessive axial stress from the neighboring tensile component reducing the opening shear strength (Mogili & Kuang, 2018; Zhang et al., 2017). Because of this, a modest opening shear stress is more dangerous for the RC roof corner joint's than closing shear stress of comparable size. Nevertheless, this aspect is mostly disregarded by the design code of practices.

The major structural design codes (NZS 3101, 2006; EN1998 Eurocode 8: Design of Structures for Earthquake Resistance, 2001; The People's Republic of China National Standard, 2010; Architectural Institute of Japan, 1999) do not provide any explicit provisions for roof corner joints. As a result, the roof corner joints are typically subject to external joint provisions. In contrast, the nominal shear strength of the roof corner joint is suggested by a single expression by codes like (ACI-318, 2019) and (ACI352R, 2002). Whereas the need for opening and closing expression of the roof corner joint is made evident by the above discussion. Additionally, the axial load on a conventional roof corner joint's column might be reduced by using a discontinuous column with beams in one or two directions. However, the supported beams may result in a significant shear and moment transmission to the roof corner joints. Thus, the theoretical foundation of current design methodologies has to be changed in order to predict the behavior of the roof corner joint that is realistic.

Numerous computational and analytical models for interior and exterior joints have been put out in the literature. An analytical study (Hwang & Lee, 1999) has recommended the modified compression field theory as one of several ideas. Additionally, further study (Hsu, 1988) introduced the modified rotating angle softened truss theory. Also available are refined models (Wong & Kuang, 2011; Pang & Hsu, 1996) using the strut-and-tie analogy. In contrast, there aren't many analytical models available for the roof corner joint's shear strength. Few researchers (Zhang, 2017; Mogili et al., 2020) have proposed analytical roof corner joint models. Due to their different force-resisting processes, these models can estimate the opening and closing shear strengths of the roof corner joint independently. A complicated finite element technique is also suggested by other researchers (Angelakos, 1999) to represent the behavior of the roof corner joint at both the opening and closing moments. All of the aforementioned models, however, need a high degree of knowledge and have far more complicated computational requirements. In order to effectively forecast the roof corner joints' reaction by gathering the available experimental data to calibrate the analytical formulations, a reevaluation of the roof corner joint design theory is necessary.

This work proposes a basic genetic algorithm based model to calculate the joint's shear capacity during opening and closing actions. The suggested model is statistically confirmed. Finally, by comparing the outcomes with those of structural design codes and earlier literature, the model's correctness is shown.

## RESEARCH SIGNIFICANCE

The poor shear capacity of columns is one of the most serious flaws in RC structures that renders them vulnerable to earthquakes (Del Vecchio et al., 2017; Elood & Moehle, 2005; K. J. Elwood, 2004; K. J. Elwood & Moehle, 2005; K. Elwood & Moehle, 2003; Lynn et al., 1996; Ousalem et al., 2002; Sezen, 2002). However, only a few models have been created to quantify the shear capacity of RC columns in compliance with ductility criteria. Similar to this, joints are regarded as rigid in structural design techniques without accounting for their shear capacity when exposed to lateral force. As a result, a typical problem for the structural integrity as a whole is the shear failure of the RC joints under cyclic loads. Therefore, to improve their seismic performance and eliminate the danger of unexpected collapse, suitable joint design and details are required. To appropriately design the joints, it is critical to precisely establish the nominal shear capacity. The nominal shear capacity of the corner roof corner joint cannot yet be adequately predicted by the available models or equations. Furthermore, the current models only consider the concrete's compressive strength and disregard other important factors like the joint geometry and longitudinal reinforcement details. Compared to other common RC joints, the roof corner joint exhibits a more complicated behavior when subjected to reverse cyclic stress. A precise model for forecasting the shear strength is required to make up for the weak theoretical foundation of the design regulations.

## FACTORS AFFECTING RC ROOF CORNER JOINT

To develop a precise genetic model for the shear behavior of roof corner joints, the key factors must be determined. According to a review of the literature, several parameters, including the compressive strength of the concrete, the longitudinal tensile reinforcement ratio, the joint aspect ratio, and the ratio between various geometric properties of adjacent members, affect the joint's shear strength. Other researchers (Borghini et al., 2016; Celik & Ellingwood, 2010; Favvata et al., 2008; Favvata & Karayannisa, 2014; Jeon et al., 2015; Laterza et al., 2017; Lowes & Altoontash, 2003b; Mitra & Lowes, 2007b; A. Niroomandi et al., 2010, 2014; Omidi & Behnamfar, 2015; Pampanin et al., 2003; Shafaei et al., 2014; Sharma et al., 2011; Shayanfar et al., 2016; Shayanfar & Bengar, 2016, 2018; Shin & Lafave, 2004; H. F. Wong, 2005) have all conducted experiments to show the relevance of these characteristics(Mazzoni et al., 1991; Megget, 1998; Meggeta, 2003; Angelakos, 1999; Zhang, 2017; Zhang et al., 2017; Kim & LaFave, 2007; Shayanfar et al., 2016; Mogili et al., 2020). Some researchers (Zhang, 2017; Mogili et al., 2020) emphasized the association between concrete compressive strength and joint shear stress during joint closure. On the other side, there is a poor association between joint opening shear stress and compressive strength. This behavior of concrete's compressive strength is seen in Figures 1(a) and 1(b). In this instance, the joint shear strength under both opening and closing movements was measured by altering just the concrete compressive strength and keeping the other key factors constant. A randomly chosen sample of specimens with comparable geometrical properties was chosen. The joint shear strength during closure behavior has a positive association with the concrete compressive strength, as shown in Figure 1(a). However, when the same specimens are exposed to opening action, the joint shear strength is decreased. The opening movement creates a tensile tension in the beam and column close to the roof corner joint, which causes the concrete strut to expand (Zhang, 2017; Mogili et al., 2020). While during a closing operation, compressive pressures from the nearby beam and column elements cause the concrete strut to compress and become confined inside the joint. This information leads to the conclusion that the opening action is more important than the closing action due to the confinement loss.

The influence of the beam-to-column depth ratio has a major impact on the roof corner joint's shear strength (joint aspect ratio). According to studies (Kim & LaFave, 2007; Zhang et al., 2017; Mogili et al., 2020), the roof corner joint's shear strength varies with the joint aspect ratio as a result of the concrete compression strut's inclination Figure (c). Further study (Zhang, 2017) has expanded the research into the effects of the beam-to-column breath ratio. The relationship between the normalized horizontal joint shear strength and the beam-column breath ratio is seen in Figure 1(d). The shear strength of the roof corner joint tends to grow with an increase in the beam-column breath ratio during both opening and closing behavior, the author observed. As a result, it is thought that a key influence on the shear strength of the RC roof corner joint is the beam-column breath ratio. However, this aspect is often overlooked in the structural design guidelines. In follow-up research (Mogili et al., 2020) to this one, it was found that straying from a 45° strut angle resulted in a narrow strut, which decreased the strength of the diagonal concrete strut. The experiment was expanded (Mogili et al., 2020) to look at the impact of shear reinforcement on the strut capacity. It is shown that 45° inclined stirrups are more efficient in enhancing the roof corner joint's strut capacity than horizontal beam stirrups and vertical column ties.

Researchers (Mogili et al., 2020) have demonstrated the advantage of having sufficient anchoring in transmitting forces across the roof corner joints in their research of the effects of longitudinal reinforcement via the roof corner joints. They showed how weak anchoring may reduce the shear strength of the joint, which can cause bond slippage and early failure. Mogili et al. (2020) demonstrated better roof corner joint shear strength when columns are wider than beams as a result of indirect confinement of the joint core as part of this study's continuation, as shown in Figure 1(d). Mogili et al. (2020) made note of the effect of longitudinal steel on the closing and opening shear capacity using Kim's investigation. According to the authors' findings, as illustrated in Figures 1(e) and 1(f), the specimen with a high opening-to-closing flexural moment capacity had a reduced joint shear capacity under the closure behavior, and vice versa.

## REVIEW OF PRIOR RESEARCH

Since the majority of experimental investigations on roof corner joints previous to 1990 concentrated on the monotonic static load response, it seems that the complicated cyclic behavior of roof corner joints was not adequately investigated. Study of the cyclic behavior became crucial after the Loma Prieta Earthquake in 1989. Researchers have reevaluated the behavior of the roof corner joints as a result of the severe damage this earthquake caused to reinforced concrete buildings' roof corner joints. Due to these historical factors, major cyclic experimental research on the roof corner joint is less frequent than those on the inner and exterior joints. In this part, the primary experimental work with cyclic loading is covered.

*Figure 1a. Variability in the normalized joint shear capacity depending on the roof corner joint shape and material properties (a-d)*

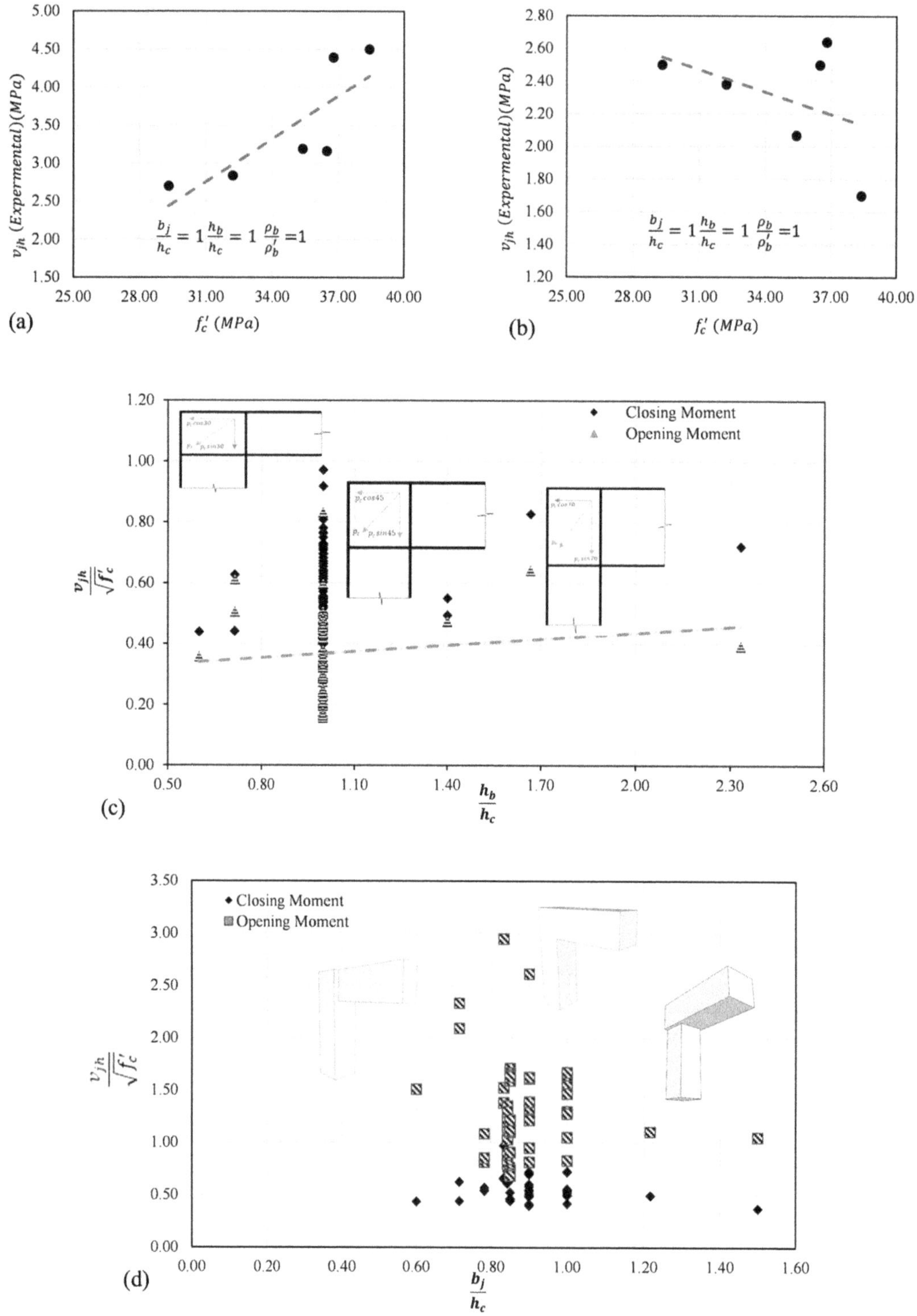

*Figure 1b. Variability in the normalized joint shear capacity depending on the roof corner joint shape and material properties (e-f)*

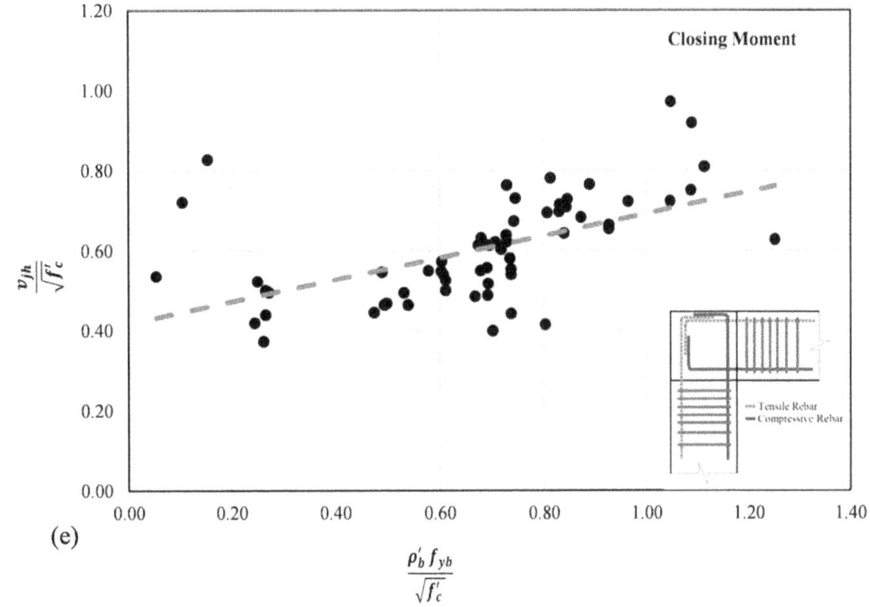


(e)

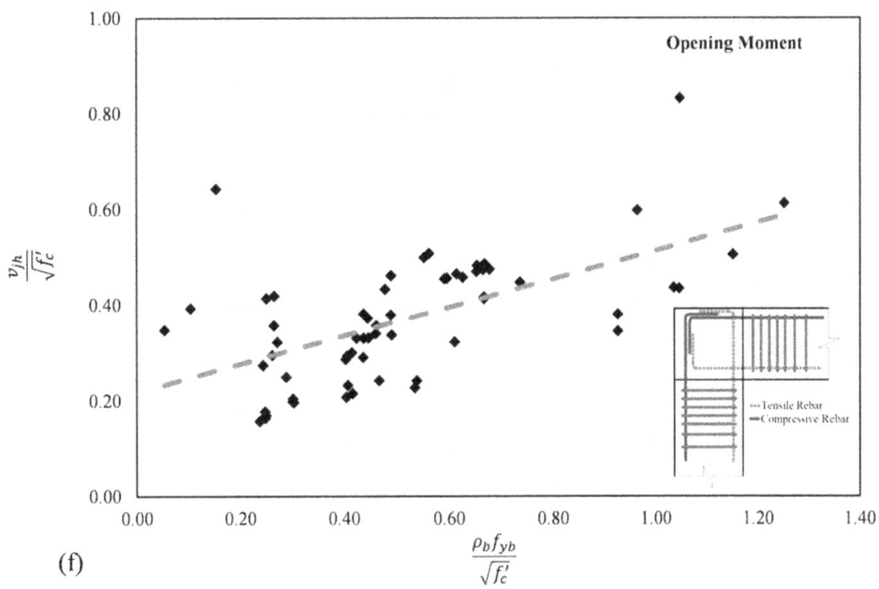


(f)

## Experimental Database

Two distinct empirical equations for evaluating the shear strength of RC roof corner joints subjected to cyclic loadings are developed using non-linear multiple genetic . In order to present a model that is utilized to assess the shear strength of the roof corner joint under cyclic stress, a database is generated from past research. The test findings of 61 tests conducted between 1991 and 2018 by various researchers make up this study. To be more precise, the random sample of 41 trials selected from the database for genetic analysis contains only specimens that have been exposed to reverse cyclic loading. The last 20 samples are utilized to validate the model. A few trials are left out due to differences in the geometric size of the beam and column, the compressive strength of the concrete, the strength of the tensile and compression reinforcement, and the strength of the shear reinforcement of the beams and columns. Table 1 summarizes the experiment database.

Three cyclic stress investigations on the roof corner joint were carried out in 1991 by Mazzoni et al. (1991). The specimens were created in accordance with the guidelines from ACI 318-19 (Standard, 2019). It was shown that none of the specimens have the joint shear capability indicated by ACI 352-85. In this regard, other researchers (Cote & Wallace, 1994; Angelakos, 1999) observed a similar anomaly in the updated structural design codes ACI 352-91 (American Concrete Institute, 1991), ACI 352-98 (American Concrete Institute, 1998), ACI 352-02 (Committee, 2010). Megget (1998) noted that, in contrast to earlier research, experimental testing on roof corner joints subjected to closure action was consistent with the shear strength recommended by (NZS 3101-2006). The test results for the initial action, however, were still far different from the suggested strength of the code. This difference in the opening motion was related to the degeneration of the bond in the cyclically loaded roof corner joint brought on by the failure of the concrete compression strut. In-depth experimental and parametric research on these joints has more recently been carried out (Angelakos, 1999; Zhang, 2017; Mogili & Kuang, 2019). The shear strength of the joint is affected by a variety of parameters, including the material, shape, and reinforcing qualities of the joint as well as those of the surrounding structural elements. The force transmission mechanisms, which constrict the joint concrete strut under the closure moment and expand the strut under the opening moment, are another finding from these studies that also contribute to the unique behavior of the roof corner joint. Figure 2 illustrates this. In light of this, it is crucial to note that the concrete compressive strength is not the only factor that significantly affects the roof corner joint's shear strength.

*Figure 2. RC roof corner joint mechanics*

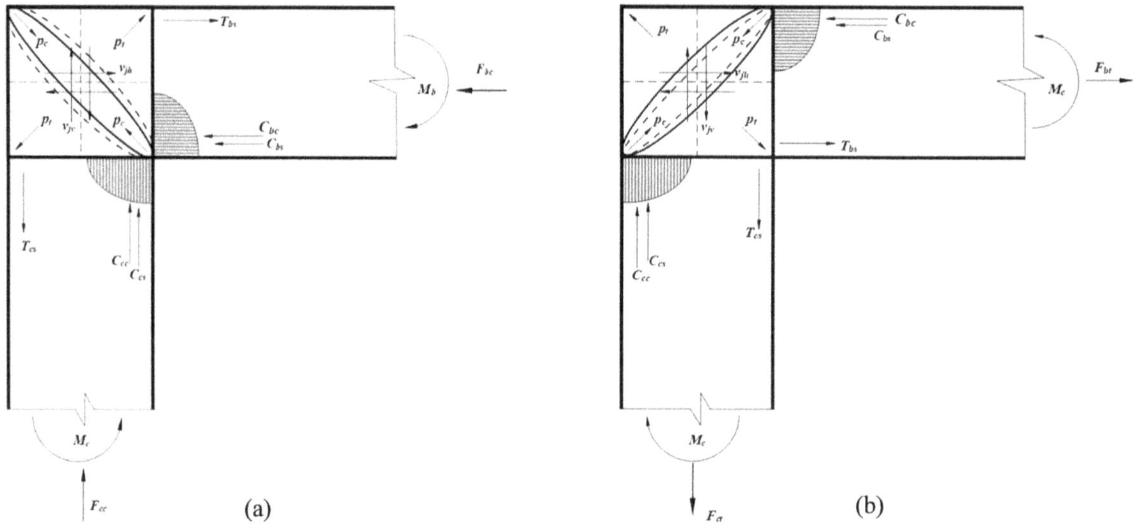

## EARLIER MODELS ASSESSING ROOF CORNER JOINT SHEAR STRENGTH

To calculate the shear strength of the roof corner joints, many researchers have published a variety of empirical formulae. When used to analyze problems with roof corner joints, these equations, which are based on experimental investigations, provide a range of answers that may not be reliable. For instance, Wallace et al. (1998), Angelakos (1999), and Kim and LaFave (2012) established the capacity equation as $0.5\sqrt{f_c^{'}}$ , although McConnell and Wallace (1995) advocated that the roof corner joint's shear capacity be $0.58\sqrt{f_c^{'}}$ . Equations $0.12f_c^{'}$ and 0.10f c' for the closure and opening shear capacities, respectively, were similarly created by Megget (1998, 2003). Priestly et al. (1996) suggested capping the primary tensile stresses at $0.29\sqrt{f_c^{'}}$ in place of forecasting the joint shear capacity. Similar to this, the following overview shows how codes of practice significantly differ in forecasting the roof corner joint shear strength:

### ACI 352R-02 (2010)

According to the safety provisions of the ACI Code, the shear capacity of joints is founded on the relation:

$$\phi V_n \geq V_u \tag{1}$$

where $V_n$ and $V_u$ represent the nominal shear capacity of the joint and the factored shear capacity, respectively. Following the strength-based design methodology, a strength reduction factor $\phi$ is multiplied with the nominal capacity. Generally, the value of this factor is taken as 0.85 for the seismic design of joints. As per the code, the concrete contribution to the nominal shear capacity of the roof corner joint is given by the least of Eq (2) and (3):

$$V_n = 0.083\gamma_c\sqrt{f_c'}A_j \tag{2}$$

$$V_n = \gamma_c\sqrt{f_c'}A_j \tag{3}$$

where $f_c'$ and $A_j$ represent the concrete compressive strength and the effective cross-sectional area of the joint, respectively. The constant term $\gamma c$ denotes the degree of confinement provided by the adjacent structural members, i.e., beam and column. For the roof corner or the corner joint, the value of $\gamma_c$ is taken as 8.

### Eurocode 8 (EN 1998-1:2004)

There exists no equation in Eurocode 8 for estimating the shear strength of the roof corner joint. Therefore, the provisions of the exterior joint are extended for the roof corner joint. The horizontal joint shear force determined from the horizontal component of the diagonal concrete strut is calculated using the following equation:

$$V_{jhd} = \lambda \eta f_{cd} \sqrt{\left(1 - \frac{v_d}{\eta}\right) b_j h_{jc}} \tag{4}$$

$$\eta = 0.6\left(1 - \frac{f_{ck}}{250}\right) \tag{5}$$

where the constant factor $\lambda$ for exterior and the interior joint is 1.0 and 0.8, respectively. $f_{cd}$ and $f_{ck}$ represents the design value of concrete compressive strength (MPa) and the characteristic cylinder compressive strength of concrete (MPa), $v_d$ is the normalized axial force and b$j$ is the effective joint width.

## New Zealand Design Standard (NZS 3101-2006)

So as to prevent the crushing of concrete in the exterior joint core, the New Zealand design code limits the horizontal shear force using

$$V_{jh} \leq \min\left[0.2 f_c' b_j h_c ; 10 b_j h_c\right] \tag{6}$$

where $h_c$ is the overall depth of the column in the direction of horizontal shear and $b_j$ represents the effective width of joint. Having no provision for the roof corner joint in this code of practice, Eq. 6 is used to calculate the shear capacity of the roof corner joint as well.

## Chinese Seismic Standards (GB 50011:2010)

The Chinese Code for Seismic Design of Buildings does not offer any provision for the exterior and the roof corner joints. Therefore, the provisions of the interior joint are extended to the roof corner joint as well as the exterior joint. The code proffers the maximum shear strength capacity equal to:

$$V_{jh} \leq \frac{1}{\gamma_{RE}}\left(0.3 \eta_j \beta_c f_c b_j h_j\right) \tag{7}$$

where $\gamma_{RE}$ is similar to the strength reduction factor and represents the seismic bearing capacity coefficient, whose value is taken as 0.85. $\eta_j$ is a constant and dependent on the joint classification. For the RC roof corner joint, its value is 1. $\beta_c$ and $f_c'$ represent the concrete strength factor and compressive strength of concrete, respectively. $b_j$ is the effective width of joint core and $h_j$ represents an effective depth of the joint core.

### Architectural Institute of Japan (AIJ)

AIJ standard also concedes the contribution of the concrete diagonal compression strut to resist the shear forces developed in the roof corner joint. Based on the ultimate strength concept, AIJ proposes the following equation for calculating the shear strength of the joint:

$$V_{ju} = k\phi F_j b_j D_j \tag{8}$$

where the factor $k$ is dependent on the shape of in-plane geometry, and its value is taken as 0.4 for the corner joint, $\phi$ factor represents the effect of out-of-plane geometry, $b_j$ and $D_j$ are the effective joint width and the effective column depth, respectively. $F_j$ represents the standard value of joint shear strength, and its value is suggested by AIJ as

$$F_j = 0.8 f_c^{'0.7} \tag{9}$$

where $f_c^{'}$ represents the concrete compressive strength. By substituting Eq. 10 in Eq.9, the roof corner joint shear is determined,

$$V_{ju} = 0.272 f_c^{'0.7} b_j D_j$$

where, $b_j$ and $D_j$ represents the joint width and the depth, respectively.

## DEVELOPED MODEL FOR CALCULATING THE HORIZONTAL JOINT SHEAR

The foregoing explanation makes it evident that, in contrast to what several design guidelines opine, the joint shear strength is not entirely influenced by the compressive stress of the concrete. The present work thus develops a genetic -based model to assess the influence of additional influencing characteristics on the roof corner joint shear capacity ($v_{jh}$). A database of 61 experimental trials on the RC roof corner beam-column joints is summarized in Table A1. Figure 1 illustrates the relationship between joint shear stresses under the opening and closing moments and eight different influencing factors to help with idea clarification. Compressive stresses on the samples in the experimental database range from 26.8 to 49.7 MPa. In this database, the starting moment's longitudinal tensile reinforcement ratio ranges from 0.001-0.012 through the closing moment's 0.005-0.020. The effective joint breath ratio varies from 0.71 to 1.50, whereas the joint aspect ratio is between 0.6-2.33. Non-linear multiple genetic analyses are used to create two distinct equations, one for the opening moment and one for the closing moment, using a random sample of 41 trials from the database. The flow chart in Figure 3 illustrates the phases of the genetic approach used to create the shear strength equations. First, the factors influencing the shear strength of the roof corner joints are identified using information from the experimental database. The shear strength model is created in the next stage using a non-linear multiple genetic analysis, and the model's ability to forecast is assessed. Up until the dataset reaches the necessary accuracy, the model is updated using a variety of genetic models from the literature. The model's performance is then examined in a validation study using a random sample of 20 trials that were not utilized in the model's creation.

The following closing moment equation is used for the final calculation based on the procedure described above:

$$v_{jh}^{c} = 1.5 f_c^{X_1} \frac{\left[ \left(\frac{h_b}{h_c}\right)^{X_2} \left(\frac{h_b}{b_b}\right)^{X_3} \left(\frac{b_j}{h_c}\right)^{X_4} \left(\frac{\rho_b' f_{yb}}{\sqrt{f_c'}}\right)^{X_5} \left(\frac{f'}{f_{yb}}\right)^{X_6} \right]^{X_7}}{\left(\frac{h_b}{h_c}\right)^{X_8} \left[1-\sqrt{1+4\left(\frac{h_c}{h_b}\right)^{X_9}}\right] \left[-0.6\left(\frac{\rho_b' f_{yb}}{\sqrt{f_c'}}\right)^{X_{10}}+1.6\right]} \quad (MPa) \tag{11}$$

and for the opening, the following:

$$v_{jh}^{o} = 60.0 f_c^{X_1} \frac{\left[ \left(\frac{h_b}{h_c}\right)^{X_2} \left(\frac{h_b}{b_b}\right)^{X_3} \left(\frac{b_j}{h_c}\right)^{X_4} \left(\frac{\rho_b f_{yb}}{\sqrt{f_c'}}\right)^{X_5} \left(\frac{f_c'}{f_{yb}}\right)^{X_6} \right]^{X_7}}{\left(\frac{h_b}{h_c}\right)^{X_8} \left[1-\sqrt{1+4\left(\frac{h_c}{h_b}\right)^{X_9}}\right] \left[-0.7\left(\frac{\rho_b f_{yb}}{\sqrt{f_c'}}\right)^{X_{10}}+1.5\right]} \quad (MPa) \tag{12}$$

In these equations $X_1, X_2, X_3, X_4, X_5, X_6, X_7, X_8, X_9, X_{10}$ represent non-dimensional empirical coefficients that are derived using the iterative genetic analysis procedure and are given in Table 1. $h_b$ and $h_c$ define the respective depths of beam and column members, $b_j$ and $b_b$ represents the respective joint effective width and the beam width, $\rho_b$ ' and $\rho b$ represent the longitudinal tensile reinforcement under respective closing and opening behavior, respectively, $f_c$ ' and $f_{yb}$ define the respective concrete compressive strength and the yield strength of tensile reinforcement.

*Table 1. Empirical coefficients to determine $v_{jh_{max}}$ for opening and closing behavior*

| | $X_1$ | $X_2$ | $X_3$ | $X_4$ | $X_5$ | $X_6$ | $X_7$ | $X_8$ | $X_9$ | $X_{10}$ |
|---|---|---|---|---|---|---|---|---|---|---|
| **Closing** | 0.50 | 0.20 | -0.10 | -0.50 | 0.20 | 1.00 | -0.01 | 1.00 | 1.00 | 1.00 |
| **Opening** | -1.00 | -1.5 | -1.6 | 1.5 | 0.5 | 1.00 | 0.10 | 1.00 | 1.00 | 0.5 |

*Figure 3. Flowchart showing genetic process*

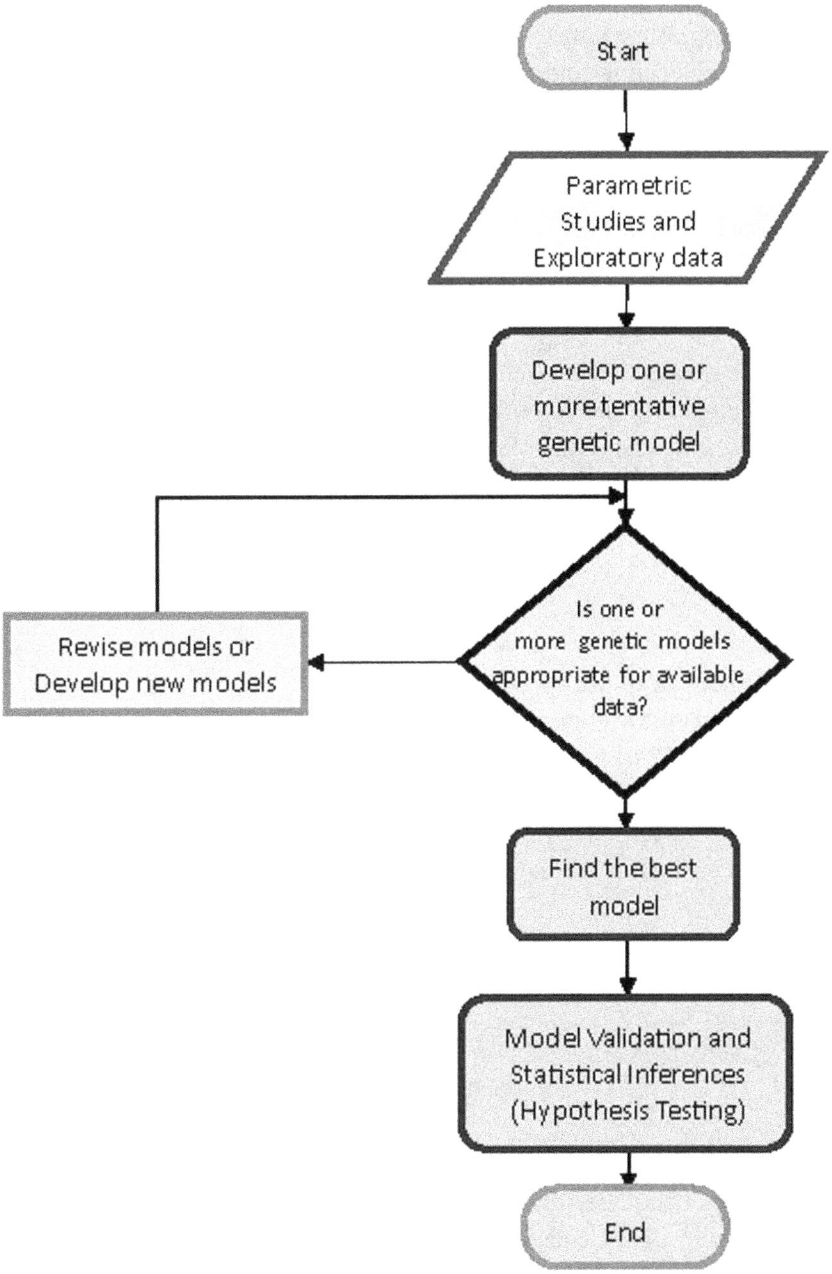


## VALIDATION OF THE SUGGESTED MODEL AND STATISTICAL ANALYSIS

Any genetic model's performance assessment depends critically on model validation. The whole dataset of 61experimental trials is split into a training data set and a validation data set for validation purposes. A random sample of 20 trials is utilized for validation, and a random sample of 41 experiments is used to derive an equation for calculating opening and closing joint shear strength. To gauge how well the suggested model fits the experimental findings, one may use variables like the mean, standard deviation, and coefficient of variation. According to Figures 4 and 5, the performance factor of the training, validation, and total data for both opening and closing behavior is very near to 1, confirming the effectiveness of both models. The coefficient of variation, or CoV, is another method for evaluating the suggested model's accuracy. CoV is defined as:

$$\mathrm{CoV}(\%)=\left\{\frac{\text{Standard Devation}(\sigma)}{\text{Mean}(\mu)}\right\}\times 100 \tag{13}$$

where the CoV's smallest value indicates that the outcome data is less dispersed.

Calculating the average absolute error (AAE), which reflects the overall error between the model findings and the experimental data, is another validation check. Therefore, the AAE is

$$AAE(\%)=\frac{1}{n}\sum\left[\frac{\left|V_{jh}^{Exp}-V_{jh}^{Est}\right|}{V_{jh}^{Exp}}\right] \tag{14}$$

where n stands for the number of test specimens.

In terms of joint closure, the average absolute error (AAE) is 11.1 percent for training data, 13.0 percent for validation data, and 12.6 percent for the whole data set. While the AAE of the opening behavior is 11.0 for the training data, 16.0 for the validation data, and 11.7 for the whole data set.

The variance dependent coefficient ($R^2$), also known as the coefficient of determination, is the most often used indicator for evaluating the dependability of a genetic model. The coefficient ($R^2$) might be expressed as

$$R^2=1-\frac{\sum\left[V_{jh}^{Exp}-V_{jh}^{Est}\right]^2}{\sum\left[V_{jh}^{Exp}-V_{jh_{(Mean)}}^{Exp}\right]^2} \tag{15}$$

where a value near 1 denotes a precise forecast.

In this regard, Figures 4 and 5 respectively display the coefficient of determination for the opening and closing statistical models. The closing event's coefficient of determination ($R^2$) is 0.98 for the training data and 0.96 for the validation data, as is clear from the figure. In contrast, the coefficient for the whole data set is 0.98. Similar to closing action, the coefficient for training and validation data, respectively, is 0.97 and 0.94. Additionally, ($R^2$) is 0.97 for the whole data under the opening moment. These findings imply that the determined coefficient of determination, which is used for both training and validation, is near the benchmark value of 1, which denotes a reliable prediction of joint shear strength.

The t-statistic is a different indicator for measuring a model's statistical significance at a certain confidence level. The null $H_0$: $\mu_D=0$, where $\mu_D$ is the assumed difference between the mean experimental and anticipated Joint Shear Stresses (MPa), and the alternative hypothesis is $H_1$: $\mu_D\neq 0$. To compare the model findings with those of experiments, the paired samples t-test is used. The t distribution's basic shape is

$$t = \frac{\bar{X} - \mu_D}{\frac{s}{\sqrt{N}}} \tag{16}$$

where

$\bar{X}$ = Sample mean difference (The mean difference between the experimental and predicted joint shear strength)

$\mu_D$= Population mean (The sample mean is tested against a value of zero, i.e., $\mu_D$=0, assuming no difference between experimental and predicted Joint Shear Stress (MPa))

s= Standard deviation

$N$= Sample size (number of observations)

$v$= Degree of freedom

The significant level $\alpha$=0.05 with $\mu$=60 degrees of freedom. The critical t-value is obtained at $t_{\frac{\alpha}{2}}$ depending on the level of significance ($\alpha$) and the degree of freedom (n–$1$). To judge the statistical insignificance of the model at the confidence level of 1-α, the calculated t value must be less than the critical t value.

As can be seen in Tables 4 and 5, the $t_{calculated}$<$t_{(0.025,60)}$, i.e., 0.60782<2.0002 and -0.2811<2.0002 for the RC roof corner joint under the closing and opening behavior, respectively, having a degree of freedom ($v$)=60 and not falling in the critical region. Consequently, the null hypothesis is accepted with the analytical model results at a 5% level of significance.

Another indicator for determining the precision and consistency of the suggested model is the slope of the best fit line. It is clear from Figures 4(a), (b), and (c) that the slope of the best fit line for the closure behavior for the training, validation, and overall data is 0.98, 0.97, and 0.98, respectively. This value is very near to the slope of the benchmark line, which is 1. Similar to the joint closure, for the joint opening, the slope of the best fit line for the training, validation, and whole data set is 0.93, 0.90, and 0.93, respectively. As demonstrated in Figures 5 (a), (b), and (c), this is likewise near to the benchmark line (c).

To address the most important factors impacting the proposed model, a parametric validation is conducted in addition to the statistical validation of the model. In Section 4, numerous affecting parameters are listed with a full discussion. In this context, the fluctuation of joint shear strength for the experimental specimen of Mogili during opening and closing behavior is shown in Figsures 6 and 7. Considering this, Table A-1 provides a summary of the specimen's geometric and material characteristics. The shear strength of the RC joint diminishes at both the opening and closing moments at a given concrete compressive strength as the beam to column depth rises (joint aspect ratio), according to Figures 6(a) and 7(a). The concrete strut's less-resistive force component is the cause of this declining tendency. For the same specimen, the joint shear strength is increased by an increase in the joint breath ratio and longitudinal reinforcement ratio. The opening and closing behavior predicted by the suggested model follows the same pattern as the parametric analysis of the actual data, as shown in Figure 1. This demonstrates that the suggested model is not largely reliant on the compressive strength of concrete and is capable of capturing the influence of change in various geometrical and material parameters.

Figure 8 represents the closing and the opening shear strength according to the proposed empirical model along with experimental joint shear strength. In this setting, the ratio of the observed shear strength and the predicted shear strength is termed as strength ratio. In the closing and the opening behavior of the RC roof corner joint, the shear strength estimation shows a satisfactory correlation with the experimental dataset, having a mean strength ratio of 1.04 and 1.01 under the opening and the closing behavior, respectively. Clearly, this establishes the accuracy of the proposed model for estimating the shear strength of the roof corner joint.

*Figure 4. Comparison of closing joint shear strength values: (a) training data, (b) validation data, (c) all data*

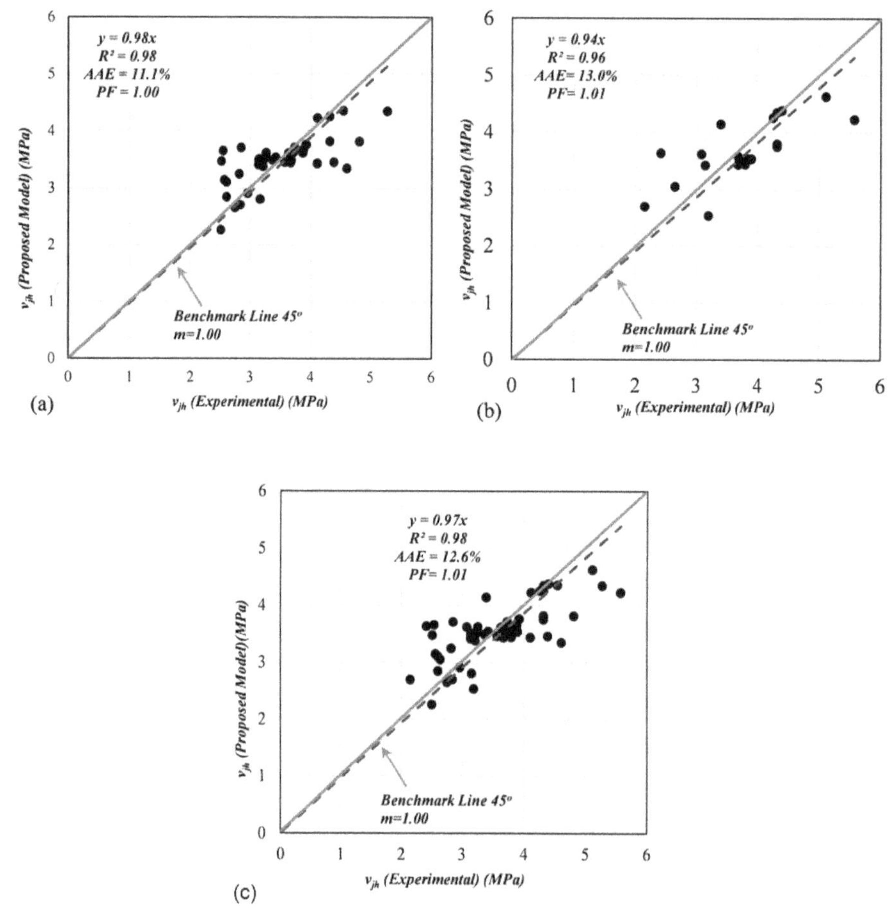

*Figure 5. Comparison of closing joint shear strength values: (a) training data, (b) validation data, (c) all data*

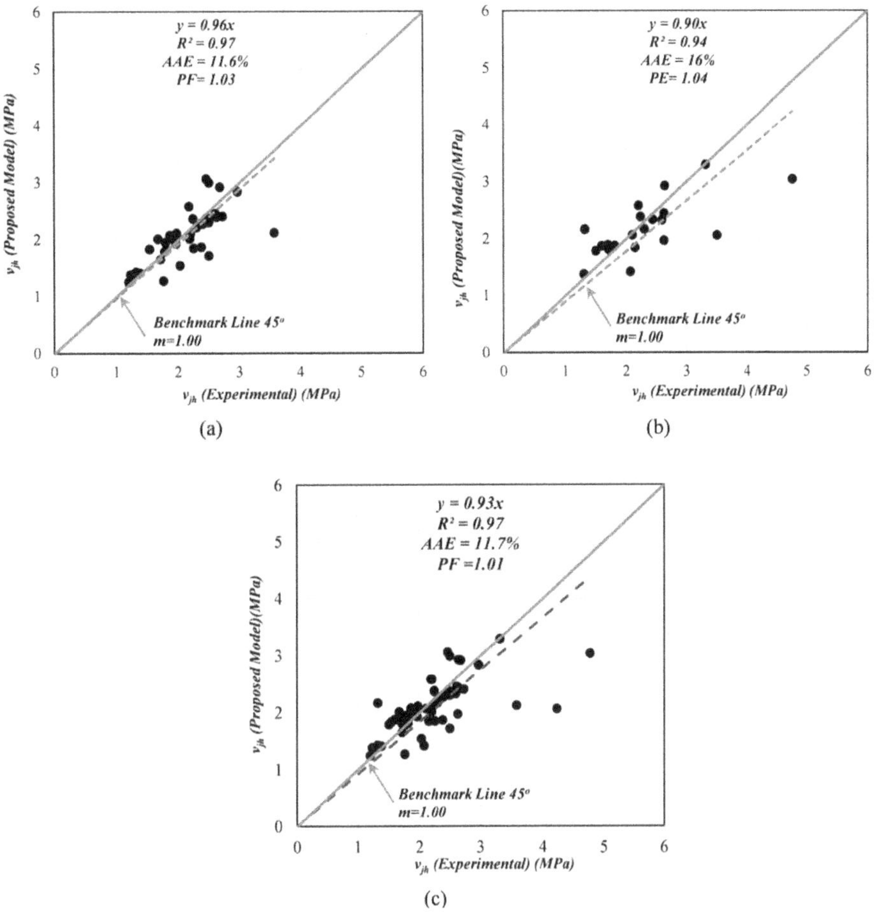


## RESULTS AND DISCUSSIONS

This section contrasts the suggested non-linear genetic model with the empirical and analytical models that were previously established. By comparing the model findings and different codes of practice, it is also possible to see the power of the suggested model. In Tables 2 and 3, for the closure and opening moments, respectively, the joint shear strength is reported and used as the foundation for the discussion. Tables 2 and 3 give statistical comparisons of different models to emphasize the performance checks described in Section 8. Tables 2 and 3 provide a statistical comparison of several models to highlight the performance checks mentioned in Section 8. The suggested model's performance indicators are calculated using a comprehensive data set of 61 trials, whose statistical analysis is detailed in detail below.

*Figure 6. Parametric study of proposed model under joint closing*

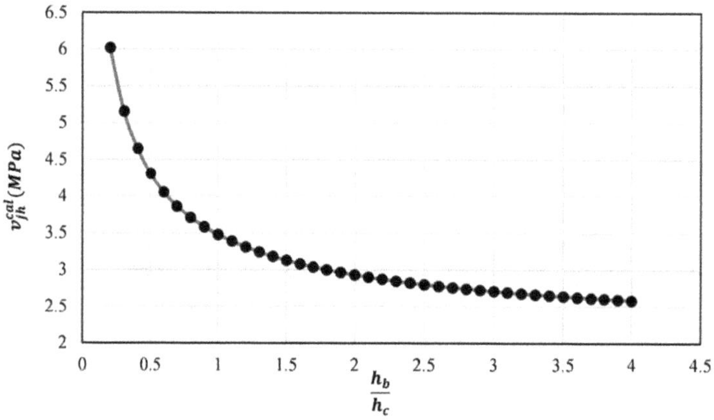


(a)

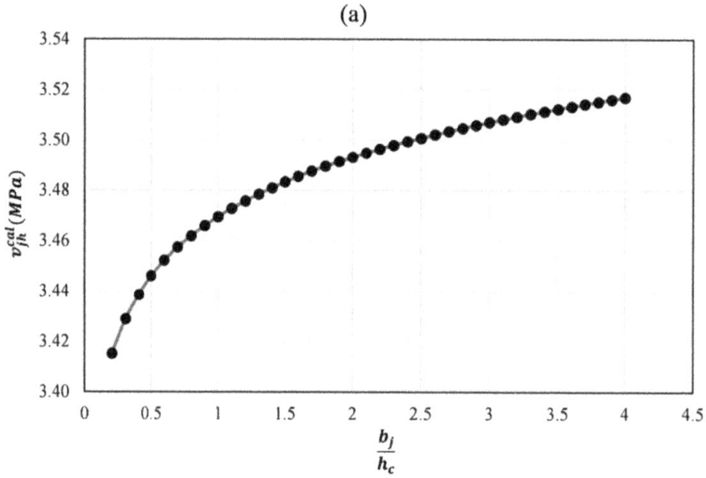


(b)

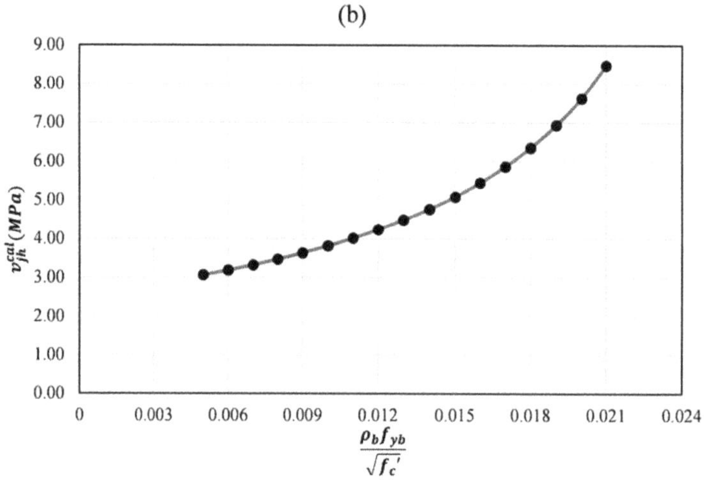


(c)

*Figure 7. Parametric study of proposed model under joint opening*

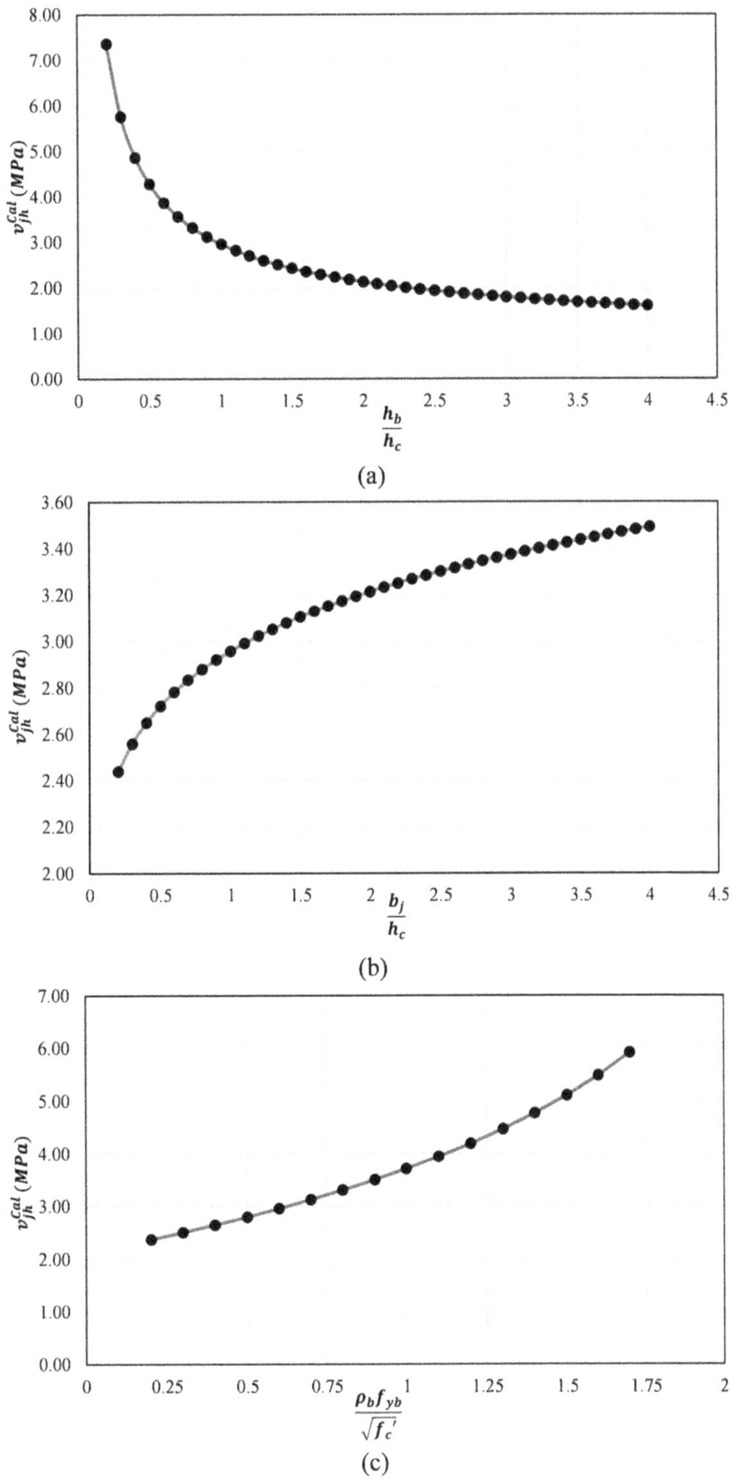

*Figure 8. Strength ratio variation in RC roof corner joint under*

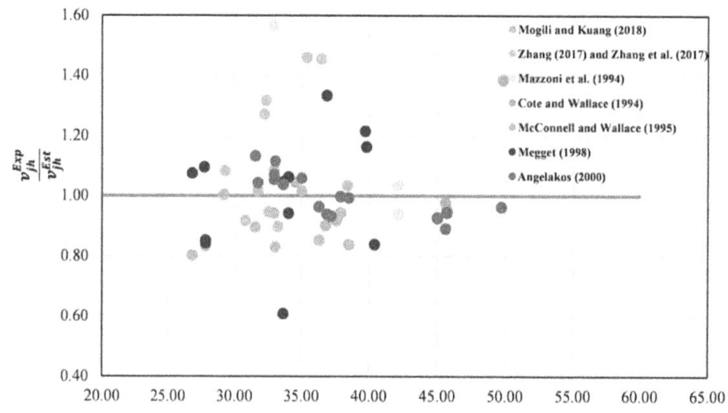


(a)

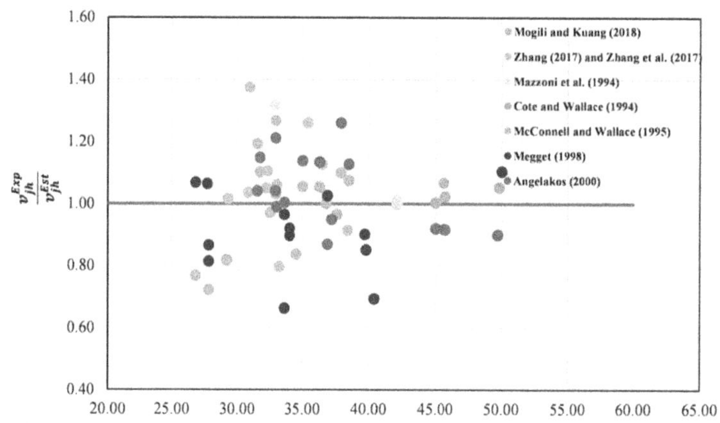


(b)

## Roof Corner Joint Closing

Table 2 displays a statistical comparison of several models. It is important to note that the suggested model is more trustworthy than the other models since its coefficient of determination ($R^2$=0.98) is greater and the slope of the best-fit line ($m$=0.97) is most similar to the benchmark line's slope. For contrast, the coefficient of determination ($R^2$=0.97) of the model of Mogili et al. (2020) is only slightly lower than that of the suggested model, however the slope of the best-fit line (m=0.86) is substantially lower than that of the proposed model. It's obvious that the majority of single factor-dependent models diverge greatly from the one being suggested. Although the coefficient of determination and slope of the best-fit line provided by the ACI 352R-02 and AIJ models change slightly from those of the suggested model, the average absolute error (AAE) is noticeably different. According to Figure 9, the suggested model has the lowest average absolute error (AAE) of all the models provided in Table 2 at 11.7 percent, demonstrating its capacity to compute the joint shear strength more accurately. Similar to how the performance factor (PF) derived from the suggested equation is slightly over 1, it attests to the model's effectiveness. The suggested model, as illustrated in Fig 10, produces less dispersion than previous models (CoV=14.6 percent).

## Roof Corner Joint Opening

Table 3 displays a statistical comparison of the opening behavior of several models. It's interesting to note that the suggested model once again has a greater coefficient of determination ($R^2$=0.97) than any of the other models presented in Table 3. Although the slope (m=0.96) of the suggested model's best-fit line is somewhat less steep than that of Mogili et al. (2020)'s model, other signs point to the proposed model as being somewhat more accurate and trustworthy. The statistical parameters for every alternative model provided in Table 3 are quite different from the suggested model. The suggested model is statistically well-founded and well-organized when all the statistical parameters shown in this table are taken into consideration.

Thus, it can be argued that the suggested model is capable of accounting for numerous geometrical aspects, leading to an appropriate calculation of joint shear strength under both closure and opening moments.

*Table 2. Statistical evaluation under a closing moment*

| Author | $PF = \frac{v_{jh}^{Exp}}{v_{jh}^{Est}}$ | Std. Deviation | COV (%) | AAE (%) | $R^2$ | m |
|---|---|---|---|---|---|---|
| | Mean | | | | | |
| Mogili et al. (2020) | 1.13 | 0.190 | 17.3 | 15.0 | 0.97 | 0.86 |
| Kim and LaFave (2012)] | 1.28 | 0.28 | 28.0 | 21.0 | 0.95 | 0.75 |
| ACI 352R-02 (2002) | 0.90 | 0.180 | 20.0 | 22.0 | 0.96 | 1.07 |
| EN 1998-1-2004 | 0.44 | 0.101 | 23.0 | 141.8 | 0.94 | 2.23 |
| GB 50010-2011 | 0.34 | 0.083 | 24.5 | 212.6 | 0.94 | 1.93 |
| NZS 3101-2006 | 0.51 | 0.125 | 24.5 | 108.4 | 0.94 | 2.89 |
| AIJ (1999) | 0.98 | 0.210 | 23.0 | 19.7 | 0.95 | 0.98 |
| Megget (1998) | 0.34 | 0.240 | 24.0 | 30.4 | 0.94 | 1.16 |
| McConnell and Wallace (1995) | 1.03 | 0.224 | 22.0 | 17.8 | 0.95 | 0.80 |
| Wallace et al. (1998) | 1.20 | 0.260 | 22.0 | 19.0 | 0.96 | 0.80 |
| Angelakos (1999) | 1.20 | 0.260 | 22.0 | 19.0 | 0.96 | 0.80 |
| Priestly et al. (1996) | 1.26 | 0.274 | 22.0 | 20.9 | 0.95 | 0.76 |
| **Proposed** | **1.01** | **0.146** | **14.6** | **11.7** | **0.98** | **0.97** |

*Figure 9. Comparison of the accuracy of the various models*

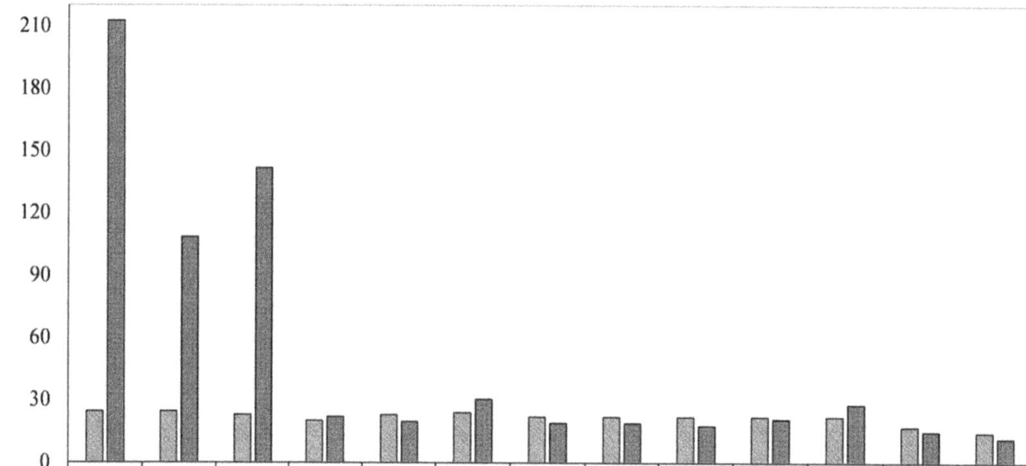


| | GB 50010-2010 | NZS 3101-2006 | EN 1998-1: 2004 | ACI 352R-02 | AIJ | Megget | Wallace et.al | Angelakos | McConnel et al. | Priestly et al. | Kim et al. | Mogili et al. | Proposed Model |
|---|---|---|---|---|---|---|---|---|---|---|---|---|---|
| COV (%) | 24.5 | 24.5 | 23.0 | 20.2 | 23.0 | 24.0 | 22.0 | 22.0 | 22.0 | 22.0 | 22.0 | 17.3 | 14.6 |
| AAE (%) | 212.6 | 108.4 | 141.8 | 22.0 | 19.7 | 30.4 | 19.0 | 19.0 | 17.8 | 20.9 | 28.0 | 15.0 | 11.7 |

## CONCLUSION

In this work, predictive shear strength models for RC roof corner joints are created using genetic analysis. The suggested models seek to characterize the shear capacity of the roof corner joints during opening and closure moments.

This study used genetic -based non-linear empirical data from earlier experiments to account for the factors that affect the shear capacity of the roof corner joint, such as concrete compressive strength, steel tensile strength, longitudinal tensile reinforcement ratio, geometric properties of structural elements, and joint aspect ratio. The model produces more precise estimates of joint shear capacity than the models put out by prior researchers. More specifically, the average absolute error (AAE) of the present model for the roof corner joint under the closure and opening behaviors is 11.7 and 12.6 percent, respectively. Similar to this, results obtained by the suggested model are less scattered when CoV values of 14.6 percent and 22 percent are used. The suggested model is more dependable since it has a larger coefficient of determination than those previously anticipated models, which are less, at 97 percent and 98 percent for the opening and close, respectively. It is determined that the roof corner joint's performance factor for the opening and closing behaviors is, respectively, 1.04 and 1.01, which is not far from the reference value 1.

Compared to models suggested by several important design codes, the present model provides findings that are more accurate when estimating the shear capacity of the roof corner joint. When compared to past test data and shear strength prediction equations of various design codes, the results are drastically different. Despite the overestimation of the roof corner joint's shear strength, a few codes, such ACI 352R-02 and AIJ, demonstrate a comparatively better calculation of the closing shear strength alone. These calculations continue to overestimate the joint shear capacity for the opening shear strength. When the RC roof corner joint's shear capacity is calculated using EN 1998-1:2004, NZS 3101-2006, and GB50010-2010, an overestimation is shown.

*Figure 10a. Shear Strength Prediction for RC Roof corner Joint Under Closing Behavior Using Different Proposed Equations (a-d)*

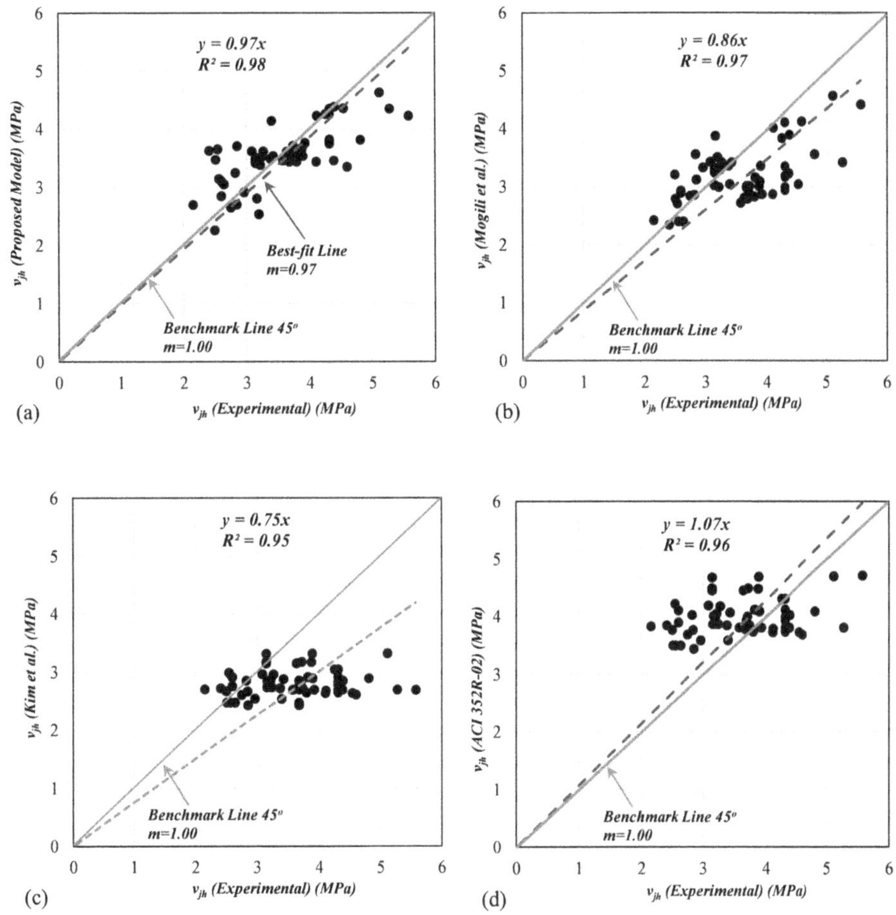

*Figure 10b. Shear Strength Prediction for RC Roof corner Joint Under Closing Behavior Using Different Proposed Equations (e-h)*

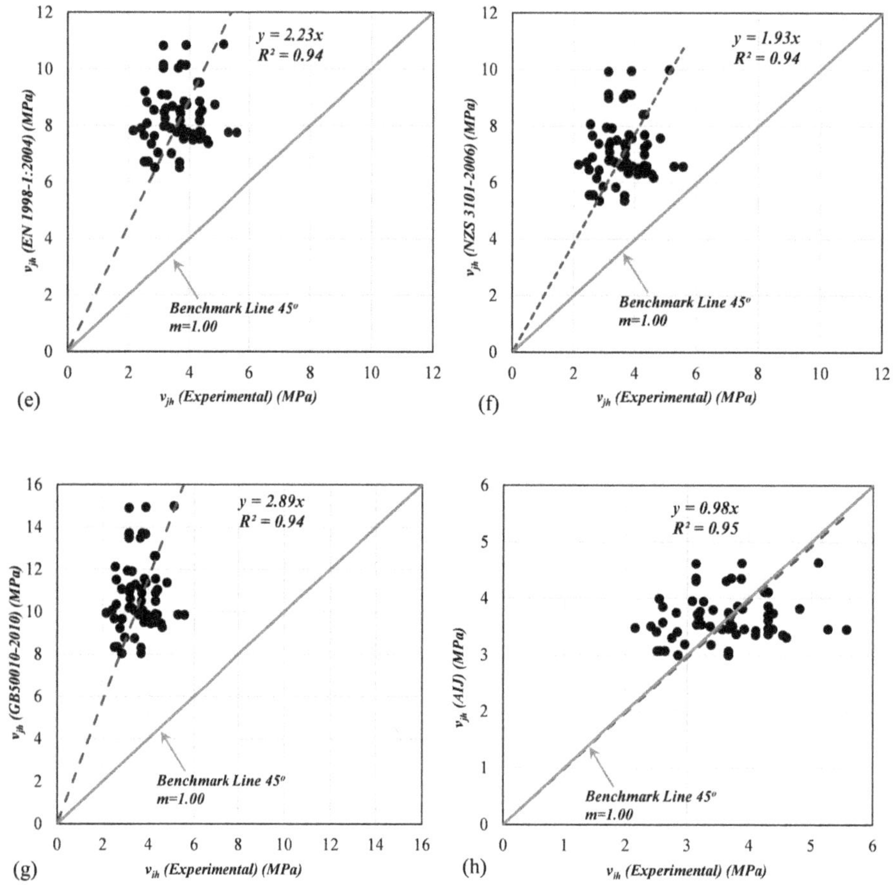

*Figure 10c. Shear Strength Prediction for RC Roof corner Joint Under Closing Behavior Using Different Proposed Equations (i-l)*

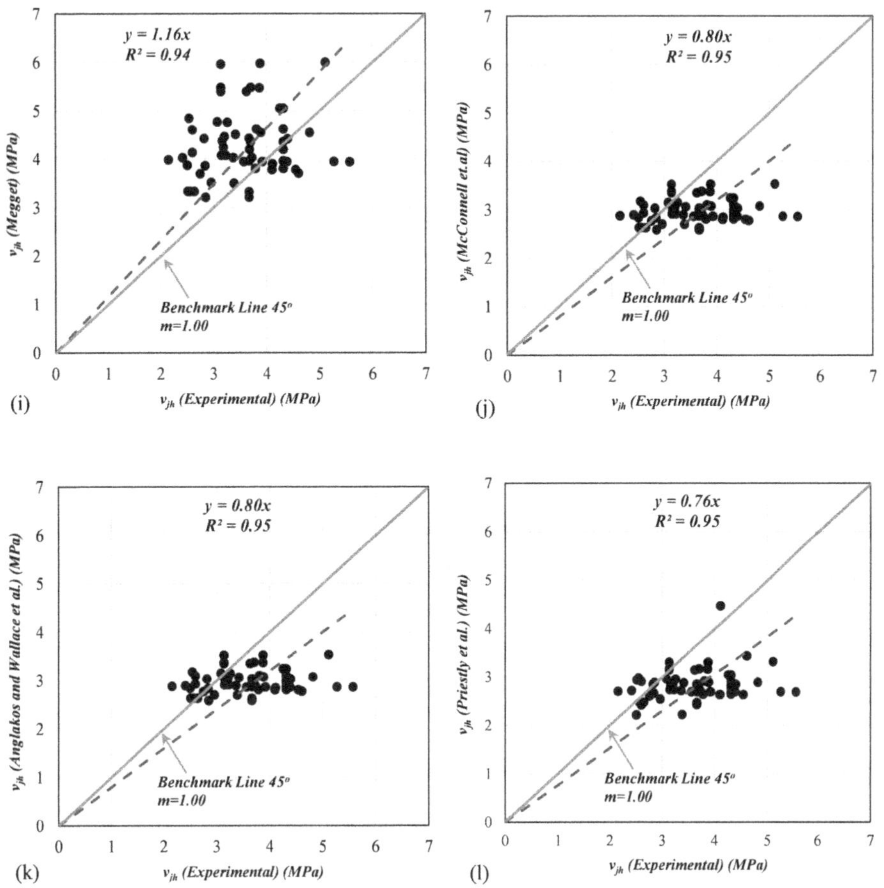

*Table 3. Different shear strength prediction models' statistical performance in predicting opening behavior*

| Author | $PF = \frac{v_{jh}^{Exp}}{v_{jh}^{Est}}$ | Std. Deviation | COV (%) | AAE (%) | $R^2$ | m |
|---|---|---|---|---|---|---|
| | Mean | | | | | |
| Mogili et al. (2020) | 0.98 | 0.210 | 22.0 | 18.0 | 0.94 | 0.96 |
| Kim and LaFave (2012) | 0.79 | 0.260 | 31.0 | 46.0 | 0.89 | 1.17 |
| ACI 352R-02 (2002) | 0.55 | 0.160 | 29.0 | 101.0 | 0.90 | 1.67 |
| EN 1998-1-2004 | 0.27 | 0.097 | 36.0 | 323.0 | 0.87 | 3.45 |
| GB 50010-2011 | 0.21 | 0.079 | 37.0 | 450.3 | 0.86 | 4.45 |
| NZS 3101-2006 | 0.31 | 0.113 | 36.0 | 266.8 | 0.86 | 2.97 |
| AIJ (1999) | 0.61 | 0.200 | 33.0 | 87.3 | 0.88 | 1.52 |
| Megget (1998) | 0.62 | 0.214 | 34.0 | 81.4 | 0.86 | 1.48 |
| McConnell and Wallace (1998) | 0.63 | 0.192 | 31.0 | 73.6 | 0.89 | 1.44 |
| Wallace et al. (1998) | 0.73 | 0.223 | 31.0 | 51.4 | 0.89 | 1.44 |
| Angelakos (1999) | 0.73 | 0.223 | 31.0 | 51.4 | 0.89 | 1.44 |
| Priestly et al. (1996) | 0.77 | 0.250 | 32.4 | 47.1 | 0.89 | 1.19 |
| **Proposed** | **1.04** | **0.230** | **22.0** | **12.6** | **0.97** | **0.93** |

*Figure 11. Comparison of the model's accuracy in terms of opening moment's coefficient of variation (CoV) and average absolute error (AAE)*

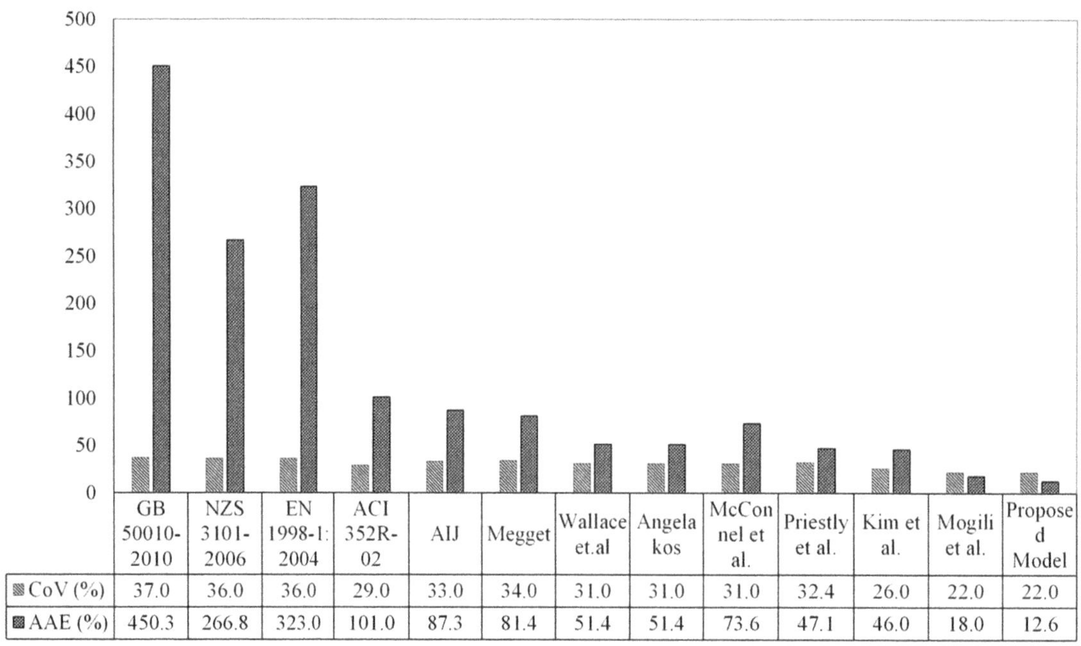

*Figure 12a. Shear strength prediction for RC roof corner joint under opening behavior using different proposed equations (a-d)*

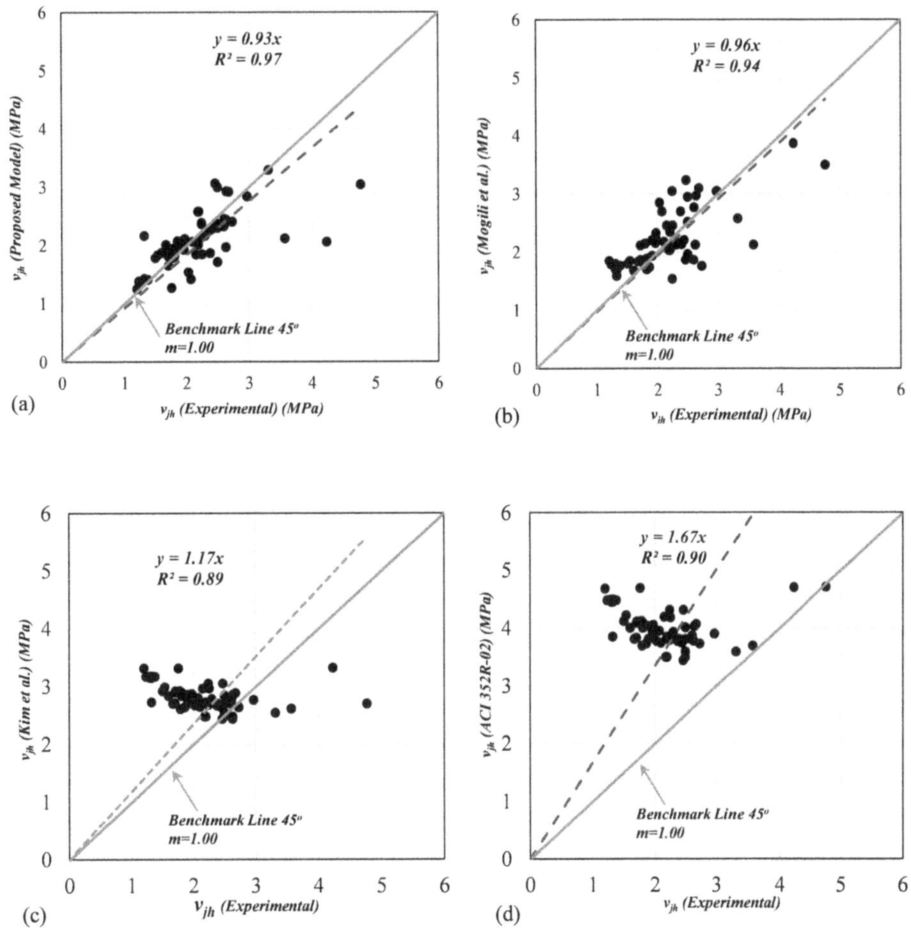

*Figure 12b. Shear strength prediction for RC roof corner joint under opening behavior using different proposed equations (e-h)*

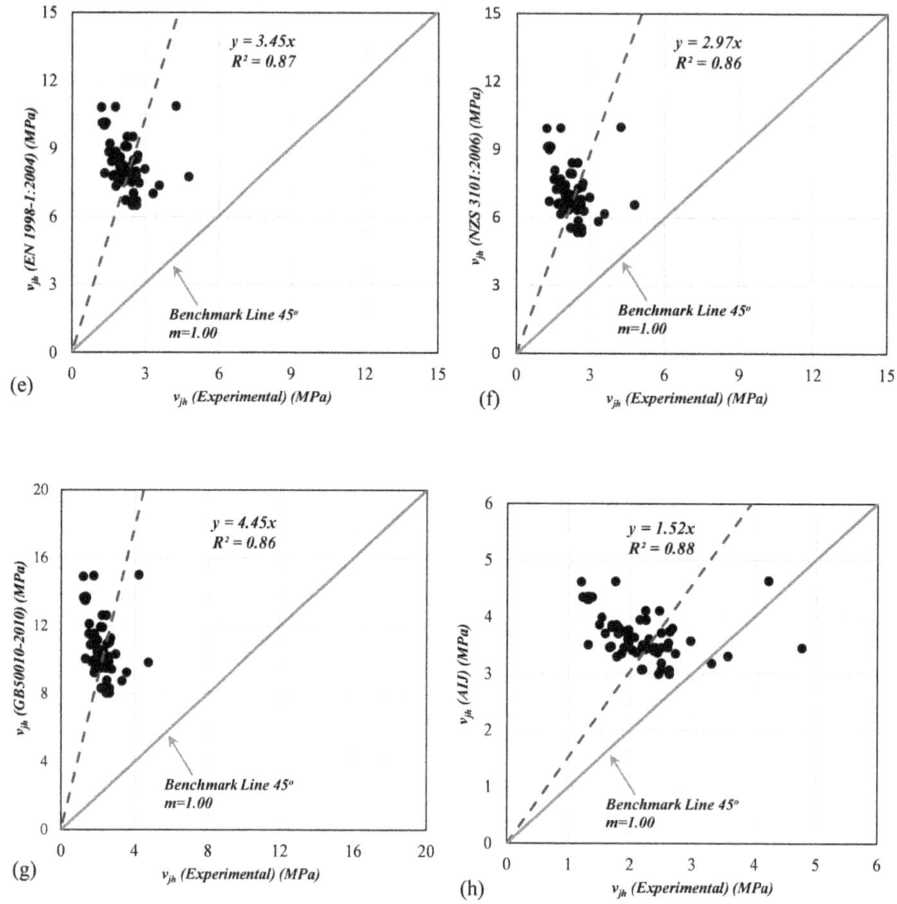

*Figure 12c. Shear strength prediction for RC roof corner joint under opening behavior using different proposed equations (i-l)*

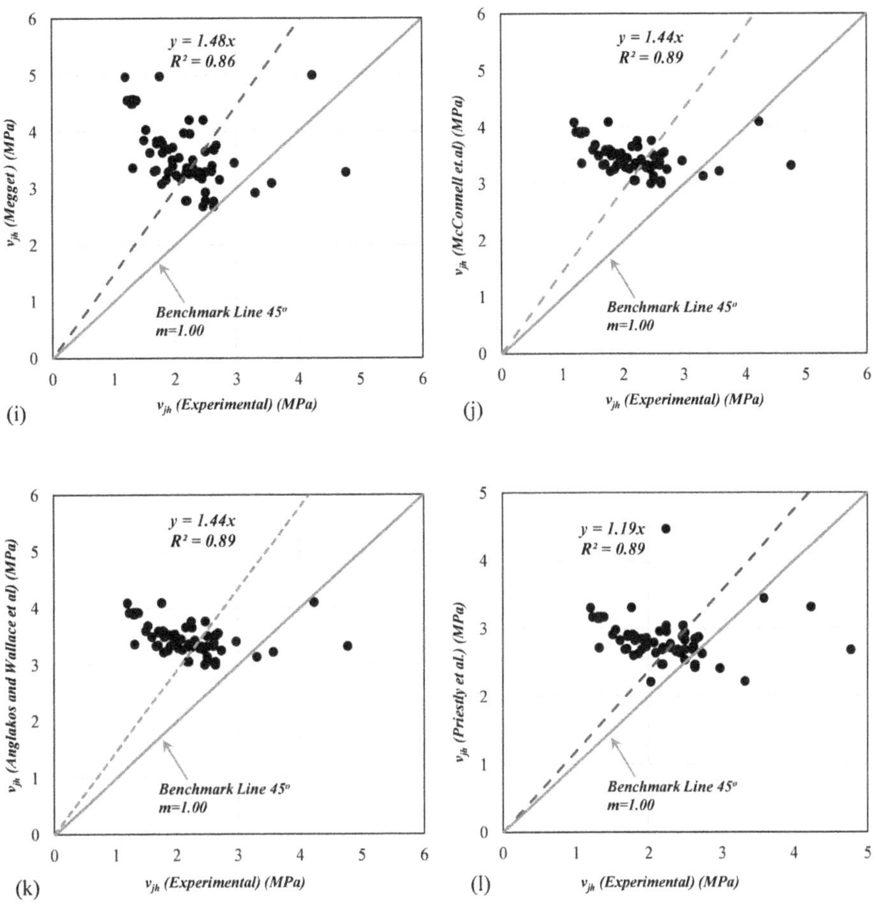


*Table 4. t-Statistics results*

| Description | Opening | | Closing | |
|---|---|---|---|---|
| | $v_{jh}^{Exp}$ (MPa) | $v_{jh}^{Est}$ (MPa) | $v_{jh}^{Exp}$ (MPa) | $v_{jh}^{Est}$ (MPa) |
| Mean | 2.16 | 2.17 | 3.58 | 3.54 |
| Variance | 0.38 | 0.38 | 0.56 | 0.23 |
| Observations | 61 | 61 | 61 | 61 |
| Pearson Correlation | 0.76 | | 0.73 | |
| Hypothesized Mean Difference | 0.00 | | 0.00 | |
| Df | 60 | 60 | 60 | 60 |
| t-Stat | -0.28 | | 0.61 | |
| t-Critical two-tail | 2.00 | 2.00 | 2.00 | 2.00 |

## REFERENCES


Akguzel, U., & Pampanin, S. (2010). Effects of variation of axial load and bidirectional loading on seismic performance of GFRP retrofitted reinforced concrete exterior beam-column joints. *Journal of Composites for Construction*, *14*(1), 94–104. doi:10.1061/(ASCE)1090-0268(2010)14:1(94)

Akguzel, U., & Pampanin, S. (2012). Assessment and design procedure for the seismic retrofit of reinforced concrete beam-column joints using FRP composite materials. *Journal of Composites for Construction*, *16*(1), 21–34. doi:10.1061/(ASCE)CC.1943-5614.0000242

American Concrete Institute. (1991). *ACI committee 352, Recommendations for design of beam-column joints in monolithic reinforced concrete structures (ACI 352R-91)*. ACI.

American Concrete Institute. (1998). *ACI committee 352, Recommendations for design of beam-column joints in monolithic reinforced concrete structures (ACI 352R-98)*. ACI.

American Concrete Institute. (2010). *ACI committee 352, Recommendations for design of beam-column joints in monolithic reinforced concrete structures (ACI 352R-10)*. ACI.

Anderson, M., Lehman, D., & Stanton, J. (2008). A cyclic shear stress-strain model for joints without transverse reinforcement. *Engineering Structures*, *30*(4), 941–954. doi:10.1016/j.engstruct.2007.02.005

Angelakos B. (1999). *The behavior of reinforced concrete roof corner joints under earthquake loads* [Ph.D dissertation]. University of Toronto.

Architectural Institute of Japan (AIJ). (1999). *Structural design guidelines for reinforced concrete buildings*. AIJ.

Ascheim, M., & Moehle, J. P. (1992). *Shear Strength and Deformability of RC Bridge Columns Subjected to Inelastic Cyclic Displacements*. Calif. Dept. Transp. Rep. No. UCBIEERC-92J04. Earthq. Eng. Res. Center, University of California, Berkeley.

Bakir, P. G., & Boduroğlu, H. M. (2002). A new design equation for predicting the joint shear strength of monotonically loaded exterior beam-column joints. *Engineering Structures*, *24*(8), 1105–1117. doi:10.1016/S0141-0296(02)00038-X

Borghini, A., Gusella, F., & Vignoli, A. (2016). Seismic vulnerability of existing R.C. buildings: A simplified numerical model to analyze the influence of the beam-column joints collapse. *Engineering Structures*, *121*, 19–29. doi:10.1016/j.engstruct.2016.04.045

Celik, O. C., & Ellingwood, B. R. (2010). Seismic fragilities for non-ductile reinforced concrete frames - Role of aleatoric and epistemic uncertainties. *Structural Safety*, *32*(1), 1–12. doi:10.1016/j.strusafe.2009.04.003

[] Cote, P.A., & Wallace, J.W. (1994). *A study of reinforced concrete roof corner-joints subjected to cyclic lateral loading*. Rep No CU/CEE-94/4. ASCE Library.

Del Vecchio, C., Del Zoppo, M., Di Ludovico, M., Verderame, G. M., & Prota, A. (2017). Comparison of available shear strength models for non-conforming reinforced concrete columns. *Engineering Structures*, *148*, 312–327. doi:10.1016/j.engstruct.2017.06.045

Elwood, K. J. (2004). Modelling failures in existing reinforced concrete columns. *Canadian Journal of Civil Engineering*, *31*(5), 846–859. doi:10.1139/l04-040

Elwood, K. J., & Moehle, J. P. (2005). Axial capacity model for shear-damaged columns. *ACI Structural Journal*, *102*(4), 578–587. doi:10.14359/14562

Elwood, K., & Moehle, J.P. (2003). *Shake table tests and analytical studies on the gravity load collapse of reinforced concrete frames.* University of California, Berkeley. https://doi.org/ doi:10.16953/deusbed.74839

EN 1998-1, Eurocode 8. (2001). *Design of structures for earthquake resistance - Part 1: General rules, seismic actions and rules for buildings.* CENEuropean Committee for Standardization.

Favvata, M. J., Izzuddin, B. A., & Karayannis, C. G. (2008). Modelling exterior beam-column joints for seismic analysis of RC frame structures. *Earthquake Engineering & Structural Dynamics*, *37*(13), 1527–1548. doi:10.1002/eqe.826

Favvata, M. J., & Karayannisa, C. G. (2014). Influence of pinching effect of exterior joints on the seismic behavior of RC frames. *Earthquakes and Structures*, *6*(1), 89–110. doi:10.12989/eas.2014.6.1.089

Ho, J. C. M., & Pam, H. J. (2003). Inelastic design of low-axially loaded high-strength reinforced concrete columns. *Engineering Structures*, *25*(8), 1083–1096. doi:10.1016/S0141-0296(03)00050-6

Hsu, T. T. C. (1988). Softened truss model theory for shear and torsion. *ACI Structural Journal*, *85*(6), 624–635. doi:10.14359/2740

Hwang, S. J., & Lee, H. J. (1999). Analytical model for predicting the shear strength of exterior reinforced concrete beam-column joints. *ACI Structural Journal*, *96*, 846–858.

Jeon, J. S., Lowes, L. N., DesRoches, R., & Brilakis, I. (2015). Fragility curves for non-ductile reinforced concrete frames that exhibit different component response mechanisms. *Engineering Structures*, *85*, 127–143. doi:10.1016/j.engstruct.2014.12.009

Kim, J., & LaFave, J.M. (2007). Key influence parameters for the joint shear behavior of reinforced concrete (RC) beam-column connections. *Eng Struct, 29*, 2523–39. https://doi.org/ruct.2006.12.012 doi:10.1016/j.engst

Kim, J., & LaFave, J. M. (2012). A simplified approach to joint shear behavior prediction of RC beam-column connections. *Earthquake Spectra*, *28*(3), 1071–1096. doi:10.1193/1.4000064

Koza, J. R. (1994). Genetic programming as a means for programming computers by natural selection. *Statistics and Computing*, *4*(2), 87–112. doi:10.1007/BF00175355

Laterza, M., D'Amato, M., & Gigliotti, R. (2017). Modeling of gravity-designed RC sub-assemblages subjected to lateral loads. *Eng Struct J, 130*, 242–60. doi:10.1016/j.engstruct.2016.10.044

Lowes, L. N., & Altoontash, A. (2003). Modeling reinforced-concrete beam-column joints subjected to cyclic loading. *Journal of Structural Engineering*, *129*(12), 1686–1697. doi:10.1061/(ASCE)0733-9445(2003)129:12(1686)

Lynn, A. C., Moehle, J. P., Mahin, S. A., & Holmes, W. T. (1996). Seismic evaluation of existing reinforced concrete building columns. *Earthquake Spectra*, *12*(4), 715–739. doi:10.1193/1.1585907

Mazzoni, S., Moehle, J. P., & Thewalt, C. R. (1991). *Cyclic response of RC beam-column roof corner joints test and retrofit.* Rep. No. UCB/EERC-91/14. Research Gate.

McConnell, S. W., & Wallace, J. W. (1995). *Behavior of Reinforced Concrete Beam-Column Roof corner-Joints Subjected to Reversed Cyclic Loading.* Rep. No. CU/CEE-95/07. NIH.

Megget, L. M. (1998). The seismic behavior of small reinforced concrete beam-column roof corner joints. *Bull New Zeal Natl Soc Earthq Eng*, *31*(4), 215–245. doi:10.5459/bnzsee.31.4.215-245

Meggeta, L. M. (2003). The Seismic Design and Performance of Reinforced Concrete Beam-Column Roof corner Joints in Buildings. *Earthquake Spectra*, *19*(4), 863–895. doi:10.1193/1.1623782

Mitra, N., & Lowes, L. N. (2007). Evaluation, calibration, and verification of a reinforced concrete beam-column joint model. *Journal of Structural Engineering*, *133*(1), 105–120. doi:10.1061/(ASCE)0733-9445(2007)133:1(105)

Mogili, S., & Kuang, J. S. (2018). Cyclic Vulnerability of RC Roof corner Connections Under Relatively High Opening Shear Stresses. In *16th Eur Conf Earthq Eng.* NICEE.

Mogili, S., & Kuang, J.S. (2019). Reversed cyclic performance of reinforced concrete roof corner joints under variable closing and opening stresses. *Eng Struct, 178,* 116–27. https://doi.org/ struct.2018.10.025 doi:10.1016/j.eng

Mogili, S., Kuang, J. S., & Hwang, S. J. (2020). Predicting Shear Strength of Reinforced Concrete Roof corner Joints in Closing and Opening Actions. *Journal of Structural Engineering*, *146*(6), 1–11. doi:10.1061/(ASCE)ST.1943-541X.0002633

Moretti, M., & Tassios, T. P. (2007). Behavior of short columns subjected to cyclic shear displacements: Experimental results. *Engineering Structures*, *29*(8), 2018–2029. doi:10.1016/j.engstruct.2006.11.001

Moretti, M. L., Tassios, T. P., & Vintzileou, E. (2014). Behavior and design of corner joints under opening bending moment. *ACI Structural Journal*, *111*(1), 3–13. doi:10.14359/51686514

Navidi, W. (2019). *Statistics for Engineers and Scientists* (5th ed.). McGraw-Hill. doi:10.1017/CBO9781107415324.004

Niroomandi, A., Maheri, A., Maheri, M.R., & Mahini, S.S. (2010). Seismic performance of ordinary RC frames retrofitted at joints by FRP sheets. *Eng Struct, 32*, 2326–36. https://doi.org/.engstruct.2010.04.008 doi:10.1016/j

Niroomandi, A., Najafgholipour, M. A., & Ronagh, H. R. (2014). Numerical investigation of the affecting parameters on the shear failure of Nonductile RC exterior joints. *Engineering Failure Analysis*, *46*, 62–75. doi:10.1016/j.engfailanal.2014.08.003

Niroomandi, A., Pampanin, S., & Dhakal, R. (2015). Comparison of alternative assessment procedures to predict seismic performance of RC columns. In *Tenth Pacific Conf. Earthq. Eng. (PCEE 2015),* (p. 1–10). AEES.

Nouali, A., & Matallah, M. (2017). A simplified approach to assess the size effect on the shear-flexure interaction in RC elements. *Engineering Structures*, *144*, 151–162. doi:10.1016/j.engstruct.2017.04.057

NZS 3101, Code of practice for the design of concrete structure (NZS 3101-2006), Standard association Newzealands, 2006.

Omidi, M., & Behnamfar, F. (2015). A numerical model for simulation of RC beam-column connections. *Engineering Structures*, *88*, 51–73. doi:10.1016/j.engstruct.2015.01.025

Osorio, E., Bairán, J. M., & Marí, A. R. (2017). Analytical modeling of reinforced concrete columns subjected to bidirectional shear. *Engineering Structures*, *138*, 458–472. doi:10.1016/j.engstruct.2017.02.029

Ousalem, H., Kabeyasawa, T., Tasai, A., & Ohsugi, Y. (2002). Experimental study on seismic behavior of reinforced concrete columns under constant and variable axial loadings. *Proc Japan Concr Inst*, *24*, 229–234.

Pampanin, S., Calvi, G. M., & Moratti, M. (2002). Seismic behavior of RC beam-column joints designed for gravity loads. In *12th Eur Conf Earthq Eng.* (vol. 726, pp. 1–10). University of Canterbury.

Pampanin, S., Magenes, G., & Carr, A. (2003). Modelling of shear hinge mechanism in poorly detailed RC beam-column joints. In *Proc. Fib Symp. 2003 Concr. Struct. Seism. Reg.* (pp. 126–7). Research Gate.

Pang, X. B., & Hsu, T. T. C. (1996). Fixed angle softened truss model for reinforced concrete. *ACI Structural Journal*, *93*, 197–207.

Park, H. G., Choi, K. K., & Wight, J. K. (2007). Strain-based shear strength model for slender beams without web reinforcement. *ACI Structural Journal*, *104*, 638–639.

Park, H. G., Kang, S., & Choi, K.-K. (2013). Analytical model for shear strength of ordinary and prestressed concrete beams. *Engineering Structures*, *46*, 94–103. doi:10.1016/j.engstruct.2012.07.015

Park, H. G., Yu, E. J., & Choi, K. K. (2012). Shear-strength degradation model for RC columns subjected to cyclic loading. *Engineering Structures*, *34*, 187–197. doi:10.1016/j.engstruct.2011.08.041

Pauletta, M., Di Luca, D., & Russo, G. (2015). Exterior beam column joints - Shear strength model and design formula. *Engineering Structures*, *94*, 70–81. doi:10.1016/j.engstruct.2015.03.040

Priestley, M. J. N., Seible, F., & Calvi, G. (1996). *Seismic Design and Retrofit of Bridge* (1st ed.). John Wiley & Sons, Inc. doi:10.1002/9780470172858

Priestley, M.J.N., Verma, R., & Xiao, Y. (1994). Seismic shear strength of reinforced concrete columns. *J Struct Eng, 120*, 2310–29. https://doi.org/120:8(2310) doi:10.1061/(ASCE)0733-9445(1994)

Sezen, H. (2002). *Seismic behavior and modeling of reinforced concrete building columns*. University of California at Berkeley.

Sezen, H., & Moehle, J. P. (2004). Shear strength model for lightly reinforced concrete columns. *Journal of Structural Engineering*, *130*(11), 1692–1703. doi:10.1061/(ASCE)0733-9445(2004)130:11(1692)

Shafaei, J., & Nezami, S. A. (2019). Effect of different size of joint enlargement on seismic behavior of gravity load designed RC beam-column connections. *Struct Des Tall Spec Build, 28,* 1–27. doi:10.1002/tal.1653

Shafaei, J., Zareian, M. S., Hosseini, A., & Marefat, M. S. (2014). Effects of joint flexibility on lateral response of reinforced concrete frames. *Engineering Structures*, *81*, 412–431. doi:10.1016/j.engstruct.2014.09.046

Sharma, A., Eligehausen, R., & Reddy, G. R. (2011). A new model to simulate joint shear behavior of poorly detailed beam-column connections in RC structures under seismic loads, Part I: Exterior joints. *Engineering Structures*, *33*(3), 1034–1051. doi:10.1016/j.engstruct.2010.12.026

Shayanfar, J., Akbarzadeh Bengar, H., & Niroomandi, A. (2016). A proposed model for predicting non-linear behavior of RC joints under seismic loads. *Materials & Design*, *95*, 563–579. doi:10.1016/j.matdes.2016.01.098

Shayanfar, J., & Bengar, H. A. (2016). Numerical model to simulate shear behavior of RC joints and columns. *Computers and Concrete*, *18*(6), 877–901. doi:10.12989/cac.2016.18.6.877

Shayanfar, J., & Bengar, H.A. (2018). A practical model for simulating non-linear behavior of FRP strength ened RC beam-column joints. *Steel Compos Struct, 27,* 49–74. https://doi.org/.2018.27.1.049 doi:10.12989/scs

Shin, M., & Lafave, J. M. (2004). Modeling of cyclic joint shear deformation contributions in RC beam-column connections to overall frame behavior. *Structural Engineering and Mechanics*, *18*(5), 645–669. doi:10.12989em.2004.18.5.645

Standard, A. A. C. I. (2019). *An ACI Standard*. ACI.

The People's Republic of China National Standard. (2010). GB50010-2010: Code for Seismic Design of Buildings. Author.

Wallace, J. W., McConnell, S. W., Gupta, P., & Cote, P. A. (1998). Use of headed reinforcement in beam-column joints subjected to earthquake loads. *ACI Structural Journal*, *95*(5), 590–606. doi:10.14359/574 PMID:31500274

Wang, G. L., Dai, J. G., & Teng, J. G. (2012). Shear strength model for RC beam-column joints under seismic loading. *Engineering Structures*, *40*, 350–360. doi:10.1016/j.engstruct.2012.02.038

Wong, H. F. (2005). *Shear strength and seismic performance of non-seismically designed reinforced concrete beam-column joints*. Hong Kong Univ Sci Technol. doi:10.14711/thesis-b914043

Wong, S.H.F., & Kuang, J.S. (2011). Predicting shear strength of RC exterior beam-column joints by modified rotating-angle softened-truss model. *Comput Concr, 8*, 59–70. https://doi.org/ac.2011.8.1.059 doi:10.12989/c

Zhang, N. (2017). Seismic Performance and Shear Strength of Reinforced Concrete Beam-column Roof corner Joints [Ph.D. Dissertation]. The Hong Kong University of Science and Technology.

Zhang, N., Kuang, J. S., & Mogili, S. (2017). Cyclic Behavior of Reinforced Concrete Beam-Column Roof corner Joints. In *16th World Conf. Earthq. Eng. 16WCEE*. Semantic Scholar.

# Chapter 6
# Predicting the Fundamental Period of Light-Frame Wooden Buildings by Employing Bat Algorithm-Based Artificial Neural Network

**Mehdi Nikoo**
*Department of Building, Civil and Environmental Engineering, Concordia University, Montréal, Canada*

**Ghazanfarah Hafeez**
https://orcid.org/0000-0002-5045-6968
*Department of Building, Civil and Environmental Engineering, Concordia University, Montréal, Canada*

**Ghasan Doudak**
*Department of Civil Engineering, University of Ottawa, Ottawa, Canada*

**Vagelis Plevris**
https://orcid.org/0000-0002-7377-781X
*Department of Civil and Architectural Engineering, Qatar University, Doha, Qatar*

## ABSTRACT

*The study utilizes an artificial neural network model for determining the fundamental period of Light-Frame Wooden Buildings, employing the Bat algorithm on a data set of 71 measured periods of wooden buildings. The number of stories, floor area, storey height, maximum length, and maximum width are selected as input parameters to estimate the fundamental period of light-frame wooden buildings. The accuracy and the competitiveness of the developed model were evaluated by comparing it with a similar particle swarm optimization (PSO)- ANN scheme, the formulas provided in the National Building Code of Canada, an equation obtained from the Eureqa software, and a non-linear regression (NLR) model. The results of the research show that the bat-ANN model exhibited the best overall performance with the lowest RMSE and MAE error values and the highest values of the Coefficient of determination, $R^2$, in comparison to the other examined models. Therefore, the proposed Bat-ANN model can be considered as a reliable, robust, and accurate tool for predicting the fundamental period of wooden buildings.*

DOI: 10.4018/978-1-6684-5643-9.ch006

## INTRODUCTION AND BACKGROUND

Reliable seismic analysis and design of a structure require precise knowledge about the fundamental period of the building, among other dynamic characteristics. Building codes around the world provide simplified formulas for estimating the fundamental periods of buildings constructed with different materials, structural systems and geometries (Panthi et al., 2021). The National Building Code of Canada (NBCC) provides empirical formulas to calculate the fundamental lateral period of vibration of different structural systems, such as shear walls, steel moment-resisting frames, reinforced concrete moment resisting frames, and steel braced frames (NRC/IRC, 2015). The formula for shear walls can also be applied to what is termed in the code as "other structures" and it is only a function of building height above the base, $h_n$, as presented in Eq. (1).

$$T_a = 0.05h_n^{3/4} \tag{1}$$

Proper estimates of the fundamental period is imperative in seismic analysis and design as it is the basis for the calculation of the design base shear and any error in estimating the fundamental period could have significant consequences in over- or under- estimating the seismic forces, leading to an overly conservative or unsafe design. An alternative method for calculating the building period, which is permitted by the NBCC is using numerical methods, such as Rayleigh's method or eigenvalue analysis. The drawback of using such methods is their association with uncertainty in input parameters, especially pertaining to the stiffness of the structural system, as well as some limitations related to lower bound limits imposed on the base shear.

It has been well recognized that considering the building height as the only variable describing the dynamic characteristics of a building and to approximate the period based only on the height is not sufficient. Several investigations have been conducted to evaluate the adequacy of these equations on concrete, steel and wooden buildings, and several authors have suggested improved equations. The subsequent section provides the state-of-the-art knowledge on building periods established based on field measurements and using finite element analysis methods. In the present study, an ANN model employing the Bat Algorithm is developed to provide a better prediction of the fundamental period of light-frame wooden buildings including several additional parameters to the building height. While similar approaches can be found in the literature, the methodologies used in these studies are limited to mainly concrete structures (Plevris & Solorzano, 2021; Solorzano & Plevris, 2021). In the current study, the model exhibiting the best performance is selected from a set of examined ANN models and subsequently compared with other models, such as ANN-Particle Swarm Optimization, Nonlinear regression, the NBCC (NRC/IRC, 2015), as well as equations derived from the Eureqa software (Dubčáková, 2011; Schmidt & Lipson, 2009).

## LITERATURE REVIEW

The majority of the available literature on experimental studies with aim to investigate the fundamental period of existing buildings have been focused on concrete structures, while limited studies are available for steel and wooden buildings. The proposed equations have primarily depended on the seismic force resisting system (SFRS) typology, material types which relates to the building mass, and overall building geometry. Farsi and Bard (Farsi & Bard, 2004), Lagomarsino (Lagomarsino, 1993) and Gilles (Gilles, 2011) performed ambient vibration tests (AVT) on several reinforced concrete shear wall buildings and suggested improved relationships for estimating the building periods required for the calculation of the base shear in seismic design. The proposed equations were limited to linear expressions with the height of the buildings. The period equations suggested by other authors for reinforced concrete shear wall buildings, such as Goel and Chopra (Goel Rakesh & Chopra Anil, 1997), Lee et al. (Lee et al., 2000), Morales (Dominguez Morales, 2000) provided expressions with multiple variables and were not limited to building height.

The study undertaken by Goel and Chopra (Goel Rakesh & Chopra Anil, 1997) assessed the code period equation for shearwall buildings specified in the US building codes (NEHRP 94, SEAOC 96 and UBC 1997) by investigating the fundamental periods measured during strong ground motions. The authors reported a significant scatter in the measured period and suggested an improved expression based on regression analyses of experimental results from nine shear wall buildings that included variables related to shear wall dimensions, such as height and equivalent shear area (Eq. (3) in Table 1). Similarly, Lee et al. (Lee et al., 2000) focused on evaluating the period formula in the Korean building code by employing fundamental period data-set of fifty reinforced concrete shear wall buildings. The periods were measured using ambient vibration testing, and the proposed relationship for estimating the period derived for a uniform cantilever shear beam is shown in Eq. (4).

Morales (Dominguez Morales, 2000) evaluated the code period equation for concrete shear walls and frame buildings and proposed an equation that accounted for the moment of inertia (equivalent second moment of area) and height of the building. The equation was linear regarding the height but nonlinear with the moment of inertia and demonstrated a reasonable correlation with the measured periods. The proposed equation (Eq. (6) in Table 1) was derived from the database of the buildings whose fundamental periods were measured during strong earthquake motions by the California Strong Motion Instrumentation Program (CSMIP) and by the National Oceanic and Atmospheric Administration (NOAA). It should be noted that the dataset used in this study was mostly similar to that employed by Goel and Chopra (Goel Rakesh & Chopra Anil, 1998) in establishing Eq. (3).

Gilles et al. (Gilles et al., 2011) created a database of 27 reinforced concrete buildings and proposed an improved relationship for the fundamental period of concrete shear wall buildings as a function of the building height. The improved equation was found to provide better estimates when compared to the equations provided in the code and other improved models available in the literature. Models to estimate the vibration period of torsion, second modes and models that could account for the uncertainty associated with fundamental period were also proposed in this study.

Efforts have also been made to investigate the applicability of code equations to wooden shear wall buildings. Camelo (Camelo et al., 2001) proposed a period formula based on a non-linear expression as a function of the building height. The proposed equation was capable of predicting the period of wooden buildings more accurately than the period formula provided in Uniform Building Code (UBC 97) (Paz & Leigh, 2004). Hafeez et al. (Hafeez et al., 2018, 2019; Hafeez et al., 2014) developed a comprehensive database of measured dynamic properties of light-frame wooden buildings based on ambient vibration tests. The study assessed the adequacy of the NBCC code formula for approximating the fundamental period of wooden shear wall buildings. The suggested equation was non-linear with building height, shear wall length and floor area.

Table 1 summarizes the experimental studies that provided alternative expressions to calculate the fundamental period of shear wall buildings for different materials. The studies reported in Table 1 emphasize the need to include more sophisticated approaches with several parameters in order to obtain a better estimate of the fundamental period. One of the issues associated with testing several buildings in order to develop an approximation of their behaviour is that the buildings are very complex systems that even when multiple parameters are included in the analysis, it does not guarantee the adequacy of the proposed model. Also, typically the effect of the individual parameters is not assessed separately. This has made assessment procedures like ANN a useful alternative as will be demonstrated in this study.

*Table 1. Equations proposed for estimating the fundamental period of shear wall buildings*

| (Author, Year) | Material | Proposed Equation | Eq. | Parameters | Method |
|---|---|---|---|---|---|
| (Lagomarsino, 1993) | Concrete | $T_1$=0.018$H$ | (2) | H: Height of the building | Ambient vibration |
| (Goel Rakesh & Chopra Anil, 1997) | Concrete | $T = \frac{0.0062h}{\sqrt{A_e}}$ | (3) | $A_e$: Area of the floor<br>h: Height of the building | Earthquake Records |
| (Lee et al., 2000) | Concrete | $T = 0.4\left[\frac{h^{0.2}}{\sqrt{L_w}}\right]$ | (4) | h: Height of building (m)<br>$L_w$: Total wall width per unit plan area ($m^{-1}$) | Ambient vibration |
| (Farsi & Bard, 2004) | Concrete | $T_1$=0.01$h$ | (5) | h: Height of the building | Ambient vibration |
| (Dominguez Morales, 2000) | Concrete | $T = 0.13\frac{h}{I^{0.25}} - 0.4$ | (6) | h: height of the building<br>I: Moment of inertia | FEA and linear regression |
| (Gilles et al., 2011) | Concrete | $T$=0.020$h$ | (7) | h: Height of the building | Ambient vibration |
| (Hafeez et al., 2018) | Wood | $T = 0.045(\frac{h}{l}A)^{0.36}$ | (8) | H: Building height<br>A: Building area<br>l: Shear wall length | Ambient vibration |
| (Camelo et al., 2001) | Wood | $T = 0.032h_n^{0.55}$ | (9) | $h_n$: Building height (ft) | Forced vibration |

During the last years, there has been an increasing trend in employing ANN for the solution of complex problems in various scientific fields, including economics, engineering, and many others. In the field of structural engineering particularly, ANN has been successfully applied to solve various engineering problem (Lagaros & Papadrakakis, 2004; Plevris & Tsiatas, 2018). Examples of ANN having been successfully applied includes the modeling of masonry failure (P.G. Asteris & V. Plevris, 2013; Panagiotis G. Asteris & Vagelis Plevris, 2013; Asteris & Plevris, 2017; Plevris & Asteris, 2015; V. Plevris & P.G. Asteris, 2014; Vagelis Plevris & Panagiotis G. Asteris, 2014; Plevris et al., 2017; Plevris et al., 2019), predicting the compressive strength of concrete containing recycled aggregate (Dabiri et al., 2022; Kandiri et al., 2021), modeling the corrosion rate in cables of suspension bridges (Ben Seghier et al., 2021), predicting the properties of FRP-Confined Concrete Cylinders (Ahmad et al., 2020), determining the nominal shear capacity of steel fiber reinforced concrete beams (Ahmadi et al., 2020), predicting the bond stress of corroded steel reinforcing bars in concrete members (Ahmadi et al., 2021), and predicting the capacity of concrete walls (Sharib et al., 2021). ANN has also been successfully used for modeling and analyzing wooden structures, such as for predicting the compression strength of heat treated woods (Tiryaki & Aydın, 2014), predicting the mechanical properties of wood (Fathi et al., 2020) and investigating wood bonding quality based on pressing conditions (Bardak et al., 2016), among other interesting and innovative applications.

## ARTIFICIAL INTELLIGENCE AND OPTIMIZATION ALGORITHMS

### Artificial Neural Networks

An ANN consists of an input layer and an output layer and between them there exist a number of hidden layers that execute factual processing through weighted connections between the neurons of each layer. In a fully connected ANN model, each neuron of a layer is connected to every neuron of the next layer. The final, output layer is used to obtain the processing outcomes (Plevris, 2009).

A feed-forward neural network is an ANN in which nodes do not form a directed cycle. The information in this network flows solely in one direction from the input nodes to the output nodes, passing through the nodes of the hidden layers. The feed-forward network comprises several layers of computing units, coupled in a feed-forward manner. Each neuron in one layer is connected to neurons in the adjacent layer via directed connections. In many applications, these networks' units use the sigmoid, tangent sigmoid, purelin (Linear), poslin (Positive linear), and log sigmoid transfer functions as activation functions (Pham & Liu, 1995).

### Bat Algorithm

The bat optimization technique was inspired by the tracking properties of little bats seeking prey and specifically the behavior of bats when they reflect sound. The Bat algorithm begins by establishing a population of bats and determining their global optimum position in the search space using a fitness cost function. Each bat's position is decided by a random step toward the global optimum position. If a bat receives a signal from the global best location during an iteration, it will migrate towards that location. In each iteration, if the calculated new position of each bat improves the value of the fitness function for that bat, then the new position is saved. The algorithm iterates until a stop criterion is reached (Dehghani & Bogdanovic, 2018). This approach is also suitable for training an ANN. The network's weights and biases are considered as the design variables of the optimization problem, i.e. the position vector of a bat, and so each bat population represents a phase of the ANN training. The network's prediction error is assigned as the bat's cost function value, which needs to be minimized. In this sense, the bat algorithm's final solution represents a fully trained network (Dehghani & Bogdanovic, 2018).

The Bat algorithm initially generates a random population of bats and then changes its frequency using Eq. (10) (Aalimahmoody et al., 2021; Dehghani & Bogdanovic, 2018):

$$f_i = f_{min} + (f_{max} - f_{min})\beta \tag{10}$$

where $f_i$ denotes the frequency of the $i$-th bat, $f_{min}$ denotes the minimum frequency, $f_{max}$ denotes the maximum frequency, and $\beta$ is a random value between 0 and 1. The bats' position and velocity are updated using Eqs (11) and (12) (Srivastava & Sahana, 2019):

$$V_i^{t+1} = V_i^t + \left(x_i^t - \mathrm{x}^*\right) f_i \tag{11}$$

$$x_i^{t+1} = x_i^t + V_i^{t+1} \tag{12}$$

where $V_i^t$ is the velocity of the $i$-th bat at iteration $t$, $x_i^t$ is the position of the $i$-th bat at iteration $t$, and $x^*$ is the global best position of the entire bat population.

The technique then uses Eq. (18) to relocate some bats towards the best global location (Shadbahr et al., 2021).

$$x_{new} = x_{old} + \varepsilon \mathrm{A} t \tag{13}$$

where A represents the loudness and ε is a random value between 0 and 1.

The cost function value of each bat's new position must be smaller than the one of the previous iterations. Following that, the algorithm modifies the pulse rate and volume using Eqs (20) and (21) (Dehghani & Bogdanovic, 2018):

$$A_i^{t+1} = \alpha A_i^t \tag{14}$$

$$r_i^{t+1} = r_i^0 \left(1 - \exp\left(-\gamma t\right)\right) \tag{15}$$

where α is a constant between 0 and 1, $r_i^0$ is the initial pulse rate, and $\gamma$ is a constant value.

## Particle Swarm Optimization

Particle swarm optimization (PSO) is a metaheuristic algorithm inspired by the behavior of swarms of animals (e.g., flock of birds) (Kennedy & Eberhart, 1995). Generally, random solutions are used to initialize the algorithm. The algorithm's potential solutions, referred to as particles, possess particular positions and velocity vectors in the design space. The particles have a "memory" as they are able to "remember" the best cost function value associated with the locations they have individually visited (local best solution). Additionally, the algorithm tracks the best solution produced thus far by the entire swarm, referring to it as the global best solution (Plevris & Papadrakakis, 2011). The particle swarm optimization algorithm changes the velocity and the position of particles throughout the iterations thus taking them closer to their local and global optimal solutions. A random weight term is used to account for the influence of local best and global best solutions. In the ANN-PSO approach, the PSO optimizer is used for training the ANN, i.e. finding the values of the network's weights and biases that lead to the minimization of the prediction error.

## Performance Metrics

To determine the model's accuracy, statistical metrics (error metrics) need to be used. These metrics help identify the best models which exhibits the least error and has the most generalization capabilities. The root mean squared error (RMSE) and the mean absolute error (MAE) are the statistical metrics used in this study to assess the accuracy of the various models, as presented in Eqs. (16) and (17) (Li & Heap, 2008):

$$RMSE = \left[ \frac{1}{n} \sum_{i=1}^{n} \left( C_i - O_i \right)^2 \right]^{\frac{1}{2}} \tag{16}$$

$$MAE = \frac{1}{n} \sum_{i=1}^{n} \left| C_i - O_i \right| \tag{17}$$

where $C_i$ and $O_i$ define the calculated (model predicted), and the observed (ground truth) data, respectively, and $n$ is the number of samples in the database. In addition, we also take into account the Coefficient of determination ($R^2$), a number between 0 and 1 that measures how well a statistical model predicts an outcome (Plevris et al., 2022).

# METHODOLOGY

## Dataset

The measured period database used in this study is obtained from past studies on the estimation of the period of wooden buildings (Hafeez et al., 2018, 2019; Hafeez et al., 2014). The authors of these studies investigated the buildings' response under low amplitude motion, where the responses of several buildings with various geometric configurations were measured using ambient vibration method. The recorded motions of the buildings were analyzed in the frequency domain to extract the modal parameters including frequency, mode shape and damping ratio. This study employs 71 samples with various input parameters to establish an expression for the fundamental period of light-frame wood buildings. The input parameters include number of stories, total building height, plan area and the maximum length and width on the building plan (where the plan dimensions are irregular). The statistical properties related to the input parameters are shown in Table 2.

Figure 1 provides an overview of the structural period distribution in the range of 0.11 s to 0.45 s, which shows that approximately 79% of the period data is less than 0.3 seconds. Furthermore, Figure 2 shows a normal distribution curve demonstrating that the distribution of the fundamental period ranges mostly between 0.15 and 0.3 seconds.

*Table 2. Statistical properties of the light-frame wooden building parameters*

| Statistical Index | Unit | Type | Min | Max |
|---|---|---|---|---|
| Stories | Number | Input | 1 | 6 |
| Total height | m | Input | 3.00 | 21.60 |
| Max length | m | Input | 6.00 | 96.00 |
| Max width | m | Input | 4.90 | 80.00 |
| Area | $m^2$ | Input | 29.89 | 7680.00 |
| Period | s | Output | 0.11 | 0.45 |

*Figure 1. The experimental data's fundamental period distribution*

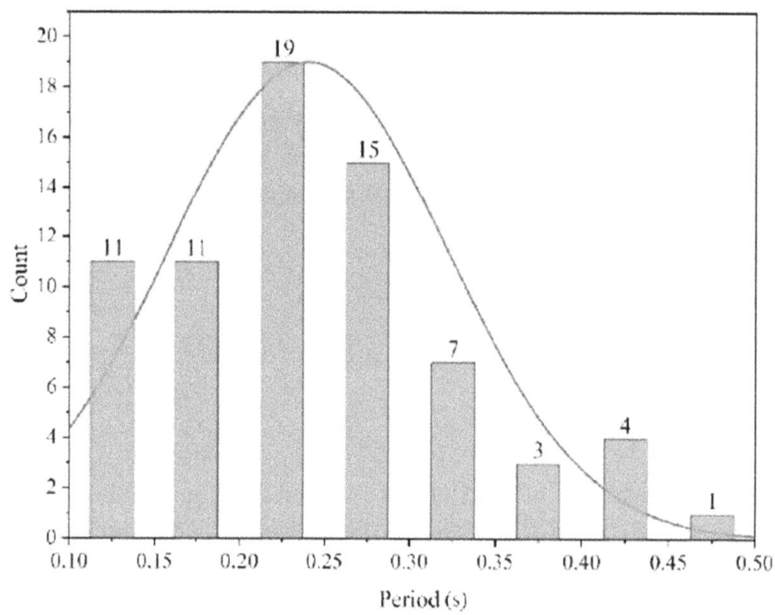


The correlation matrix of the input and output variables used in this study is presented in Figure 3, indicating a greater influence of the building height and the number of stories on the period of the building.

The subsequent section provides the details about various NN architectures developed employing optimization algorithms (Bat, PSO) used in this study, followed by the regression model (NLR), equations based on Eureqa software and the NBCC code model. Finally, the comparison among various best-performed models is discussed, and the conclusions of the study are presented.

Unlike a correlation plot, a Variation Inflation Factor (VIF), displayed in Eq. (18), is a measure to identify the multicollinearity of variables within the dataset. When correlation is considered among more than one variable, VIF will be the preferred method for estimating the variance of a regression coefficient. The value of VIF varies from 1, 1-5 and >10, indicating no, moderate and high correlation between the variables. Table 3 displays the evaluation of the correlation between variables with the VIF metric. As shown in the table, in most cases variables appear to be somehow correlated, having VIF values greater than 1, although in some cases the value is very close to 1 denoting practically no correlation.

$$VIF = \frac{1}{1 - R^2} \tag{18}$$

*Figure 2. Box normal plot for the fundamental period*

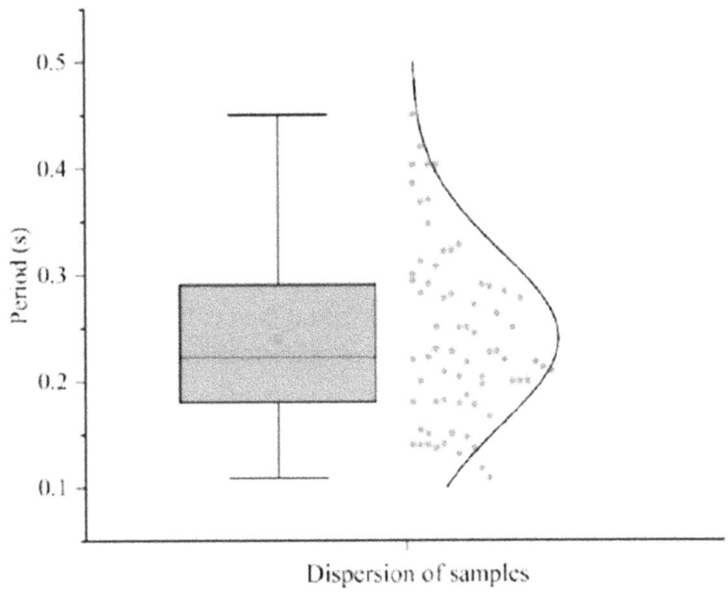


*Figure 3. Correlation matrix for the input and output variables*

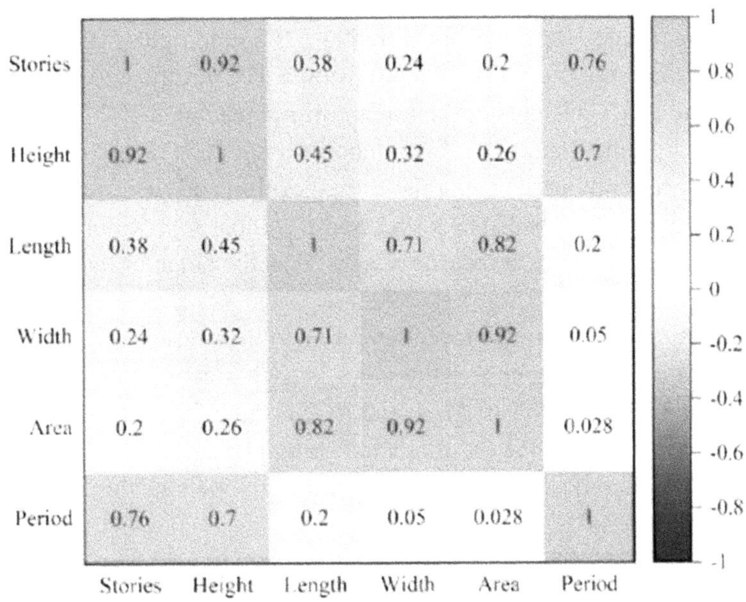

*Table 3. Evaluation of correlation between variables with Variation Inflation Factor (VIF)*

| | Stories | Height | Max length | Max width | Area | Period |
|---|---|---|---|---|---|---|
| **Stories** | | | | | | |
| **Height** | 6.641 | | | | | |
| **Max length** | 1.171 | 1.258 | | | | |
| **Max width** | 1.060 | 1.114 | 1.990 | | | |
| **Area** | 1.041 | 1.073 | 3.110 | 6.170 | | |
| **Period** | 2.342 | 1.968 | 1.042 | 1.003 | 1.001 | |

## ANN Model Combined With Bat Algorithm

A Feed-Forward type ANN is employed for calculating the fundamental period of the 71 samples of light-frame wooden buildings. The data were arbitrarily separated into two sets, as follows: 80% (57 samples) of the samples data were dedicated to the network training while the remaining 20% (14 samples) was reserved for network performance testing. In the development of the ANN model, the number of hidden layers and the number of neurons were selected based on the complexity of the problem, and Eq. (19) was used to obtain an estimate of the total number of neural cells of an ANN (Bowden et al., 2005).

$$N_H \leq \min(2N_I + 1) \quad (19)$$

where $N_H$ denotes the number of nodes of the hidden layer, and $N_I$ denotes the number of inputs. Given the five inputs, according to Eq. (19) the minimum number of nodes at each hidden layer should be 11. A number of different ANN architectures with two hidden layers are examined, as presented in Table 4. The first hidden layer has two to six neurons, and the second hidden layer has two to five neurons. For all ANNs, different transfer functions, including sigmoid, tangent sigmoid, purelin (Linear), poslin (Positive linear), and log sigmoid, have been used for the hidden and the output layers, as shown in the table. The Bat Algorithm was employed to adjust the weights and biases in order to minimize the prediction error, along with ANN training. Table 4 describes the different ANN topologies (16 in total) used in the study, while Table 5 describes the properties of the Bat Algorithm. The optimization parameters employed for reducing the errors in the ANNs weights of the Bat algorithm, shown in Table 5, were chosen based on recommendations from a previous study (Aalimahmoody et al., 2021).

All 16 ANN models were optimized using the Bat algorithm and their performances were assessed using three statistical indices, namely *MAE*, *RMSE* and $R^2$. Table 6 provides the top three models and the corresponding statistical assessment measures, as well as the slope of the straight line, used for the training and the testing sets.

In Table 6, the inclusion of "2L" in the topology refers to the presence of two hidden layers and the numbers in parentheses indicate the number of neurons in each hidden layer. It can be observed that the Bat-ANN 2L (4-3) network has the lowest *MAE* and *RMSE* index in the training and testing phases. It also has the highest $R^2$ value in both the training and testing phase (0.96 and 0.95, respectively) which indicates that the model is the most accurate among the 16 models of this study. The corresponding Bat-ANN 2L (4-3) network topology is shown in Figure 4.

*Table 4. Different topologies used in ANN training*

| No | Topology | Hidden and Output Activations | No | Topology | Hidden and Output Activations |
|---|---|---|---|---|---|
| 1 | 5-6-5-1 | TANSIG-PURELIN | 9 | 5-4-5-1 | LOGSIG-PURELIN |
| 2 | 5-6-4-1 | PURELIN-PURELIN | 10 | 5-4-4-1 | TANSIG-TANSIG |
| 3 | 5-6-3-1 | POSLIN-PURELIN | 11 | 5-4-3-1 | TANSIG-PURELIN |
| 4 | 5-6-2-1 | LOGSIG-PURELIN | 12 | 5-4-2-1 | PURELIN-PURELIN |
| 5 | 5-5-5-1 | TANSIG-TANSIG | 13 | 5-3-5-1 | POSLIN-PURELIN |
| 6 | 5-5-4-1 | TANSIG-PURELIN | 14 | 5-3-4-1 | LOGSIG-PURELIN |
| 7 | 5-5-3-1 | PURELIN-PURELIN | 15 | 5-3-3-1 | TANSIG-TANSIG |
| 8 | 5-5-2-1 | POSLIN-PURELIN | 16 | 5-3-2-1 | TANSIG-PURELIN |

*Table 5. Parameters of the Bat algorithm*

| Parameter | Value | Parameter | Value |
|---|---|---|---|
| Population size | 100 | Max Generations | 200 |
| Loudness | 0.9 | Pulse Rate | 0.5 |
| Min Frequency | 0 | Max Frequency | 2 |
| Alpha | 0.99 | Gamma | 0.01 |

*Table 6. Performance metrics for the top three ANNs combined with the Bat algorithm*

| Num | Topology | Train | | | | Test | | | |
|---|---|---|---|---|---|---|---|---|---|
| | | *MAE* | *RMSE* | $R^2$ | y=ax+b | *MAE* | *RMSE* | $R^2$ | *y=ax+b* |
| 1 | Bat-ANN 2L(5-5) | 0.006 | 0.030 | 0.88 | y = 0.9942x + 0.0052 | 0.017 | 0.038 | 0.82 | y = 1.0933x – 0.0091 |
| 2 | Bat-ANN 2L(3-2) | 0.020 | 0.028 | 0.87 | y = 0.9176x + 0.0185 | 0.023 | 0.026 | 0.87 | y = 0.7933x + 0.0417 |
| 3 | Bat-ANN 2L(4-3) | 0.010 | 0.015 | 0.96 | y = 0.9862x + 0.0029 | 0.010 | 0.016 | 0.95 | y = 0.9555x + 0.0127 |

To illustrate the performance of the Bat-ANN 2L(4-3) model, Figure 5 displays the calculated values of the model compared to their experimental counterparts (ground truth data) for the training set, while Figure 6 presents a similar trend for the testing set. Figure 7 shows the corresponding results for all data, training and testing, for the 71 samples. In each diagram the line of perfect agreement (i.e., $y=x$) is also displayed. It can be observed that most data points are close to the line of perfect agreement, which indicates the model's ability to predict the data with reasonable accuracy.

*Figure 4. The architecture of the ANN with topology 5-4-3-1*

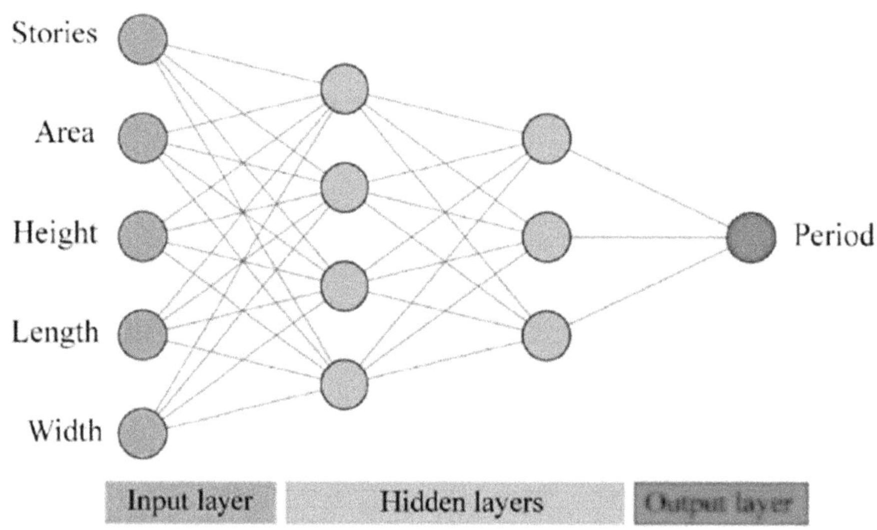


*Figure 5. Experimental (ground truth) vs calculated values of the Bat-ANN 2L(4-3) model's fundamental period (training data)*

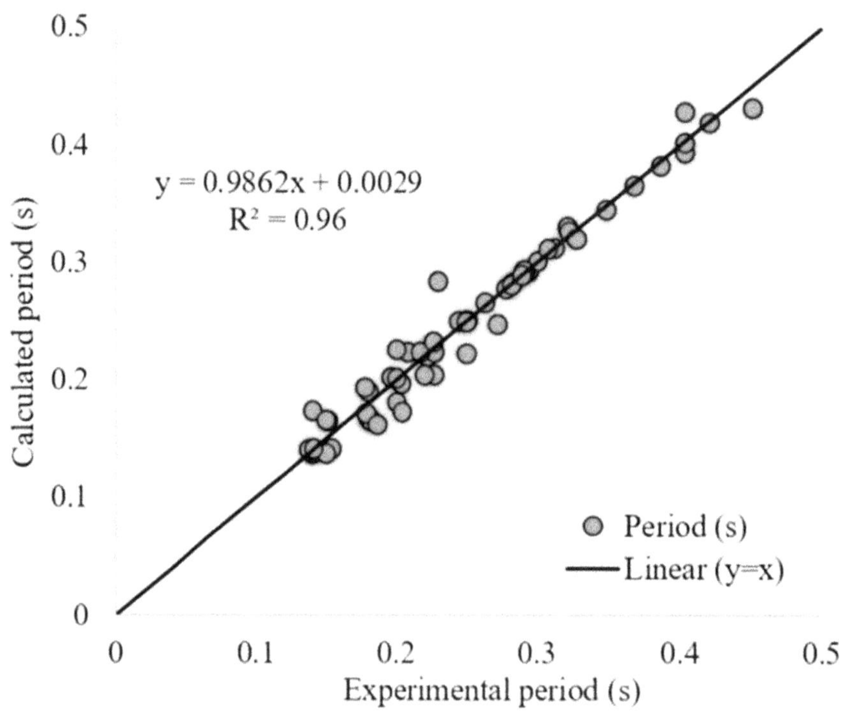

*Figure 6. Experimental (ground truth) vs calculated values of the Bat-ANN 2L(4-3) model's fundamental period (testing data)*

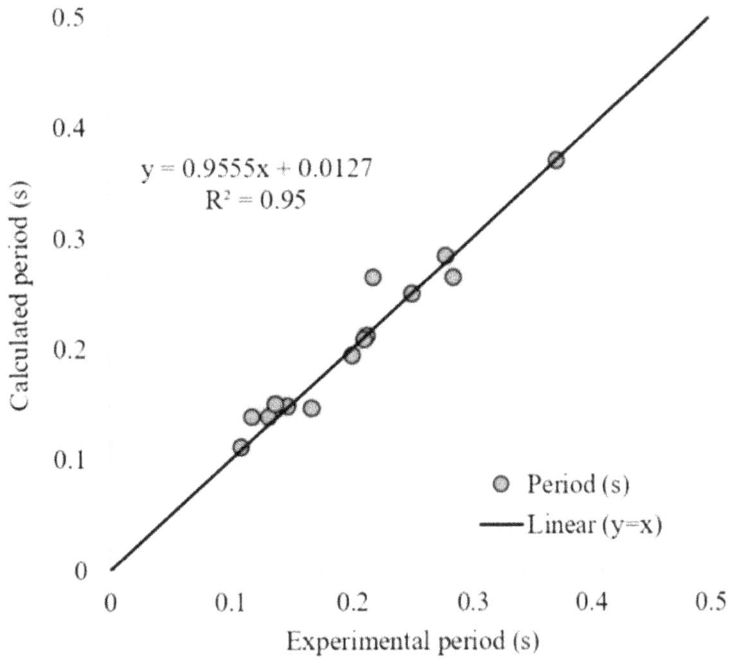


*Figure 7. Experimental (ground truth) vs calculated values of the Bat-ANN 2L(4-3) model's fundamental period (all data)*

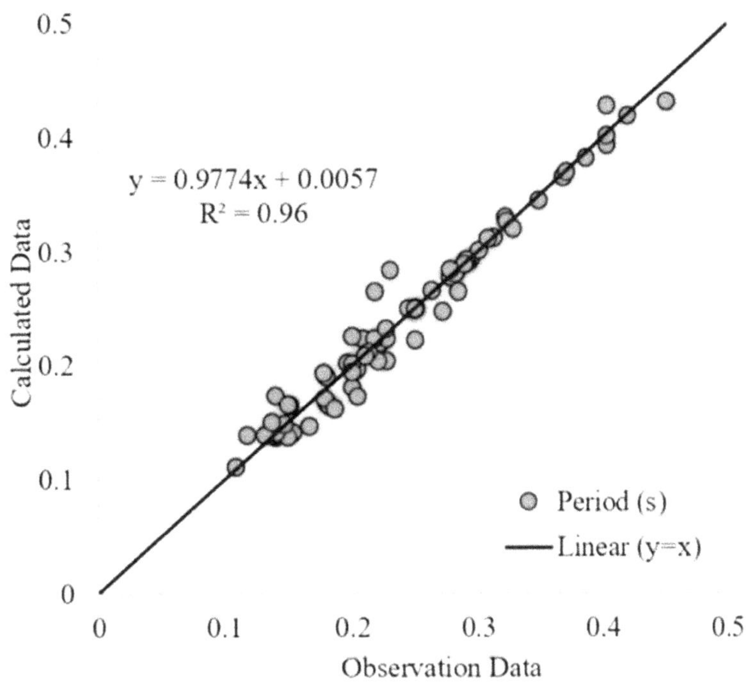

## ANN Model Combined With PSO Algorithm

The PSO technique was used to train the same 16 ANN architectures specified in Table 4 in order to identify the best-performing PSO-ANN scheme. Table 7 provides the PSO algorithm parameters used in the study. The optimization parameters employed for reducing the errors in the ANNs weights of the PSO algorithm, shown in Table 7, were selected based on recommendations from a previous study (Sadowski et al., 2019). Table 8 presents the top three models and the corresponding statistical assessment measures, as well as the slope of the straight line, used for all data sets.

*Table 7. The PSO algorithm parameters*

| Parameter | Value | Parameter | Value |
|---|---|---|---|
| Max iterations | 100 | Swarm size | 200 |
| Cognition Coefficient | 2 | Social Coefficient | 2 |

*Table 8. Performance metrics for the top three ANNs combined with the PSO algorithm*

| Model | All datasets | | | |
|---|---|---|---|---|
| | *MAE* | *RMSE* | $R^2$ | $y = ax + b$ |
| PSO-ANN 2L(5-5) | 0.014 | 0.026 | 0.90 | $y = 0.9238x + 0.0159$ |
| PSO-ANN 2L(4-4) | 0.013 | 0.038 | 0.83 | $y = 1.0176x + 0.0017$ |
| PSO-ANN 2L(5-3) | 0.038 | 0.053 | 0.57 | $y = 0.568x + 0.1004$ |

The PSO-ANN model with the best performance was the PSO-ANN 2L(5-5) model with 5 neurons in each hidden layer, exhibiting an $R^2$ value equal to 0.90 and indicating reasonable accuracy. The results of this model are presented in Figure 8 where the values of the experimental periods are on the horizontal axis (ground truth values) and the PSO-ANN predicted values are plotted on the vertical axis.

## Nonlinear regression model and the Eureqa formula

Nonlinear regression (NLR) is a type of regression analysis in statistics in which observational data are modeled using a function of the form $y = e^{ax_1+bx_2+\dots}$ that is a nonlinear combination of the model parameters and is function of one or more independent variables. Successive approximations are used to fit the data. In the present work, five variables are employed as independent parameters, and the fundamental period is calculated using the DataFit software. The formula $\ln(y) = a_1 \times x_1 + a_2 \times x_2 + a_3 \times x_3 + a_4 \times x_4 + a_5 \times x_5 + a_6$ is used to evaluate several equations in this study. Eq. (20) illustrates the best case.

$$\ln y = 0.1535 \times x_1 + 0.0094 \times x_2 + 0.0001 \times x_3 - 0.023 \times x_4 + 0 \times x_5 - 1.9748 \quad (20)$$

where the dependent variable $y$ denotes the period, $x_1$ is the number of stories variable, $x_2$ is the plan area, $x_3$ is the story height, and $x_4$ and $x_5$ are the variables associated with the maximum length and maximum width of the buildings, respectively. It should be highlighted here that the variable $x_5$ (maximum width of the building) is in fact not included in the model, since it is multiplied with zero.

*Figure 8. Experimental (ground truth) vs calculated values of the fundamental period for the PSO-ANN 2L(5-5) model (all data)*

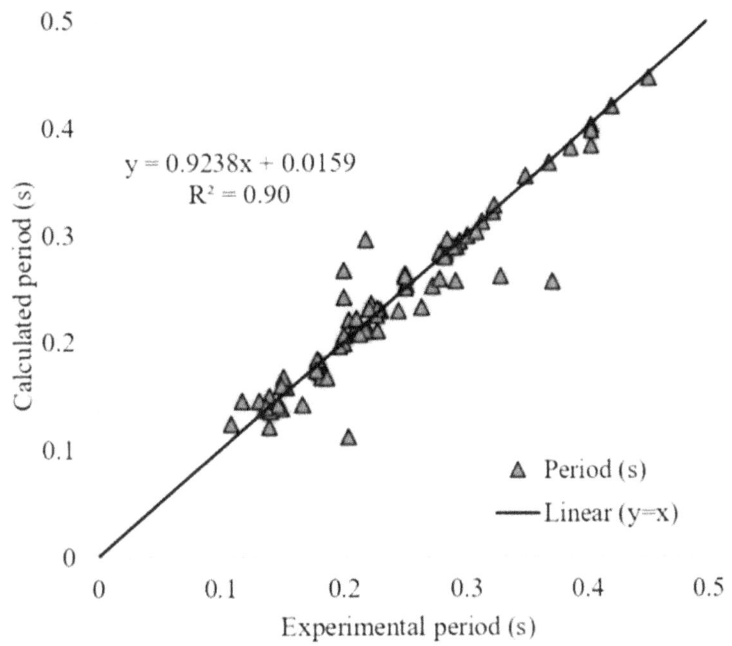


In the examined NLR model, presented in Figure 9, an $R^2$ value of 0.58 is obtained, and the straight-line slope is determined as 0.5869. It can be noted that a wider scatter is observed for the period in the range of 0.3 to 0.5 s, compared to the period range from 0.1 to 0.3 s.

Eureqa software (Dubčáková, 2011; Schmidt & Lipson, 2009) is a tool used to derive mathematical equations that represent varied collections of measurable inputs and outputs. A sequence of simple and sophisticated models is constructed using various mathematical operations such as addition, subtraction, multiplication, and trigonometric functions. The best-fitting data model with the least amount of error can then be chosen according to the procedures for developing mathematical models using the software. Prior to the search for suitable models, it is necessary to select the mathematical operators that will be utilized to build the models. In the beginning, basic mathematical operations such as addition and multiplication are selected to formulate the model for every parameter. If the model does not fit the data precisely, then additional operators including trigonometric and logarithmic functions may be added, and the procedure is repeated until the most precise model is finally achieved (Al-Subhi, 2020). This paper presents a mathematical equation obtained using the Eureqa software, to estimate the fundamental period as presented in Eq. (21).

$$y = 0.0704 + (0.0517 \times x_1) + \frac{0.0786 \times cos(0.325 \times x_3 + cos(x_2 + 0.135 \times x_4 - 0.344 \times x_1))}{x_1} \quad (21)$$

Where the dependent variable $y$ denotes the period, $x_1$ is the number of stories, $x_2$ is the area, $x_3$ is the height, and $x_4$ is the maximum length. In the mathematical model obtained from the Eureqa software (Figure 10), the $R^2$ value is equal to 0.65, and the slope of the straight line is 0.7114. The scatter trend found with Eureqa is somewhat similar to the one of the NLR model with slightly better correlation between the measured and the calculated period.

*Figure 9. Experimental (ground truth) vs calculated values of the fundamental period for the NLR model (all data)*

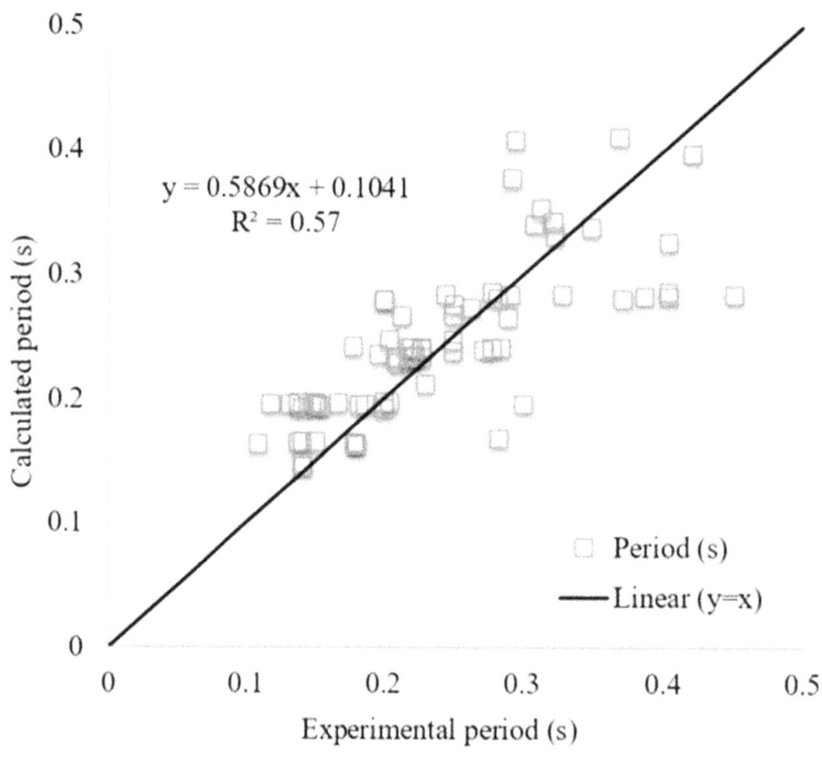


## NBCC Standard Code Formula

The suitability of contemporary building code formulas to estimate the fundamental period of a structure has been discussed in the introduction section. This section provides a measure of the ability of the empirical formula in the NBCC (NRC/IRC, 2015) to predict the period of the sample buildings used in the current study. It can be observed from Figure 11 that a wide scatter is present in the data, indicating the limitation of the building code expression being a function of the building height, only.

*Figure 10. Experimental (ground truth) vs calculated values of the fundamental period for the Eureqa model (all data)*

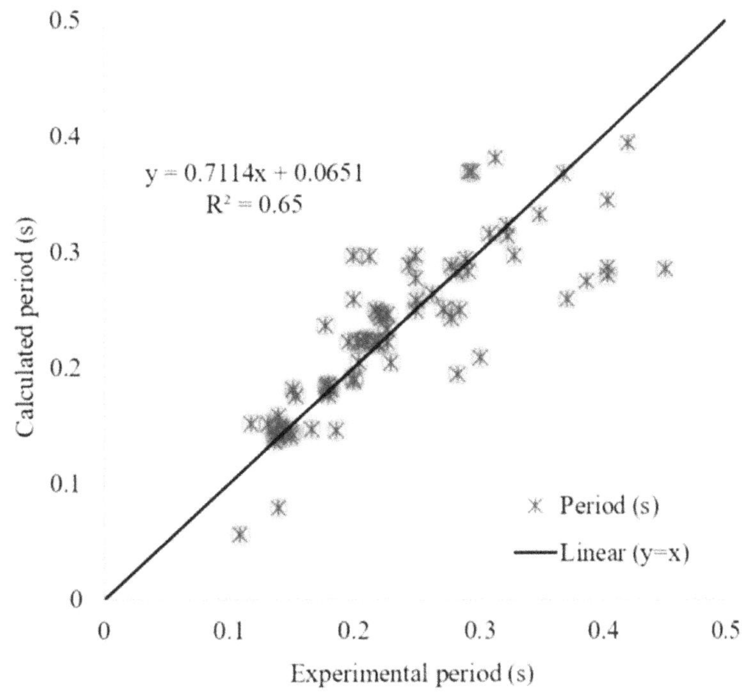


*Figure 11. Experimental vs calculated values of the fundamental period for the NBCC standard (all data)*

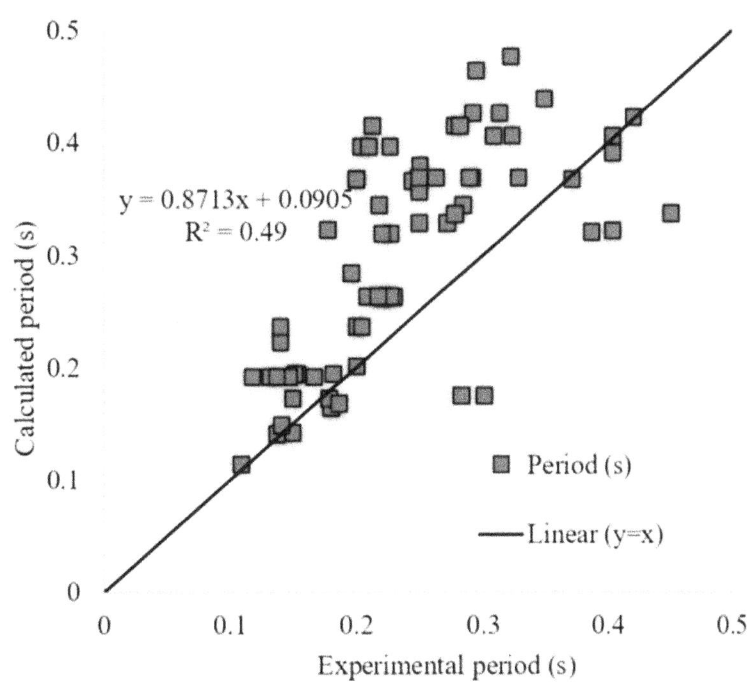

## Comparison Between the Various Models

The best performing models for each category, i.e. (i) Bat-ANN, (ii) PSO-ANN, (iii) NBCC, (iv) Eureqa equation, and (v) NLR are compared based on the *MAE* and *RMSE* error metrics values, to evaluate their overall performance. The results are presented in Table 8, and they show that the Bat-ANN model provides better performance and precision for predicting the building period followed by the PSO-ANN, Eureqa equation, NLR, and then the NBCC model.

*Table 8. Error metric values of the different models*

| No | Topology | Train set | | | Test set | | | All data | | |
|---|---|---|---|---|---|---|---|---|---|---|
| | | MAE | RMSE | $R^2$ | MAE | RMSE | $R^2$ | MAE | RMSE | $R^2$ |
| 1 | NLR | 0.037 | 0.052 | 0.58 | 0.049 | 0.054 | 0.64 | 0.039 | 0.053 | 0.58 |
| 2 | Eureqa equation | 0.030 | 0.047 | 0.66 | 0.042 | 0.053 | 0.53 | 0.032 | 0.049 | 0.65 |
| 3 | NBCC | 0.074 | 0.090 | 0.49 | 0.085 | 0.105 | 0.57 | 0.076 | 0.094 | 0.50 |
| 4 | PSO-ANN 2L(5-5) | 0.012 | 0.021 | 0.93 | 0.025 | 0.039 | 0.70 | 0.014 | 0.026 | 0.90 |
| 5 | Bat-ANN 2L(4-3) | 0.010 | 0.015 | 0.97 | 0.010 | 0.016 | 0.95 | 0.010 | 0.015 | 0.97 |

The predictions of all models in the testing phase are shown in Figure 12, where the top part of the figure compares Bat-ANN, PSO-ANN and NBCC with the ground truth (experimental) values, and the bottom part of the figure shows the comparison between Bat-ANN, Eureqa, and NLR model vs the ground truth values.

According to the results presented in Table 8 and Figure 12, it is shown that the Bat-ANN is the most accurate model, achieving predicted values of the period which are closer to the ground truth values. The Bat-ANN model exhibits the lowest *RMSE* and *MAE* metric values in the training phase, testing phase and overall data comparisons. The same conclusions are drawn also by examining the $R^2$ index where the Bat-ANN model performs best with an $R^2$ value of 0.96 for the overall dataset.

Another way to represent the results is through the Taylor diagram, which is a mathematical representation designed to graphically indicate which of several approximate models is the most realistic (Taylor, 2001). The diagram combines three statistical quantities, namely the Centered Root Mean Square Difference (*CRMSD*), the Pearson correlation coefficient $R$ and the Standard Deviation $\sigma$ in a single diagram that is easy to read and interpret (Elvidge et al., 2014). Details on the formulas used for the calculation of these three quantities can be found in the work of (Plevris et al., 2022). The Taylor diagram of the models examined in the current study is presented in Figure 13, where the ground truth results are represented with the "Reference" point. It can again be observed that the Bat-ANN model demonstrates once again the best performance, being the one closest to the reference point.

*Figure 12. Comparison between the experimental results and the various models, for the test data*

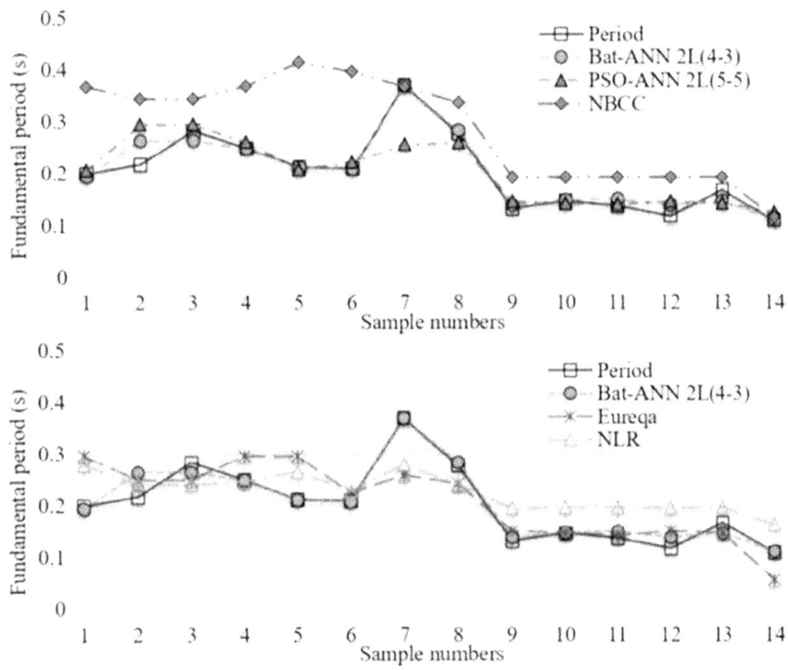


*Figure 13. Taylor diagram of the examined models for the estimation of the fundamental period*

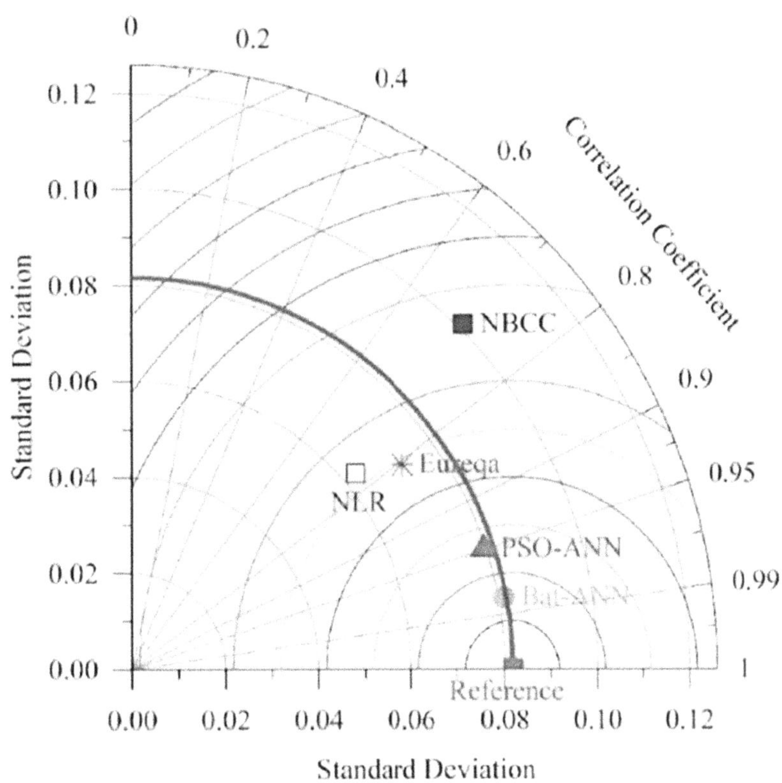

## CONCLUSION

The fundamental period of a building is an important parameter required for the base shear estimation of seismic design. This paper presents an ANN model developed to predict the fundamental period of light-frame wood buildings, implementing the Bat algorithm based on reported data-set of periods obtained from 71 buildings. Sixteen Bat-ANN models, having two hidden layers, were developed. The Bat-ANN model that demonstrated the best performance was the Bat-ANN 2L network (4-3), with a 5-4-3-1 network architecture. The selection of this specific model was predicated on its performance, providing the lowest values of mean absolute error (*MAE*) and root mean squared error (*RMSE*) among all other models. This model was compared to other models, including the PSO-ANN scheme, the formula of the National Building Code of Canada (NBCC), an equation obtained from the Eureqa software, and a Non-linear regression (NLR) model. The PSO-ANN scheme was also examined with 16 variations, and the best variation was found to be that corresponding to a 5-5-5-1 network architecture. The bat algorithm showed the best fit of the data set with $R^2$ value of 0.96 for all data, while the corresponding $R^2$ value for the best PSO-ANN scheme was 0.90 for all data. The other models, including NLR, Eureqa, and NBCC, demonstrated lower accuracy and showed wider scatter than the BAT-ANN and PSO-ANN. Therefore, the proposed Bat-ANN model can be considered as a reliable, robust and accurate tool for predicting the fundamental period of wooden buildings.

## ACKNOWLEDGMENT

This research received no specific grant from any funding agency in the public, commercial, or not-for-profit sectors.

## REFERENCES

Aalimahmoody, N., Bedon, C., Hasanzadeh-Inanlou, N., Hasanzade-Inallu, A., & Nikoo, M. (2021). BAT Algorithm-Based ANN to Predict the Compressive Strength of Concrete—A Comparative Study. *Infrastructures*, *6*(6), 80. doi:10.3390/infrastructures6060080

Ahmad, A., Plevris, V., & Khan, Q.-u.-Z. (2020). Prediction of Properties of FRP-Confined Concrete Cylinders Based on Artificial Neural Networks. *Crystals*, *10*(9), 1-22. doi: 10.3390/cryst10090811

Ahmadi, M., Kheyroddin, A., Dalvand, A., & Kioumarsi, M. (2020). New empirical approach for determining nominal shear capacity of steel fiber reinforced concrete beams. *Construction & Building Materials*, *234*, 117293. doi:10.1016/j.conbuildmat.2019.117293

Ahmadi, M., Kheyroddin, A., & Kioumarsi, M. (2021). Prediction models for bond strength of steel reinforcement with consideration of corrosion. *Materials Today: Proceedings*, *45*, 5829–5834. doi:10.1016/j.matpr.2021.03.263

Al-Subhi, A. (2020). Parameters estimation of photovoltaic cells using simple and efficient mathematical models. *Solar Energy*, *209*, 245–257. doi:10.1016/j.solener.2020.08.079

Asteris, P. G., & Plevris, V. (2013, 2013). Neural network approximation of the masonry failure under biaxial compressive stress. *ECCOMAS Special Interest Conference - SEECCM 2013: 3rd South-East European Conference on Computational Mechanics, Proceedings*. IACM. doi:10.7712/130113.4411.S2158

Asteris, P. G., & Plevris, V. (2013, 12-14 June 2013). *Neural Network approximation of the masonry failure under biaxial compressive stress* 3rd South-East European Conference on Computational Mechanics (SEECCM III), Kos Island, Greece.

Asteris, P. G., & Plevris, V. (2017). Anisotropic masonry failure criterion using artificial neural networks. *Neural Computing & Applications*, *28*(8), 2207–2229. doi:10.1007/s00521-016-2181-3 doi:10.100700521-016-2181-3

Bardak, S., Tiryaki, S., Nemli, G., & Aydın, A. (2016). Investigation and neural network prediction of wood bonding quality based on pressing conditions. *International Journal of Adhesion and Adhesives*, *68*, 115–123. doi:10.1016/j.ijadhadh.2016.02.010

Ben Seghier, M. E. A., Corriea, J. A. F. O., Jafari-Asl, J., Malekjafarian, A., Plevris, V., & Trung, N.-T. (2021). On the modeling of the annual corrosion rate in main cables of suspension bridges using combined soft computing model and a novel nature-inspired algorithm. *Neural Computing & Applications*, *33*(23), 15969–15985. doi:10.1007/s00521-021-06199-w doi:10.100700521-021-06199-w

Bowden, G. J., Dandy, G. C., & Maier, H. R. (2005). Input determination for neural network models in water resources applications. Part 1—Background and methodology. *Journal of Hydrology (Amsterdam)*, *301*(1), 75–92. doi:10.1016/j.jhydrol.2004.06.021

Camelo, V., Beck, J. L., & Hall, J. F. (2001). *Dynamic Characteristics of Woodframe Buildings*.

Dabiri, H., Kioumarsi, M., Kheyroddin, A., Kandiri, A., & Sartipi, F. (2022). Compressive strength of concrete with recycled aggregate; a machine learning-based evaluation. *Cleaner Materials, 3*, 100044.

Dehghani, H., & Bogdanovic, D. (2018). Copper price estimation using bat algorithm. *Resources Policy*, *55*, 55–61. doi:10.1016/j.resourpol.2017.10.015

Dominguez Morales, M. (2000). *Fundamental period of vibration for reinforced concrete buildings*. University of Ottawa.

Dubčáková, R. (2011). Eureqa: Software review. *Genetic Programming and Evolvable Machines*, *12*(2), 173–178. doi:10.100710710-010-9124-z

Elvidge, S., Angling, M. J., & Nava, B. (2014). On the use of modified Taylor diagrams to compare ionospheric assimilation models. *Radio Science, 49*(9), 737-745. doi:10.1002/2014RS005435

Farsi, M. N., & Bard, P.-Y. (2004). Estimation des périodes propres de bâtiments et vulnérabilité du bâti existant dans l'agglomération de Grenoble. *Revue Française de Génie Civil*, *8*(2-3), 149–179. doi:10.1080/12795119.2004.9692601

Fathi, H., Nasir, V., & Kazemirad, S. (2020). Prediction of the mechanical properties of wood using guided wave propagation and machine learning. *Construction & Building Materials*, *262*, 120848. . doi:10.1016/j.conbuildmat.2020.120848

Gilles, D., McClure, G., & Chouinard, L. E. (2011). Uncertainty in fundamental period estimates leads to inaccurate design seismic loads. *Canadian Journal of Civil Engineering*, *38*(8), 870–880. doi:10.1139/l11-055

Gilles, D. C. (2011). *In situ dynamic characteristics of reinforced concrete shear wall buildings*. McGill University.

Goel Rakesh, K., & Chopra Anil, K. (1997). Period Formulas for Moment-Resisting Frame Buildings. *Journal of Structural Engineering*, *123*(11), 1454–1461. doi:10.1061/(ASCE)0733-9445(1997)123:11(1454)

Goel Rakesh, K., & Chopra Anil, K. (1998). Period Formulas for Concrete Shear Wall Buildings. *Journal of Structural Engineering*, *124*(4), 426–433. doi:10.1061/(ASCE)0733-9445(1998)124:4(426)

Hafeez, G., Doudak, G., & McClure, G. (2018). Establishing the fundamental period of light-frame wood buildings on the basis of ambient vibration tests. *Canadian Journal of Civil Engineering*, *45*(9), 752–765. doi:10.1139/cjce-2017-0348

Hafeez, G., Doudak, G., & McClure, G. (2019). Dynamic characteristics of light-frame wood buildings. *Canadian Journal of Civil Engineering*, *46*(1), 1–12. doi:10.1139/cjce-2017-0266

Hafeez, G., Mustafa, A., Doudak, G., & McClure, G. (2014). Predicting the Fundamental Period of Light-Frame Wood Buildings. *Journal of Performance of Constructed Facilities*, *28*(6), A4014004. doi:10.1061/(ASCE)CF.1943-5509.0000519

Kandiri, A., Sartipi, F., & Kioumarsi, M. (2021). Predicting Compressive Strength of Concrete Containing Recycled Aggregate Using Modified ANN with Different Optimization Algorithms. *Applied Sciences (Basel, Switzerland)*, *11*(2), 485. doi:10.3390/app11020485

Kennedy, J., & Eberhart, R. (1995). Particle swarm optimization *IEEE International Conference on Neural Networks*, Piscataway, NJ, USA.

Lagaros, N. D., & Papadrakakis, M. (2004). Learning improvement of neural networks used in structural optimization. *Advances in Engineering Software*, *35*(1), 9–25. doi:10.1016/S0965-9978(03)00112-1

Lagomarsino, S. (1993). Forecast models for damping and vibration periods of buildings. *Journal of Wind Engineering and Industrial Aerodynamics*, *48*(2), 221–239. doi:10.1016/0167-6105(93)90138-E

Lee, L.-H., Chang, K.-K., & Chun, Y.-S. (2000). Experimental formula for the fundamental period of RC buildings with shear-wall dominant systems. *The Structural Design of Tall Buildings, 9*(4), 295-307.

Li, J., & Heap, A. D. (2008). *A Review of Spatial Interpolation Methods for Environmental Scientists*. G. Australia.

NRC/IRC. (2015). *National Building Code of Canada 2015*. National Research Council of Canada, Institute for Research in Construction.

Panthi, B., Dahal, P., Shrestha, P., & Thapa, K. B. (2021). Fundamental period of RC buildings with infill walls in Nepal. *Asian Journal of Civil Engineering*, *22*(5), 983–993. doi:10.100742107-021-00359-y

Paz, M., & Leigh, W. (2004). Uniform Building Code 1997: Equivalent Lateral Force Method. In M. Paz & W. Leigh (Eds.), *Structural Dynamics: Theory and Computation* (pp. 699–729). Springer US. doi:10.1007/978-1-4615-0481-8_23

Pham, D. T., & Liu, X. (1995). *Neural Networks for Identification, Prediction and Control.* Springer. doi:10.1007/978-1-4471-3244-8

Plevris, V. (2009). *Innovative computational techniques for the optimum structural design considering uncertainties.* National Technical University of Athens.

Plevris, V., & Asteris, P. (2015). Anisotropic failure criterion for brittle materials using Artificial Neural Networks. 5th International Conference on Computational Methods in Structural Dynamics and Earthquake Engineering (COMPDYN 2015), Crete Island, Greece. doi:10.7712/120115.3537.3648

Plevris, V., & Asteris, P. G. (2014). Modeling of masonry compressive failure using Neural Networks OPT-i 2014. *1st International Conference on Engineering and Applied Sciences Optimization, Proceedings.* ASPETE.

Plevris, V., & Asteris, P. G. (2014). Modeling of Masonry Failure Surface under Biaxial Compressive Stress Using Neural Networks. *Construction & Building Materials*, *55*, 447–461. doi:10.1016/j.conbuildmat.2014.01.041

Plevris, V., Bakas, N. P., Markeset, G., & Bellos, J. (2017, June 15-17). Literature review of masonry structures under earthquake excitation utilizing machine learning algorithms *6th ECCOMAS Thematic Conference on Computational Methods in Structural Dynamics and Earthquake Engineering (COMPDYN 2017)*, Rhodes, Greece. 10.7712/120117.5598.18688

Plevris, V., & Papadrakakis, M. (2011). A Hybrid Particle Swarm – Gradient Algorithm for Global Structural Optimization. *Computer-Aided Civil and Infrastructure Engineering*, *26*(1), 48–68. doi: 10.1111/j.1467-8667.2010.00664.x

Plevris, V., & Solorzano, G. (2021, 28-30 June). *Prediction of The Eigenperiods of MDOF Shear Buildings using Neural Networks.* 8th ECCOMAS Thematic Conference on Computational Methods in Structural Dynamics and Earthquake Engineering (COMPDYN 2021), Athens, Greece. doi:10.7712/120121.8755.20415

Plevris, V., Solorzano, G., & Bakas, N. P. (2019, 24-26 June). *Literature review of historical masonry structures with machine learning.* 7th International Conference on Computational Methods in Structural Dynamics and Earthquake Engineering (COMPDYN 2019), Crete, Greece. doi:10.7712/120119.7018.21053

Plevris, V., Solorzano, G., Bakas, N. P., & Ben Seghier, M. E. A. (2022, 5-9 June). *Investigation of performance metrics in regression analysis and machine learning-based prediction models.* 8th European Congress on Computational Methods in Applied Sciences and Engineering (ECCOMAS 2022), Oslo, Norway. doi:10.23967/eccomas.2022.15510.23967/eccomas.2022.155

Plevris, V., & Tsiatas, G. (2018). Computational Structural Engineering: Past Achievements and Future Challenges. *Frontiers in Built Environment*, *4*(21), 1–5. doi:10.3389/fbuil.2018.00021

Sadowski, Ł., Nikoo, M., Shariq, M., Joker, E., & Czarnecki, S. (2019). The Nature-Inspired Metaheuristic Method for Predicting the Creep Strain of Green Concrete Containing Ground Granulated Blast Furnace Slag. *Materials (Basel)*, *12*(2), 293. doi:10.3390/ma12020293 PMID:30658508

Schmidt, M., & Lipson, H. (2009). Distilling Free-Form Natural Laws from Experimental Data. *Science*, *324*(5923), 81–85. doi:10.1126/science.1165893 doi:10.1126cience.1165893 PMID:19342586

Shadbahr, E., Aminnejad, B., & Lork, A. (2021). Determining post-fire residual compressive strength of reinforced concrete shear walls using the BAT algorithm. *Structures, 32*, 651-661. doi: 10.1016/j.istruc.2021.03.002

Sharib, S., Ahmad, N., Plevris, V., & Ahmad, A. (2021, 28-30 June 2021). *Prediction Models for Load Carrying Capacity of RC Wall through Neural Network.* 14th ECCOMAS Thematic Conference on Evolutionary and Deterministic Methods for Design, Optimization and Control (EUROGEN 2021), Athens, Greece. doi:10.7712/140121.7956.18529

Solorzano, G., & Plevris, V. (2021). *Design of Reinforced Concrete Isolated Footings under Axial Loading with Artificial Neural networks.* 14th ECCOMAS Thematic Conference on Evolutionary and Deterministic Methods for Design, Optimization and Control (EUROGEN 2021), Streamed from Athens, Greece. doi:10.7712/140121.7955.18448

Srivastava, S., & Sahana, S. K. (2019). Application of Bat Algorithm for Transport Network Design Problem. *Applied Computational Intelligence and Soft Computing*, *2019*, 9864090. doi:10.1155/2019/9864090

Taylor, K. E. (2001). Summarizing multiple aspects of model performance in a single diagram. *Journal of Geophysical Research*, *106*(D7), 7183–7192. doi:10.1029/2000JD900719

Tiryaki, S., & Aydın, A. (2014). An artificial neural network model for predicting compression strength of heat treated woods and comparison with a multiple linear regression model. *Construction & Building Materials*, *62*, 102–108. doi:10.1016/j.conbuildmat.2014.03.041

# Chapter 7
# Shear Capacity of RC Elements With Transverse Reinforcement Through a Variable-Angle Truss Model With Machine-Learning-Calibrated Coefficients

**Dario De Domenico**
*University of Messina, Italy*

**Qingcong Zeng**
*Zhejiang University, China*

**Giuseppe Quaranta**
https://orcid.org/0000-0001-8295-0912
*Sapienza University of Rome, Italy*

**Giorgio Monti**
*Sapienza University of Rome, Italy*

## ABSTRACT

*This chapter addresses the shear capacity prediction of reinforced concrete elements with transverse reinforcement through a hybrid approach in which a mechanical model (i.e., the variable-angle truss model incorporated in the Eurocode 2) is enhanced with two machine-learning-calibrated corrective coefficients aimed at improving the accuracy of the numerical predictions. Genetic programming is adopted to derive closed-form expressions of the corrective coefficients, thus making the proposed formulation suitable for design purposes and usable by practitioners. The model predictive performance and the improvements over alternative code-based formulations is demonstrated through a wide database of experimental results of reinforced concrete beams and columns with plain and hollow sections failing in shear under both monotonic and cyclic loading conditions. The proposed approach leads to numerical-to-experimental shear capacity ratios having mean value close to one and coefficients of variation equal to 32%, 28% and 24% for beams, columns with plain and hollow sections, respectively.*

DOI: 10.4018/978-1-6684-5643-9.ch007

## INTRODUCTION

The prediction of the shear capacity of reinforced concrete (RC) elements with transverse reinforcement is a critical topic that has attracted the interest of several researchers over the last four decades (ASCE-ACI Committee 445, 1998). Many existing buildings are characterized by considerably lower shear reinforcement than recommended in current design codes and often exhibit a shear-dominated failure. Hence, in order to perform an accurate vulnerability assessment of such structures, it is of the utmost importance to develop reliable, unbiased, and precise numerical formulations capable of predicting the actual shear strength of RC elements with stirrups.

Some formulations proposed in international codes were found to be overly conservative compared to experimental findings but also characterized by excessive dispersion, which is usually accounted for within design procedures through the adoption of large safety coefficients (Cladera & Marí 2007; De Domenico & Ricciardi 2019). In this context, the development of machine learning techniques has been increasingly exploited to solve structural engineering problems (Azadi Kakavand 2021; Quaranta et al. 2020; Seghier et al. 2021a, 2021b; Mai et al. 2022; Quaranta et al. 2022). Instead of adopting a pure data-driven approach to derive a new empirical formula as proposed by some researchers (Fiore et al. 2016; Feng et al. 2021), the strategy proposed in this contribution intends to use artificial intelligence tools to improve one of the most popular mechanical models adopted in international codes for the shear capacity prediction of RC elements with stirrups, i.e., the variable-angle truss model (European Committee for Standardization 2004). More specifically, genetic programming is used to calibrate two empirical coefficients governing the concrete contribution of the truss resisting mechanism so as to increase its predictive accuracy. In this way, the mechanical basis of the resisting mechanism is preserved, but the final predictions are improved thanks to a hybrid approach in which mechanical principles are assisted by machine learning. The effectiveness and potentials of the proposed formulation is demonstrated by comparisons with experimental results taken from a large database comprising RC beams and columns with solid and hollow sections failing in shear under both monotonic and cyclic loading conditions. A comparative analysis is also made with an ample set of expressions from reference building codes to demonstrate that the proposed formulation leads to more accurate and economical outcomes for practical design purposes.

## A SHORT REVIEW ABOUT SOME CODE-CONFORMING SHEAR CAPACITY EQUATIONS FOR RC BEAMS AND COLUMNS

Starting from the truss resisting mechanism postulated by Ritter and Mörsch in the early 1900s, two approaches were developed to reduce the conservativeness of the original truss model: 1) the additive approach, in which an additional concrete contribution (generally having empirical nature) is considered along with the truss contribution with compression struts inclined at 45° (e.g., ACI 318 Building Code (2019) and the pre-standard version of the Eurocode 2 (1991)); 2) the variable-angle truss model with compression diagonals inclined of angles generally less than 45° (e.g., Model Code 90 (1993), the Eurocode 2 (2004) and other national building codes in Germany and Italy). The reason for such "variable" inclination is related to the observation of some physical phenomena occurring during shear failure of RC members with stirrups, such as aggregate interlock, dowel forces and residual tensile stress, that induce a strut rotation mechanism to cross adjacent cracks (Walraven et al. 2013).

These classes of mechanical models are based on simple equilibrium equations (and the theory of plasticity) without any explicit consideration for compatibility conditions. An alternative group of mechanics-based models incorporates compatibility equations and material constitutive relationships in addition to equilibrium equations in order to determine the strut inclination angle, e.g., the modified compression field theory (MCFT) (Vecchio & Collins 1986; Bentz et al. 2006). The MCFT inspired the development of simplified code expressions incorporated into the AASHTO standards (2012) and the Canadian Building Code A23.3-04 (2004). Finally, the Model Code 2010 (2013) includes different levels of approximation ranging from a fixed-angle truss model (in which the compression diagonals inclination is less than 45°) without concrete contribution (level I) up to an additive approach with concrete contribution calculated through compatibility conditions in a similar manner to the simplified MCFT (Sigrist et al. 2013).

## GENETIC-PROGRAMMING-BASED IMPROVEMENT OF THE VARIABLE-ANGLE TRUSS MODEL

### Outline of the Proposed Methodology

An overview of the available technical codes for practice shows that the shear capacity of RC members is based on different formulations, and various sets of parameters governing the resisting mechanism are involved. The accuracy of the various models is highly dependent on the manner in which such parameters are determined. In almost all models, the concrete contribution is inspired by a resisting mechanism, but some underlying parameters have an empirical (or partly empirical) basis and were calibrated upon a set of experimental data.

In the literature, several models followed a pure data-driven approach, wherein the adjective "pure" implies that the resulting capacity equation is not linked to any physical resistant mechanism but, rather, it is solely based on data (e.g., Mansour et al. (2004); Naderpour & Mirrashid (2020)). Although this approach might result in more accurate predictions, developing a capacity model on a mechanical ground is also a very important goal in view of the possible extensions to other cases as well as considering the inherent advantage that the role of the governing (tuning) coefficients can be interpreted through some mechanical justifications. Starting from this consideration, a hybrid novel approach is herein proposed. Compared to existing literature studies, the relevance of the proposed approach relies on the fact that, here, machine learning is not intended to replace a mechanics-informed model able to explain how structural elements resist to applied loads. Actually, its application is meant at improving the accuracy and alleviating the inherent simplifications of the resulting capacity equations through the data-driven definition of the involved corrective parameters. The distinctive peculiarities of the adopted "gray-box" framework make the proposed approach different from both white box (i.e., fully mechanics-based) and black box (i.e., pure data-driven) modeling techniques.

The mechanical model taken into consideration as the starting point is that of the Eurocode 2 (EC2) (2004) grounded on the truss analogy, which, as overviewed above, stems from the refinement of the historical shear resisting mechanical model proposed by Ritter and Mörsch dating back to the beginning of last century. The mechanical basis of this model is fully preserved but, at the same time, is enriched and improved by two corrective coefficients whose expressions are identified by means of genetic programming (GP), to better match experimental data. The governing equations of the proposed variable-angle truss model with machine-learning-calibrated coefficients are expressed as follows:

$$V = \min\{V_{Rs}, V_{Rc}\} \tag{1}$$

$$V_{Rs} = \frac{A_{sw}}{s} z\, f_{ysw} \cot\theta \tag{2}$$

$$V_{Rc} = \eta_c \cdot b_w \cdot z \cdot \eta \cdot f_c \cdot \frac{\cot\theta}{1+\cot^2\theta} \tag{3}$$

where $V_{Rs}$ and $V_{Rc}$ are the shear capacity ascribed to two failure mechanisms, namely the yielding of the steel transverse reinforcement and the crushing of the concrete struts, respectively, $A_{sw}$ is the cross-sectional area of the transverse reinforcement, $s$ is the spacing of the stirrups, $z$ is the inner lever arm (with $z$=0.9•$d$, $d$ being the effective depth of the cross section), $f_{ysw}$ is the yield strength of the shear reinforcement, $\theta$ is the angle between the concrete compression struts and the longitudinal axis of the RC member, b$w$ is the (minimum) width of the concrete cross section and f$c$ is the concrete (cylinder) compressive strength. The terms $\eta$ and $\eta$c are the two tuning coefficients to be determined through the machine learning approach. Indeed, these equations (1)-(3), apart from the coefficients $\eta$ and $\eta$c, coincide with those of the EC2, and they can be obtained in a fully analytical fashion starting from the idealization of the variable-angle truss resistant mechanism. More specifically, $\eta$ and $\eta_c$ are introduced in the proposed formulation in place of $v$ and $\alpha_c$ governing the concrete contribution in the original EC2 (2004) formulation. The coefficient $\eta$, like $v$, represents a strut efficiency factor, whereas the term $\eta_c$ incorporates not only the effect of applied compressive stress on the resulting shear capacity (like $\alpha_c$) but also the effect of cyclic loading. The methodology pursued in this chapter aims at finding the best tuning (corrective) coefficients $\eta$ and $\eta_c$ to match a series of experimental data collected from the literature. To this aim, an experimental database is compiled as detailed in the sequel of the paper.

The determination of the best expression for $\eta$ is carried out for each sample of the database (training) under the assumption $\eta_c$=1, i.e., by selectively excluding the effect of axial compressive stress (only RC beams are considered) and that of the cyclic loading (only monotonic loading is considered). In mathematical terms, the optimal value for $\eta$ is obtained for each sample of the database by solving the following constrained optimization problem:

$$\min_{\eta} \frac{\left|V_{num}(\eta \mid \eta_c = 1) - V_{\exp}\right|}{V_{\exp}} \quad s.t. \begin{cases} \eta_{\min} \le \eta \le \eta_{\max} \\ \theta_{\min} \le \theta \le \theta_{\max} \end{cases} \tag{4}$$

where $V_{num}(\eta|\eta_c=1)$ represents the numerical prediction of the shear capacity through Eqs. (1)-(3) depending on the coefficient $\eta$ and assuming $\eta_c$=1. Based on Eq. (4), the optimal value of $\eta$ is sought so that the numerical shear capacity approaches, in the best possible fashion, the corresponding experimental result, while satisfying certain bounding limits given by $\eta_{min}$ and $\eta_{max}$, $\theta_{min}$ and $\theta_{max}$.

Once the optimal value of $\eta$ is determined for each sample of the database, namely $\eta_{opt}$, a machine learning technique is adopted to obtain an analytical expression depending on some dominant variables $\vartheta$, namely $\eta=f(\vartheta)$, such that its predictions fit the retrieved optimal values $\eta_{opt}$ in the best possible way.

In a similar fashion, the determination of the best expression for $\eta c$ is carried out for each $_s$ample of the database including RC columns under cyclic loading (thus including the effects of both compressive stress and cyclic loading) under the previously obtained expression for $\eta$, namely $\eta$=f($\vartheta$), which in mathematical terms is expressed as the following constrained optimization problem:

$$\min_{\eta_c} \frac{\left|V_{num}\left(\eta_c \mid \eta = f(\vartheta)\right) - V_{exp}\right|}{V_{exp}} \quad s.t. \begin{cases} \eta_{c,\min} \le \eta_c \le \eta_{c,\max} \\ \theta_{\min} \le \theta \le \theta_{\max} \end{cases} \tag{5}$$

where $V_{num}(\eta_c|\eta=f(\vartheta))$ represents the numerical **p**rediction of the shear capacity through Eqns. (1)-(3) depending on the coefficient $\eta_c$. Based on Eq. (5), the optimal value of $\eta_c$ is sought so that the numerical shear capacity approaches, in the best possible fashion, the corresponding experimental results, while satisfying certain bounding limits given by $\eta_{c,min}$ and $\eta_{c,max}$, $\theta_{min}$ and $\theta_{max}$.

Once the optimal value of $\eta_c$ is determined for each sample of the database, namely $\eta_{c,opt}$, a machine learning technique is adopted to obtain an analytical expression depending on some dominant variables $\varphi$, namely $\eta_c=g(\varphi)$, such that its predictions fit the **retrieved** optimal values $\eta c$,opt in the best possible way.

The functional relationships f and g identified by GP are searched w*i*thin *t*he class of function sets including standard arithmetic operators only, so that the resulting expressions are suitable for design purposes and usable by practitioners. It is worth noting that the expression of $\eta_c$ in Eq. (3) is aimed at incorporating the effects of compressive stress and cyclic loading; however, its value also inherently corrects the concrete contribution by accounting for a larger database of results, as the expression of $\eta$ appears not only for RC beams (based on which the calibration of $\eta$ in Eq. (4) was conducted) but also for RC columns, which are, in general, characterized by different ranges of mechanical and geometrical data.

Lower and upper bounds for $\eta$ and $\eta_c$ in Eq. (4) and (5), respectively, are assumed equal to $\eta_{min}=1.001\omega_{sw}$ (where $\omega_{sw}$ is the mechanical ratio of transverse reinforcement), $\eta_{max}=1.0$, $\eta_{c,min}=0.3$ and $\eta_{c,max}=3.0$. On the other hand, $\theta_{max}=45°$ in agreement with EC2, since this upper bound is motivated by mechanical considerations and is also supported by experimental data (Biskinis et al. 2004). With regard to the lower bound $\theta_{min}$, Biskinis et al. (2004) pointed out that it can be lower than the value of 21.8° recommended in the EC2 formulation. With the range of variation of $\theta$ recommended in Eurocode 2 ($1.0\le \cot\theta \le 2.5$), the corresponding effective concrete compress*i*ve strength is assumed around 0.5-0.55$f_c$ owing to the presence of the efficiency factor $v=0.6(1-f_c/250)$, and the corrective term $\alpha_c$ takes on *v*alues of around 1-1.25. However, the range of variation of $\eta$ and $\eta_c$ allowed in the proposed formulation is broader than that of the original coefficients in the EC2 formulations (v and $\alpha_c$), which implies that a wider class of plastically admis*s*ible stress fields could be allowed accordingly. Based on these considerations, experimental data by Biskinis et al. (2004) and other truss models proposed in the literature (De Domenico & Ricciardi 2019), the lower bound for $\theta$ is set as $\theta_{min}=11.31°$ (i.e., $1\le\cot\theta\le5$).

## Data Preparation and Results

In this Section, the experimental database used for the machine learning approach is presented. In particular, experimental results concerning RC beams are collected from three databases published in the literature by Mansour et al. (2004), including 194 samples, by Zhang et al. (2016), including 194 samples, and by Reineck et al. (2014, 2017), including 170 samples. Moreover, experimental results concerning RC columns are collected from the NEES ACI 369 rectangular column database compiled by Ghannoum et al. (2015), with recent extensions by Azadi Kakavand et al. (2019) in the PRJ-2526 database. These databases are selected because they contain detailed information on the specimens and testing configurations, transparent criteria for data filtering, are quite popular in the field of shear capacity predictive models of RC elements and are adopted by other authors. In particular, the database of RC beams was adopted in another study (De Domenico & Ricciardi 2019) aimed at improving the Eurocode 2 truss model on a mechanical basis, while the database of RC columns was adopted in another study developing a pure data-driven approach (Azadi Kakavand et al. 2019).

Excluding replications of samples among different databases and specimens not failing in shear, additional filters were applied to ensure consistency with the underlying hypotheses of the truss resisting mechanism considered in this work. The applied filters considered in this study are: 1) shear span-to-effective depth ratio $a/d \geq 2.2$, so as to include only slender RC members; 2) mechanical ratio of transverse reinforcement $\omega_{sw} \leq 0.25$; 3) applied compressive stress ratio $\sigma_c/f_c \leq 0.50$. The distribution of the main mechanical and geometrical data of the samples collected from the literature is plotted in the histograms of Figure 1 for RC beams (373 samples) and RC columns (119 samples).

Several machine learning approaches were developed in the literature, which can be used for the determination of novel expressions for $\eta$ and $\eta_c$. To derive closed-form expressions of the corrective coefficients that are suitable for design purposes and usable by practitioners, in this work a standard GP with a tree-based representation of the candidate solution is adopted. Once the optimal solution for each data point is identified, the optimal model is searched to maximize the accuracy of the numerical predictions given by either $\eta = f(\vartheta)$ or $\eta_c = g(\varphi)$ with respect to the previously identified optimal solutions ($\eta_{opt}$ or $\eta_{c,opt}$). The accuracy measure to maximize is the Pearson coefficient $R^2$. Dimensional and non-dimensional variables are taken into account in the derivation of the final expressions, i.e., $\vartheta = \{f_c, b_w/d, a/d, \omega_{sw}\}$ and $\varphi = \{f_c, b_w/d, a/d, \omega_{sw}, \sigma_c/f_c, \mu\}$, where $\mu$ is the displacement ductility demand appearing in the case of RC columns tested under cyclic loading.

A standard GP approach, whose schematic flowchart is shown in Figure 2, has been implemented to determine the optimum expressions for the corrective coefficients $\eta$ and $\eta_c$, including fitness evaluation, crossover and mutation operation$_s$, iteratively. More details of the algorithms of the GP can be found in the literature (Banzhaf et al. 1998; Langdon et al. 2008).

In the GP, the following parameters are set: population size equal to 1,000 tournament selection (tournament size equal to 10); subtree crossover (with crossover rate equal to 0.90, functions and leaves being selected as crossover point 90% and 10% of the times, respectively), point mutation (with mutation rate equal to 0.15) and reproduction (by duplicating 5 candidate solutions within the current population to the next without changes). When the iterative procedure does not lead to further improvements after 100 generations, a stop criterion is established. The entire database is divided into training (80% of the samples) and testing (20% of the samples) datasets.

*Figure 1. Distribution of the main variables within the considered experimental databases for RC beams (top) and columns with solid cross-section (bottom)*

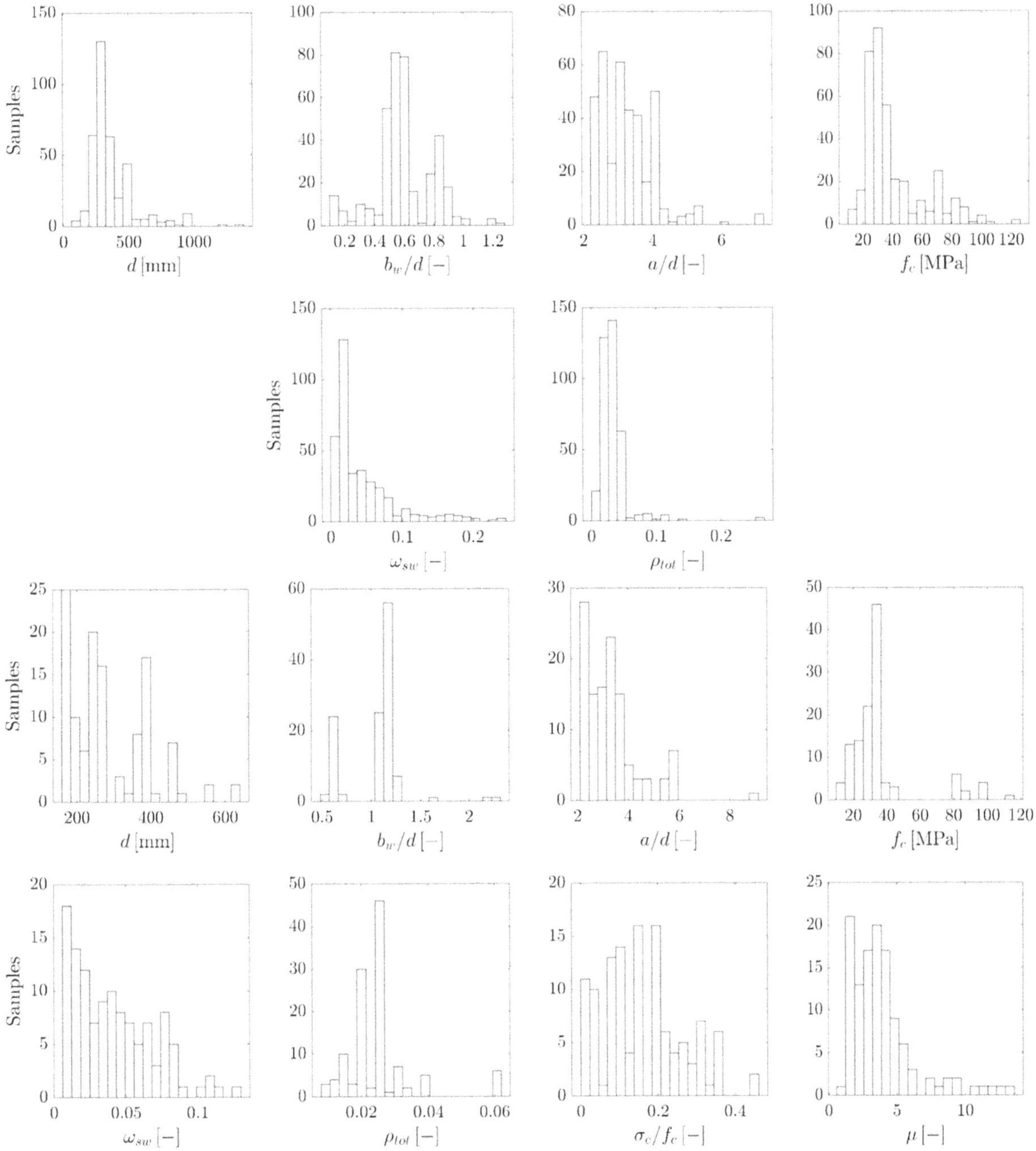

*Figure 2. Schematic flowchart of the GP approach implemented to determine the corrective coefficients*

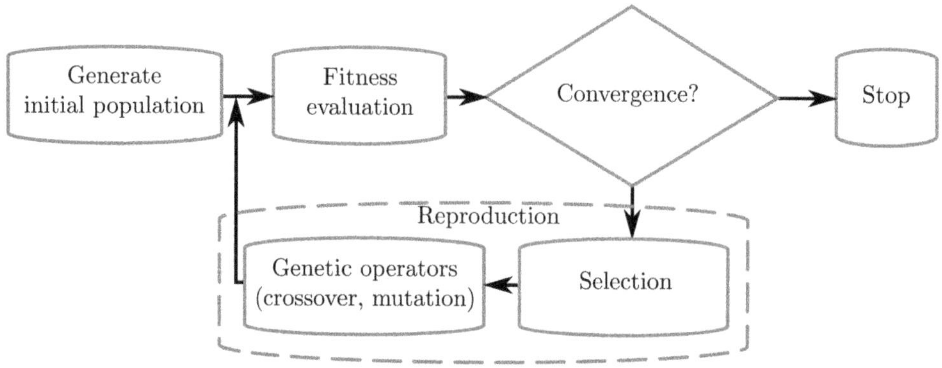


The final expressions of the corrective factors $\eta=f(\vartheta)$ or $\eta_c=g(\varphi)$ are:

$$\eta = 0.12 + \frac{3.86 + 3.94(b_w/d)}{8.21 - 0.08(b_w/d) - 0.08 f_c + f_c(b_w/d)} \quad \text{(with fc in[MPa])} \tag{6}$$

$$\eta_c = \begin{cases} 1 \\ 0.37 + \dfrac{0.30 + 0.75(\sigma_c/f_c) - 1.39(a/d)\omega_{sw}}{a/d}\left(3.79 + 57.52\dfrac{\sigma_c/f_c}{\mu^2}\right) \end{cases}$$ for beams, monotonic loading (top): for columns, cyclic loading (bottom) (7)

These expressions are valid within the range of the mechanical and geometrical parameters in the compiled databases (see again Figure 1). Moreover, the following constraints are applied to the corrective coefficients: $0.1 \leq \eta \leq 1.0$, $1/3 \leq \eta_c \leq 2.6$. The $R^2$ values resulting from the calibration procedure of $\eta$ are quite low, equal to 22.47% and 19.76% for the training and complete dataset, respectively, whereas those pertinent to the calibration procedure of $\eta_c$ are much larger, equal to 75.64% and 73.52% for the two datasets.

By analyzing Eq. (6), it can be noted that the corrective coefficient $\eta$ decreases with increasing $f_c$, in line with the inverse relationship existing between $v$ and $f_c$ in the original EC2 formulation. This trend is plotted in Figure 3, in which it is also observed that the coefficient $\eta$ generally decreases as the $b_w/d$ ratio increases. This result may be justified, from a mechanical standpoint, by the fact that the effective compressive stress of compression struts decreases as the flexural inertia of the concrete diagonals decreases, an aspect that is not incorporated in the coefficient $v$ of the EC2 formulation.

On the other hand, as expected, the corrective coefficient $\eta_c$ given by Eq. (7) is influenced by the applied compression in term of the ratio $\sigma_c/f_c$: similar to the $\alpha_c$ expression of the original EC2 formulation, the value of $\eta_c$ increases with increasing compressive stresses $\sigma_c/f_c$. An additional term that appears in Eq. (7) is the displacement ductility demand $\mu$: by looking again at Figure 3, the coefficient $\eta_c$ decreases with increasing $\mu$. This result is a major novelty with respect to the formulation of $\alpha_c$ in the original EC2 model, which does not explicitly consider the influence of ductility demand on the shear capacity of RC members. The decreasing trend (gradient) of the shear strength with increasing ductility becomes less pronounced for higher values of $\mu$, in line with other models from the literature, such as the model proposed by Biskinis et al. (2004) incorporated in the Eurocode 8 (2005). Based on the obtained results and plots, the threshold of $\mu$ above which the decreasing trend of the shear strength tends to vanish depends upon the value of the compressive stress. Further inspection of Eq. (7) reveals that the corrective coefficient $\eta c$ generally decreases with $\omega sw$ and with a/d; these two factors are not involved in the original EC2 formulation.

*Figure 3. Parametric study of the new parameters ruling the concrete contribution within the variable-angle truss model obtained by genetic programming*

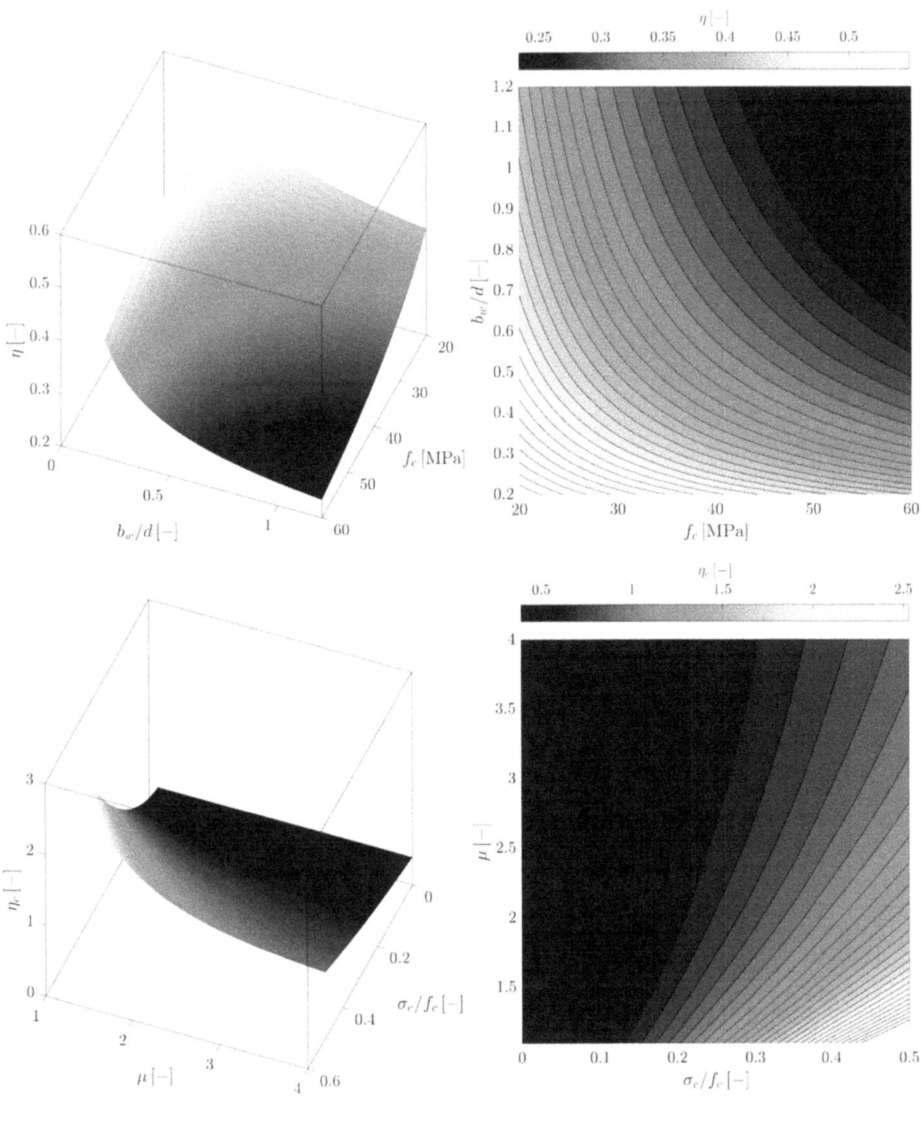

*Figure 4. Comparison between experimental shear strength values of RC beams and corresponding predictions obtained by means of different capacity equations*

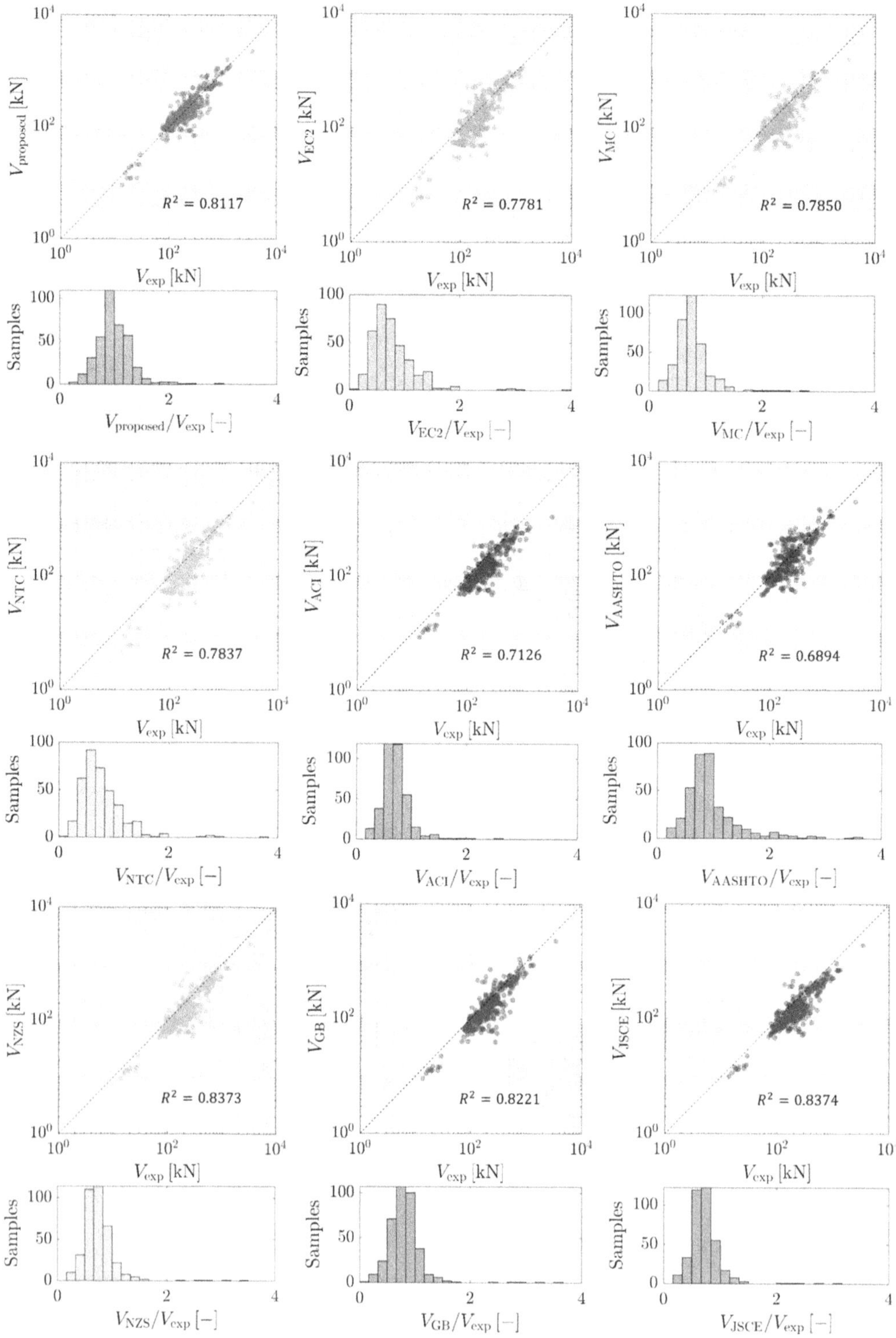

## NUMERICAL INVESTIGATION

### Application to RC beams

Since the present work deals with a variant of a code-based formulation, more specifically, of the EC2 truss model, the accuracy of the proposed expressions is discussed in a comparative manner by including other shear capacity equations from alternative technical codes proposed by international organisms, national/federal regulatory agencies or standardization bodies.

In particular, in addition to the proposed model (i.e., Eqs. (1)-(3) with the corrective coefficients provided by Eqs. (6) and (7)), the following models are considered in this comparative assessment (abbreviated nomenclature is indicated within brackets): 1) original EC2 formulation (EC2); 2) Model Code 2010 (MC); 3) Italian Technical Code 2018 (NTC); 4) ACI 318 Building Code (ACI); 5) provisions from the American Association of State Highway and Transportation Officials (AASHTO); 6) New Zealand Standards for concrete structures (NZS); 7) Codes for Design of Concrete Structures (China) GB50010-2010 (GB); 8) Japan Society of Civil Engineering Guidelines for Concrete no. 15 (JSCE).

*Figure 5. Assessment of different capacity equations for RC beams*

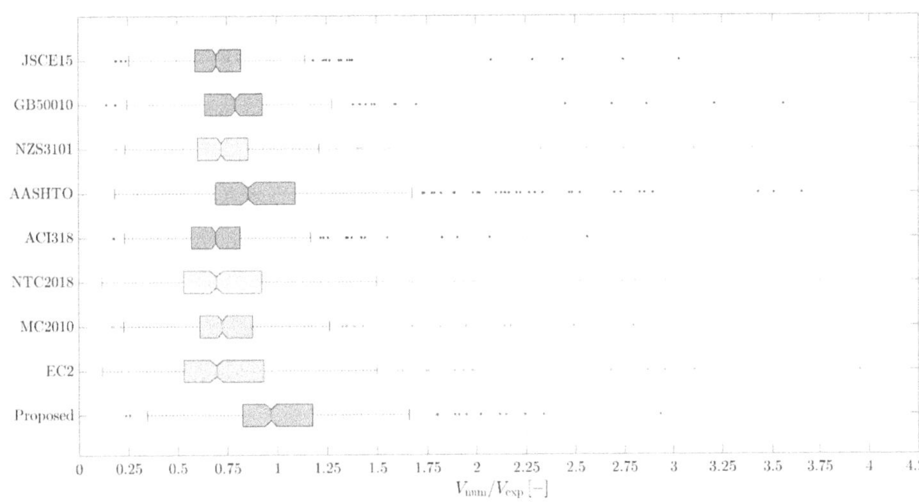


For the RC beams under monotonic loading collected in the present study, the graphs in Figure 4 illustrate the experimental shear strength along the horizontal axis versus the numerically predicted shear strength along the vertical axis. The same axis scale is adopted in the two axes so that the perfect agreement between the two sets of results is obtained along the 45° line. By comparatively analyzing the various subplots, it is clear that the proposed model is associated with less dispersion (distance of the data points from the 45° line) than other code-based formulations, and a general agreement with the experimental findings. A more synthetic assessment of the various shear capacity equations for RC beams is illustrated in Figure 5 through a box-and-whisker plot of the Vnum/Vexp ratio, where Vnum represents the numerical prediction and $V_{exp}$ the experimental counterpart. In particular, the mean value of the $V_{num}/V_{exp}$ is equal to 1.0034 for the proposed model (coefficient of variation CoV equal to 32.45%), whereas that of the original EC2 formulation is 0.7926 (CoV equal to 54.16%). Similar improvements are noted by comparing the proposed model with other code formulations. Almost all the existing code-conforming formulations underestimate, on average, the shear capacity of RC beams (median value of the $V_{num}/V_{exp}$ ratio lower than one) and are characterized by higher dispersion of results (CoV values ranging from 35.92% to 54.16%). This comparison demonstrates that the machine-learning improvement of the corrective coefficients $\eta$ and $\eta_c$ in the proposed formulation remarkably enhances the predictive accuracy of the variable-angle truss model of the EC2.

## Application to RC Columns With Solid and Hollow Cross-Section

Similar to the RC beams, the accuracy of the proposed model is also assessed for RC columns with solid and hollow cross-section under cyclic loading. Figure 6 reports the numerical versus experimental plots concerning the RC columns with solid cross section for the various code-based formulations.

In this case, most of the examined code-based formulations overestimate the actual value of the shear strength, as the data points lie, in most cases, above the 45° line of the subplots. On the contrary, the proposed shear capacity equation leads to the most accurate results, with the majority of data points falling close to the 45° perfect-prediction line. The results and main statistical metrics are summarized in the box-and-whisker plot of Figure 7: the proposed formulation shows a significantly smaller interquartile range than all the other models. In particular, the mean value of the $V_{num}/V_{exp}$ ratio is equal to 1.0324 for the proposed model (coefficient of variation CoV equal to 28.07%), whereas that of the original EC2 formulation is 1.1816 (CoV equal to 46.12%).

Finally, the performance of the model is also assessed considering RC columns with hollow cross section. A series of experimental results were collected from the literature, mainly based on the tests discussed by AlAjarmeh et al. (2020), Calvi et al. (2005), Delgado (2008) and Mo & Nien (2002). In this case, the thickness of the hollow cross section tw plays the role of the (minimum) width of the solid cross section $b_w$. Pertinent results, obtained by limiting the analysis to values of $t_w/d>0.1$ (corresponding to $b_w/d>0.2$, which is in line with the minimum value observed in the database for the RC columns), are illustrated in Figure 8: once again, the model exhibits very good accuracy, with a mean value of the $V_{num}/V_{exp}$ ratio equal to 1.02 and a CoV equal to 24.71%. The Italian technical code (NTC2018) would provide a similar mean value, but a much higher CoV equal to 48.84%. In Figure 8 (right) we also report a comparative assessment with other four technical formulations taken from codes or literature papers, previously analyzed by Cassese et al. (2017) for the same database considered here: this comparison shows once again a very good predictive performance of the proposed model, especially in terms of mean value. This final comparison demonstrates that the achieved calibration of the corrective coefficients allows one, without any further refinement or adjustments, to accurately predict the shear strength of RC members. Notably, the proposed model represents the best equation for predicting the shear strength for all typologies of RC members (i.e., RC beams under monotonic loading and RC columns under cyclic loading with solid or hollow cross-section).

*Figure 6. Comparison between experimental shear strength values of RC columns with solid cross-section and corresponding predictions obtained by means of different capacity equations*

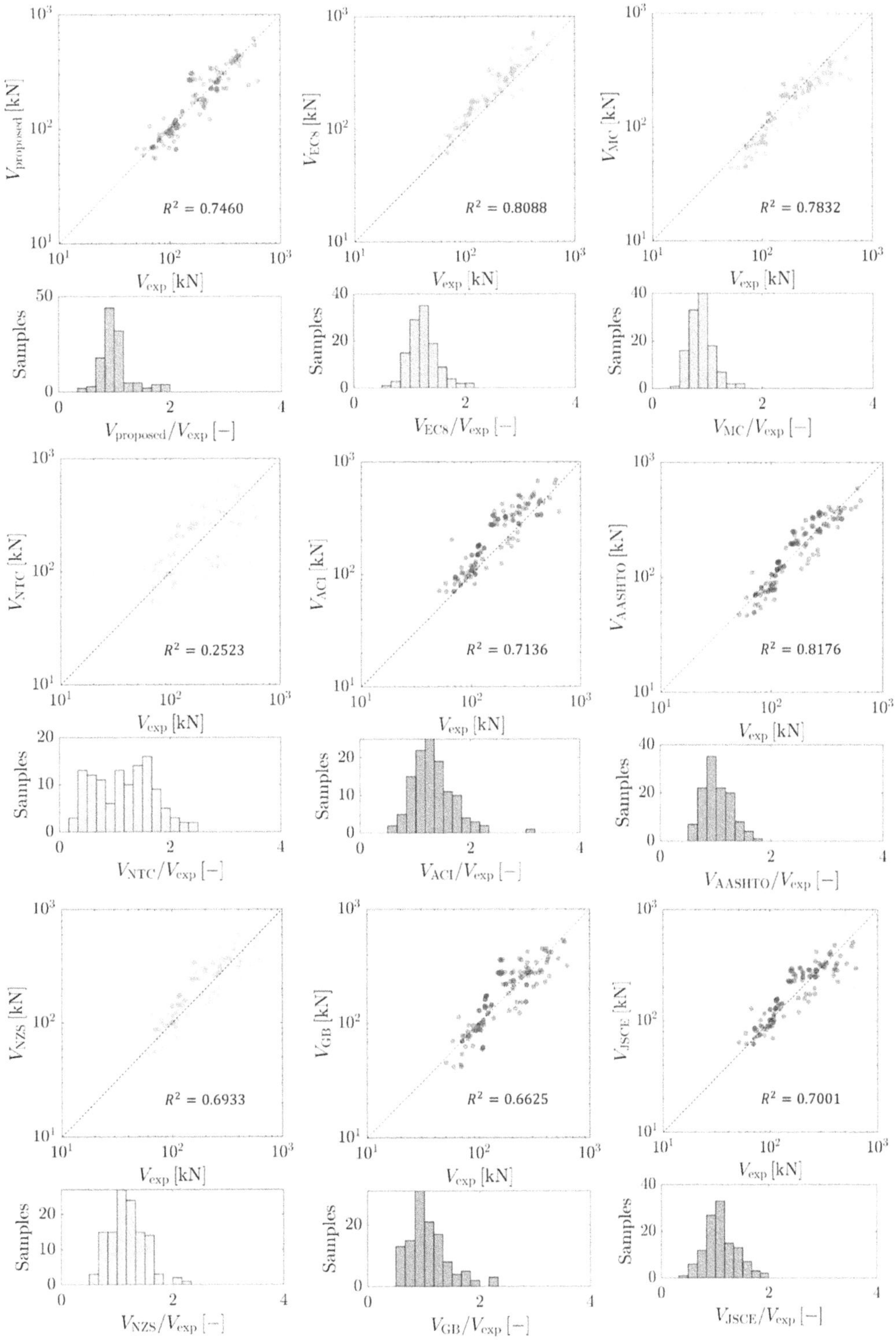

*Figure 7. Assessment of different capacity equations for RC columns with solid cross-section*

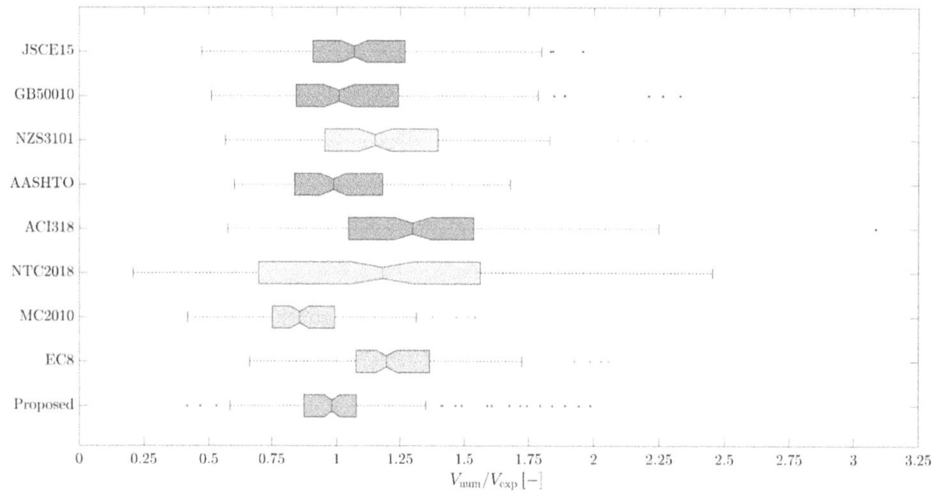


*Figure 8. Comparison between experimental shear strength values of RC columns with hollow cross-section and corresponding predictions obtained by means of the proposed equation (left) and assessment of different capacity equations for RC columns with hollow cross-section*

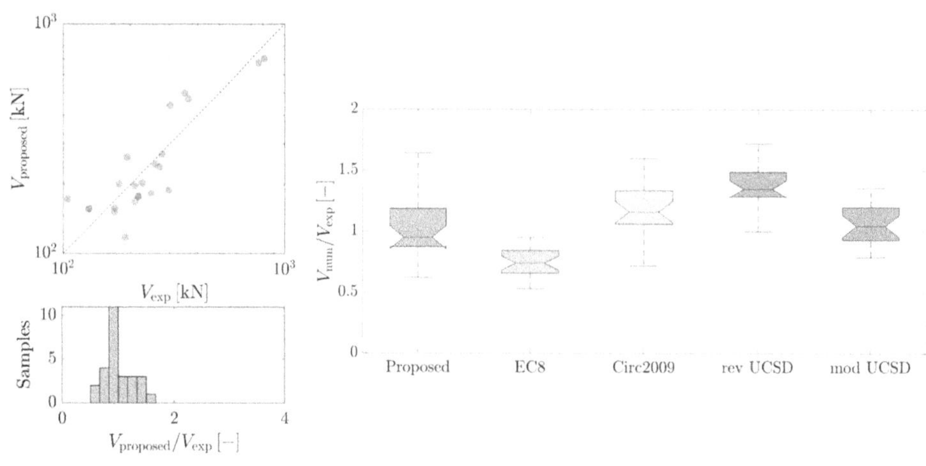


## CONCLUSION

The shear capacity prediction of RC members with transverse reinforcement has been addressed in this chapter. A gray-box modeling approach has been adopted, which is different from alternative black-box modeling (i.e., pure data-driven) techniques and white-box modeling (i.e., mechanics-based) models. In the proposed approach, a mechanical model, namely the variable-angle truss model of the EC2 formulation, is improved through two corrective coefficients governing the concrete contribution, whose expressions have been determined through a data-driven technique, namely genetic programming. The proposed model avoids typical overfitting problems of pure data-driven approaches and improves the accuracy of traditional mechanics-based models.

The proposed approach has allowed the derivation of inline equations for two corrective coefficients governing the concrete contribution, called $\eta$ and $\eta_c$, that replace the v and $\alpha$c coefficients of the original EC2 formulation. These expressions are quite compact, thus making the proposed formulation suitable for design purposes and usable by practitioners. While the machine learning technique has been adopted to identify suitable expressions of the involved corrective parameters, the capacity estimation is ultimately governed by a mechanics-based resisting mechanism, which allows interpreting and discussing the role of the various parameters involved in the detected expressions. In particular, the coefficient $\eta$ (replacing the efficiency factor v of the EC2 formulation) depends not only on the compressive strength of concrete $f_c$, but also on the cross-section shape factor $b_w/d$, somehow suggesting that the effective concrete compressive strength in the truss model decreases as the flexural inertia of the concrete diagonals cross-section decreases. Moreover, the coefficient $\eta_c$ (replacing $\alpha_c$ of the EC2 formulation) depends not only on the applied compressive strength on the RC member $\sigma_c/f_c$, but also on the displacement ductility $\mu$, suggesting that shear strength of RC columns under cycling loading condition decreases as the displacement ductility demand increases, in line with available experimental findings.

The methodology illustrated in this Chapter is quite general and can be applied to derive other capacity equations of interest in other structural engineering problems. Further work is underway to apply the proposed technique to the shear capacity prediction of RC members with externally bonded fiber-reinforced cementitious matrix (FRCM) sheets.

## ACKNOWLEDGMENT

Giuseppe Quaranta and Giorgio Monti acknowledge the support from the Laboratories University Network of Seismic Engineering (ReLUIS) through the research project DPC/ReLUIS 2022-2024 – UR RM1 WP11.

## REFERENCES

AASHTO. L. R. F. D. (2012). AASHTO LRFD bridge design specifications. American Association of State Highway and Transportation Officials: Washington, DC, USA.

ACI Committee 318 (2019). Building code requirements for structural concrete (ACI 318-*19) and commentary (ACI 318r-19). ACI.*

*AlAjarmeh, O. S., Manalo, A. C., Benmokrane, B., K*arunasena, K., Ferdous, W., & Mendis, P. (2020). Hollow concrete columns: Review of structural behavior and new designs using GFRP reinforcement. Engineering Structures, 203, 109829. doi:10.1016/j.engstr*uct.2019.109829*

*ASC*E-*ACI* Committee 445 on Shear and Torsion. (1998). Recent approaches to shear design of structural concrete. Journal of Structural Engineering, 124(12), 1375–1417. *doi:10.1061/(ASCE)0733-9445(1998*)*124:*12(1375)

Azadi Kakavand, M. R., Sezen, H., & Taciroglu, E. (2019). Rectangular column database. designsafe-ci.

Azadi Kakavand, M. R., Sezen, H., & Taciroglu, E. (2021). Data-driven models for predicting the shear strength of rectangular and circular reinforced concrete columns. Journal of Structural Engineering, 147(1), 04020301. doi:10.1061/(ASCE)ST.1943-541X.0002875

Banzhaf, W., Nordin, P., Keller, R. E., & Francone, F. D. (1998). Genetic programming: an introduction: on the automatic evolution of computer programs and its applications. Morgan Kaufmann Publishers Inc.

Bentz, E. C., Vecchio, F. J., & Collins, M. P. (2006). Simplified modified compression field theory for calculating shear strength of reinforced concrete elements. ACI Structural Journal, 103(4), 614.

Biskinis, D. E., Roupakias, G. K., & Fardis, M. N. (2004). Degradation of shear strength of reinforced concrete members with inelastic cyclic displacements. Structural Journal, 101(6), 773–783.

Calvi, G. M., Pavese, A., Rasulo, A., & Bolognini, D. (2005). Experimental and numerical studies on the seismic response of RC hollow bridge piers. Bulletin of Earthquake Engineering, 3(3), 267–297. doi:10.100710518-005-2240-0

Canadian Standards Association. (2004). Design of Concrete Structures. A23, 3–04.

Cassese, P., Ricci, P., & Verderame, G. M. (2017). Experimental study on the seismic performance of existing reinforced concrete bridge piers with hollow rectangular section. Engineering Structures, 144, 88–106. doi:10.1016/j.engstruct.2017.04.047

Cladera, A., & Marí, A. R. (2007). Shear strength in the new Eurocode 2. A step forward? Structural Concrete, 8(2), 57–66. doi:10.1680tco.2007.8.2.57

Comité Euro-International du Béton. (1993). CEB-FIP model code 1990: Design code. Comité Euro-International du Béton.

De Domenico, D., & Ricciardi, G. (2019). Shear strength of RC beams with stirrups using an improved Eurocode 2 truss model with two variable-inclination compression struts. Engineering Structures, 198, 109359. doi:10.1016/j.engstruct.2019.109359

Delgado, P. (2008). Avaliação da Segurança Estrutural em Pontes. [PhD thesis, FEUP, Porto]

European Committee for Standardization (1991). Eurocode 2: Design of concrete structures - Part 1-1: General rules and rules for buildings. ECS.

European Committee for Standardization (2004). Eurocode 2: Design of concrete structures – Part 1-1: General rules and rules for buildings. ECS.

European Committee for Standardization (2005). Eurocode 8: Design of structures for earthquake resistance – Part 3: Assessment and retrofitting of buildings. ECS.

Fédération Internationale du Béton (fib) (2013). Model Code for concrete structures 2010. FIB.

Feng, D. C., Wang, W. J., Mangalathu, S., Hu, G., & Wu, T. (2021). Implementing ensemble learning methods to predict the shear strength of RC deep beams with/without web reinforcements. Engineering Structures, 235, 111979. doi:10.1016/j.engstruct.2021.111979

Fiore, A., Quaranta, G., Marano, G. C., & Monti, G. (2016). Evolutionary polynomial regression–based statistical determination of the shear capacity equation for reinforced concrete beams without stirrups. *Journal of Computing in Civil Engineering*, *30*(1), 04014111. doi:10.1061/(ASCE)CP.1943-5487.0000450

Ghannoum, W., Sivaramakrishnan, B., Pujol, S., Catlin, A. C., Fernando, S., Yoosuf, N., & Wang, Y. (2015). *NEES: ACI 369 rectangular column database*. National Science Foundation.

Langdon, W. B., Poli, R., McPhee, N. F., & Koza, J. R. (2008). Genetic programming: An introduction and tutorial, with a survey of techniques and applications. *Computational intelligence. Compendium (Newtown, Pa.)*, 927–1028.

Mai, S. H., Ben Seghier, M. E. A., Nguyen, P. L., Jafari-Asl, J., & Thai, D. K. (2022). A hybrid model for predicting the axial compression capacity of square concrete-filled steel tubular columns. *Engineering with Computers*, *38*(2), 1–18. doi:10.100700366-020-01104-w

Mansour, M. Y., Dicleli, M. U. R. A. T., Lee, J. Y., & Zhang, J. J. E. S. (2004). Predicting the shear strength of reinforced concrete beams using artificial neural networks. *Engineering Structures*, *26*(6), 781–799. doi:10.1016/j.engstruct.2004.01.011

Mo, Y. L., & Nien, I. C. (2002). Seismic performance of hollow high-strength concrete bridge columns. *Journal of Bridge Engineering*, *7*(6), 338–349. doi:10.1061/(ASCE)1084-0702(2002)7:6(338)

Naderpour, H., & Mirrashid, M. (2020). Bio-inspired predictive models for shear strength of reinforced concrete beams having steel stirrups. *Soft Computing*, *24*(16), 12587–12597. doi:10.100700500-020-04698-x

Quaranta, G., De Domenico, D., & Monti, G. (2022). Machine-learning-aided improvement of mechanics-based code-conforming shear capacity equation for RC elements with stirrups. *Engineering Structures*, *267*, 114665. doi:10.1016/j.engstruct.2022.114665

Quaranta, G., Lacarbonara, W., & Masri, S. F. (2020). A review on computational intelligence for identification of nonlinear dynamical systems. *Nonlinear Dynamics*, *99*(2), 1709–1761. doi:10.100711071-019-05430-7

Reineck, K. H., Bentz, E., Fitik, B., Kuchma, D. A., & Bayrak, O. (2014). ACI-DAfStb Databases for Shear Tests on Slender Reinforced Concrete Beams with Stirrups. *ACI Structural Journal*, *111*(5). doi:10.14359/51686819

Reineck, K. H., & Dunkelberg, D. (Eds.). (2017). *ACI-DAfStb databases 2015 with shear tests for evaluating relationships for the shear design of structural concrete members without and with stirrups*. Beuth Verlag GmbH.

Seghier, M. E. A. B., Gao, X. Z., Jafari-Asl, J., Thai, D. K., Ohadi, S., & Trung, N. T. (2021a). Modeling the nonlinear behavior of ACC for SCFST columns using experimental-data and a novel evolutionary-algorithm. *Structures*, *30*, 692–709. doi:10.1016/j.istruc.2021.01.036

Seghier, M. E. A. B., Ouaer, H., Ghriga, M. A., Menad, N. A., & Thai, D. K. (2021b). Hybrid soft computational approaches for modeling the maximum ultimate bond strength between the corroded steel reinforcement and surrounding concrete. *Neural Computing & Applications*, *33*(12), 6905–6920. doi:10.100700521-020-05466-6

Sigrist, V., Bentz, E., Ruiz, M. F., Foster, S., & Muttoni, A. (2013). Background to the fib Model Code 2010 shear provisions–part I: Beams and slabs. *Structural Concrete*, *14*(3), 195–203. doi:10.1002uco.201200066

Vecchio, F. J., & Collins, M. P. (1986). The modified compression-field theory for reinforced concrete elements subjected to shear. *ACI J.*, *83*(2), 219–231.

Walraven, J., Belletti, B., & Esposito, R. (2013). Shear capacity of normal, lightweight, and high-strength concrete beams according to Model Code 2010. I: Experimental results versus analytical model results. *Journal of Structural Engineering*, *139*(9), 1593–1599. doi:10.1061/(ASCE)ST.1943-541X.0000742

Zhang, T., Visintin, P., & Oehlers, D. J. (2016). Shear strength of RC beams with steel stirrups. *Journal of Structural Engineering*, *142*(2), 04015135. doi:10.1061/(ASCE)ST.1943-541X.0001404

# Chapter 8
# Groundwater Modelling of the Saq Aquifer Using Artificial Intelligence and Hydraulic Simulations

**Ghufran Ahmed Pasha**
*Department of Civil Engineering, University of Engineering and Technology, Taxila, Pakistan*

**Abdul Razzaq Ghumman**
*Department of Civil Engineering, College of Engineering, Qassim University, Buraidah, Saudi Arabia*

**Mohammed R. Almuhaylan**
*Department of Civil Engineering, College of Engineering, Qassim University, Buraidah, Saudi Arabia*

**Md. Shafiquzzaman**
*Department of Civil Engineering, College of Engineering, Qassim University, Buraidah, Saudi Arabia*

**Ibrahim Saleh Al-Salamah**
*Department of Civil Engineering, College of Engineering, Qassim University, Buraidah, Saudi Arabia*

**Husnain Haider**
*Department of Civil Engineering, College of Engineering, Qassim University, Buraidah, Saudi Arabia*

**Afzal Ahmed**
*Department of Civil Engineering, University of Engineering and Technology, Taxila, Pakistan*

**Riaz Akhtar Khan**
*Department of Civil Engineering, LEADS University, Lahore, Pakistan*

## ABSTRACT

*Water resources are directly related to the economic conditions of a region. Precise estimation of groundwater is an important step toward better planning and management. This book chapter is dedicated to modelling groundwater in terms of both quantity and quality utilizing ANN (artificial neural networks), ANFIS (adaptive neuro-fuzzy inference system), and the numerical-hydraulic modeling by MODFLOW (modular three-dimensional finite-difference groundwater flow model). The model performance was determined using mean of square error and Nash-Sutcliffe efficiency of model. The pumping data of the*

DOI: 10.4018/978-1-6684-5643-9.ch008

*area was used to determine the parameters of Saq Aquifer (Qassim, Saudi Arabia) including specific storage and transmissivity. It has been found that the ANFIS model is the most effective for qualitative and quantitative modelling of the aquifer. Following sensitivity analysis, different future scenarios for sustainable groundwater pumping were examined. This book chapter presents research findings that will be useful for engineers, planners, and managers of water systems in arid areas.*

## INTRODUCTION

Groundwater hydrology is the branch of hydrology that studies the occurrence, distribution, and transportation of water found below the surface of the earth. It also describes the physical and biological interactions between groundwater and the surrounding environment. Humans have used artificial means such as infiltration galleries or wells to extract groundwater for thousands of years, but the exact date is unknown. Groundwater was considered a water supply throughout the 19th and early 20th centuries due to the increase in population. In many parts of the country, the groundwater resources are severely strained. In most of the transboundary aquifers, similar to the Saq, common in ownership for Saudi Arabia, Jordan, and Iraq, groundwater is depleting at distressing rate of approximately 6.8 millimeters per annum. Figure 1 shows a decreasing trend in groundwater levels from 1980 to 2020. The figure also suggests that the pumping rate was almost half of the available groundwater in the early years. However, currently, the pumping rate exceeds almost 50%. In order to meet the demand of excessive population growth, people mostly rely on the extraction of groundwater for their homes and agriculture, even at low annual rainfall (Almuhaylan et al., 2020; Fallatah, 2020). It is of the utmost significance to make accurate projections of future groundwater levels to ensure the environmental management and improvement of groundwater resources across the earth (Ahmed, 2020; Li et al., 2021; Mallick et al., 2021; Su et al., 2020). Several variables greatly influence the fluctuation of groundwater levels. The duration and intensity of rainfall, topography, aquifer properties, subsoil properties etc., all contribute to groundwater level variation. The interaction between groundwater and surface water can also affect these responses. An understanding of the complex groundwater system may be provided by field studies or remote sensing data obtained by remote sensing techniques about aquifer characteristics. In most cases, there are limitations in measuring groundwater data spatially and temporally; thus, a complete understanding of the system remains inadequate. This is because the groundwater system is extremely complex.

Groundwater models can provide some insight into the complicated dynamics of the groundwater system. Once a model has been developed and validated correctly, it can help predict groundwater behaviour. These predictions can aid decision-makers in exploring additional methods to manage water resources efficiently (Barthel et al., 2021; Li et al., 2021; Shiri et al., 2021; Shuler & Mariner, 2020). The chapter of this book includes several objectives. Out of the key objectives of this research one objective is to develop groundwater models to predict water levels for the Saq Aquifer. However, because all the numerical models have some uncertainty, no single model can deliver results that absolutely satisfy the researcher. This research has therefore examined a range of data-driven models. Three models have been used in this study to investigate the groundwater level. There are a) ANN (Artificial Neural Networks), b) ANFIS (Adaptive Neuro-Fuzzy Inference System), and c) Numerical-hydraulic modelling by MODFLOW (Modular Three-Dimensional Finite-Difference Groundwater Flow Model).

All the hydraulic models that simulate transboundary aquifers in particular places with unknown boundary conditions necessitate extremely costly data collecting strategies. The "general head boundary condition" used in this study has overcome this problem; however, it is still necessary to examine additional modelling strategies. For the purpose of forecasting groundwater levels in the neighborhood of critical locations such as pumping stations, data-driven models can be an effective option (Malekinezhad & Banadkooki, 2018; Roshni et al., 2020). In Table 1, the hydraulic and artificial intelligence model types have been compared. There are several Data-dependent models that are classified into numerous groups. Khedri et al. (2020) evaluated and analyzed five distinct data-dependent models for predicting levels of groundwater. They studied the effectiveness of ANN, ANFIS, Support Vector Regression, Group Data Handling, and Fuzzy Logic. Khedri et al. (2020) suggested that ANN is a highly popular and effective technique. In order to model the flow of groundwater, Majumder & Eldho (2020) used ANN in conjunction with an optimization strategy. In addition, they found that ANN in combination with optimization produced improved convergence.

Another researcher, Suprayogi et al. (2020) used ANN to predict variation in groundwater levels in Meskom, Indonesia accurately. They found that the ANN performed well in forecasting groundwater levels over a short time and observed strong correlation coefficient values. Wunsch et al. (2021) demonstrated the ability of recurrent ANNs to predict groundwater levels using data from 17 existing groundwater wells located in the Upper Rhine Graben, which is one of Central Europe's major groundwater resources. Solgi et al. (2021) examined the working of ANN for predictions of short and long term changes in levels of groundwater. They used groundwater data of more than 17 years located in San Antonio, Texas, United States of America. The water quality of groundwater is another important aspect that needs to be predicted in addition to the groundwater levels (Li et al., 2021; Pal & Chakrabarty, 2021; Shiri et al., 2021). Fadipe et al. (2021) have investigated quality of groundwater in the vicinity of Boluwaduro, Ofatedo, located in the Osun State. The amount of iron, lead, and nitrate in the groundwater has been studied using the ANN structure. Using a variety of approaches that use artificial intelligence, Al-Adhaileh & Alsaade (2021) investigated the groundwater in terms of its quality in several aquifers located in different parts of India. A comprehensive review of the relevant literature reveals that modelling groundwater flow is a complex phenomenon. In data-driven models, there are many aspects that should be examined to select the most excellent model for groundwater accurate forecasts.

ANN research includes the selection of architecture, the investigation of the effects of activation functions, and the selection of an effective training technique. Considering the activation function, number of hidden layers, and number of neurons present in each hidden layer, the ANN may have one of several different architectures (Almuhaylan et al., 2020; Kulisz et al., 2021). When different training functions are utilized, the operation of these models changes (Almuhaylan et al., 2020; Solgi et al., 2021). Various researchers have investigated the performance of different training functions. Coulibaly et al. (2001) examined the effectiveness of three different ANN models with different training functions in predicting groundwater levels in the Gondo plain of Burkina Faso. They evaluated the recurrent neural networks, generalized radial basis function models, and input delay neural networks. Based on level prediction; recurrent neural networks were the most effective. Nordin et al. (2021) used two types of ANN architectures to examine groundwater modelling. The number of various effluents present in groundwater was estimated, and the numerical model was tested using actual groundwater data from Pakistan's largest industrial city, Faisalabad. Daliakopoulos et al. (2005) used seven types of ANN architectures and training algorithms to predict groundwater levels in the Messara Valley, Crete, Greece. It was found that the standard feed-forward ANN modelled with the Levenberg–Marquardt structure produced accurate predictions for up to 18 months. Yoon et al., (2011) used ANN with a back-propagation technique to train the model for finding short-term groundwater level readings in the seaside town of Mukho, Donghae city, Korea. They found that in the results of ANN, there is some degree of uncertainty and proposed that these models should always be meticulously prepared. A back-propagation technique was employed by Krishan G (2015) to predict changes in groundwater level for various districts of Punjab in India. With ANN-backpropagation method, they were able to predict groundwater levels for the next six months with acceptable accuracy. In advanced types of ANN, the learning process is facilitated by a training method comparable to optimization (Li et al., 2021). The basic goal of ANN learning/training is reducing the discrepancy between estimated and measured levels of groundwater and contaminant concentrations in each iteration.

*Figure 1. Pumping rates and depletion of groundwater regarding Saq Aquifer*

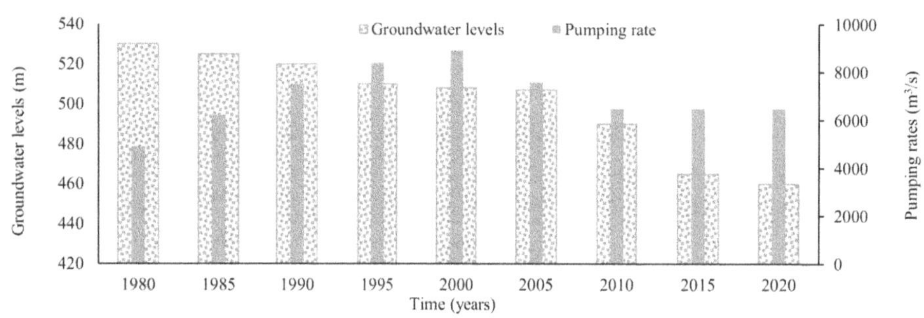


With multiple training trials, weights and biases are changed in several training trials efficiently to achieve a global minimum. Multilayer perceptron has around twelve distinct training functions. If the most effective training function is used, there are changes that ANN will perform best in a short period. In previous investigations, just a few out of 12 outcomes were documented. This study evaluates the performance of the 5 significant training functions to estimate the quality and quantity of subsurface water for pumping locations in Buraydah, Qassim, Saudi Arabia.

In order to determine the quality of groundwater, multiple parameters/contaminants must be assessed. Nevertheless, the primary objective of this book chapter is to evaluate the performance of multiple ANN and ANFIS models. Therefore, few quality factors have been selected to make the comparison more transparent and relatively simple. The data of TDS (total dissolved solids), Fe (iron), and Mn (manganese) were readily accessible. Consequently, these three pollutants were chosen for the quality evaluation. Additionally, it is important to note that the available data comprises annual contaminant concentrations and groundwater levels. Therefore, seasonal variation research falls outside the scope of this work.

*Table 1. Data Dependent Models and Numerical Hydraulic Modelling (A review)*

| Characteristics | ANN/ANFIS | Hydraulic Model | Remarks |
|---|---|---|---|
| Efforts/costs involved in computation expenses (need high-speed computers and time for computations) | *x* | About 10 times *x* | Hydraulic models like MODFLOW necessitate high-performance processors compared to data-driven models like ANNs and ANFIS (Chen et al., 2020). However, this has not been taken into account because of the ease with which computers are easily available. |
| Conditions required for processing models and data | *x* | About 3 times *x* | There is a need for data on topography, water levels, contamination concentrations, pumping rates, and hydraulic characteristics. Therefore, the hydraulic model (MODFLOW) needs highly accurate aquifer data. In addition, a mesh with very high resolution and smaller simulation time intervals is required that satisfies the limiting ratio of time interval to the distance between nodes. For data-driven models such as ANN and ANFIS, merely a decent data set can be adequate (Bizhanimanzar et al., 2020; Das et al., 2019; Emamgholizadeh et al., 2014). |
| Model Category | Data-driven (Black-box/ semi-black box) | Distributed (Based on laws of physics) | For accurate prediction of groundwater levels and contamination parameters, hydraulic models require values of parameters related to the processes of physics and equations based on the laws of nature. On the other hand, all the data-driven models like ANFIS and ANN, do not require equations based on physics. Therefore, these models can only be trained, tested, and validated using the data that has been previously recorded (Bizhanimanzar et al., 2019, 2020; Das et al., 2019; Emamgholizadeh et al., 2014). |
| Model Prediction Biases | *x* | About 10 times *x* | Biases in hydraulic model outputs can range widely, however, biases in data-driven model (ANFIS/ANN) predictions are typically restricted to a specific range. The ANFIS/ANN, on the other hand, lacks generality (Bizhanimanzar et al., 2019, 2020; Das et al., 2019; Emamgholizadeh et al., 2014). |
| Future Predictions for a Long Time. | Highly-challenging | Much easier once the model is calibrated | After the hydraulic model has been calibrated and validated, it is quite simple to utilise it for future forecasts. On the other hand, to apply ANFIS/ANN for long-term future predictions, one must have a significant amount of experience and competence in a specific phenomenon (Bizhanimanzar et al., 2019, 2020; Das et al., 2019; Emamgholizadeh et al., 2014). |

## MATERIALS AND METHODS

### Important Hydraulic Properties of Groundwater and Subsoil Strata

*Porosity:* It is defined as the volume of subsoil openings divided by the entire sample volume.
*Moisture Content:* It is the proportion of water in a sample to its overall weight or volume.
*Aquifer*: It is subsoil strata which can store and transmit water.
*Hydraulic Head:* The water level height from a reference plus pressure head.
*Specific Storage*: In a unit volume of the saturated permeable matrix, the amount of water taken into storage under a unit rise in pressure or the volume of water obtained under a unit reduction in pressure is referred to as specific storage.
*Storage Coefficient:* T The storage coefficient or storativity of an aquifer is equal to the specific storage multiplied by the saturated thickness.
*Hydraulic Conductivity:* It refers to how easily water is able to travel through an aquifer.
*Darcy Law:* The velocity flux in an aquifer is directly proportional to the hydraulic gradient.

*Transmissivity:* It is obtained by multiplying the hydraulic-conductivity and the fully saturated aquifer-thickness.

## Methodological Framework

In this chapter, the prediction of quantity and quality of groundwater for the Saq Aquifer is done using two approaches, a) MODFLOW, a hydraulic model, and b) ANN and ANFIS, data-driven models. Table 1 provides a conceptual comparison of the two strategies. A comprehensive approach framework for groundwater modelling is explained in Figure 2. It is worth noting that five ANN models (ANN-1, ANN-2, ANN-3, ANN-4 & ANN-5) presented in Figure 2 are due different training functions only. The actual number of ANN models are much more than 5 because of 5, 10 and 15 neurons in double and triple layer ANN models. This research involves three essential steps. In the first step, data related to pumping rates, groundwater levels, the locations of wells, contaminants concentration, and the properties of the aquifer were collected. There are a huge number of data points. Data from a total of 44 public wells has been collected. Water levels and pumping rates data from 1980 to 2018 and contaminants (TDS, Fe & Mn) data from 2010 to 2018 has been taken from relevant authorities. The data of pumping rates, water level, and contaminant concentration have been normalized for use in the ANFIS and ANN models. In the second step, (hydraulic/ ANFIS/ ANN), calibration/ training/ testing/ validation was done, considering necessary model requirements. 60% data was used for training/ calibration, while the other 40% was used for testing/validation. The calibration and validation were carried out against the measured data of both quality and quantity of groundwater. In addition, adjustments were made to the hydraulic parameters, so that the level obtained by the model becomes closer to the actual groundwater level in a predetermined model run time. Pollutant transport model calibration is primarily based on adjusting transport parameters (dispersivity coefficient, etc.) to ensure that the model's predicted concentration values match the observed concentrations. Six wells along with their location coordinates and past groundwater level records were identified and used as observation wells for concentration.

For the hydraulic model (MODFLOW), the data distribution was made in two parts, i.e., 60% and 40% for calibration and validation, respectively. Whereas, for data-driven models (ANN/ANFIS), the distribution of data was made in three parts, i.e., 60 percent for training, 20 percent for validation, and 20% for testing. The ANFIS/ANN models were generated using MATLAB to fulfil the required goals.

An analysis for sensitivity of the variables was conducted in order to determine the effect of various model parameters on results of the model. The third step consisted of choosing the best model for future forecasts of groundwater quality and levels. The use of groundwater is on the rise as a result of the extraordinary growth rise in population and high quality of life (Gonand, 2016). Eight future scenarios with varying pumping rates (mentioned in Table 2) have been analyzed to investigate aquifer's depletion in terms of quality and quantity. Table 2 provides an analysis of the probable possibilities between the two extremes, as well as the most optimistic and pessimistic scenarios.

*Figure 2. Framework for groundwater modeling to explore the effects of withdrawals of groundwater on aquifer diminution*

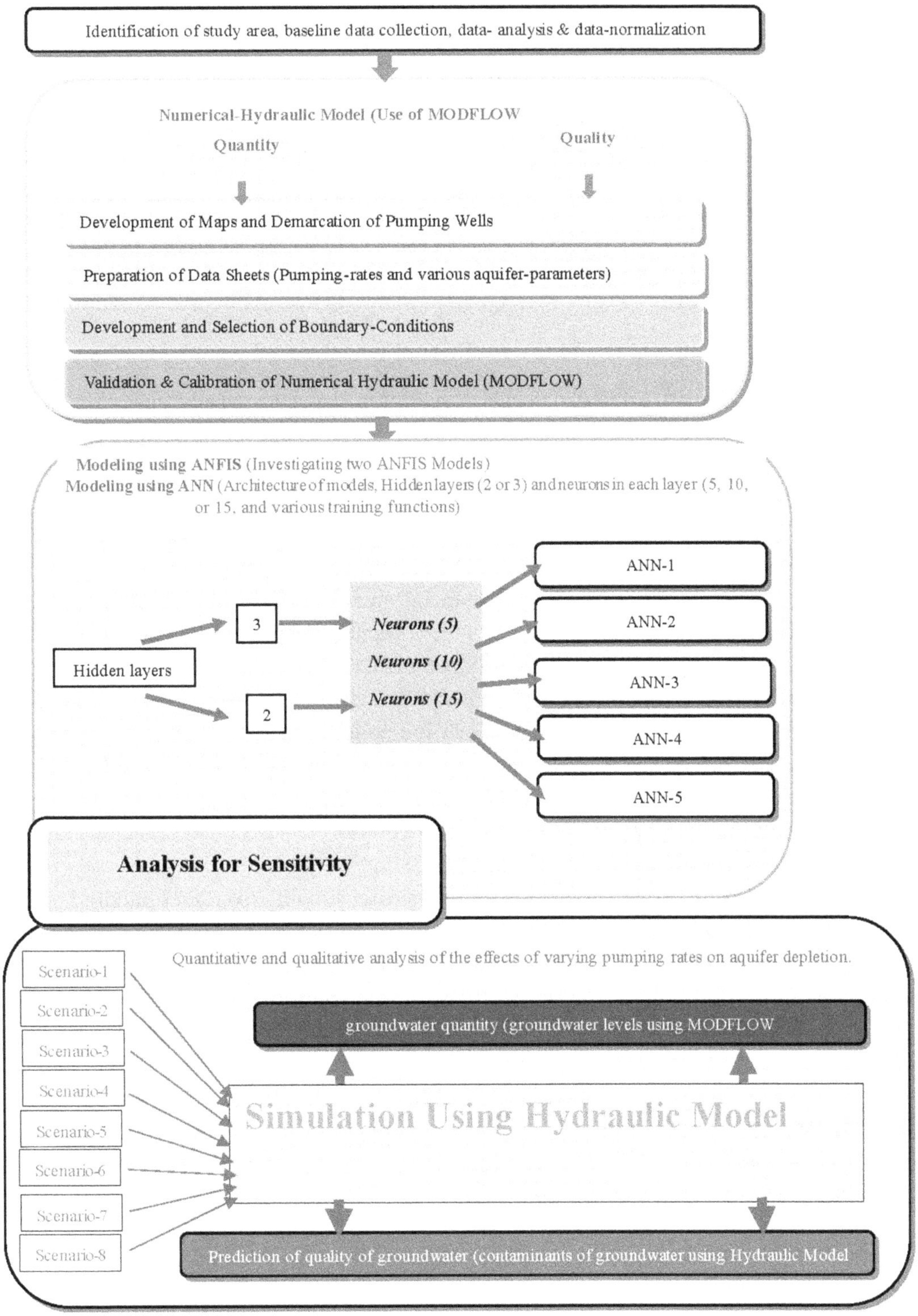

*Table 2. Future pumping rate scenarios*

| No (Scenario) | Explanation | No (Scenario) | Explanation |
|---|---|---|---|
| 1 | Assuming an annual drop of 1% in pumping rate | 5 | Assuming an annual increase of 3% in pumping rate |
| 2 | Assuming the current groundwater withdrawal rate remains constant for the whole duration of the period | 6 | Assuming an annual increase of 4% in pumping rate |
| 3 | Assuming an annual increase of 1% in pumping rate | 7 | Assuming an annual increase of 5% in pumping rate |
| 4 | Assuming an annual increase of 2% in pumping rate | 8 | Assuming Scenario 2 with artificial recharging |

## Study Area

In this research the Saq Aquifer, located in the Qassim Region, was selected as a study area (Figure 3). It is an important region of the Saudi Arabia and is an important arid region in the country. It lies between the latitudes of 25° 00' and 27° 00' North and the longitudes of 42° 30' and 45° 00' East and encompasses approximately 80,000 km$^2$ of the land area (Almuhaylan et al., 2020; Mallick et al., 2021). Its altitude ranges from 600 to 850 meters above sea level. As on 2017 estimates. The total population of Qassim region is 1,423,935 per square kilometer and it is the second-largest sectors in the field of agricultural in Saudi Arabia. The people have an easy access to the groundwater from the Saq aquifer. Furthermore, the region mostly includes deserts environment, with hot and dry summers and cold and wet winters. The mean yearly evaporation is almost 4,500 millimeters. The maximum daily temperature during the summer is 49°C, while the minimum nightly temperature is 36°C. However, winter temperatures may occasionally drop below 0 degrees Celsius. The yearly precipitation in Qassim is approximately 125 mm (Almuhaylan et al., 2020). The Saq Aquifer, one of the most significant aquifers in the Kingdom, serves as the region's primary source of groundwater supplies (Al-Omran et al., 2017; Mallick et al., 2021). It has a very extensive outcrop, which spans over 1200 kilometers in Saudi Arabia alone and joins the Jordanian border as fat a longitude 45°00' E and latitude 24°30' N.

The large aquifer is naturally constrained, with thicknesses ranging from four hundred meters in its southern to 700 meters in its northern portions. The aquifer's portion being studied has an average thickness of 500 m (Almuhaylan et al., 2020). The strata is composed of sandstone with a medium to coarse grain size. A few localized locations contain sandstone of fine grain size. Most of the public and private tube wells are only partially-penetrating, reaching a depth of around 650 meters with 125-meter screens (Almuhaylan et al., 2020). A large valley known as Wadi Rummah runs through the Saq portion of the Qassim region. Compared to the region farther from Wadi Rummah, the Saq Aquifer is more salinized in the vicinity. Wadi recharge, effluent from sewage treatment plants, and agricultural activities, are crucial elements that may add to the relatively increased saltiness of the waters in the Wadi Rummah region (Mallick et al., 2021).

*Figure 3. The map of the study area with the locations of the pumping locations, the Saq outcrop, and the Qassim Area*

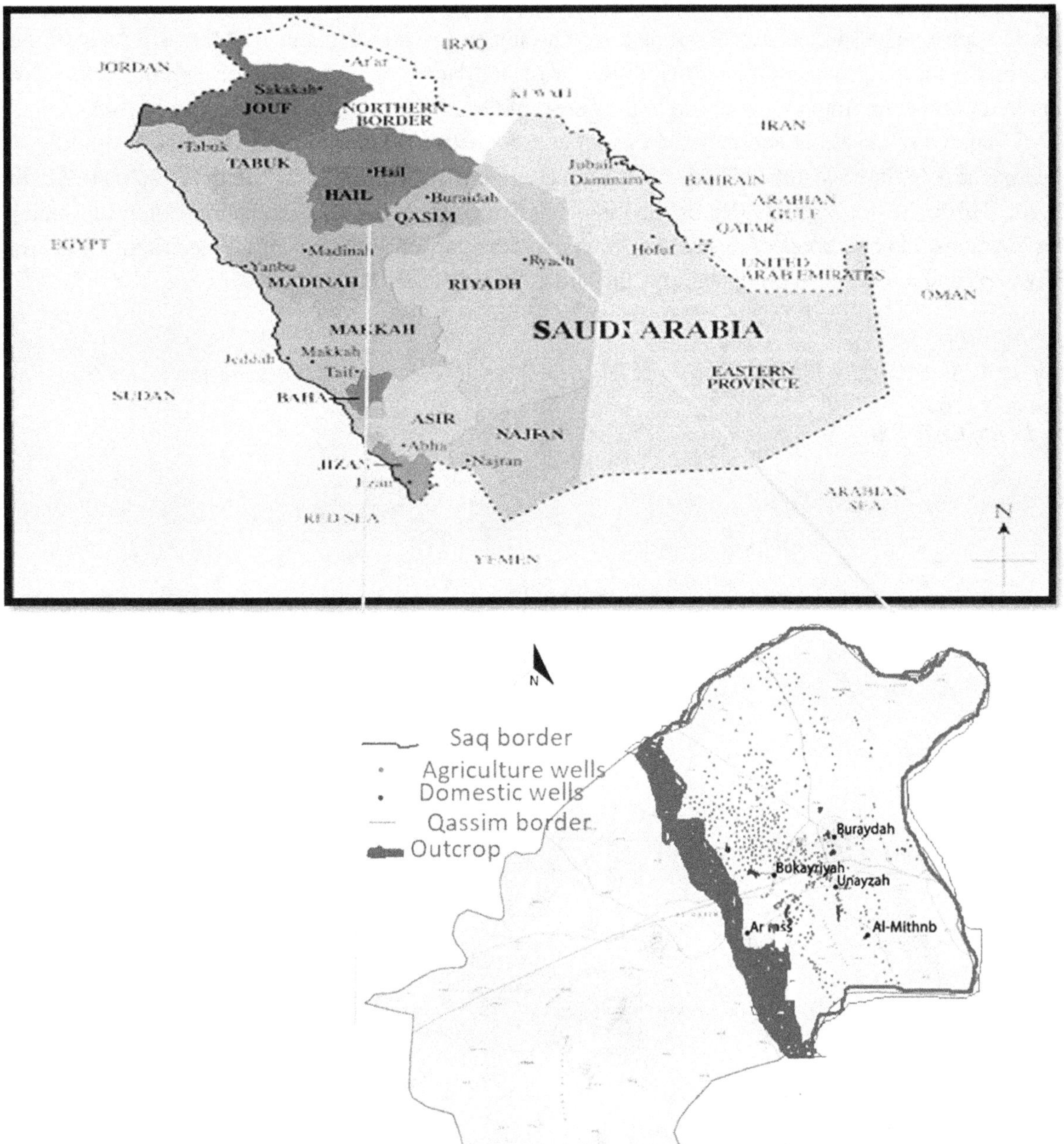

## Saq-Aquifer: Significant Geographic Elements and Hydraulic Characteristics

The Saq-Aquifer begins at 24°30'N-latitude and extends to 45°E-longitude at the Jordanian border. It has an approximate surface area of 65000 $km^2$. The subsurface area is about 160000 $km^2$. As described previously, the Saq sandstone outcrops to the West and North of Tabuk Region continues for seven hundred kilometers from the west side of the desert of Great Nafud towards the border of Jordanian. At Jabal Saq, the thickness of aquifer reaches six hundred metres (Saq-Mountain). In the Qassim Region, the thickness of the Sandstone of Saq slowly increases from 400 metres in the South to 700 metres in the North. This evidence suggests that the Saq is a massive aquifer with thousands of groundwater pumping locations. In some areas of this aquifer, water was available at shallow depths (see Figure 4 below). However, due to excessive pumping now, the situation is different.

*Figure 4. A well at the Tannumah town in Alasyah, with its elevated head of artesian water from Saq-aquifer (1964)*
*(Al-Jerash, 1968)*

There are localized patches of fine sandstone throughout the medium to coarse sandstone. The type of rock is quartz sandstone that is coarse to well sorted. Abdel-Aal et al. (1997) state that the conductivity is on the rise (it grew to 2.76 dS/m in 1987 from 1.93 dS/m in 1983). The TDS (total dissolved solids) grew from 1395 mg/L in 1983 to 1992 mg/L in 1987, according to their findings. The Saq Aquifer was mineralized to acceptable levels. As per Sharaf & Hussein (1996) total dissolved solids (TDS) ranged from 300 to 1,000 milligrams per litre.

The distinctive characteristic of this Saq is that it has boundaries which can be well identified only at its ends, and at various places its boundary conditions keep on changing on daily basis because of continuously pumping of water from the aquifer whereas the recharge is negligible (Almuhaylan et al., 2020). To encompass the real boundary conditions, one must gather vast data over an extended period. In addition, it is necessary to collect information on the number of pumping wells. Therefore, the only possibility is to deal with a limited area with unknown borders. In the present work, a generalized head boundary has been examined.

The data point frequency for Saudi Arabia's main Saq Sandstone and related aquifers, which cover an area of around 370000 km$^2$, is nearly 01 measurement per ten thousands of km$^2$ (Lloyd, 2007). Therefore, the pumping test data collected by the authors (Figure 5) were used to assess the hydraulic parameters of the aquifer. The hydraulic conductivity k, transmissivity, T and storage coefficient S, were calculated using the following equations.

$$T = \frac{2.3Q}{4\pi\left[\Delta\left(h_o - h\right)\right]} \tag{1}$$

$$S = \frac{2.25Tt_o}{r^2} \tag{2}$$

$$K = T/B \tag{3}$$

Where $Q$ is the well's discharge, $r$ represents the distance in-between the pumping-well and observation-well, $t_o$ is the intercept of a line joining various points on graph and extended towards the x-axis, and $\Delta(ho\text{-}h)$ is the drawdown for one log time cycle, depicted in Figure 5.

*Figure 5. Pumping test drawdown (m) against time (min) plotted on semi-logarithmic paper using the Jacob Method of analysis*

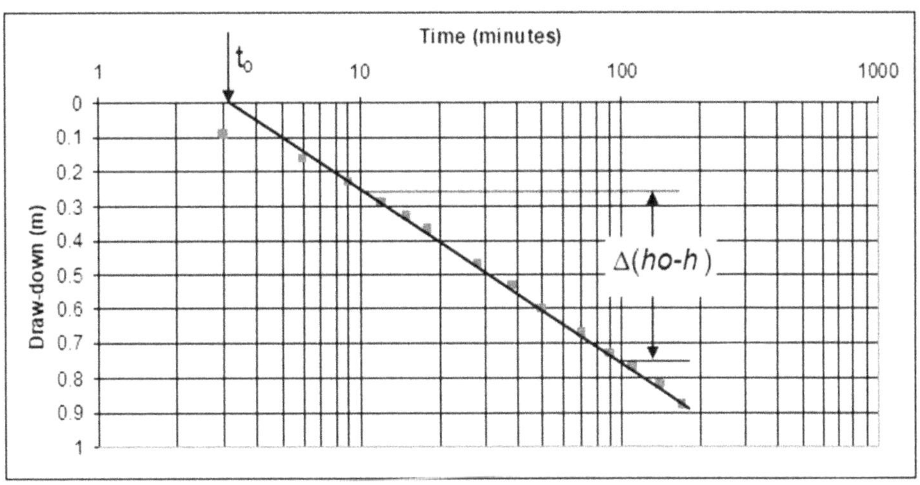

The conductivity (hydraulic) of the aquifer was determined by ratio of the transmissivity to the aquifer's fully-saturated thickness. In this study, the hydraulic conductivity was determined to be 3.60 metres per day. Further the aquifer-specific-storage was determined by the ratio of aquifer-storage-coefficient to the aquifer's saturated thickness. The particular storage was determined to be 6e-7 $m^{-1}$.

## Model Formation

In MODFLOW, a discretized domain composed of several nodes and related finite difference blocks replaces the aquifer system (cells). The study area is represented by a total of 60 columns and rows each, with additional columns and rows added where wells are located to reduce the consequences of a mistake caused by big grid components. The locations of the wells are depicted in Figures 6. In the model, three layers i.e., 600, 250 and 250 m, were considered. A 600-meter-thick impermeable stratum covers the 500-meter-thick aquifer. To adequately represent the partially penetrated wells, the aquifer was separated into two strata of 250 m thickness each. The length and width of the modelling domain were 30,000 m and 30,000 m, respectively. In the model, "CONFINED" layer type was selected. The regional decline was estimated by assuming the effective average transmissivity (T) was assumed to be 1.25 $m^2$/min.

The hydraulic conductivity in all three dimensions, i.e., x, y and z, was calculated using the formula $K=T/B$, where $B$ is the aquifer's thickness in metres; therefore, $K_x = K_y = K_z = 3.6$m/day. In the model, no water recharge from the surface was considered. Also, the boundary conditions of the study analysis area were not known and varied over time; therefore, these conditions were calibrated.

## Description of Hydraulic Model (MODFLOW)

### Levels of Groundwater and Flow Modeling

The modelling of the flow of groundwater is an extremely difficult endeavor (Younes et al., 2021). In order to represent groundwater flow, partial differential equations are used, such as the one shown below (Mezouary & Mansouri, 2021).

$$\frac{\partial}{\partial x}\left(K_{xx}\frac{\partial h}{\partial x}\right)+\frac{\partial}{\partial y}\left(K_{yy}\frac{\partial h}{\partial y}\right)+\frac{\partial}{\partial z}\left(K_{zz}\frac{\partial h}{\partial z}\right)=S_s\frac{\partial h}{\partial t}-W \tag{4}$$

Where "$K_{xx}$", "$K_{yy}$", and "$K_{zz}$" represent the conductivities (hydraulic) in the x, y, and z directions, respectively; "$h$" represents the groundwater head; "$S_s$" represents the specific storage of the aquifer; "$W$" represents the source/sink; and "$t$" denotes time. At the locations of wells, a denser mesh was built to obtain accurate predictions (Behera et al., 2022; Jafari et al., 2021; Mamo et al., 2021; Namitha et al., 2019).

Visual MODLOW is widely applied in modeling groundwater flow problems. It has graphical user interface that is used in commercial applications. As an input, Excel files, Surfer grids, data from GIS and AutoCAD, and other formats can be used instead of the traditional text file format. In addition, the results can be displayed graphically, such as on charts and contour maps. Due to these features, the amount of time needed for execution is reduced, thus making the analysis of findings more simple predictions (Jafari et al., 2021; Mamo et al., 2021; Namitha et al., 2019). In order to solve Equation 1, in addition to finite element methods, there are many explicit and implicit finite differences methods. A finite difference approach is used implicitly by MODFLOW to solve Equation 1. Table 3 presents the input data for MODFLOW. In large aquifers such as Saq, choosing the size of mesh and correct boundary conditions requirement is a tough task. A 30 km × 20 km rectangular mesh was built by producing 70 columns and 80 rows; however, in one case, a larger area was considered for modelling to examine the region surrounding Wadi Rumah. Dense mesh was created to get more accurate results. In order to forecast groundwater (both quality and quantity) as precisely as possible, the number of rows and columns for the mesh surrounding the pumping wells was increased to create a highly dense mesh. The vertical depth of the aquifer was estimated to be around 500 metres, with an impervious cover of roughly 628 metres above the 500-meter-thick layer of Saq aquifer (Al-Omran et al., 2017; Almuhaylan et al., 2020). As previously stated, a 125 m long screen partially penetrates the pumping wells. Therefore, the constrained 500 m thick aquifer has been separated into four layers of similar thickness to accurately simulate the impacts of the well's screen (125 m each). As a result, five layers with a thickness of 631, 125, 125, and 125 m have been used, with the top impermeable layer of 631 m thickness (Figure 6).

## Quality Modeling

For investigating the quality of groundwater, the movement and concentrations of pollutants are modelled. Consequently, an extra equation is needed to simulate water quality depending on the movement of pollutants. Equation 2 describes the three-dimensional transport of contaminant concentrations in a system of groundwater flow (Kamali & Niksokhan, 2017). It is a partial differential equation as given below.

$$\frac{\partial C}{\partial t}=\frac{\partial}{\partial x_i}\left(D_{ij}\frac{\partial C}{\partial x_i}\right)-\frac{\partial}{\partial x_i}\left(\upsilon_i C\right)+\frac{q_s}{\theta}C_s+\sum_{1}^{N}R_k \tag{5}$$

In above Equation 2, "$C$" represents the dissolved concentration of pollutant, "$k$", "$\theta$" is the porosity of the subsurface strata, "$D_{ij}$" is dispersion coefficient (hydrodynamic), "$x_i$" is the distance lengthwise over the particular coordinate axis, "$vi$" is linear pore water velocity, "$q_s$" is flow rate(volumetric), "$R_k$" is the term related to the chemical reaction and "$C_s$" is the source-concentration (Jabbari et al., 2020; Kamali & Niksokhan, 2017).

Several solutions exist for Equation 4 when combined with Equation 5. The most used methods are finite element and finite difference. As it is difficult to simulate Equations 4 and 5, various commercial softwares are available to perform this task. In this research, MODFLOW is used to predict the quantity and quality of groundwater.

For simulations of contaminant transport, all the above factors specified in preceding section utilized to estimate groundwater levels were held constant (quality modeling). As a transport boundary condition, the constant concentration options have been specified in the Visual MODFLOW. As a starting point for the model, the mean values for Fe, TDS, Mn in 2010 were used. To apply initial concentration values, "Assign" was selected from the MODFLOW menu, and the starting concentration values were entered in the "Assign Window." The concentration of pollutants in groundwater and their transport path were simulated using Visual MODFLOW's MT3DMS module. This module analyzes Equation 2 paired with Equation 1 using the technique of finite differences. The standard parameters for dispersion have been adopted. In addition, the fully-saturated constant density option was selected. The calibration of the contaminant (Fe, TDS, and Mn) transport model was carried out by carefully adjusting the pollutant transport attributes (dispersivity coefficient, etc.) till the observed concentration values of Fe, TDS, and Mn coincide with the predicted concentration values of the model for these contaminants. As explained earlier, the six wells were selected as observation wells for contaminant concentration. The model was given the locations of these wells, TDS, Fe, and Mn concentrations, and historical data related to groundwater levels.

*Table 3. Input data values for MODFLOW*

| Required Data | Data-Description/Estimation | Adopted Values of Data |
|---|---|---|
| Transmissivity & Conductivity | To determine the hydraulic-conductivity of the Saq-Aquifer located in the Qassim Area, the Cooper Jacob technique was performed to the pumping test (Al-Omran et al., 2017; Cooper Jr & Jacob, 1946) | $T$=0.024 to 1.62 $m^2$/minute, $K_{yy}= K_{xx}=K_{zz}$ =2.8 m/day |
| No of aquifer layers | Research publications (Al-Omran et al., 2017; Almuhaylan et al., 2020). | Layers-adopted = 5, having thicknes of 631m, 125m, 125m, 125m, and 125 m, respectively |
| Mesh size | selected the best with the help of modelling with several sizes of mesh. | Highly dense. Around the wells, it was made denser further |
| Pumping rates | According to the data taken from the Ministry of Environment, Agriculture and Water, Saudi Arabia | Data of private and public tube-wells (1980 to 2018) |
| Water levels (a) | According to the data taken from the Ministry of Environment, Agriculture and Water, Saudi Arabia | Data of private and public tube-wells (1980 to 2018) |
| Initial conditions | According to the data taken from the Ministry of Environment, Agriculture and Water, Saudi Arabia | Levels of groundwater at the beginning of simulations |
| Boundary conditions | Calibration was made by assuming the conditions of general head boundary-condition | A daunting task |
| Storage coefficient | Method of Cooper-Jacob was used for pumping-test-data to calculate the hydraulic conductivity of Saq Aquifer in Qassim Area (Al-Omran et al., 2017; Cooper Jr & Jacob, 1946) | 2.5 x $10^{-3}$ to 6.3 x $10^{-5}$ |
| Contaminant concentrations | Measured data | Measured data |

*Figure 6. Saq Aquifer Sub-soil Strata used for MODFLOW*
*(Almuhaylan et al., 2020)*

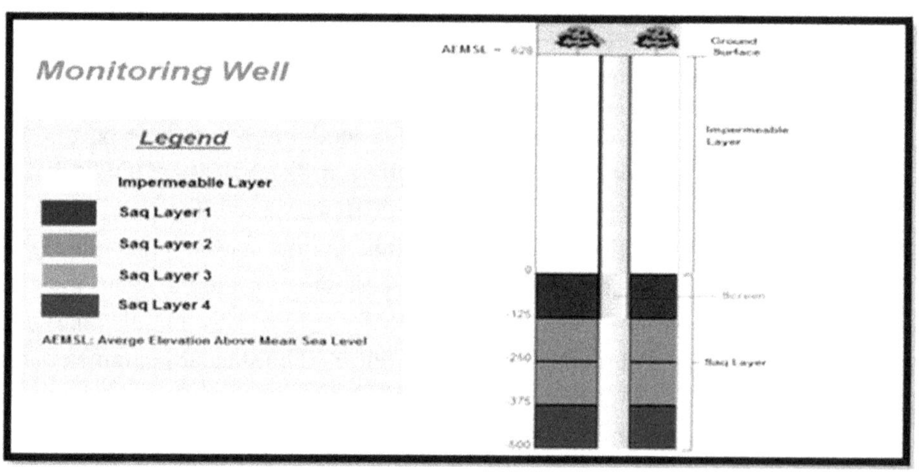

## Estimation of Parameters of Model

### Models Based on Artificial Intelligence

In modelling groundwater flow, the implementation of artificial intelligence (AI)-based models is becoming increasingly common and beneficial (Al-Adhaileh & Alsaade, 2021; Almuhaylan et al., 2020; Li et al., 2021; Wunsch et al., 2021). All the domains of Water resources, including development, planning, management, hydrology, quality of water forecasts, and reservoir operations, have all benefitted from AI approaches (Behera et al., 2022; Bozorg-Haddad et al., 2021; Mattas et al., 2021; Singha et al., 2021). Furthermore, there has been an increase in the adoption of AI models since they can generalise relationships for a complicated phenomenon through the learning process (Abd-Elmaboud et al., 2021; Tabari et al., 2021). Nevertheless, the application of AI models in the field of groundwater quality and quantity prediction specialization is still a crucial and critical research problem despite the limited scope of this application of AI models (Abd-Elmaboud et al., 2021; Tabari et al., 2021; Wu et al., 2020). Figures 7 a&b demonstrate a typical AI model configuration.

### Models Based on ANN

In ANN-type models, neurons play a crucial part in defining the values of the input and output variables compared to the target values. Figure 7a shows that there are primarily three layers named as output, input, and hidden. The internal-structural-method does the calculations. The hidden layers inside the internal-structural technique operate as the backpropagation (BP) and the feed-forward mechanism for providing outputs with sufficient precision. These hidden layers are considered as the backbone of predictions. When there are numerous hidden layers, the "Multi-Layer Perceptron" is formed (MPL). One of the many types of ANN models is one that includes a training function (Stylianoudaki et al., 2022). Thin research used five different types of training in this study (Table 4). A total of five different artificial neural network models have been created, each with its own name, i.e., ANN1, ANN2, ANN3, ANN4 and ANN5. Tests have been conducted on both the double (DL) and triple (TL) hidden layers. It is possible that the hidden layers include a varied number of neurons, which will need to be determined in order to achieve the best outcomes feasible (Arifin et al., 2019; Ogunbo et al., 2020; Sheela & Deepa, 2013). In this research, the strategy of "hit and trial" was utilized for the purpose of selecting the optimal number of neurons. Models are said to be accurate when their predictions are in line with the observed data. Modern ANN algorithms do this by combining an efficient optimization mechanism that adjusts the weights of the ANN process in order to find the maximum possible precision.

*Table 4. Different training functions chosen for models based on ANN*

| Model-Name | Used Function | Training Function |
|---|---|---|
| ANN-1 | trainscg | Scaled conjugate gradient BP |
| ANN-2 | trainlm | Levenberg-Marquardt BP |
| ANN-3 | trainbr | Bayesian regularization |
| ANN-4 | trainbfg | BFGS Quasi-Newton BP |
| ANN-5 | trainrp | Resilient backpropagation (Rprop) |

## Adaptive Neuro-Fuzzy Inference Systems

Groundwater quality and quantity can be accurately predicted using a hybrid AI model and an effective optimization subroutine i.e., "Adaptive Neuro-Fuzzy Inference Systems (ANFIS)". Many researchers have used ANFIS in solving Hydraulic phenomena with high precision and accuracy (Almuhaylan et al., 2020; Stylianoudaki et al., 2022). Figure 7b illustrates the structure of the ANFIS, which consists of the five levels detailed below (Table 5).

*Table 5. Structural layers for ANFIS*

| Layer No | Description |
|---|---|
| 1 | Every intersection of a fuzzy-layer is referred to as an adaptive-node. All the member functions can be viewed as nodal functions, similar to the Generalized-Bell, and Gaussian functions can be used as nodal functions. |
| 2 | A product layer where every nodal output denotes the firing strength of a rule. |
| 3 | A normalized layer where every node denotes the firing strength of a rule in its normalized form. |
| 4 | A de-fuzzy layer in which each node denotes an adaptive node with a nodal-function that specifies the impact of the rules on the output as a whole. |
| 5 | A total output layer represents the output values after de-fuzzification. |

## Model Performance Evaluation

In this study, the performance of models is characterized by two statistical indicators: the NSE (Nash–Sutcliffe Efficiency of the Model), and MSE (Mean Square Error). Both of the indicators are determined by a weighted difference between the measured and predicted variables; Mn, TDS, Fe, and groundwater levels. From 1980 to 2018, nearly thirty-eight years of data for pumping rates, TDS, Mn, Fe, and groundwater levels, were used to assess the model's performance at different run-time phases (calibration/training, validation, and testing). Similar methods have been used in numerous earlier investigations (Afzaal et al., 2020; Almuhaylan et al., 2020; Stylianoudaki et al., 2022).

The NSE (Nash–Sutcliffe efficiency of the model) is represented as follows:

$$NSE = 1.0 - \frac{\sum_{i=1}^{n}(GWL_i^o - GWL_i^p)^2}{\sum_{i=1}^{n}(GWL_i^o - GWL_{avg})^2} \tag{6}$$

where the symbol "$(GWL_i^o$ and $GWL_i^p)^2$" represent the measured and estimated levels of groundwater respectively for $i^{th}$ value of data, and "$n$" represents the total number of measured or estimated data points. The model-performance is deemed "very good" for values of NSE between 1.0 and 0.75, "good" for its values between 0.75 and 0.65, "satisfactory" for values of NSE between 0.65 and 0.5, and "acceptable" for its values between 0.5 and 0.4 (Lin et al., 2017; McCuen et al., 2006).

*Figure 7. AI Modeling technique; (a) Flow-chart representing the MLP (a feed-forward-ANN), (b) ANFIS; 5 functional-blocks*

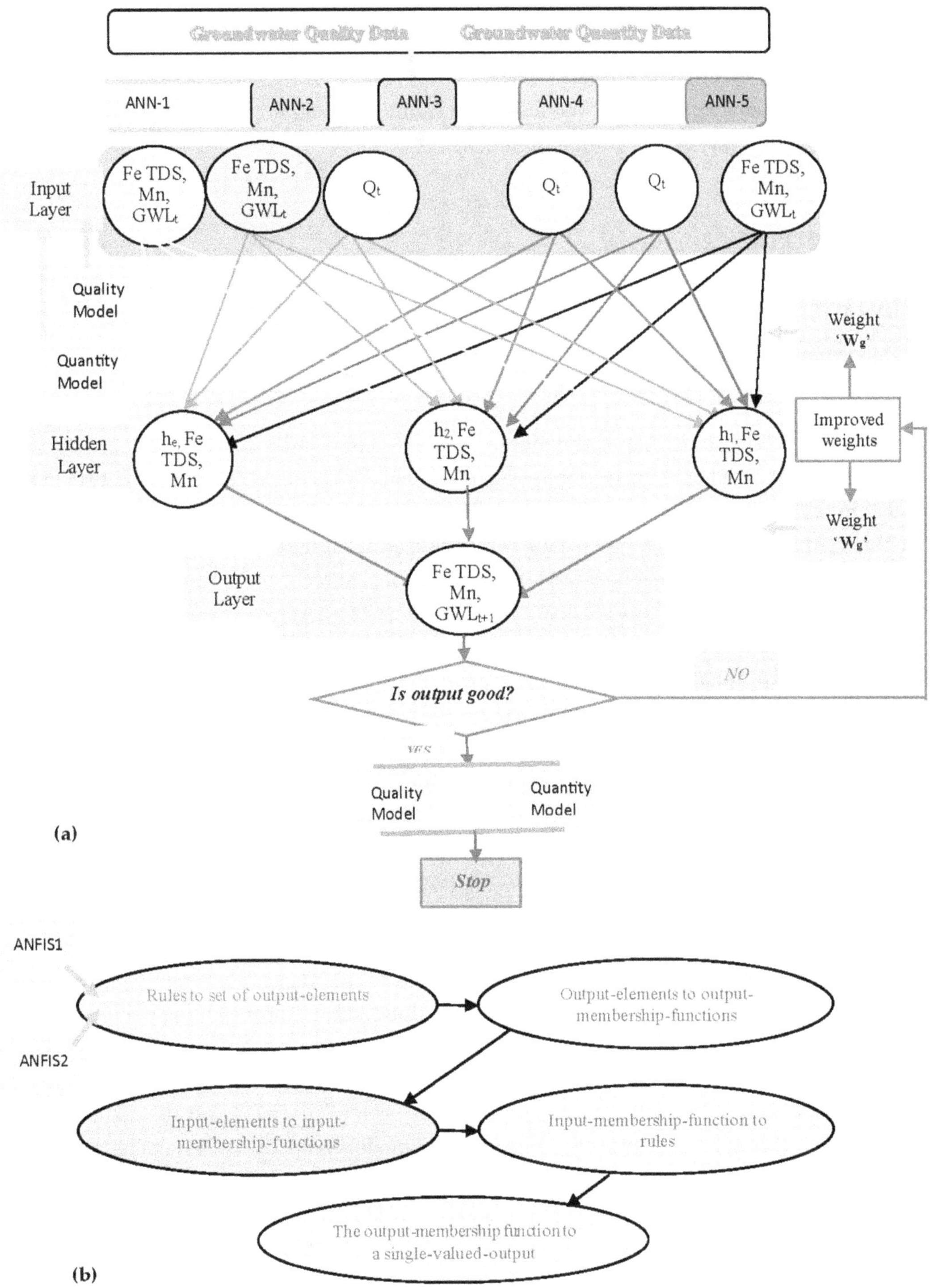

The Mean Square Error (*MSE*) is represented as follows:

$$MSE = \frac{1}{n}\sum_{i=1}^{n}\left(GWL_i^o - GWL_i^p\right)^2 \quad (7)$$

For the case of groundwater quality modelling, in Equations 6 and 7, the term *GWL* will be replaced by the values of TDS or the concentration of Mn and Fe.

## RESULTS AND DISCUSSION

### Modeling for Quantity

#### Results of Models Based on AI and Hydraulic Principles

Figures 8a&b show the results for model-calibration & validation and general head boundary condition. Figure 8a shows that an NSE value of 0.9 has been attained. In terms of the criteria outlined by earlier studies, it is in the "very good" range 4 (Lin et al., 2017; McCuen et al., 2006). Figure 8b depicts a graph that compares simulated data to recorded data. This graph is also relatively close to the line indicating 1:1 (the 1:1 line reflects an optimum fit between simulated and measured groundwater levels.) Because of the NSE of 0.87, the validation findings are also considered satisfactory. Modifying general head boundary conditions for the hydraulic model, as detailed in section 2.3, is one of the most challenging and novel elements of this study. The outcomes of this task, which are illustrated in Figures 9a and 9b, reveal that it has been completed plausibly. A few simulated groundwater depths have been slightly underestimated, while others have been slightly exaggerated. These findings are supported by prior studies by Lyons et al. (2020) and Mohanty et al. (2013). Such results are appropriate for modelling groundwater flow. This aspect is true for several reasons. To begin, one of the primary contributors to the unpredictability of the model findings is the selection of the hydraulic parameters used in the model. These factors can include hydraulic conductivity and specific storage. Secondly, the data quality in technologically developing nations is insufficient to generate very high groundwater model efficiency. Thirdly, parameter estimations may be significantly impacted by the local grid refinement, which may have an impact on the modelling outcomes. In addition, the precision of the results depends on the sophistication of the optimization technique utilized in MODFLOW. Our findings show that traditional techniques (physical and numerical modeling) depend on a wide range of parameter inputs and that the underlying mechanisms are typically too complex to understand. Consequently, data-driven methods should also be evaluated.

Figures 9a and 9b present the outcomes obtained by ANN. The hit-and-trial method was used to decide the optimum number of neurons for all five ANN models named: ANN1, ANN2, ANN3, ANN4 and ANN5. ANNs with ten hidden neurons performed best in the two- and three-layer architectures. Although, the performance with five or fifteen neurons is also "very good". A neuron in an ANN is a mathematical operation that simulates the operation of a biological neuron. There is an abundance of information about the functions of neurons. It is also interesting to learn how ANN behavior varies as the count of neurons increases. This exercise is further presented in the section of quality modelling. However, engineers are concerned with locating the optimal solution to a practical issue. Therefore, the exercise done in this section is enough to resolve the present groundwater quantity modelling. So, to compare ANN for different layers, i.e., double layer (DL) and triple layer (TL), ten neurons have been used. This exercise was performed solely for model ANN1. In the case of DL, the NSE values for training, testing, and validation were 0.95, 0.94, and 0.96. For TL, the values for training, testing, and validation were 0.97, 0.81, and 0.92. NSE values between 0.75 and 1.0 can be considered 'very good' in terms of the model's performance. Additionally, the MSE error is quite negligible, values ranging between 0.002 and 0.004. As a result, the performance of both the models is considered as "very good." When compared to the ANN with DL, the ANN with TL has far better performance. Deep comprehension of the behaviour of ANN with DL and TL requires extensive reading of the literature on ANN. However, for engineers working in groundwater hydrology, the findings of this section may be sufficient. Stylianoudaki et al. (2022) and Almuhaylan et al. (2020) have obtained results that are comparable. Therefore, a neural network with 10 neurons and TL has been employed for further study.

*Figure 8. Results of MODFLOW; (a) The NSE values, (b) Comparison of observed vs predicted depths of groundwater and 1:1 line*

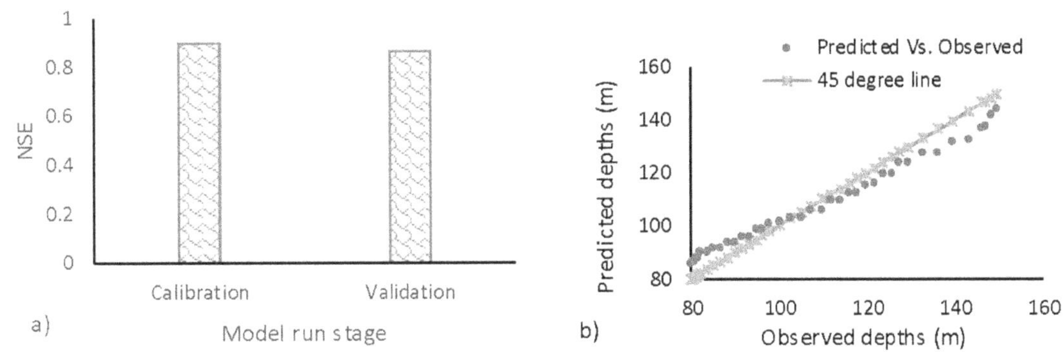


*Figure 9. Results of ANN-1; (a) The NSE values, and (b) The MSE values*

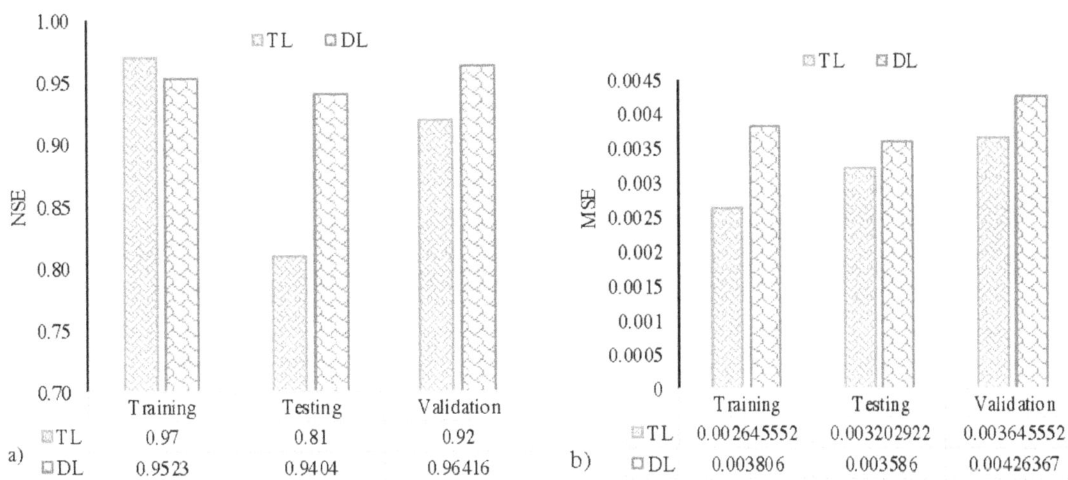

Figures 10a & b illustrate the performance of five ANN models. Figure 10a demonstrates that ANN1, which employs the Scaled Conjugate Gradient training function, performs better than the other ANN models (see Table 4 for the description of ANN models). The Scaled Conjugate Gradient (SCG) is an efficient training process with a fast convergence rate and good accuracy. The superior performance of ANN1 is most likely attributable to the adoption of a highly effective optimization process in the Scaled Conjugate Gradient in order to reach the best global minimum achievable (Almuhaylan et al., 2020; Lyons et al., 2020; Mohanty et al., 2013). Figure 10a reveals that for almost all ANN models, the NSE values vary from 0.65 to 0.97. As a result, the performance of all five models falls somewhere between the categories of "good" and "very good." An MSE of 0.02 to 0.03 (Figure 10b) supports these findings even more strongly. Previous studies have made very similar predictions (Almuhaylan et al., 2020; Djurovic et al., 2015; Jeong & Park, 2019; Khedri et al., 2020; Stylianoudaki et al., 2022). However, the majority of these articles fall short in some way or another in terms of covering a broad spectrum of variables related to groundwater flow. As a result, the current work can be considered both an exhaustive investigation and innovative addition.

*Figure 10. Results of ANN model; (a) The NSE values, and (b) The MSE values*

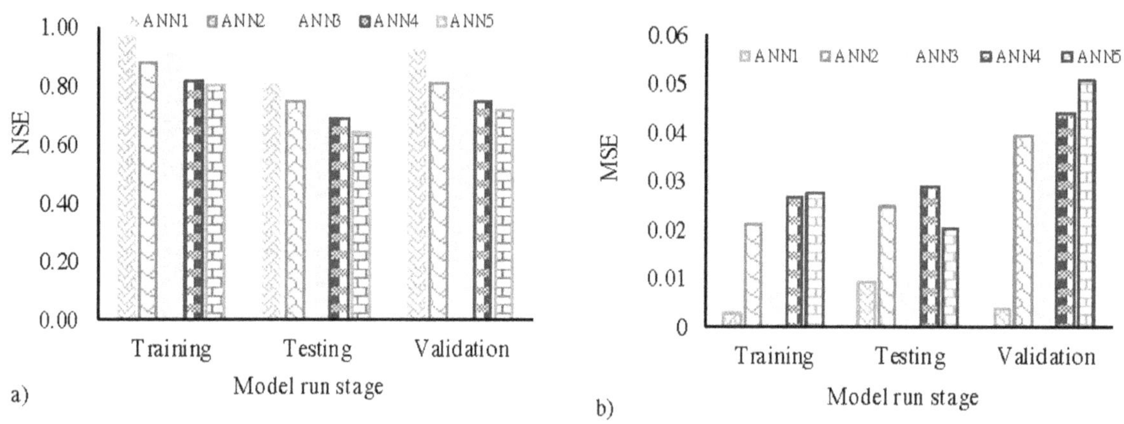


The results of modelling the quantity of groundwater using different methods are displayed in Figures 11 a,b, and c. The overall NSE values for training, testing, and validation are displayed in this graph. The results show that the hydraulic model performed better than ANN and ANFIS. It has been observed that the NSE-values range from 0.998 for ANFIS1 and 0.997 for ANFIS2, 0.997 for ANN1 and 0.995 for ANN2, 0.996 for ANN3 and 0.995 for ANN4, and 0.995 for ANN5 to 0.997 for ANN and 0.9 for the hydraulic model (Figure 11a). Our simulations are also supported by the results of recent publications (Bizhanimanzar et al., 2019, 2020; Chen et al., 2020; Emamgholizadeh et al., 2014). Figures 11b and 11c support this conclusion. In the case of both ANN and ANFIS, the simulated versus observed graph is near the 1:1-line (as it is explained above, the line 1:1 denotes the best match of predicted and observed levels of groundwater). However, the hydraulic model initially underestimated slightly the depths of groundwater (Figure 11b). This has already been explained in section 3.1.1.

*Figure 11. Results of comparative analysis of the ANN, ANFIS, and Hydraulic models. (a) NSE values, (b) predicted vs. measured depths of groundwater, and (c) time-dependent comparisons of measurements and simulations by various model*

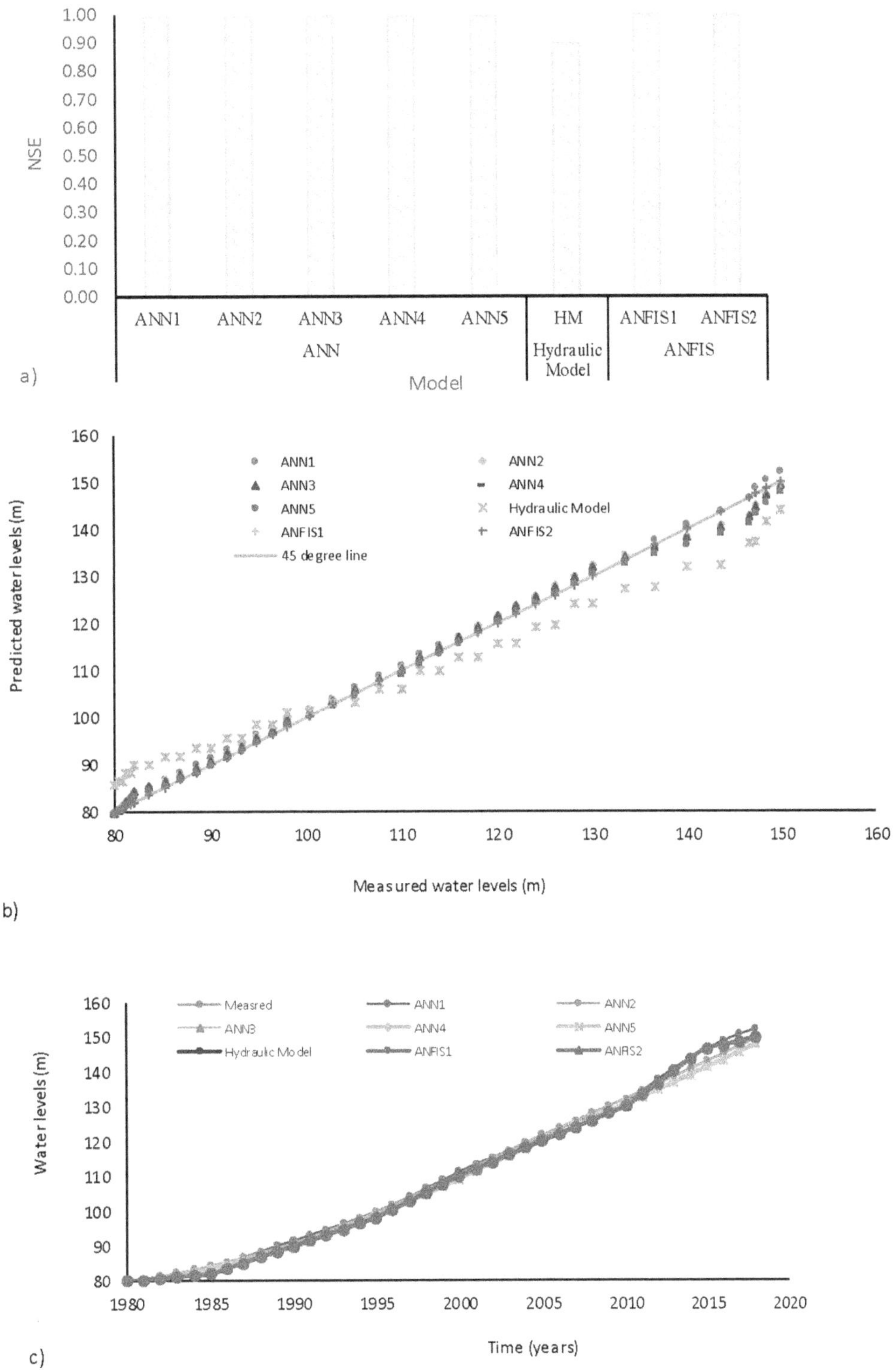

## Quality Modeling

### Results of AI and Numerical Hydraulic Models

Figure 12 compares the overall outcomes of the ANFIS, ANN, and Numerical Hydraulic models. Figure 12 a shows performance of DL-ANN-1 for simulating TDS with different number of neurons in hidden layers. It is observed that performance with 10 neuron is the best one. Although, the performance with neuron in the range of 5 to 20 is also plausible. ANN models with double and triple layers perform approximately identically when it comes to quality modelling. Furthermore, the NSE values for DL architecture and TL architecture are not significantly different (Figures 12 b and c allow for a visual comparison, and Tables 6 and 7 show the precise NSE values). The MSE values presented in Figures 12d and 12e also support this conclusion.

Among the five ANN models, ANN1 has the best performance. The NSE values for TDS, Fe, and Mn in the case of the TL architecture of the model are 0.99, 0.756, and 0.859, respectively, while in the case of the DL architecture of the model, they are 0.989, 0.769, and 0.843, respectively. Nonetheless, the performance of the remaining four ANN models is likewise satisfactory (Figures 12 and Tables 6, 7). Among the thirty NSE values, only five or six are less than 0.7.

Figures 12 b, c, d, and d demonstrate that ANFIS-2 and ANFIS-1 perform better than ANN. Meanwhile, the performance of the numerical hydraulic and ANN models can also be described as good to exceptional, with NSE values ranging between 0.886 and 0.998. NSE for ANN1 is 0.859, 0.99, and 0.756 for Mn, TDS, and Fe respectively. For ANN2, the values of NSE for Mn, TDS, and Fe, are 0.738, 0.995, and 0.642 respectively. The values of NSE in case of ANN-3, are 0.638, 0.91, and 0.729 for, Fe, TDS, and Mn respectively. Similarly in case of ANN-4, these values of NSE are 0.939, 0.73, and 0.84 for TDS, Fe and Mn. In case of ANN-5, the values of NSE for Mn, TDS, and Fe are 0.655, 0.937, and 0.71 respectively. On the other hand, the behaviour depicted in Figures 12 f, g, and h as well as the Taylor Diagrams shown in Figures 13 a, b, c, and d are distinct. According to the NSE and MSE, the performance of the models is not all that great because there is a significant amount of range scatter from the 1:1 line for the projected values. As a result, researchers shouldn't judge the effectiveness of models based just on the NSE and MSE values alone all of the time. In addition to this, the Taylor's Diagram, graphs comparing what was predicted to what was measured, and a few more statistical parameters ought to be investigated. The Taylor's Diagram confirms this fact (Figures 13 a,b,c&d). There could be a variety of causes for the disparity in model results. It may depend upon the code selected. Simplification of assumptions in conceptualization can be another reason. The temporal and geographical resolution also affects the results, and data quality can be the other criteria which may have impact on model forecasts.

The values of the standard deviation values for various parameters are presented in Table 8. It has been determined that groundwater's maximum concentrations of TDS, Fe, and Mn are 3553, 4.29, and 1.49 mg/L, respectively. The respective minimal concentrations are 456 mg/L, 0.78 mg/L, and 0.19 mg/L. The standard deviation values show a significant degree of fluctuation in the contaminant concentrations, the minimum and highest TDS values range, and the concentrations of Fe and Mn. In addition, the maximum concentrations of TDS, Fe, and Mn surpassed the WHO drinking-water standard limit of 0.1, 700, 0.3 mg/L, respectively, for Mn, TDS, and Fe. As a result, the groundwater that was considered for investigation needs to be treated before it can be used for drinking.

*Figure 12a. Results of ANFIS, ANN, and Numerical-Hydraulic models; a) Performance of DL-ANN-1 for simulating TDS with different number of neurons in hidden layers, b) NSE values for ANN-DL*

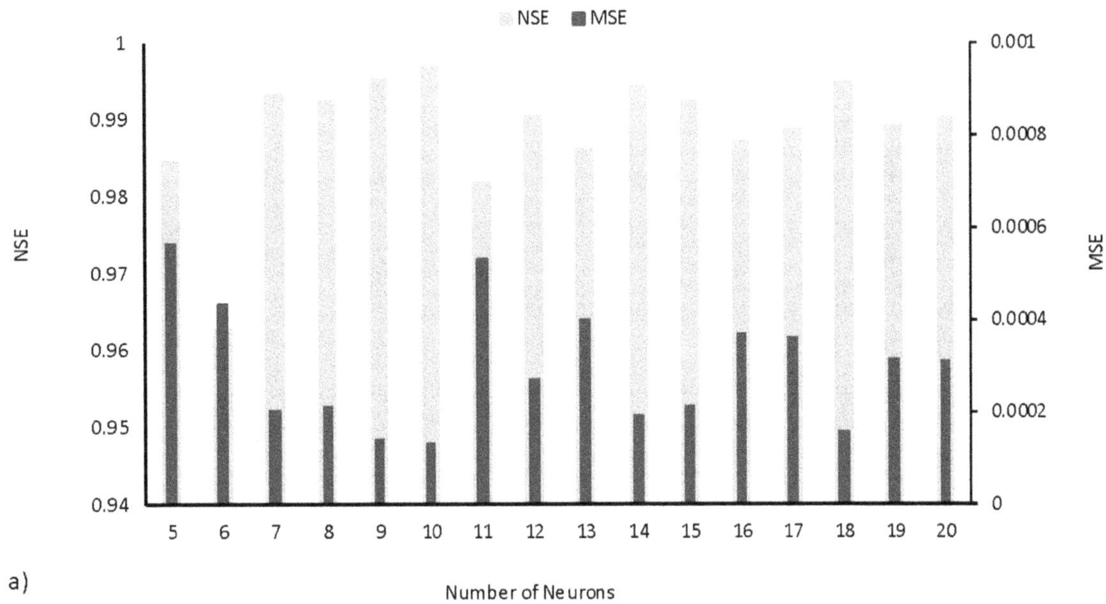


a)

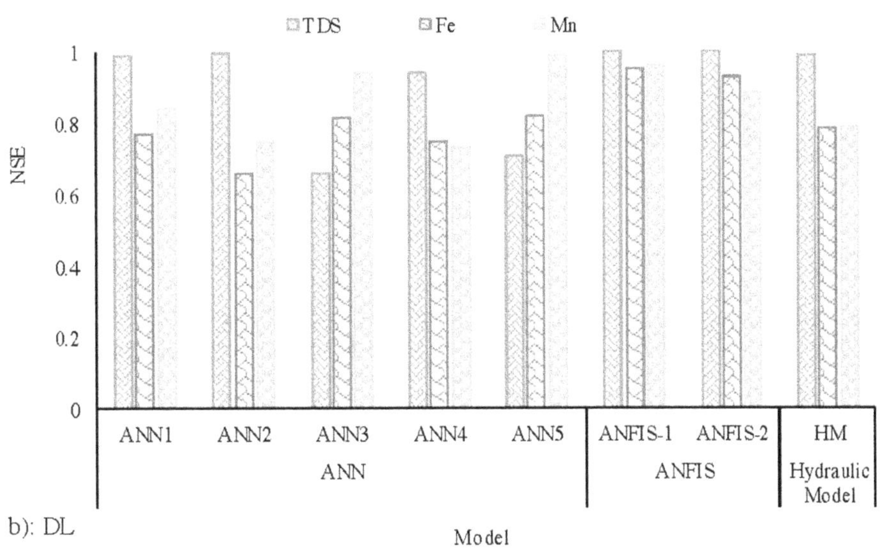


b): DL

*Figure 12b. Results of ANFIS, ANN, and Numerical-Hydraulic models; c) NSE values for ANN-TL, d) MSE values for ANN-DL*

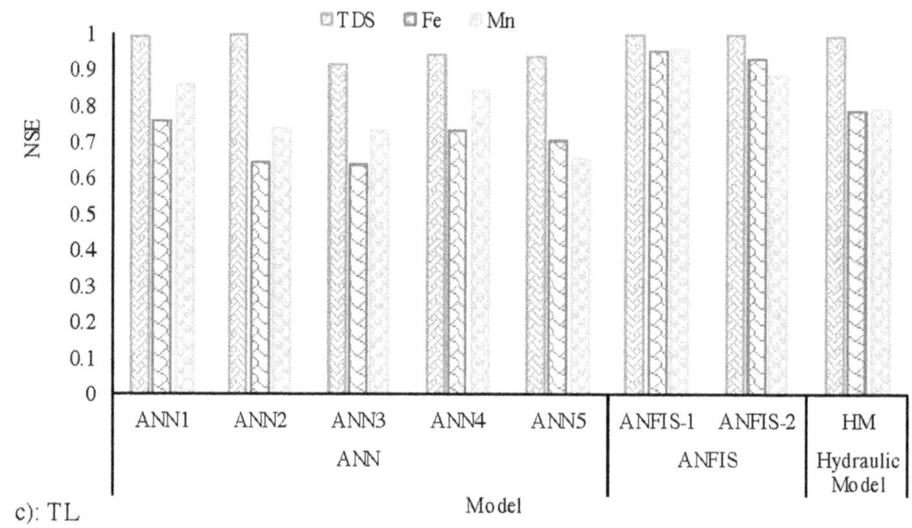


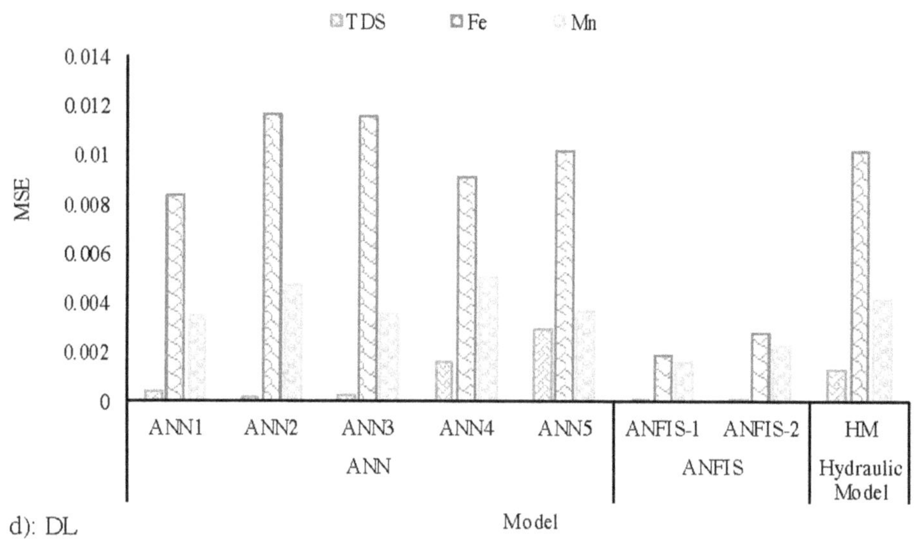


## Results of Sensitivity Analysis

The results of sensitivity analysis show that the the most important element is hydraulic conductivity as compared to other characteristics. A slight change in hydraulic conductivity substantially affects the expected levels of groundwater. It is the most important element in modelling groundwater quantity. Therefore, it must be calculated with extremely high precision through pumping tests or by employing a potent optimization strategy in the model calibration.

*Figure 12c. Results of ANFIS, ANN, and Numerical-Hydraulic models; e) MSE values for ANN-TL, f) Estimated vs measured TDS*

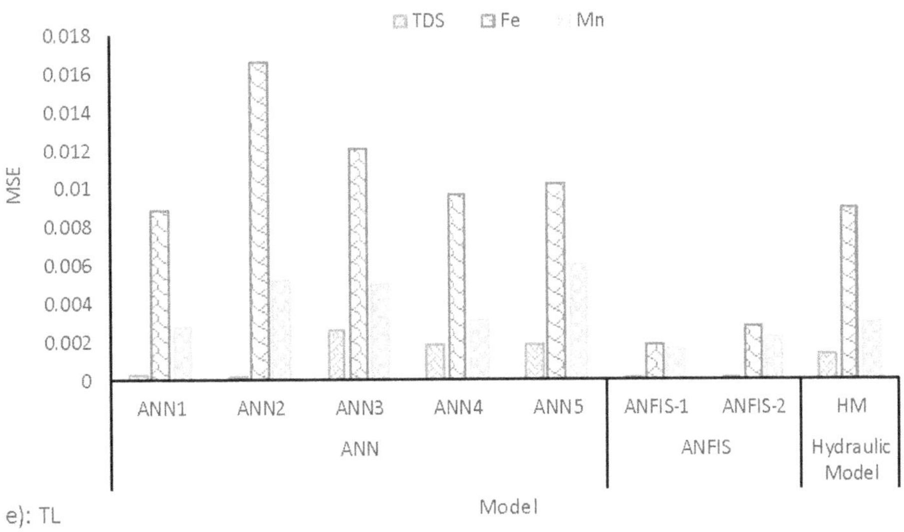


e): TL

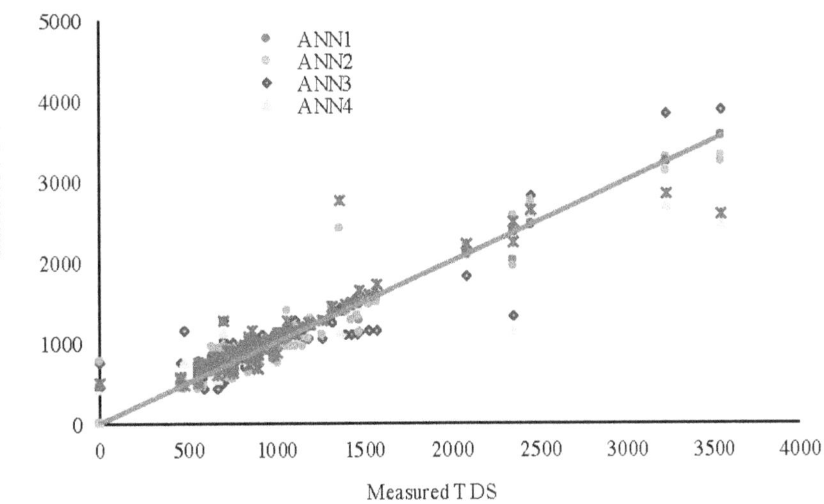


f)

The sensitivity analysis of the parameters of the pollutant transport model showed that from small to large, the most sensitive parameters were: hydraulic conductivity, porosity, and dispersion coefficient. Sensitivity analysis yields realistic results. The dispersion coefficient is an essential metric for contaminant transport.

*Figure 12d. Results of ANFIS, ANN, and Numerical-Hydraulic models; g) Estimated vs measured Fe, h) Estimated vs measured Mn*

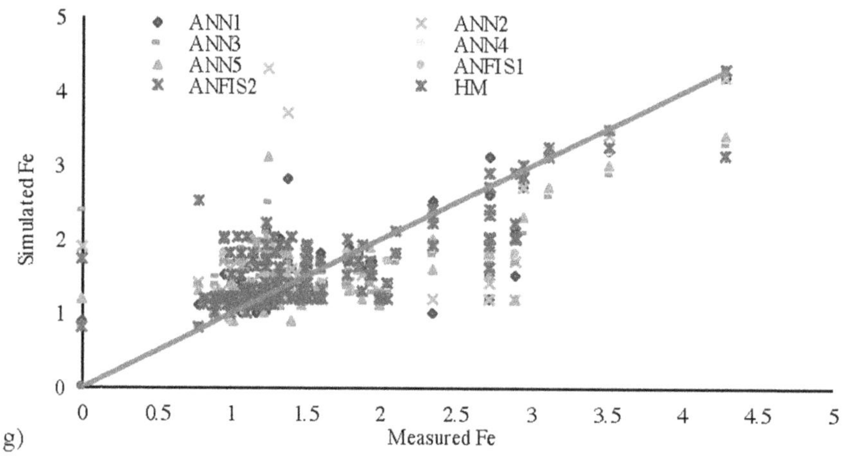


g)

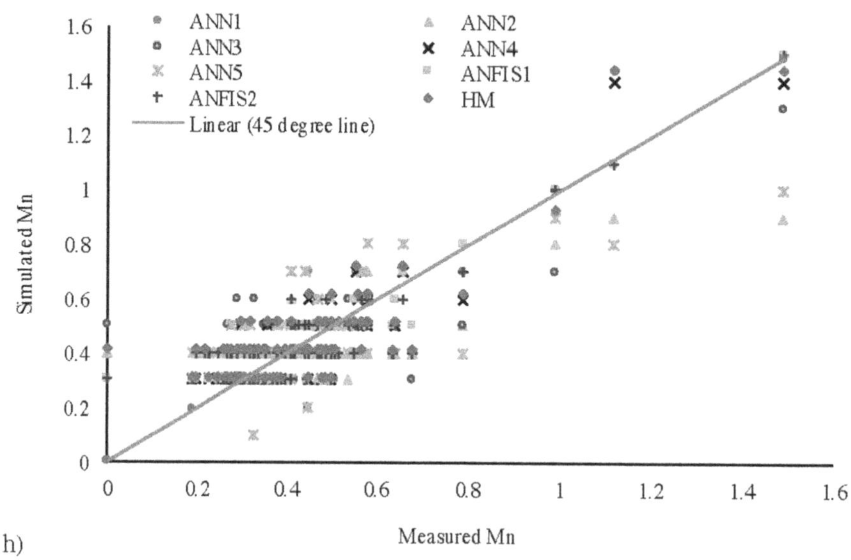


h)

## Results of Future Forecasts From 8 Scenarios

According to Figures 14 a&b, MODFLOW has predicted the depths of the groundwater for eight possible scenarios. Based on the groundwater levels in 2020, the drawdown values for 2070 are 52, 70, 82, 90, 105, 118, 135 and 66, respectively, for scenarios 1 through 8. (Figure 14a). These scenarios have been generated based on the data of 2020 reference pumping rates. Results show that the drawdown is still very substantial (70 m) even if pumping rates are not increased proportionally to population growth. Even the reduction in pumping rates resulted in a 52m drawdown. This dangerous condition concerns water scarcity and foretells a severe pollution crisis (Figure 14b). The Saq aquifer is confined; above this, there is another unconfined aquifer called Qassim Aquifer. It contains extremely contaminated groundwater at shallow depths. Under the conditions of extremely large drawdowns in the Saq Aquifer, the aquifer may generate extremely low pressures, allowing dirty water from the overlying Qassim-Aquifer to migrate downwards, which can result in a catastrophic ecological calamity. This necessitates adopting highly efficient sustainable planning, development, and management solutions.

*Table 6. Values of NSE for ANN -TL and ANN-DL models*

| Model | TL NSE values | | | DL NSE values | | |
|---|---|---|---|---|---|---|
| | TDS | Fe | Mn | TDS | Fe | Mn |
| ANN-1 | 0.9906 | 0.7559 | 0.8592 | 0.9896 | 0.7698 | 0.8431 |
| ANN-2 | 0.9947 | 0.6424 | 0.7387 | 0.9949 | 0.6594 | 0.7501 |
| ANN-3 | 0.9131 | 0.6384 | 0.7292 | 0.6574 | 0.8161 | 0.9408 |
| ANN-4 | 0.9389 | 0.7303 | 0.8404 | 0.9408 | 0.7467 | 0.7334 |
| ANN-5 | 0.9374 | 0.7056 | 0.6550 | 0.7118 | 0.8181 | 0.9906 |
| ANFIS-1 | 0.9982 | 0.9531 | 0.9600 | 0.9982 | 0.9531 | 0.9600 |
| ANFIS-2 | 0.9987 | 0.9305 | 0.8866 | 0.9987 | 0.9305 | 0.8866 |
| HM | 0.9890 | 0.7867 | 0.7934 | 0.9890 | 0.7867 | 0.7934 |

*Table 7. Values of MSE for ANN -TL and ANN-DL models*

| Model | Quality parameters | | | | | |
|---|---|---|---|---|---|---|
| | ANN-TL (Values of NSE) | | | ANN-DL (Values of NSE) | | |
| | TDS | Fe | Mn | TDS | Fe | Mn |
| ANN-1 | 0.000272 | 0.008811 | 0.002763 | 0.000387 | 0.008324 | 0.003474 |
| ANN-2 | 0.000166 | 0.016562 | 0.005208 | 0.000145 | 0.011591 | 0.004689 |
| ANN-3 | 0.002585 | 0.012132 | 0.004986 | 0.000213 | 0.011575 | 0.003553 |
| ANN-4 | 0.001848 | 0.009636 | 0.003139 | 0.001645 | 0.009123 | 0.005025 |
| ANN-5 | 0.001816 | 0.010232 | 0.006052 | 0.002927 | 0.01019 | 0.00364 |
| ANFIS-1 | $5.24 \times 10^{-5}$ | $1.87 \times 10^{-3}$ | $1.64 \times 10^{-3}$ | $5.24 \times 10^{-5}$ | $1.87 \times 10^{-3}$ | $1.64 \times 10^{-3}$ |
| ANFIS-2 | $9.01 \times 10^{-5}$ | 0.00273 | 0.002249 | $9.01 \times 10^{-5}$ | 0.00273 | 0.002249 |
| HM | 0.001296 | 0.008911 | 0.002963 | 0.001296 | 0.01019 | 0.00414 |

## CONCLUSION AND RECOMMENDATIONS

This research focuses on predicting groundwater in terms of its quality and quantity in the Saq Aquifer, Qassim Region using the hydraulic model 'MODFLOW'. A comparison of the hydraulic model has also been made with two Artificial Intelligence (AI) models i.e., ANN and ANFIS. In ANN, multiple architectures with double and triple layers were considered. Each layer comprised 5, 10, and 15 hidden neurons. Furthermore, five distinct types of training functions have been investigated. In the second AI model, two versions of ANFIS have been effectively applied to find the depletion of an aquifer in terms of quantity and quality. Three important quality criteria (TDS, Fe, and Mn) have been examined for predicting groundwater quality.

*Figure 13. Correlation, root mean square error, and standard deviation; represented by blue contours in Taylor Diagram for; a) TDS, b) Fe, c) Mn, d) Groundwater Levels*

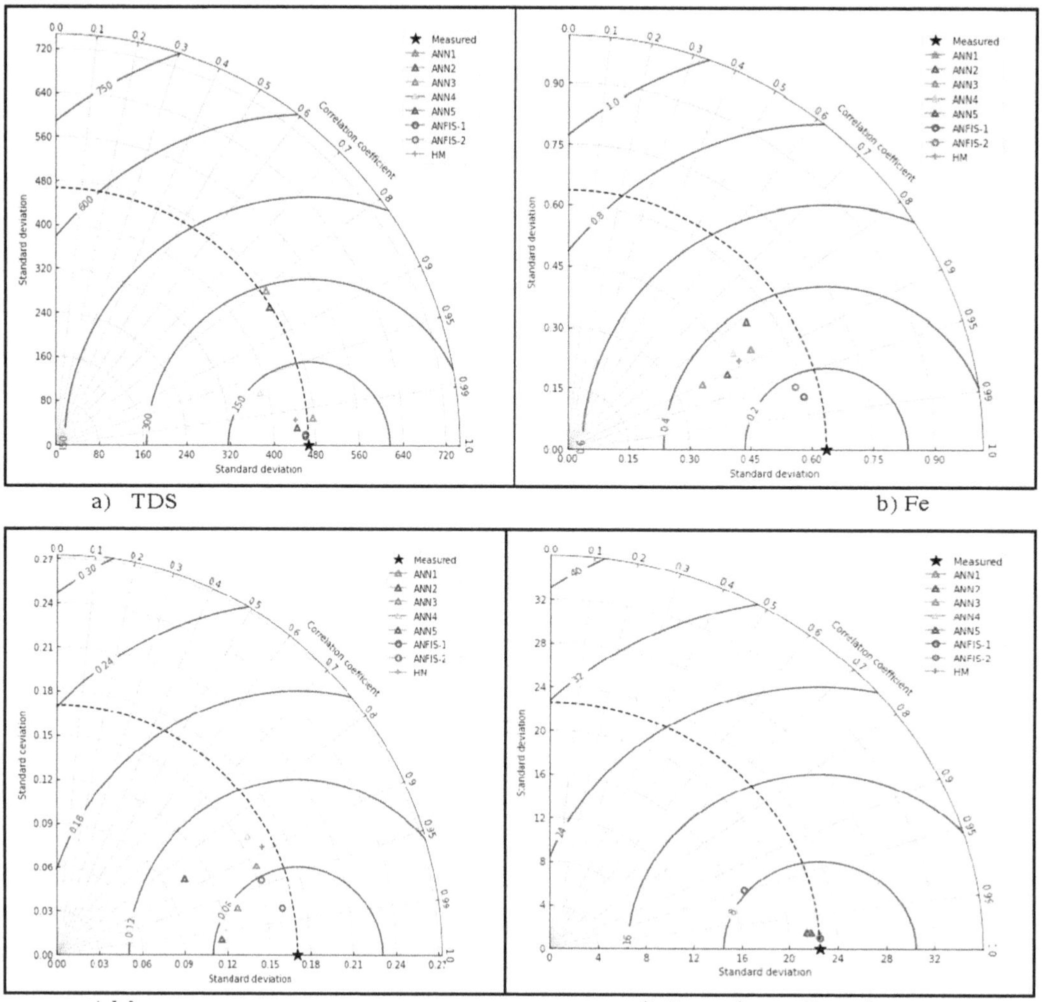


a) TDS

b) Fe

c) Mn

d) Groundwater Drawdowns

The results show that ANFIS models (ANFIS1 and ANFIS2) can accurately estimate groundwater levels and contaminants concentration with NSE values as high as 0.99. In the case of ANN, the model with ten neurons in the hidden layer showed the best findings. Moreover, the results were better in three layers ANN model with ten hidden neurons than in two layers ANN model. Among various training functions of the ANN model, the Scaled Conjugate Gradient training is better at predicting groundwater levels and contaminants than Levenberg–Marquardt, Bayesian Regularization, BFGS Quasi-Newton

BP, and Resilient backpropagation. The hydraulic model performs well for a particular pumping well and contamination data set, but not as well as the ANN or ANFIS models. Nevertheless, the hydraulic model is the most strong and dependable as it is based on physical laws. However, for contaminant and flow transport, the ANN and ANFIS models are relatively less complicated than the hydraulic model.

*Table 8. Standard deviation values (results from various models)*

| Model | TDS | Fe | Mn |
|---|---|---|---|
| Measured | 466.312 | 0.63592 | 0.17047 |
| ANN-1 | 447.122 | 0.51336 | 0.15312 |
| ANN-2 | 445.261 | 0.53890 | 0.10372 |
| ANN-3 | 475.065 | 0.36769 | 0.13123 |
| ANN-4 | 385.524 | 0.46943 | 0.15703 |
| ANN-5 | 464.100 | 0.43246 | 0.11656 |
| ANFIS-1 | 459.899 | 0.59282 | 0.16182 |
| ANFIS-2 | 459.301 | 0.57971 | 0.15217 |
| HM | 442.941 | 0.47413 | 0.16174 |

For different pumping rate scenarios, the Saq Aquifer will experience significantly severe drawdowns (varying from 70 - 135 m relative to the water levels in 2020). Even with the current pumping rate, which does not consider the requirements of the ever-increasing population in the area, the drawdown is predicted to be 70 metres over the next fifty years. In addition, the level of contaminants (TDS, Fe, and Mn) is rising dramatically. If preemptive measures are not taken, a massive ecological catastrophe is possible. However, a constant pumping rate may result in a longer lifespan for the Saq Aquifer compared to the other pumping scenarios. Compared to pumping at greater rates to meet the growing population's water needs, the drawdown will be reduced by 33% during the next 50 years. The Saq Aquifer receives a small amount of recharge, which could lead to a severe water shortage in the future. In order to achieve the aims of the Kingdom's Vision 2030, it is necessary to develop alternative water sources.

The findings and models established in this research can greatly determine the pumping rates necessary for a more environmentally friendly future. For instance, by maintaining constant pumping rates (i.e., without increasing pumping), the Saq-life Aquifer's span could be significantly extended. By applying groundwater preservation measures, it is possible to achieve a 33% reduction in groundwater depletion over the next 50 years, compared to a continuous increase in pumping. The findings of this study will be helpful in the planning, development, and management of strained water resources in the Gulf Region, Saudi Arabia, and other arid regions.

It is strongly recommended that a substantial amount of research be carried out to protect existing water supplies and establish new water sources. For the study area, it is of the utmost need to have very effective technology, data that is both incredibly exact and sufficient, and a delineation of the position of outcrops.

*Figure 14. Results of different scenarios of pumping rates, (a) future forecasts of groundwater depths (depletion of aquifer), (b) future forecasts of TDS*

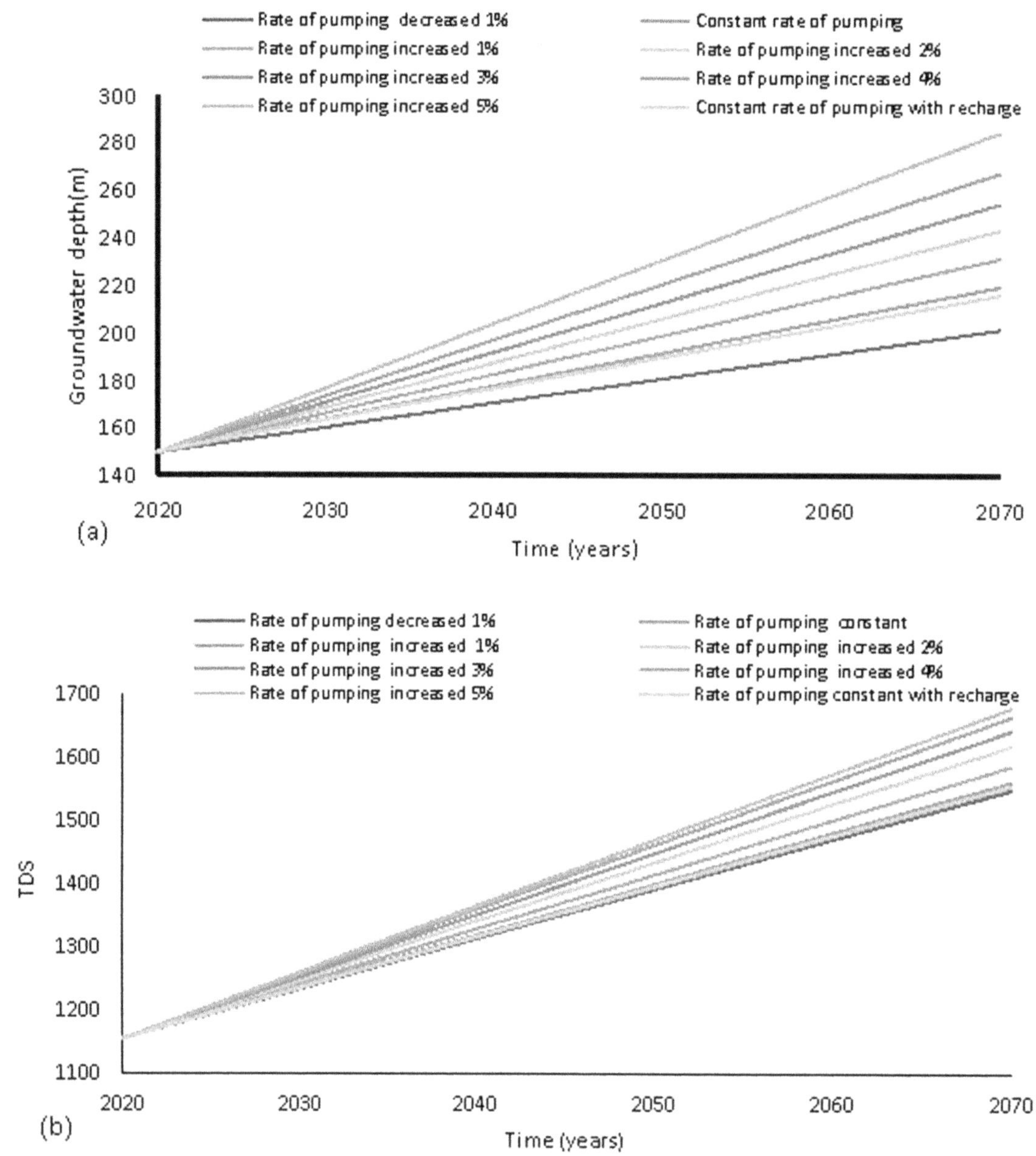


## FUNDING STATEMENT

This research was not funded by any external funding agency.

## ACKNOWLEDGMENT

The authors acknowledge the Ministry of Environment, Water, and Agriculture in Saudi Arabia for data sharing. The authors are thankful to the Deanship of Scientific Research for their Fast track Publication Program. The authors are grateful to the Administration of the College of Engineering (Dean and Vice Dean Administration) and the Head of the Civil Engineering Department for their continued support in this research.

## REFERENCES

Abd-Elmaboud, M. E., Abdel-Gawad, H. A., El-Alfy, K. S., & Ezzeldin, M. M. (2021). Estimation of groundwater recharge using simulation-optimization model and cascade forward ANN at East Nile Delta aquifer, Egypt. *Journal of Hydrology. Regional Studies*, *34*, 100784. doi:10.1016/j.ejrh.2021.100784

Abdel-Aal, S. I., Sabrah, R. E. A., Rabie, R. K., & Abdel Magid, H. M. (1997). Evaluation of groundwater quality for irrigation in Central Saudi Arabia. *Arab Gulf Journal of Scientific Research*, *15*(2), 361–377.

Afzaal, H., Farooque, A. A., Abbas, F., Acharya, B., & Esau, T. (2020). *Groundwater estimation from major physical hydrology components using artificial neural networks and deep learning*. Water., doi:10.3390/w12010005

Ahmed, M. (2020). Sustainable management scenarios for northern Africa's fossil aquifer systems. *Journal of Hydrology (Amsterdam)*, *589*, 125196. doi:10.1016/j.jhydrol.2020.125196

Al-Adhaileh, M. H., & Alsaade, F. W. (2021). *Modelling and prediction of water quality by using artificial intelligence*. Sustainability., doi:10.3390u13084259

Al-Jerash, M. A. (1968). *Soils and agricultural development in the region of Al-Qassim*. Durham University.

Al-Omran, A. M., Aly, A. A., Al-Wabel, M. I., Al-Shayaa, M. S., Sallam, A. S., & Nadeem, M. E. (2017). Geostatistical methods in evaluating spatial variability of groundwater quality in Al-Kharj Region, Saudi Arabia. *Applied Water Science*, *7*(7), 4013–4023. doi:10.100713201-017-0552-2

Almuhaylan, M. R., Ghumman, A. R., Al-Salamah, I. S., Ahmad, A., Ghazaw, Y. M., Haider, H., & Shafiquzzaman, M. (2020). *Evaluating the impacts of pumping on aquifer depletion in arid regions using MODFLOW, ANFIS and ANN*. Water., doi:10.3390/w12082297

Arifin, F., Robbani, H., Annisa, T., & Ma'Arof, N. (2019). Variations in the Number of Layers and the Number of Neurons in Artificial Neural Networks: Case Study of Pattern Recognition. *Journal of Physics: Conference Series*, *1413*(1), 12016. doi:10.1088/1742-6596/1413/1/012016

Barthel, R., Stangefelt, M., Giese, M., Nygren, M., Seftigen, K., & Chen, D. (2021). Current understanding of groundwater recharge and groundwater drought in Sweden compared to countries with similar geology and climate. *Geografiska Annaler. Series A. Physical Geography*, *0*(0), 1–23. doi:10.1080/04353676.2021.1969130

Behera, A. K., Pradhan, R. M., Kumar, S., Chakrapani, G. J., & Kumar, P. (2022). Assessment of Groundwater Flow Dynamics Using MODFLOW in Shallow Aquifer System of Mahanadi Delta (East Coast), India. *Water (Basel)*, *14*(4), 611. doi:10.3390/w14040611

Bizhanimanzar, M., Leconte, R., & Nuth, M. (2019). Modelling of shallow water table dynamics using conceptual and physically based integrated surface-water-groundwater hydrologic models. *Hydrology and Earth System Sciences*, *23*(5), 2245–2260. doi:10.5194/hess-23-2245-2019

Bizhanimanzar, M., Leconte, R., & Nuth, M. (2020). *Catchment-scale integrated surface water-groundwater hydrologic modelling using conceptual and physically based models: A model comparison study.* Water., doi:10.3390/w12020363

Bozorg-Haddad, O., Baghban, S., & Loáiciga, H. A. (2021). Assessment of global hydro-social indicators in water resources management. *Scientific Reports*, *11*(1), 17424. doi:10.103841598-021-96776-9 PMID:34465799

Chen, C., He, W., Zhou, H., Xue, Y., & Zhu, M. (2020). A comparative study among machine learning and numerical models for simulating groundwater dynamics in the Heihe River Basin, northwestern China. *Scientific Reports*, *10*(1), 3904. doi:10.103841598-020-60698-9 PMID:32127583

Cooper, H. H. Jr, & Jacob, C. E. (1946). A generalized graphical method for evaluating formation constants and summarizing well-field history. *Eos (Washington, D.C.)*, *27*(4), 526–534.

Coulibaly, P., Anctil, F., Aravena, R., & Bobée, B. (2001). Artificial neural network modeling of water table depth fluctuations. *Water Resources Research*, *37*(4), 885–896. doi:10.1029/2000WR900368

Daliakopoulos, I. N., Coulibaly, P., & Tsanis, I. K. (2005). Groundwater level forecasting using artificial neural networks. *Journal of Hydrology (Amsterdam)*, *309*(1-4), 229–240. doi:10.1016/j.jhydrol.2004.12.001

Das, U. K., Roy, P., & Ghose, D. K. (2019). Modeling water table depth using adaptive Neuro-Fuzzy Inference System. *ISH Journal of Hydraulic Engineering*, *25*(3), 291–297. doi:10.1080/09715010.2017.1420497

Djurovic, N., Domazet, M., Stricevic, R., Pocuca, V., Spalevic, V., Pivic, R., Gregoric, E., & Domazet, U. (2015). Comparison of Groundwater Level Models Based on Artificial Neural Networks and ANFIS. *TheScientificWorldJournal*, *2015*, 1–13. doi:10.1155/2015/742138 PMID:26759830

Emamgholizadeh, S., Moslemi, K., & Karami, G. (2014). Prediction the Groundwater Level of Bastam Plain (Iran) by Artificial Neural Network (ANN) and Adaptive Neuro-Fuzzy Inference System (ANFIS). *Water Resources Management*, *28*(15), 5433–5446. doi:10.100711269-014-0810-0

Fadipe, O. O., Abidoye, L. K., Adeosun, J. O., Oguntola, B. B., Adewusi, O., & Okeowo, T. E. (2021). Simulation of Groundwater Quality Characteristics using Artificial Neural Network. *Nigerian Journal of Technology*. doi:10.4314/njt.v40i1.21

Fallatah, O. A. (2020). Groundwater Quality Patterns and Spatiotemporal Change in Depletion in the Regions of the Arabian Shield and Arabian Shelf. *Arabian Journal for Science and Engineering*, *45*(1), 341–350. doi:10.100713369-019-04069-1

Gonand, F. (2016). *Impacts of Higher Energy Effciency on Growth and Welfare Across Generations in Saudi Arabia.* Kapsarc. https://www.kapsarc.org/wp-content/uploads/2016/10/KS-1657-DP051A-Impact-of-Higher-Energy-Efficiency-on-Growth-and-Welfare-Across-Generations-in-Saudi-Arabia.pdf

Jabbari, E., Fathi, M., & Moradi, M. (2020). Modeling groundwater quality and quantity to manage water resources in the Arak aquifer, Iran. *Arabian Journal of Geosciences*, *13*(14), 663. doi:10.100712517-020-05681-4

Jafari, T., Kiem, A. S., Javadi, S., Nakamura, T., & Nishida, K. (2021). Fully integrated numerical simulation of surface water-groundwater interactions using SWAT-MODFLOW with an improved calibration tool. *Journal of Hydrology. Regional Studies*, *35*, 100822. doi:10.1016/j.ejrh.2021.100822

Jeong, J., & Park, E. (2019). Comparative applications of data-driven models representing water table fluctuations. *Journal of Hydrology (Amsterdam)*, *572*, 261–273. doi:10.1016/j.jhydrol.2019.02.051

Kamali, A., & Niksokhan, M. H. (2017). Multi-objective optimization for sustainable groundwater management by developing of coupled quantity-quality simulation-optimization model. *Journal of Hydroinformatics*, *19*(6), 973–992. doi:10.2166/hydro.2017.007

Khedri, A., Kalantari, N., & Vadiati, M. (2020). Comparison study of artificial intelligence method for short term groundwater level prediction in the northeast Gachsaran unconfined aquifer. *Water Science and Technology: Water Supply*, *20*(3), 909–921. doi:10.2166/ws.2020.015

Krishan, G. L. A. (2015). Groundwater Level Simulation Using Artificial Neural Network in Southeast, Punjab, India. *Journal of Geology & Geosciences*. doi:10.4172/2329-6755.1000206

Kulisz, M., Kujawska, J., Przysucha, B., & Cel, W. (2021). Forecasting water quality index in groundwater using artificial neural network. *Energies*, *14*(18), 5875. doi:10.3390/en14185875

Li, J., Lu, W., & Fan, Y. (2021). Groundwater Pollution Sources Identification Based on Hybrid Homotopy-Genetic Algorithm and Simulation Optimization. *Environmental Engineering Science*, *38*(8), 777–788. doi:10.1089/ees.2020.0117

Lin, F., Chen, X., & Yao, H. (2017). Evaluating the use of Nash-Sutcliffe efficiency coefficient in goodness-of-fit measures for daily runoff simulation with SWAT. *Journal of Hydrologic Engineering*, *22*(11), 5017023. doi:10.1061/(ASCE)HE.1943-5584.0001580

Lloyd, J. W. (2007). The difficulties of regional groundwater resources assessments in arid areas. *Arabian Journal for Science and Engineering*, *32*(1C), 35.

Lyons, R. T., Peralta, R. C., & Majumder, P. (2020). Comparing single-objective optimization protocols for calibrating the birds nest aquifer model—A problem having multiple local optima. *International Journal of Environmental Research and Public Health*, *17*(3), 853. doi:10.3390/ijerph17030853 PMID:32019060

Majumder, P., & Eldho, T. I. (2020). Artificial Neural Network and Grey Wolf Optimizer Based Surrogate Simulation-Optimization Model for Groundwater Remediation. *Water Resources Management*, *34*(2), 763–783. doi:10.100711269-019-02472-9

Malekinezhad, H., & Banadkooki, F. B. (2018). Modeling impacts of climate change and human activities on groundwater resources using modflow. *Journal of Water and Climate Change*, *9*(1), 156–177. doi:10.2166/wcc.2017.147

Mallick, J., Singh, C. K., Almesfer, M. K., Singh, V. P., & Alsubih, M. (2021). Groundwater quality studies in the kingdom of saudi arabia: Prevalent research and management dimensions. In Water (Switzerland). doi:10.3390/w13091266

Mamo, S., Birhanu, B., Ayenew, T., & Taye, G. (2021). Three-dimensional groundwater flow modeling to assess the impacts of the increase in abstraction and recharge reduction on the groundwater, groundwater availability and groundwater-surface waters interaction: A case of the rib catchment in the Lake Tana s. *Journal of Hydrology. Regional Studies*, *35*, 100831. doi:10.1016/j.ejrh.2021.100831

Mattas, C., Dimitraki, L., Georgiou, P., & Venetsanou, P. (2021). Use of factor analysis (Fa), artificial neural networks (anns) and multiple linear regression (mlr) for electrical conductivity prediction in aquifers in the gallikos river basin, Northern Greece. *Hydrology*, *8*(3), 127. doi:10.3390/hydrology8030127

McCuen, R. H., Knight, Z., & Cutter, A. G. (2006). Evaluation of the Nash–Sutcliffe efficiency index. *Journal of Hydrologic Engineering*, *11*(6), 597–602. doi:10.1061/(ASCE)1084-0699(2006)11:6(597)

Mezouary, L. El, & Mansouri, B. El. (2021). *Groundwater flow equation, overview, derivation, and solution. i*, 2–5.

Mohanty, S., Jha, M. K., Kumar, A., & Panda, D. K. (2013). Comparative evaluation of numerical model and artificial neural network for simulating groundwater flow in Kathajodi-Surua Inter-basin of Odisha, India. *Journal of Hydrology (Amsterdam)*, *495*, 38–51. doi:10.1016/j.jhydrol.2013.04.041

Namitha, M. R., Devi Krishna, J. S., Sreelekshmi, H., & Muhammed Ashik, P. (2019). Ground water flow modelling using visual modflow. *Journal of Pharmacognosy and Phytochemistry*, *8*(1), 2710–2714.

Nordin, N. F. C., Mohd, N. S., Koting, S., Ismail, Z., Sherif, M., & El-Shafie, A. (2021). Groundwater quality forecasting modelling using artificial intelligence: A review. *Groundwater for Sustainable Development*, *14*, 100643. doi:10.1016/j.gsd.2021.100643

Ogunbo, J. N., Alagbe, O. A., Oladapo, M. I., & Shin, C. (2020). N-hidden layer artificial neural network architecture computer code: Geophysical application example. *Heliyon*, *6*(6), e04108. doi:10.1016/j.heliyon.2020.e04108 PMID:32566777

Pal, J., & Chakrabarty, D. (2021). Effects of input/output parameters on artificial neural network model efficiency for breakthrough contaminant prediction. *Water Science and Technology: Water Supply*, *21*(7), 3614–3628. doi:10.2166/ws.2021.125

Roshni, T., Jha, M. K., & Drisya, J. (2020). Neural network modeling for groundwater-level forecasting in coastal aquifers. *Neural Computing & Applications*, *32*(16), 12737–12754. doi:10.100700521-020-04722-z

Sharaf, M. A., & Hussein, M. T. (1996). Groundwater quality in the Saq aquifer, Saudi Arabia. *Hydrological Sciences Journal*, *41*(5), 683–696. doi:10.1080/02626669609491539

Sheela, K. G., & Deepa, S. N. (2013). Review on methods to fix number of hidden neurons in neural networks. *Mathematical Problems in Engineering, 2013*, 2013. doi:10.1155/2013/425740

Shiri, N., Shiri, J., Yaseen, Z. M., Kim, S., Chung, I. M., Nourani, V., & Zounemat-Kermani, M. (2021). Development of artificial intelligence models for well groundwater quality simulation: Different modeling scenarios. *PLoS One*, *16*(5), e0251510. doi:10.1371/journal.pone.0251510 PMID:34043648

Shuler, C. K., & Mariner, K. E. (2020). Collaborative groundwater modeling: Open-source, cloud-based, applied science at a small-island water utility scale. *Environmental Modelling & Software*, *127*, 104693. doi:10.1016/j.envsoft.2020.104693

Singha, S., Pasupuleti, S., Singha, S. S., Singh, R., & Kumar, S. (2021). Prediction of groundwater quality using efficient machine learning technique. *Chemosphere*, *276*, 130265. doi:10.1016/j.chemosphere.2021.130265 PMID:34088106

Solgi, R., Loáiciga, H. A., & Kram, M. (2021). Long short-term memory neural network (LSTM-NN) for aquifer level time series forecasting using in-situ piezometric observations. *Journal of Hydrology (Amsterdam)*, *601*, 126800. doi:10.1016/j.jhydrol.2021.126800

Stylianoudaki, C., Trichakis, I., & Karatzas, G. P. (2022). Modeling Groundwater Nitrate Contamination Using Artificial Neural Networks. *Water (Basel)*, *14*(7), 1173. doi:10.3390/w14071173

Su, Y., Guo, B., Zhou, Z., Zhong, Y., & Min, L. (2020). *Spatio-temporal variations in groundwater revealed by GRACE and its driving factors in the Huang-Huai-Hai Plain, China*. Sensors., doi:10.339020030922

Suprayogi, I., Sandhyavitri, A., Anon, N., Anon, J., Anon, W., & Anon, A. (2020). Groundwater level forecasting model in tropical peatland using artificial neural network. *International journal of civil engineering and technology (IJCIET)*. doi:10.34218/IJCIET.11.2.2020.008

Tabari, M. M. R., Azari, T., & Dehghan, V. (2021). A supervised committee neural network for the determination of aquifer parameters: A case study of Katasbes aquifer in Shiraz plain, Iran. *Soft Computing*, *25*(6), 4785–4798. doi:10.100700500-020-05487-2

Wu, H.-N., Shen, S.-L., Chen, R.-P., & Zhou, A. (2020). Three-dimensional numerical modelling on localised leakage in segmental lining of shield tunnels. *Computers and Geotechnics*, *122*, 103549. doi:10.1016/j.compgeo.2020.103549

Wunsch, A., Liesch, T., & Broda, S. (2021). Groundwater level forecasting with artificial neural networks: A comparison of long short-term memory (LSTM), convolutional neural networks (CNNs), and non-linear autoregressive networks with exogenous input (NARX). *Hydrology and Earth System Sciences*, *25*(3), 1671–1687. doi:10.5194/hess-25-1671-2021

Yoon, H., Jun, S. C., Hyun, Y., Bae, G. O., & Lee, K. K. (2011). A comparative study of artificial neural networks and support vector machines for predicting groundwater levels in a coastal aquifer. *Journal of Hydrology (Amsterdam)*, *396*(1-2), 128–138. doi:10.1016/j.jhydrol.2010.11.002

Younes, A., Fahs, M., & Ackerer, P. (2021). *Modeling of Flow and Transport in Saturated and Unsaturated Porous Media*. Multidisciplinary Digital Publishing Institute. doi:10.3390/w13081088

# Chapter 9
# Reliability Analysis of RC Code for Predicting Load-Carrying Capacity of RCC Walls Through ANN

**Afaq Ahmad**
https://orcid.org/0000-0001-9427-4296
*University of Engineering and Technology, Taxila, Pakistan*

**Demitrios M. Cotsovos**
*Heriot Watt University, UK*

## ABSTRACT

*Over the past couple of decades, a significant rise in utilization of artificial neural network (ANN) in the field of civil engineering has been observed. ANNs have been proven to be very helpful for researchers working in concrete technology. Reinforced cement concrete (RCC) shear walls play an important role in the stability of high-rise reinforced concrete structures. Current study is focused on using ANN-based design technique as an alternative to conventional design codes and physical models to estimate the ultimate load carrying capacity of RCC shear walls. In this study, database of 95 RCC wall samples has been collected from previously published literature. Various critical parameters considered for current research are; length of web portion of the wall (Lw), thickness of wall boundary member (bw), effective depth of wall (d), height of wall (H), shear span ratio (av/d), vertical steel ratio (ρv), horizontal steel ratio (ρh), yield strength of vertical and horizontal steel (fy), compressive strength of concrete (fc), and the ultimate load carrying capacity (Vexp).*

DOI: 10.4018/978-1-6684-5643-9.ch009

# 1. INTRODUCTION

## 1.1 Background

The vertical elements that are used to resist lateral forces are termed as shear wall, as illustrated by Figure 1. The shear wall is utilized where the presence of wind and earthquake load is significantly high. Unfortunately, the common structural elements cannot handle the lateral forces to the extreme limits due to which the shear walls are constructed (Lefas and Kotsovos 1990, Kotsovos, Cotsovos et al. 2011, Altin, Kopraman et al. 2013, Christidis, Vougioukas et al. 2013). The mechanism of handling the load is completely parallel from the orientation to the wall, which provides the ability to resist the shear forces developed in the structure and make it stable in windy conditions. Everywhere there are millions of dollars of damage done by the earthquake to the buildings, which can be significantly reduced by constructing the RC shear walls as they can handle the lateral forces adequately and act as safety structural elements in the structure (Lefas and Kotsovos 1990, Kotsovos, Cotsovos et al. 2011, Altin, Kopraman et al. 2013, Christidis, Vougioukas et al. 2013).

The most critical element in determining the type of failure through which the shear wall will go after the load exceeds its bearing capacity. The various types of configuration are utilized to construct the shear walls so that the required mood of failure can be implemented in the structure when earthquake loading exceeds any specified limit (Altin, Kopraman, & Baran, 2013). However, there are always some deficiencies existing from the perspective of improper handling of the reinforcement and design, which creates more damage to the structures. The most common interpretation of the modelling technique is for the evaluation of seismic performance, as it is one of the critical performance factors in measuring the efficiency of a shear wall (Alarcon, Hube et al. 2014, Christidis, Vougioukas et al. 2014). It is also an undeniable reality that significant development is taking place from the modelling of shear walls because it is the critical component to resist the earthquake loading. The finite element approach is better, but it has its disadvantages from the limitations corresponding to the mesh size. However, the latest techniques significantly develop more important outcomes in developing a better analysis of the shear walls (Christidis, Vougioukas et al. 2013, Barkhordari and Tehranizadeh 2021). If the slenderness ratio of the initial wall is to hide, then most of the time, the failure which will occur is ductile, and it is very critical in terms of engaging the proper safety factors in the building. However, the ideal shear controlled failures mechanism is complicated to achieve because it significantly depends on the geometry and orientation of the reinforcement within the RC shear wall. Bending is also important in this case because the slenderness ratio is getting too high, then there are high chances that the lateral forces can significantly better the shear wall at any plane, and it can produce unwanted consequences for the overall structure (Mangalathu, Jang et al. 2020).

The researcher (Raza, Khan et al. 2020) has used artificial neural network (ANN) to perform reliability analysis of existing strength models used to predict effect of confinement on carbon fibre reinforced polymer (CFRP) concrete cylinders. For this study, in addition to using currently available equations, new empirical equations were also proposed based on available published literature. Critical parameters were defined from a database comprising 708 samples. For concrete cylinders studied, critical parameters defined were; height and diameter of cylinders, thickness and elastic modulus of carbon fibre reinforced polymers used, concrete of CFRP (Ef), unconfined concrete strength and ultimate strength of confined concrete. It was observed that proposed analytical strength model performed well for strength prediction of CFRP confined concrete cylinders with R2 and RMSE values of 0.90 and 0.20, respectively. Whereas, the performance of proposed ANN model was even better having R2 and RMSE values of 0.94 and 0.29, respectively. Comparison of statistical parameters has shown that proposed ANN model was better than previously used models for strength prediction of CFRP confined concrete cylinders.

*Figure 1. RCC Shear wall with composite columns*
*(Mangalathu, Jang et al. 2020)*

The other researcher (Barkhordari and Tehranizadeh 2021, Barkhordari, Tehranizadeh et al. 2021) investigated the efficiency of RCC shear walls in buildings under earthquake loadings. They studied seismic response of G+10 storey building by using response spectrum method as per the code. The ten (10) storey building is modelled, and seismic analysis is performed with and without shear wall using response spectrum analysis. ETABS finite element analysis software is used for this purpose. It is concluded that, lateral stiffness of building can be increased by providing shear wall in structure. Moreover, reinforcement percentage in columns can also be reduced by using shear wall. So, in this way performance of building with shear walls enhanced if compared to the one without shear walls

## 1.2 RCC Shear Walls

Shear walls have gained significant attention because they are enough to handle the development of the lateral forces in the structure and provide the ability to increase service life substantially. The RCC shear wall is utilized where the presence of wind and earthquake load is significantly high. Unfortunately, the common structural elements cannot handle the lateral forces to the extreme limits due to which the shear walls are constructed (Liao, Han et al. 2009). The mechanism of handling the load is completely parallel from the orientation to the wall, which provides the ability to resist the shear forces developed in the structure and make it stable in windy conditions. Everywhere there are millions of dollars of damage done by the earthquake to the buildings, which can be significantly reduced by constructing the RC shear walls as they can handle the lateral forces adequately and act as safety structural elements in the structure (Liao, Han et al. 2009).

The shear walls are mostly connected to the other structural elements to properly transfer the load and help it mitigate the negative impact of earthquake or wind load, as illustrated by Figure 2. Regardless of the construction material, it is entirely possible to construct the shear walls, but RC shear walls are more common in the frame structures, which are significantly prone to earthquake damage or any other load condition that can damage it by creating the lateral moment (Kotsovos, Cotsovos et al. 2011, Christidis, Vougioukas et al. 2013, Christidis, Vougioukas et al. 2014). Depending on the condition and the magnitude of the load, it entirely depends on the shear wall's orientation and design to fail. The most common failure mechanisms present in the shear walls are related to the type of shear developed in the different parts of the wall. However, it should be understood that the shear wall systems are not working from the perspective of an isolated condition. However, they can be wholly integrated into the existing structural frame to provide a coupled system behaviour that is essential to handle the negative impact of the earthquake forces (Kotsovos, Cotsovos et al. 2011).

*Figure 2. Layout of Reinforcement In Shear Walls*

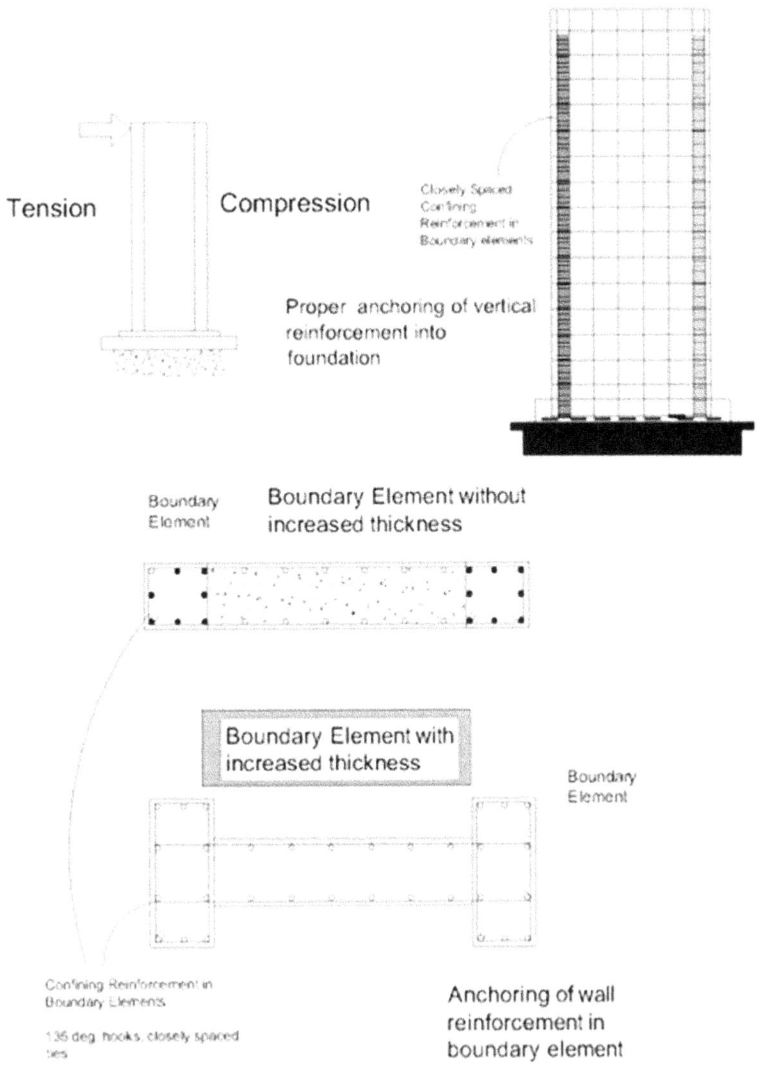

## 1.3 Failure of RCC Shear Wall

Failure of shear wall is mainly of two types either in flexure or in shear. Near the bottom part of wall at tensioned zone flexure cracks appear. If reinforcement ratio in wall web is small then the type of failure would be shear. Figure 3 shows different types of failure in RCC Shear Wall. Shear wall aspect ratio affects a lot in determining the failure type. Moderate to short walls are subjected to shear failure whereas high rise wall usually undergoes flexure failure (a). As per ACI-318 (ACI 2014) a shear wall having aspect ratio greater three (3) is classified as thin wall and its failure is controlled by bending, whereas a wall with aspect ratio less than 1.5 is classified as low wall and its failure will generally be controlled by shear. And behaviour of walls having aspect ratio between 1.5 and 3 is affected both by bending and shear. According to EC2 (EC2 2004) ductile failure is generally observed in well-designed and relatively tall walls having aspect ratio higher than three (3) because such walls are having higher shear strength.

*Figure 3. Typical failure mechanism of RCC Walls*

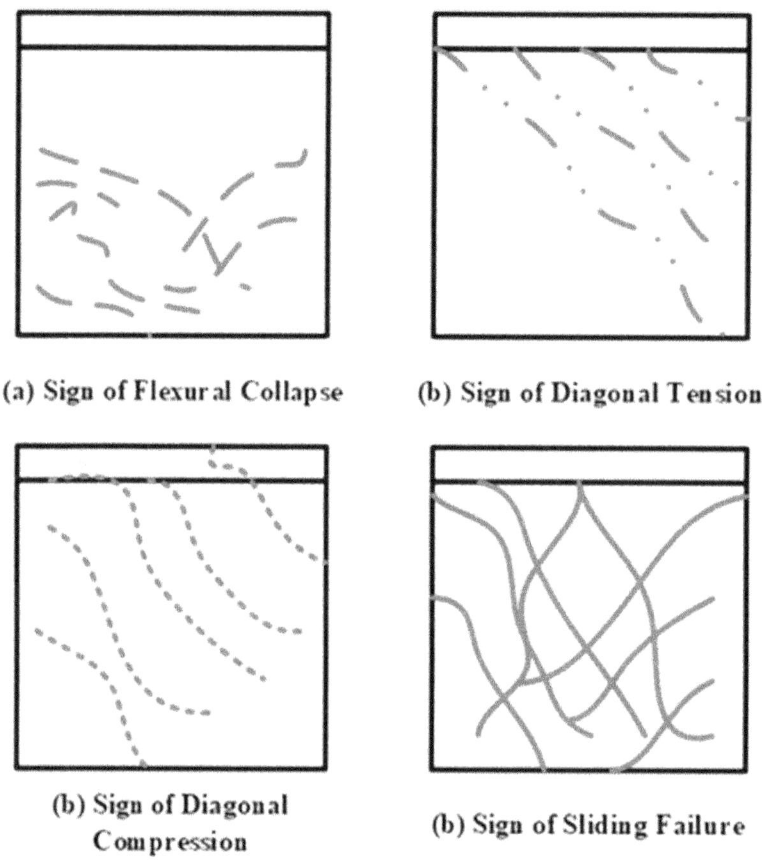

### 1.3.1 Sliding Shear Failure

The sliding shear failure is one of the important failure modes present in the RC shear wall, and it acts as a critical factor in achieving a sustainable service life. It is found from any studies that to promote the diagonal kind of failure, and horizontal reinforcement must be provided so that it could resist the moment of the different planes of the shear wall in the diagonal direction (Tang and Zhang 2011). Furthermore, there are different indications of the complete shear failure in the wall from the sliding perspective if long cracks are evident in the bottom part of the RC shear walls. When any sliding failure occurs in RC shear walls, the concrete cannot hear the compressive forces to their ultimate limit and, therefore, easily fail. It is an important indication that the best concrete results in sliding failure of the overall shear wall because the high pressure cannot be handled by the concrete, the cause of sliding friction (Greifenhagen and Lestuzzi 2005). Most of the time, there are multiple kinds of failure knocking at the same time because of the proper materials used in the RC shear walls and further difficulties that exist from the perspective of the complication that arises due to the improper reinforcement design.

### 1.3.2 Relationship Between Shear Failure and Ultimate Drift

It is commonly believed that the horizontal reinforcement ratio is significantly critical for increasing the sustainability of the RC share wall. However, from the latest studies, it is evident that if the shear wall is not reinforced too much, the value of the horizontal reinforcement ratio is not essential. For the more, the deformation capacity significantly increases in the RC shear walls in which the amount of reinforcement is less (Kotsovos, Cotsovos et al. 2011). A close relationship exists between the shear failure and the ultimate drift because the behaviour of the ultimate and drift in the shear wall is entirely dependent on the shear failure limitations.

### 1.3.3 Mitigation of Failure in Shear Wall

However, it is important to note that the different kinds of failure in the RC shear wall can be easily mitigated by selecting the properly designed reinforcement and improving the compressive strength of the concrete because it could prevent compression failures due to high axial loading. In some cases, the letter load is being applied on the shear wall from both directions due to which state could easily produce multiple times for failure at the same time. Regardless of any failure mode (Tang and Zhang 2011, Christidis, Vougioukas et al. 2013, Christidis, Vougioukas et al. 2014). It is always desired that the RCC shear walls fail in the ductile mode so that the service life of the certificate can be maintained and the safety factor can increase. Fortunately, the proper combination of the design factors and the reinforcement utilized in the construction of RC shear walls to handle the earthquake loading is improving, and multiple possibilities exist from the perspective of the modelling in order to acquire the data that is sufficient for the overall analysis (Tang and Zhang 2011). Therefore, it is important to understand the field mechanisms and the overall structural stability of the RCC shear walls when they are supposed to handle the earthquake loading. Any modelling technique that is adapted to determine the efficiency of the shear walls can significantly depend on the data acquired from multiple sources and analytically interpreted to provide accurate results in predicting the failure of the RCC shear walls

## 1.4 Application of ANN Models in Structural Engineering

This section discusses the available soft computing methods commonly used for problem solving with main focus on the use of ANN and its biological counterparts (Ahmad, Cotsovos et al. 2020). The ANN structure, its development and applications are also discussed. Based on available literature, an attempt has been made for establishing guidelines regarding ANN models development for precise predictions of RCC shear walls behaviour approaching ultimate limit state (ULS). MATLAB has been used to develop a code for Feed-Forward Back Propagation (FFBP) ANNs. This approach provides freedom to users for changing critical parameters required for ANN models development. Finally a detailed discussion is made on: (a) compilation and analysis of published data for formation of data samples for the study, i.e., data compilationa and normalization followed by selection of critical parameters for RCC shear wall design (b) training process adopted for developed ANN models, i.e., distribution of data sets, training of the models, criteria adopted for learning modules and their convergance, number of hidden neurons (HNs) and hidden layers (HLs), and type of activation function used, (c) finally, post-processing of the output data, i.e. results de-normalization, and optimized ANN model selection based on assigned error function (Cotsovos and Lagaros, Waris, Mir et al. 2020).

The recent advancements in the technologies utilized for the evaluation of the structural system are constantly being revolutionized because of the existence of new methods such as artificial neural networks and the combination of soft computing software (Malik 2005, Das, Kumar et al. 2013). The computational algorithms are an essential part of evaluating the protective system because they are based on the concepts related to two revaluations of the structure mechanics and getting output with maximum accuracy, which is not possible through any manual method. The invention of computational technologies are not new, but it is present in the world for decades. However, its implementation is significantly increasing in the structural evaluation as the results are getting more accurate because of the increasing efficiency of the computational algorithms. The artificial neural networks can increase the computational output from the multidimensional structures because the modelling becomes easy (Adeli 2001, Karamodin, Kazemi et al. 2008, Chakraborty, Kumar et al. 2012).

Furthermore, it is also easy to observe the behaviour of the individual structural element in the overall structure, which is very important in increasing the service life of the structures. It is evident from the latest studies that the multidimensional models of the structures are proposing different kinds of failure modes that cannot be analysed from the minute and four methods due to which the implementation of artificial neural networks is becoming more common and essential (Carmona and Ruiz 2014, Shuraim 2014). The finite element analysis is becoming the central part of the sector valuation because it significantly impacts the different strategies that improve the outcomes. Multiple computational resources are utilized to get more than any other approach (Ahmad and Cotsovos 2023).

Furthermore, from the perspective of the computation resources, it is evident that they matter the most from the cost perspective as it can result in decreasing the time of analysis which has a direct impact on the sustainability of the overall evaluation (Behnia, Chai et al. 2014, Dackermann, Smith et al. 2014, Yi, Lei et al. 2014). The artificial neural networks are trained from the perspective of multiple databases, due to which they can contact the conventional design codes and other physical models from the nonlinear perspective and carry out the assessments in which the behaviour of the structure is very critical. Recent studies have shown evidence of the success of utilizing the artificial neural networks in the evaluation of direct sustainability as it results in helping the engineers select the proper configuration for the different structural elements so that the desired outcome should be achieved.

In the RCC structures, the implications of the sector analysis significantly increase because they are present in large amounts worldwide, and they have a substantial impact on the environment if they fail this because of improper structural evaluation. Different calibration factors play a critical role in the success of artificial neural network models in evaluating the RC structures (Zenon Waszczyszyn 2001, Hadi 2003, Günaydın and Doğan 2004, Xu, Wu et al. 2004, Zhang and Foschi 2004, Qian and Mita 2007). Therefore, it is necessary to maintain the constant flow of information and data to meet the desirable physical conditions and significant finite element analysis. The different kinds of evaluation for the physical test data and the essential techniques necessary for the analysis are entirely included in the calibration of artificial neural networks. They ultimately help to increase the sustainable outcomes from any perspective (Ahmad, Cotsovos et al. 2015, Ahmad, Plevris et al. 2020, Waris, Mir et al. 2020).

## 1.5 Artificial Neural Networks

Biological advancements have facilitated the researchers to know how information is stored and processed in human brain. The information is stored in the form of memories in the brain. This stored information can be retrieved any time to solve relevant problems. Figure 4 shows a typical biological neuron that controls functioning of our brain. It has the ability to recall information or experiences it is exposed to and can empirically react to certain stimuli or actions. Our nervous system has more than 100 billion neurons, each of which has been connected to other neurons with 200,000 connections (Ahmad, Cotsovos et al. 2020).

*Figure 4. The Components of Biological Neurons*
*(Ahmad, Cotsovos et al. 2020)*

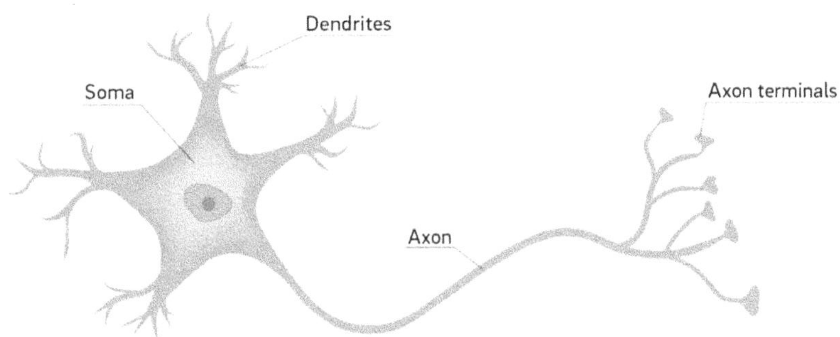


The efficient problem-solving ability of our brain is mainly due to Biological Neural Network (BNN) formed by individual neurons and their connections. Fig 4.2 shows the four basic components, i.e., (a) Dendrites (b) Soma (c) Axon and (d) Axon terminals shared by all type of neurons within our brain. The axon spreads from cell body and many smaller branches are often formed by it before its ending at nerve terminals (Ahmad, Cotsovos et al. 2020). Dendrites spreads from body of neuron cells and its main faunction is to receive messages from other neurons. And Axon terminals are actually contact points used for internal communication of neurons. On receiving or sending messages, electrical impulses are transmitted by neurons along their axons, ranging in a fraction of millimetre to a meter or even more.

It was this natural process that inspired researchers to develop soft computing techniques for obtaining solutions to complex problems using already existing relevant data, and without the requirement of any assumptions, constitutive laws, underlying theory, and complex matrices formation.

Artificial Neural Networks (ANNs) tend to replicate the functioning of biological neurons in a simple way (Rudy Setiono, Angus 1991, Krogh and Vedelsby 1995, Sarle 1995), Table 1 shows a comparison between the two. ANNs are efficiently used to predict the functions depending on a number of parameters, whose role is not clear. Based on capability to remember the data or information presented to them during training, ANNS can predict and categorize the output data. ANNS are comprised of a number of layers consisting of interconnected neurons. The values generated by neurons are multiplied by a specific weight/coefficient assigned to each link. As shown in Figure 5 values attained from neurons of a certain layer are summed with a bias value after being transferred through links. The relationship between neurons of adjacent layers is then represented by introducing the latter sum (as shown in Equations 1 and 2) into a predefined transfer function, *f*. As shown in Figure 5, the output values obtained from a certain layer's activation functions serve as input values for next layer's neurons. Initially, weights are randomly allotted to corresponding links. Then through training of available data, final values are obtained.

$$x_j = \sum(y_k w_{kj}) + b_j \quad (1)$$

$$y_j = f(x_j) \quad (2)$$

In above equations, $\boldsymbol{x}_j$ is output variable obtained through neural network, yk values are obtained using activation function $f(x_j)$, weight coefficients are represented by $\boldsymbol{w}_{kj}$, and bias value by $\boldsymbol{b}_j$, whereas, the activation, neuron number in a certain layer and number of layers, are represented by ***f***, *I*, and *k*, respectively.

*Figure 5. Analogy Between Artificial and Biological Neurons*
*(Ahmad, Cotsovos et al. 2020)*

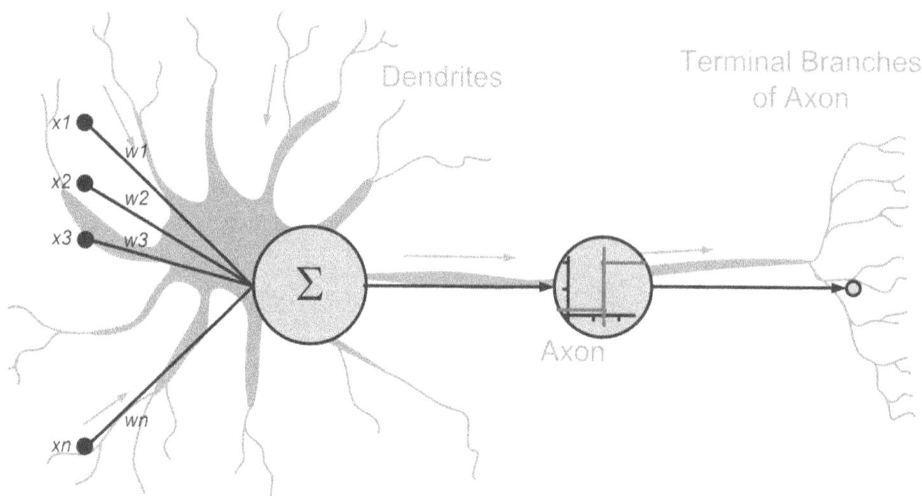

In biological neurons, weighted outputs are not simply summed up as done in artificial neural networks, neither the bias values and weights are kept changing till achievement of required output values(Rudy Setiono, Angus 1991, Krogh and Vedelsby 1995, Sarle 1995). This shows the differences between working of ANN and biological neurons as shown in Table 1. The main similarity is that just like BNN, working of ANN also highly depends on interconnectivity of neurons between different layers and their parallel working for problem solving. A wide range of problems could be solves using ANNs as computational tool(Adeli 2001). Once the models are developed, there is no need to requirement of providing a certain set of instructions during their calibration or training phases. It is the quality of available data and experience of user that affects quality of data prediction.

*Table 1. Equivalence between Biological and Artificial Neural Networks*

| Human | | Artificial |
|---|---|---|
| Neuron | **Processing Element** | Input Neurons |
| Synapses | **Weights** | Weights |
| Dendrites | **Combining Function** | Summation |
| Soma | **Transfer Function** | Activation Function |
| Axons | **Element Output** | Output Neurons |

### 1.5.1 Architecture of ANN

Considering the architecture of the artificial neural network it is evident that they are the combination of the input nodes, which are further connected to the hidden nodes through the links and connected to the various output nodes according to the expected outcomes, as shown by Figure 6. The artificial neural networks tend to ultimately create recurrent behaviour, which is also essential in some applications where that data needs to travel in various directions according to the need of the problem. Furthermore, the division between the artificial neural networks also exists because sometimes they are fixed around the perspective of the bed, and sometimes they are entirely allowed to adapt according to the environment in which they are producing the outboard based on the input variables (Rudy Setiono, Angus 1991, Krogh and Vedelsby 1995, Sarle 1995). The alternative behaviour is very critical for the artificial neural network for function, and therefore it is entirely to the supervised learning process, which is essential to get the sustainable outcomes from the overall process in the structural applications.

### 1.5.2 Architecture and Performance Criteria for ANN Models

The parameters considered while selection of an ANN architecture control the ultimate performance of ANN model. Figure 7 described the whole ANN training process in the flow chart,

1. Input variables that are selected from database
2. Division of these variables into subsets used during ANN training
3. Normalisation of input/output values
4. Selection of initial weight and bias values

*Figure 6. Multiple hidden layers in ANN*

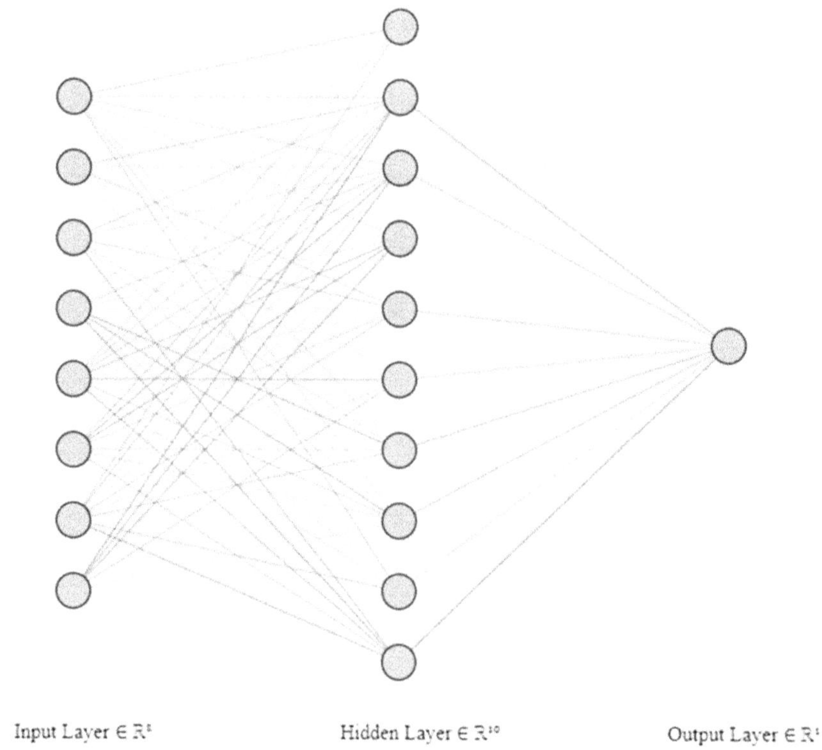


5. Number of the hidden layers
6. Number of neurons in these hidden layers
7. Type of activation functions used between different layers
8. Error function (based on the difference between target and predicted output values) used to measure model performance.
9. Training algorithm used to change bias values and weights for each connection, and contributes in better approximation of target values.

It should be kept in mind that no specific guidelines are available for selection of above-mentioned parameters. Therefore, ANN models development is usually based on trial and error method based on guidelines available in already published literature (Rudy Setiono, Angus 1991, Krogh and Vedelsby 1995, Sarle 1995) which suggest:

1. Sufficient input data should be used, so that sufficient samples should be available for model training, its validation and testing.
2. Sigmoid and hyperbolic tangent activation functions are used between first two layers and output layers, respectively.
3. Hidden layer should include the number of neurons double that the ones included into input layer.
4. Initially randomised values to be used for all bias weights may range between -0.5 and 0.5.

*Figure 7. Detail flow chart of ANN modelling*
*(Ahmad, Cotsovos et al. 2020)*

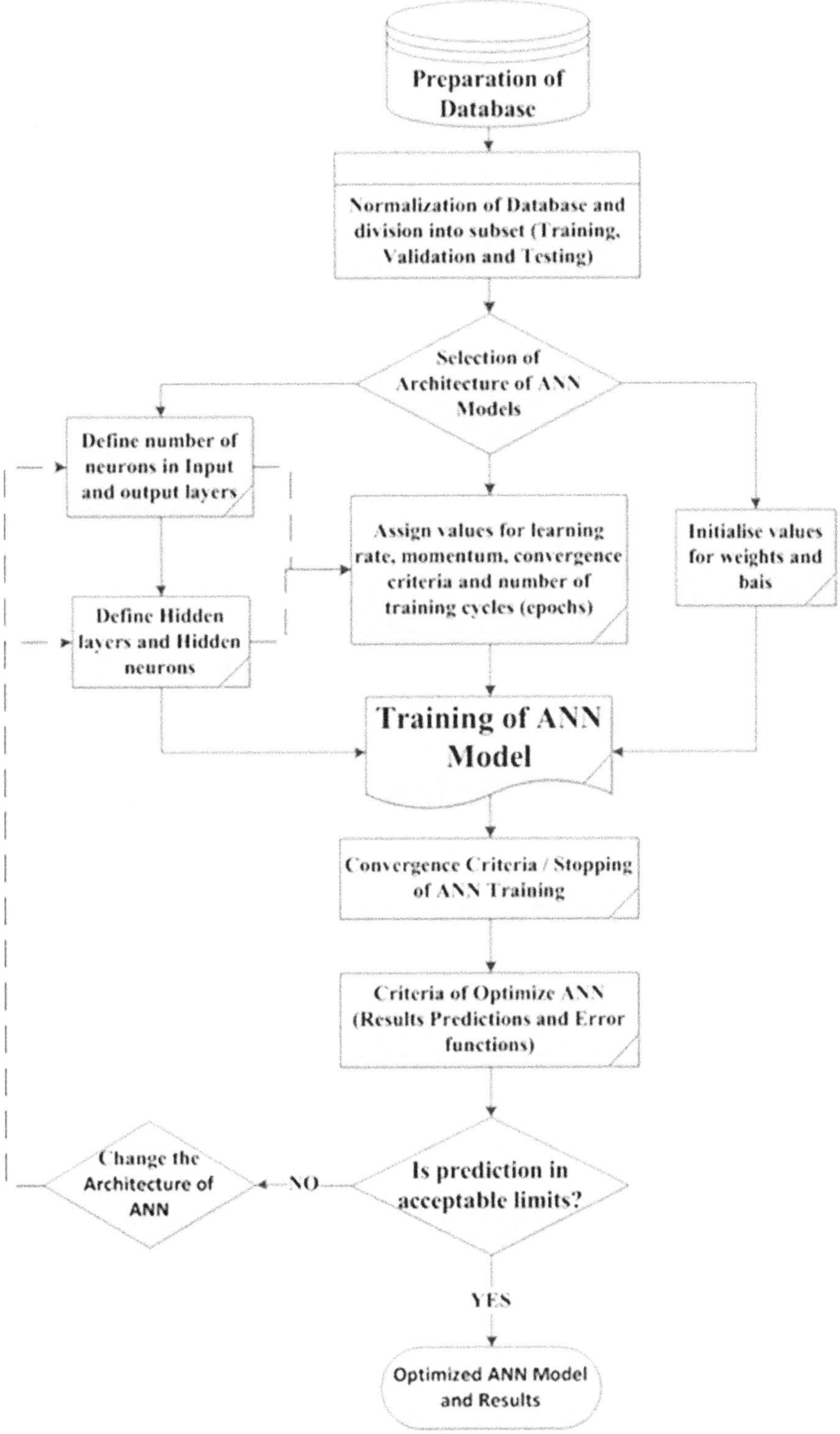

5. To avoid local minima and over-fitting problem in BPA of ANN, validation check must be done after every epoch.

## 1.6 Physical Models for RCC Walls

The design of reinforced concrete structural members are usually done by conventional design codes (CSA 2004, EC2 2004, Kotsovos 2012, ACI 2014). Majority of these codes are based on physical models (i.e. truss-analogy, strut & tie models) that describes reinforced concrete structural response at Ultimate Limit State.The main assumption is that reinforced concrete members behaves as truss in which the reinforcement and concrete in compression zone behave as struts whereas reinforcing steel in tension zone behaves as ties.Based on above physical model the load bearing capacity of RC members are determined using sectional approach. This method only utilizes the critical zones of RC members such as high values of shear force etc, so the assessment of shear and flexure capacity are carried out separately. So far as flexural capacity is concerned the experimental and predicted values are close to each other but in case of shear capacity significant difference in predicted and experimental results are observed.

*Figure 8. Typical truss model approach for RCC shear wall*
*(ACI 2014)*

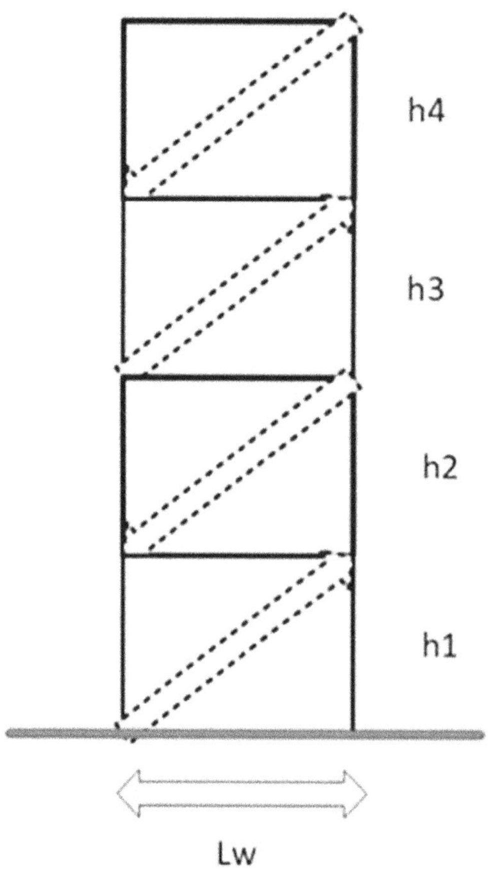

### 1.6.1 Truss Analogy

According to the assumptions of current design philosphy RC shear walls subjected to internal shear component is transferred by In the case of shear capacity, current design philosophy assumes that the internal shear component acting on an RC structural element is transferred by a mixture of: (i) direct sheart transfer (from un cracked portion of rc shear wall), (ii) Cracked portion of wall (inter-locking of aggregates and dowel (iii) Tension acted upon stirrups. Figure shows the typical truss model approach for RC shear Wall. So it is concluded by looking at the Figure 8 that failure of vertical ties (tension member) or struts (inclined compression members) shear failure is either associated with the failure of the vertical ties (acting in tension) or the inclined struts (acting in compression) of the TA model that represents the RC shear wall at ULS.

From ACI-318-14,

$$V_N = A_{Cv}(a_c\lambda\sqrt{fc'} + \rho_t fy \tag{3}$$

$$V_{s,45^\circ} = \frac{A_{sw}f_{yw}d}{s} \tag{4}$$

But not greater than

$$V_N = 0.83A_{C\omega}\sqrt{f_c'} \tag{5}$$

From EC2,

$$V_c = 0.18k(100\rho_1 f_c)^{1/3}bd \tag{6}$$

$$V_s = \frac{A_{sw}f_{yw}(0.9d)Cot\theta}{s} \tag{7}$$

$$V_s = 0.9\frac{A_{sw}f_{yw}dCot\theta}{s} \quad \text{(when } 1 \le Cot\theta \le 2.5) \tag{8}$$

$$k = \left(1 + \sqrt{\frac{200}{d}}\right) < 2 \tag{9}$$

$$\rho_l = \frac{A_s}{bd} < 0.02 \tag{10}$$

But not less than,

$$V_c = 0.035(k)^{3/2}(f_c)^{1/2}bd \tag{11}$$

### 1.6.2 Compressive-Force-Path Method

For the practical design of reinforced concrete members another method termed as Compressive Force Path (CFP) (Kotsovos 2012) as shown by Figure 9. For the practical design of RC structures this method provides a design approach in which performance within limit state philosophy is focussed. The numerical and experimental validation of predicted results determined from CFP method is carried out for the case of static as-well seismic loading (Kotsovos 2012). The fundamental assumptions upon which the CFP method is based are:

- Behaviour of concrete is three dimensional and
- Brittle in nature.
- The shear capacity is significantly affected by the compression zone.
- Shear capacity is negligibly affected by dowel action and aggregate interlocking.

## 1.7 ANN Modelling for RCC Walls

Based on author previous experience (Ahmad, Cotsovos et al. 2015, Waris, Mir et al. 2020, Ahmad and Cotsovos 2023) multilayer feed-forward (MLFF) ANN models were considered most suitable for current study. The procedure for ANN model development consists of preparing a database and its normalization, selection of ANN architecture, i.e., selection of quantity of hidden layers, number of neurons in each hidden layer and activation function, and type of learning/training algorithm and error functions. In addition to validating the previous guidelines, current study has also proposed new model for predicting ultimate load carrying capacity of RC shear walls.

The proposed ANN models were also employed to evaluate prediction accuracy as obtained by using current design codes (EC2 2004, ACI 2014) and compressive force path method (Kotsovos 2012). The assumptions used for model development based on CFP are different from the ones used in conventional design codes. The variation in the dataset of RCC shear wall is created by using different combinations of critical parameters, i.e., wall height (H), length of web portion of the wall ($L_w$), *wall* effective depth (*d*), thickness for wall boundary member ($b_w$), shear span ratio ($a_v/d$), concrete compressive strength ($f_c$), yield strength of horizontal and vertical steel ($f_y$), horizontal steel ratio ($\rho_h$), vertical steel ratio ($\rho_v$), ultimate load carrying capacity of the wall ($V_{exp}$).

### 1.7.1 Proportioning of RCC Shear Wall

The database developed for current study has been collected from 95 samples of RC shear walls. Plan and cross-section of typical reinforced cement concrete shear wall alongwith different geometric parameters used in the study have been shown in Figure 10 (a) and (b), respectively. In Figure 10 (a), L, $L_w$, $b_c$, $b_w$, $A_{hc}$, $A_{vc}$, $A_{hw}$ and $A_{vw}$ represent total wall length, length of web portion for wall, boundary column element width, width of wall web, boundary column area horizontal reinforcement area, boundary column area vertical reinforcement area, wall horizontal reinforcement area and wall vertical reinforcement area, respectively. Whereas H and $a_v$ in Figure 10 (b) represent height and shear span of wall, respectively.

*Figure 9. a)CFP Method for Reinforced Member and b) Comparison between CFP and Conventional Methods*
*(Kotsovos 2012)*

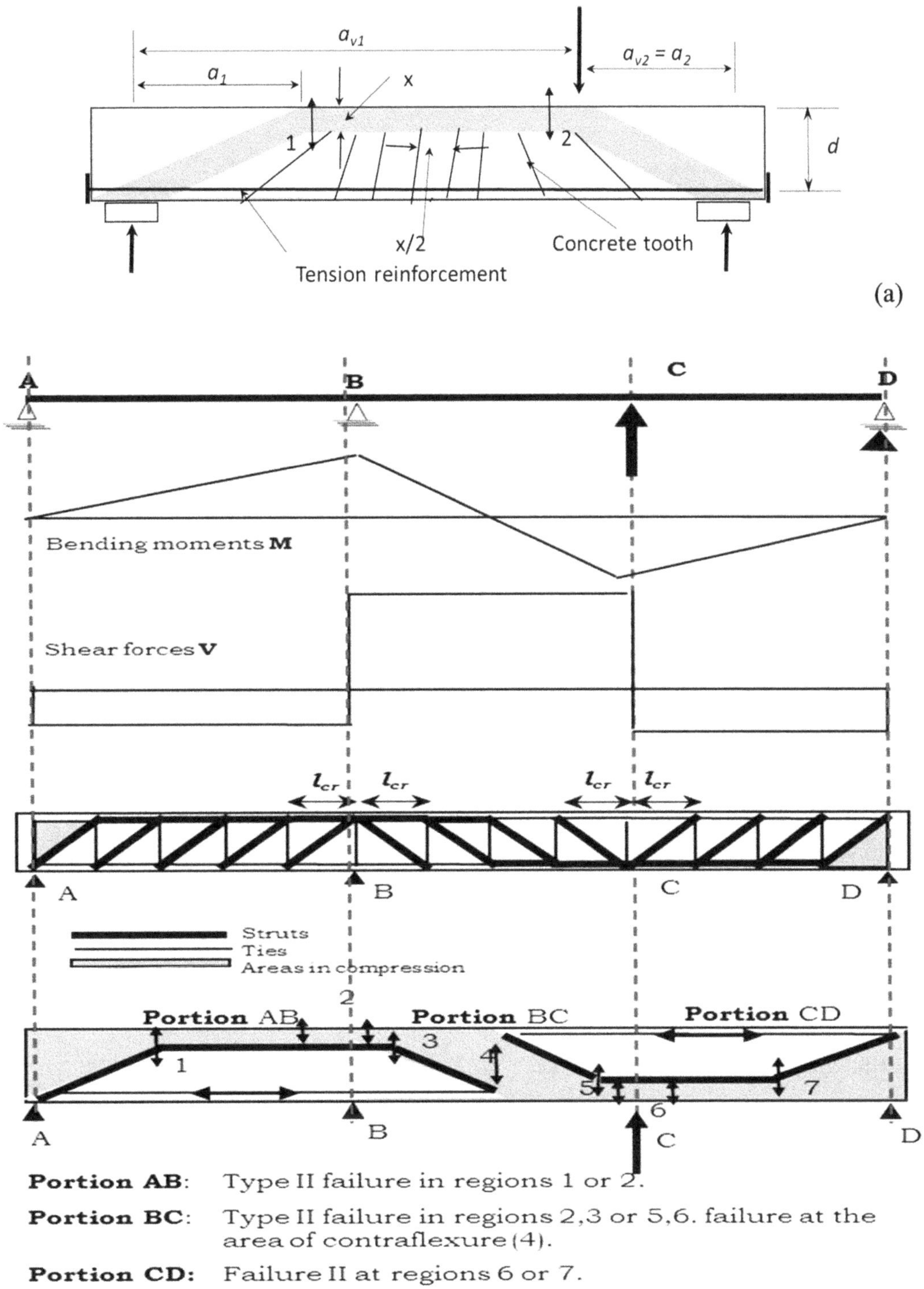

*Figure 10. RCC shear wall geometry (a) Plan (b) Cross-section*

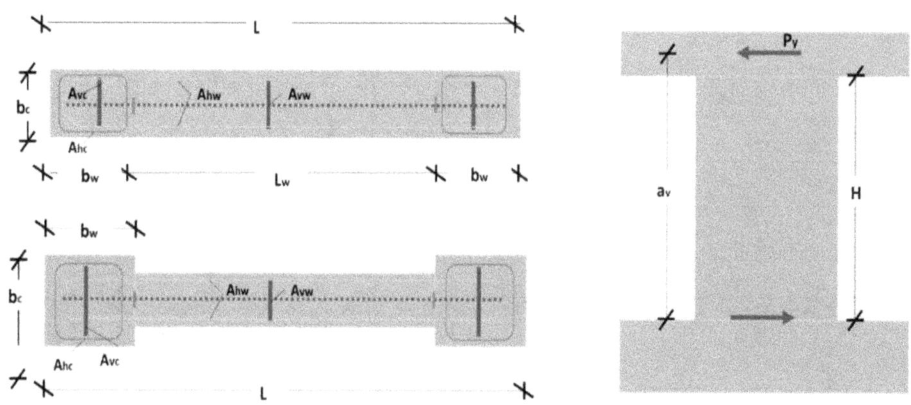


The critical parameters for RCC shear wall considered in the database and statistical information on their variation has been presented in Table 2 below. These parameters are either related to design details or structural response shown at ultimate limit state by the RCC shear walls.

*Table 2. Statistical information on variation of critical parameters*

| Description | Unit | Min | Max | Mean | St. Dev | COV |
|---|---|---|---|---|---|---|
| Wall height (H) | mm | 610 | 4570 | 1567.61 | 746.78 | 0.48 |
| Length of web portion of wall ($L_w$) | mm | 650 | 1700 | 961.68 | 302.33 | 0.31 |
| Wall effective depth (d) | mm | 520 | 1360 | 771.5 | 240.92 | 0.31 |
| Thickness for wall boundary member (*bw*) | mm | 65 | 130 | 95.58 | 18.25 | 0.19 |
| Shear span ratio (av/d) | | 0.86 | 3.04 | 2.19 | 0.62 | 0.28 |
| Concrete compressive strength ($f_c$) | MPa | 20.1 | 53.8 | 33.69 | 7.05 | 0.21 |
| Yield strength for horizontal and vertical (*fy*) | MPa | 375 | 622 | 247 | 506.59 | 63.63 |
| Horizontal steel ratio ($\rho_h$)=$Ah_w/b_w$•sh | | 0.11 | 1.57 | 0.62 | 0.32 | 0.52 |
| Vertical steel ratio (ρv)=$Av_w/b_w$•sv | | 0.35 | 3.33 | 2.98 | 1.57 | 1.02 |
| Ultimate load carrying capacity of wall (*Vexp*) | kN | 65 | 980 | 280.69 | 221.53 | 0.79 |

### 1.7.2 Sensitivity Analysis of Database

Sensitivity Analysis has been performed for determination of relationship between different parameters used in the database. Pearson correlation coefficient (r) as given in Equation (12) has been used for this purpose. Its values may vary between -1 and +1. Positive value indicates a positive correlation (increase in one parameter increases the value of other) and vice versa. The interdependency of one parameter on the other is indicated by a higher r-value. It assisted in selection of different combination of input parameters. All parameters shown in Table 1 have been used in different combinations for different

ANN models developed (Ahmad, Cotsovos et al. 2020). The variation of correlation coefficient for all the parameters used in current study has been shown in Fig. 6.2.

$$r = \frac{n\Sigma xy - (\Sigma x)(\Sigma y)}{\sqrt{n(\Sigma x^2) - (\Sigma x)^2}\sqrt{n(\Sigma y^2) - (\Sigma y)^2}} \tag{12}$$

Where r, n and x,y are Pearson correlation coefficient, total number of samples, and two variables considered, respectively. Correlation values (r) were calculated individually for each critical parameter against the experimental values and have been shown in Figure 11 below.

*Figure 11. Variation of Correlation factor for critical parameters*

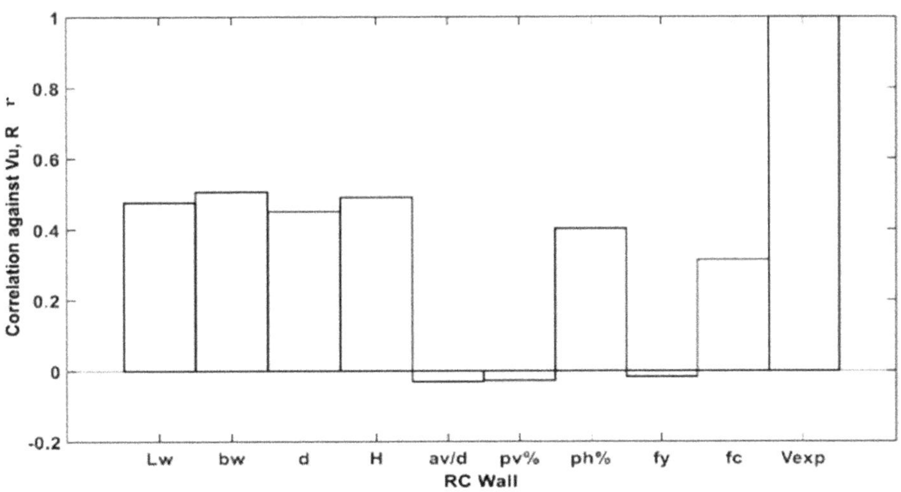


From Figure 11, it can be seen that wall geometry parameters have significant effect on wall strength having a positive linear relationship with it. Contrary to this, yield strength of reinforcement, shear span ratio and reinforcement ratio in vertical direction have negative relationship with wall strength.

## 1.7.3 Development of ANN Models

Four different ANN models, i.e., WAL-1 to WAL-4, were developed using certain combinations of input parameters for prediction of a single output parameter, i.e., $V_u$. From Table 3, it can be seen that two hidden layers were used between input and output layers with different number of neurons for each layer. Sigmoid function has been used for all the models between input and first hidden layer, whereas Tanghn function was used between two hidden layers and between second hidden layer and output layer. The input data was randomly divided into 60%, 20% and 20% for training, validation and testing, respectively.

*Table 3. Structures for different ANN models used*

| 1 | 2 | 3 | 4 | 5 | 7 | 8 |
|---|---|---|---|---|---|---|
| | **Database** | **Hidden Layers** | **Hidden Neurons** | **Activation Function** | **Error Function** | **Division** |
| net= newff | (Input, Trg, | H1, H2 | {H1, H2 | 'logsig',' logsig', 'tansig'} | 'mse' | 'dividerand') |
| WAL-1 | Input=09 Output=1 | Double | 9,9 | STT | MSE, MAE, R | Randomly Division (60%, 20%, 20%) |
| WAL-2 | Input=08 Output=1 | Double | 8,8 | STT | MSE, MAE, R | Randomly Division (60%, 20%, 20%) |
| WAL-3 | Input=04 Output=1 | Double | 4,4 | STT | MSE, MAE, R | Randomly Division (60%, 20%, 20%) |
| WAL-4 | Input=04 Output=1 | Double | 4,4 | STT | MSE, MAE, R | Randomly Division (60%, 20%, 20%) |

Where, WAL-1 to WAL-4 are four ANN modelsH1 and H2 are first and second hidden layers, respectively S and T are Logsig and Tansig activation functions, respectivel .MSE, R and MAE are mean square e, correlation factor and mean absolute error, respectively

## 1.7.4 Selection of Best ANN Model

For precise and efficient predictions made by ANN models, the data collected is initially normalized and then divided into three subsets. These normalized data will then be used for training, validation and testing purpose. For ANN model development and distribution of data into three subsets, MATLAB (Beale, Hagan et al. 2015) has been used. The data used for training, validation and testing phases was 60%, 20% and 20%, respectively. For better performance of the models, results obtained after third analysis were compared. The combinations of input variables and number of hidden neurons for each hidden layer were different for these models, as shown in Table 4 below.

*Table 4. Different parameters for ANN models*

| Sr. No | Model | Input | Function of IP-H1 | Function of H1-H2 | Function of H2-OP | No of Hidden Neurons | Output |
|---|---|---|---|---|---|---|---|
| 1 | WAL-1 | $H, L_w, d, b_w, a_v/d, f_c, f_y, \rho_h, \rho_v$ | Sigmoid | Tanghn | Tanghn | 9-9 | $V_u$ |
| 2 | WAL-2 | $L_w, b_w, H, a_v/d, f_c, M_f/f_c b_w d^2$ | Sigmoid | Tanghn | Tanghn | 8-8 | $V_u$ |
| 3 | WAL-3 | $b_w/d, a_v/d, \rho_v f_y/\rho_h, f_c/f_y$ | Sigmoid | Tanghn | Tanghn | 4-4 | $V_u$ |
| 4 | WAL-4 | $b_w/d, a_v/d, M_f/f_c b_w d^2, f_c/f_y$ | Sigmoid | Tanghn | Tanghn | 4-4 | $V_u$ |

Where,WAL-1 to WAL-4 are four ANN models
H1 and H2 are first and second hidden layers, respectively.
IP is input layer and OP is output layer

### 1.7.5 Best Model Based on Error Functions

Figure. 12 shows the three error functions, i.e., MAE, R and MSE (Ahmad, Cotsovos et al. 2020), for four proposed ANN models as discussed before, it can be seen that model two (WAL-2) corresponds to lowest error values (MSE = 0.22‰ and MAE = 2.05%) and highest Pearson correlation coefficient (R = 98%).

*Figure 12. Performance of different ANN models*

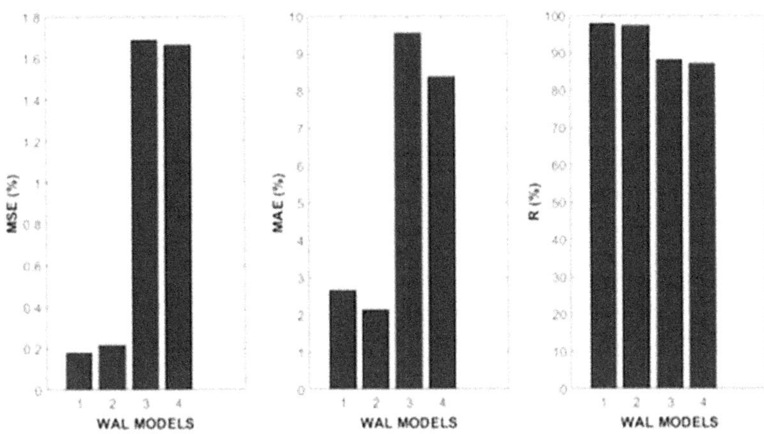


### 1.7.6 Best Model Based on Experimental Data Prediction

All these models were also used for strength prediction of RCC shear walls. Statistical parameters for actual experimental data and those predicted from models have been presented in Table 5 below.

*Table 5. Statistical parameters for different ANN models developed for strength prediction*

| *Vu* | Unit | MIN | MAX | DIFF | AVG | SD | COV |
|---|---|---|---|---|---|---|---|
| *V*exp (target value) | kN | 83 | 977 | 894 | 269.83 | 176.43 | 0.65 |
| WAL-1 prediction | kN | 93 | 950 | 857 | 262.03 | 179.36 | 0.68 |
| WAL-2 prediction | kN | 96 | 978 | 882 | 267.45 | 173.14 | 0.65 |
| WAL-3 prediction | kN | 92 | 753 | 661 | 292.81 | 182.03 | 0.62 |
| WAL-4 prediction | kN | 109 | 817 | 708 | 291.38 | 177.84 | 0.61 |

Based on minimum and maximum values, differences, standard deviation, average values and coefficient of variation as shown in Table 5, it can be seen that model 2 performed well and its statistical parameters were close to real experimental data.

*Figure 13. RCC wall strength prediction using ANN model*

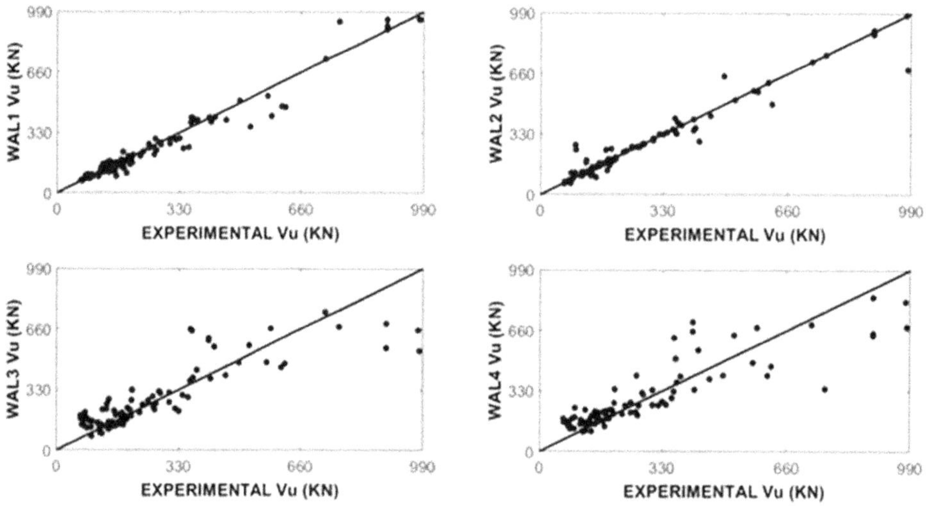


The same four models were used to predict strength of RCC shear wall as shown in Figure 13 Once again it was observed that strength predicted by second ANN model was close to real experimental values as obtained from published literature. So, second model is proposed for comparison with other physical models in upcoming sections.

## 1.8 Comparison of ANN With Physical Models

The proposed ANN model is compared with conventional design codes, i.e., ACI and EC2 (EC2 2004, ACI 2014) and compressive force path (CFP) (Kotsovos 2012) for predicting experimental values of ultimate load carrying capacity for RCC shear wall. The scatter plots are the most suitable ways to present the relationship between two variables and nature between these variables. The observed relationships might be weak or strong, linear or non-linear, and positive or negative.

### 1.8.1 Ultimate Strength Prediction

Figure 14 showed prediction of experimental values for RCC wall strength using conventional and non-conventional techniques. It has been observed that proposed ANN model and CFP model predicted the experimental values more closely, compared to their counterparts based on conventional design codes. Figure 14 (a) showed that ANN model predicted the experimental values very well, providing a closer fit to experimental data for range of strength considered for this study. Figure 14 (b) represented prediction of experimental values as made by using CFP model (Kotsovos 2012). It was observed that the model generally underestimated the values. For lower range of strength, e,g., upto 200 kN, there was a underestimation of experimental data that kept on increasing with increase in wall strength, making the use of CFP code more efficient for wall with smaller strength. Predictions of experimental values as made by ACI code has been shown in Figure 14 (c). It has been observed that prediction was relatively better for smaller wall strength values with slight underestimation and overestimation from experimental values. However, with increase in wall strength, once again higher underestimation of the experimental data was observed. Predictions for experimental values using model based on EC2 has been shown in Figure 14 (d). Once again it was observed that for relatively lower wall strength values, EC2 based model predicted the experimental values relatively better with slight variation, compared to its performance for higher strength walls. For higher strength walls, EC2 model underestimated the experimental values. It was also observed that most of the models predicted the experimental values comparatively better for lower strength walls but more data scatter was observed as wall strength was increased. Although, no such scatter was observed for data predicted by ANN for all strength ranges used in the study.

*Figure 14. Comparison of conventional and non-conventional prediction models (a) ANN, (b) CFP, (c) ACI, (d) EC2*

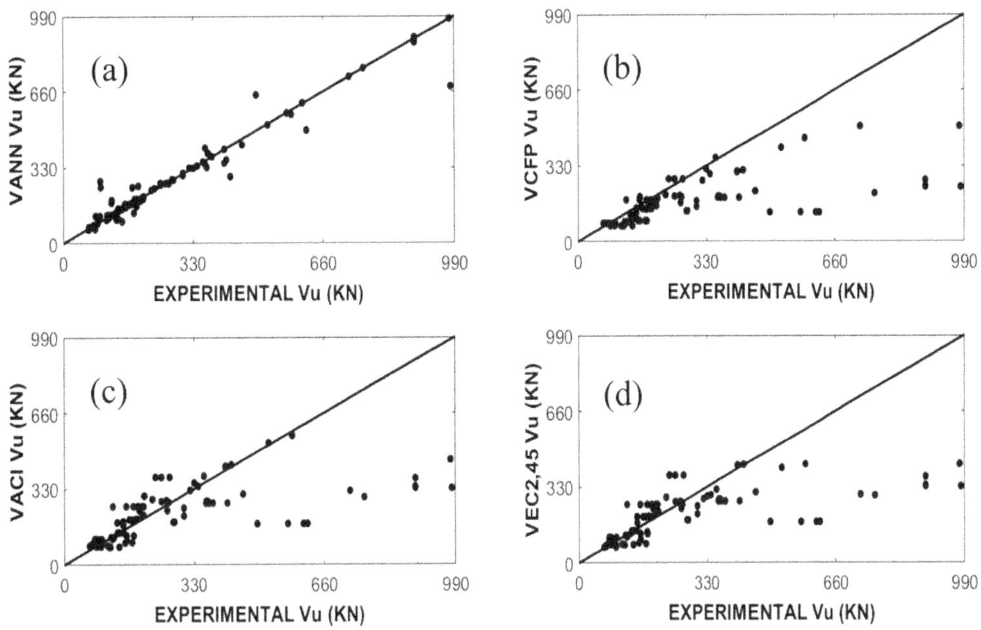


## 1.8.2 Gaussian Distribution

To analyse distribution and spread of data, normal distribution curves have been plotted for different prediction models and the same has been shown in Figure 15. It has been observed that ANN model has predicted the experimental values very well, showing narrow band of $V_{exp}/V_{pre}$ with least standard deviation, with a mean value of 1.05, compared to widespread curves observed for other prediction models. Models based on all other codes and methods could not predict the experimental values well and developed Gaussian distribution curves with peas lower than 0.5 and a wide bottom.

*Figure 15. Normal distribution curves for different prediction models*

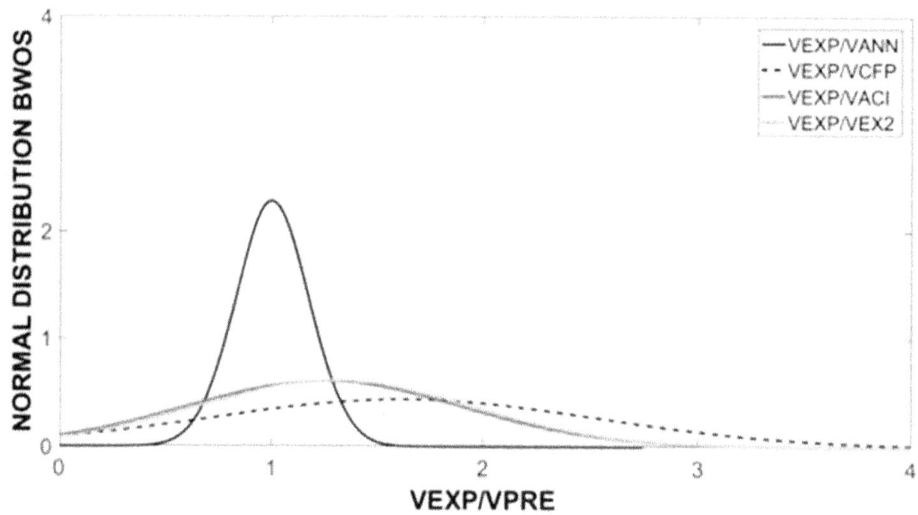


## 1.8.3 Overestimation and Underestimation of Experimental Data

In addition to standard deviation, overestimation and underestimation of ultimate load carrying capacity of wall samples for every prediction model were also calculated as shown in Figure 16. It has been observed that approximately 90% of the results obtained from ANN model were in the range of 0.755 to 1.255 ($V_{exp}/V_{pre}$ value of unity describes best prediction). The predictions made by conventional design codes and compressive force path method are mostly away from accurate range of 0.755 to 1.255 and a large number of samples were in underestimate range, predicting lower strength values than the experimental ones.

*Figure 16. ULS prediction using different models*

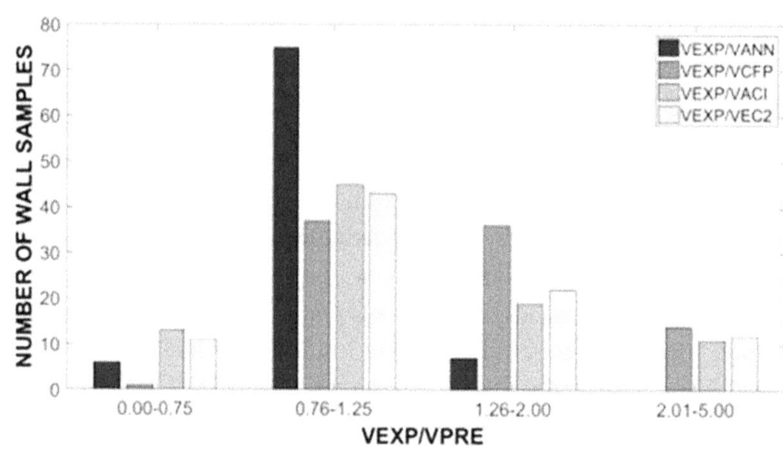

## 1.9 Parametric Studies

This section deals with ACI code, EC2 code, compressive force path method and ANN predictions to study the effect of critical parameters on ultimate load carrying capacity of RCC shear walls. Parameters considered for this study are: shear span ratio, wall depth, compressive strength, transverse steel ratio and longitudinal steel ratio and their effects have been shown in Figure 17 to Figure 21, respectively.

### 1.9.1 Shear Span Ratio

Figure 17 shows the effect of shear span ratio on $V_{EXP}/V_{PRED}$. It has been observed that values predicted by ANN model, i.e., $V_{EXP}/V_{ANN}$ ratio, are close to unit line, showing minimum scatter and good agreement with experimental values. Some $V_{EXP}/V_{ANN}$ values show deviation from unit line, especially for shear span value greater than 2.5. However, overall efficiency of ANN model was proved to be better than its counterparts. For other methods studied, the values of $V_{EXP}/V_{ACI}$, $V_{EXP}/V_{EC2}$ and $V_{EXP}/V_{CFP}$ were more scattered and located away from unit line. A careful inspection of the data suggested that data points against shear span ratio between 1 and 2, data points were relatively close to unit line compared to higher shear span ratios where predictions made by conventional models were too scattered and far away from unit line. From Figure 17, it has been observed that ANN model predicted the ultimate strength more accurately for all the shear span ratios used for current study.

*Figure 17. Prediction of RCC wall strength for walls with different shear span ratios*

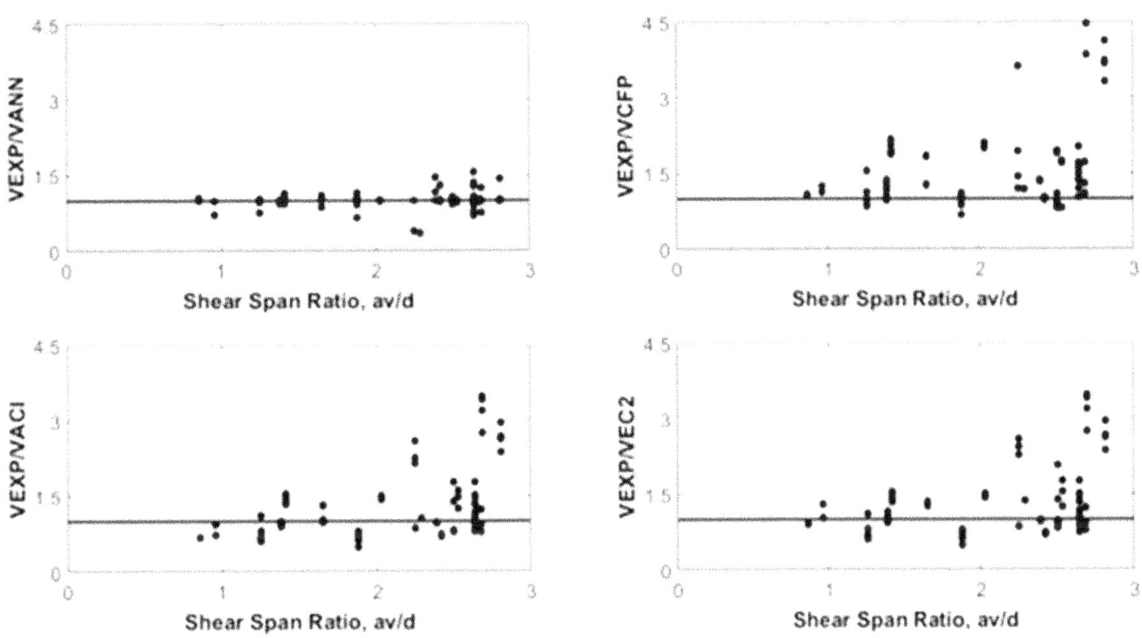

### 1.9.2 Effective Depth

Figure 18 shows the effect of depth on $V_{EXP}/V_{PRED}$. Again it has been observed that values predicted by ANN model, i.e., $V_{EXP}/V_{ANN}$ ratio, were close to unit line, showing minimum scatter and good agreement with experimental values. Some $V_{EXP}/V_{ANN}$ values were both above and below the unit line, showing underestimation and overestimation of experimental data. For other methods studied, the values of $V_{EXP}/V_{ACI}$, $V_{EXP}/V_{EC2}$ and $V_{EXP}/V_{CFP}$ were more scattered and located away from unit line. For $V_{EXP}/V_{CFP}$, most of the data points were above unit line, showing underestimation of experimental data. For $V_{EXP}/V_{ACI}$, the scatter was both above and below the unit line, showing both underestimation and overestimation of the experimental values. Whereas, the plot for ratio $V_{EXP}/V_{EC2}$, showed both underestimation and over-estimation of the experimental values with more data points above unit line, showing underestimation of experimental values. From Figure 18, it has been observed that once again ANN model predicted the ultimate strength more accurately for all the depth values used for current study than its counterparts.

*Figure 18. Prediction of RCC wall strength for walls of different depths*

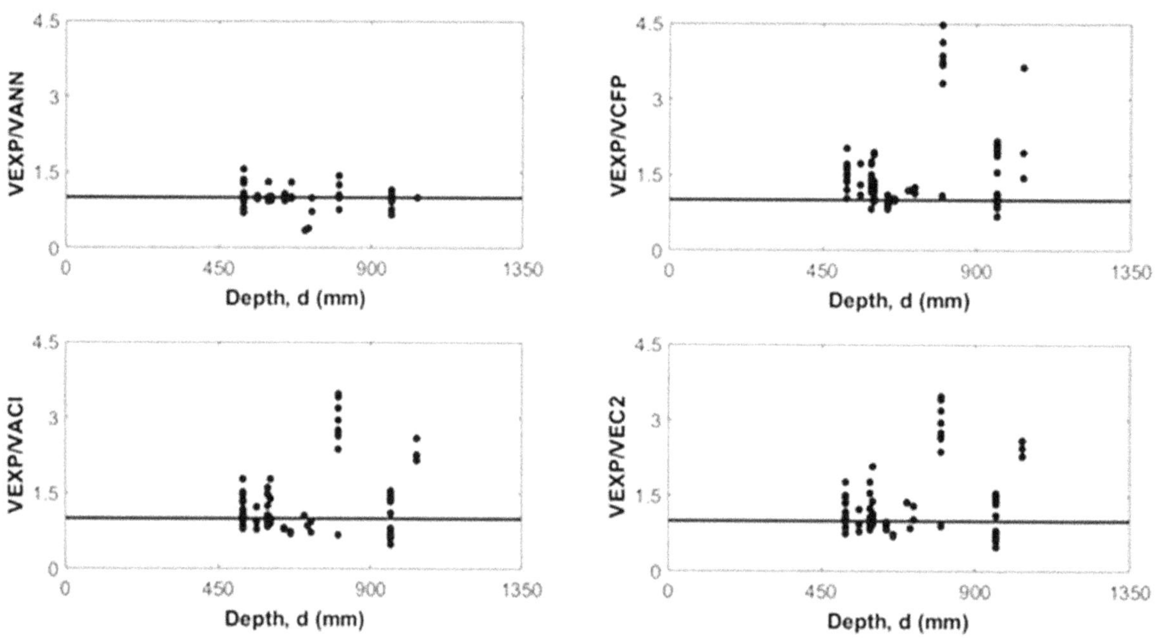


### 1.9.3 Compressive Strength

Figure 19 shows the effect of compressive strength on $V_{EXP}/V_{PRED}$. Again, it has been observed that values predicted by ANN model, i.e., $V_{EXP}/V_{ANN}$ ratio, were close to unit line, showing minimum scatter and good agreement with experimental values. Some scatter of $V_{EXP}/V_{ANN}$ values was observed but it was minimum compared to that observed by other models. For other methods studied, the values of $V_{EXP}/V_{ACI}$, $V_{EXP}/V_{EC2}$ and $V_{EXP}/V_{CFP}$ were more scattered and located away from unit line. For $V_{EXP}/V_{CFP}$, most of the data points were above unit line, showing underestimation of experimental data by CFP method. For $V_{EXP}/V_{ACI}$, most of the points were below unit line, showing overestimation of the values but they were still close to unit line compared to less data points above but far away from unit line, showing more underestimation of the data. The plot for ratio $V_{EXP}/V_{EC2}$ was quite similar to the one observed for $V_{EXP}/V_{ACI}$. From Figure 19, it has been observed that ANN model predicted the ultimate strength more accurately for all the compressive strength values used for current study than its counterparts.

*Figure 19. Prediction of RCC wall strength for walls with different compressive strengths*

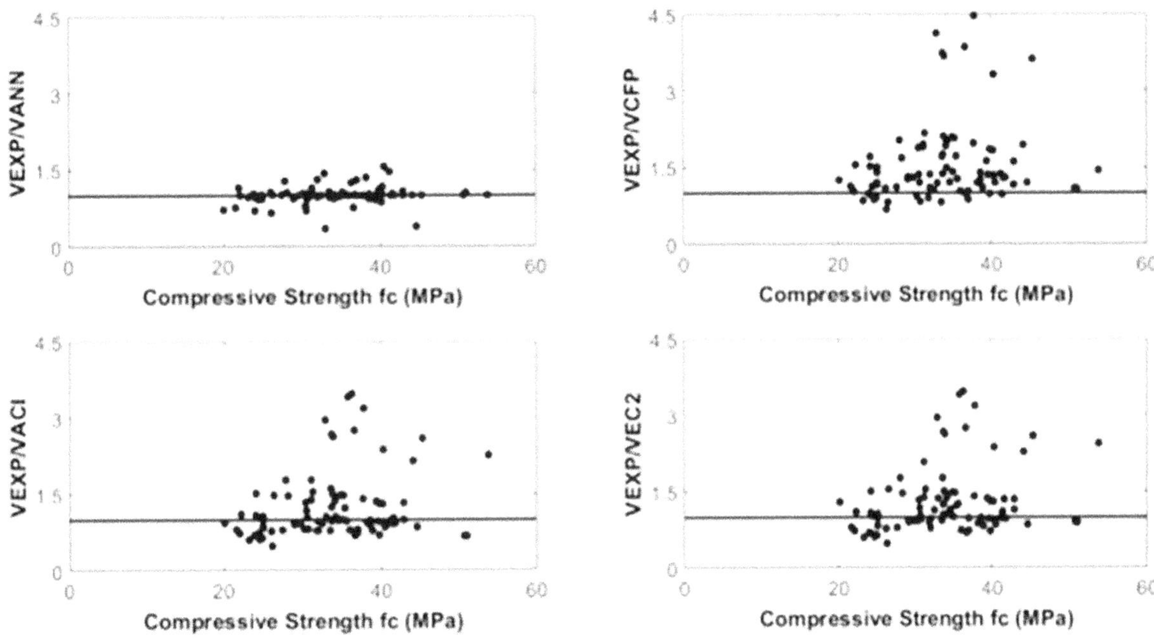


## 1.9.4 Transverse and Longitudinal Steel Ratio

Figure 20 and Figure 21 show the effect of transverse steel ratio and longitudinal steel ratio on $V_{EXP}/V_{PRED}$, respectively. For both the cases, it has been observed that values predicted by ANN model, i.e., $V_{EXP}/V_{ANN}$ ratio, were close to unit line, showing minimum scatter and good agreement with experimental values. Some scatter of $V_{EXP}/V_{ANN}$ values was observed but it was minimum compared to that observed by other models. For other methods studied, the values of $V_{EXP}/V_{ACI}$, $V_{EXP}/V_{EC2}$ and $V_{EXP}/V_{CFP}$ were more scattered and located away from unit line. For $V_{EXP}/V_{CFP}$, most of the data points were above unit line in both the figures, showing underestimation of experimental data by CFP method. For $V_{EXP}/V_{ACI}$, again most of the data points were above unit line in both cases, showing an underestimation of experimental results. The plots for ratio $V_{EXP}/V_{EC2}$ were also similar to previous two plots, having maximum points above unit line and showing underestimation of experimental data. From Figure 20 and Figure 21 it has been observed that ANN model predicted the ultimate strength more accurately for all the longitudinal and transverse steel ratios used for current study than its counterparts.

*Figure 20. Prediction of RCC wall strength for walls with different transverse steel ratio*

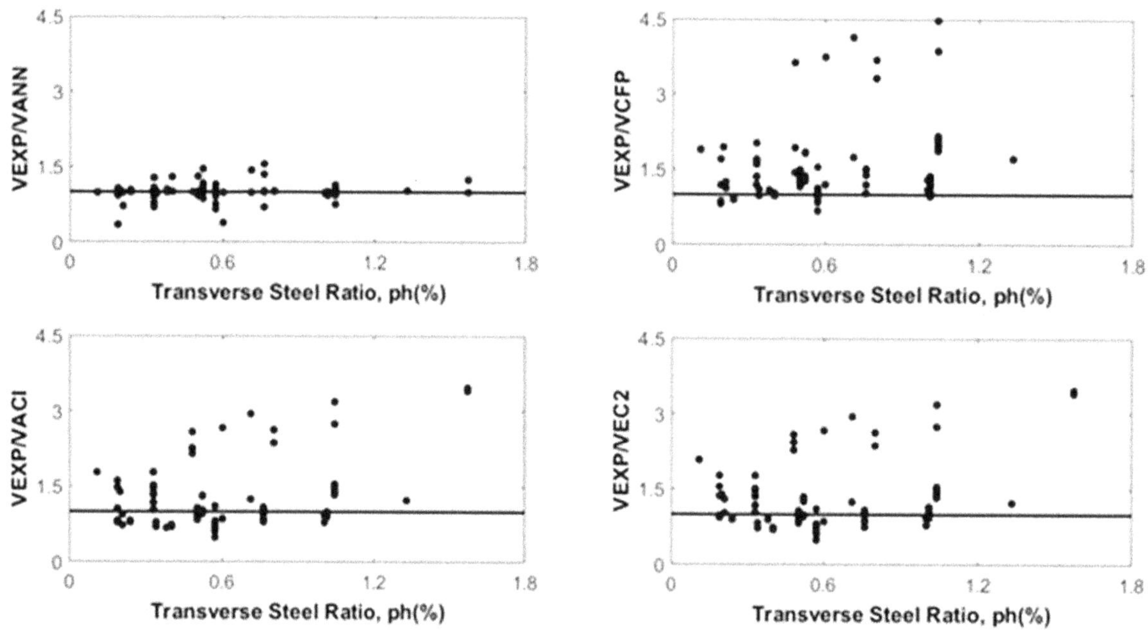


*Figure 21. Prediction of RCC wall strength for walls with different longitudinal steel ratio*

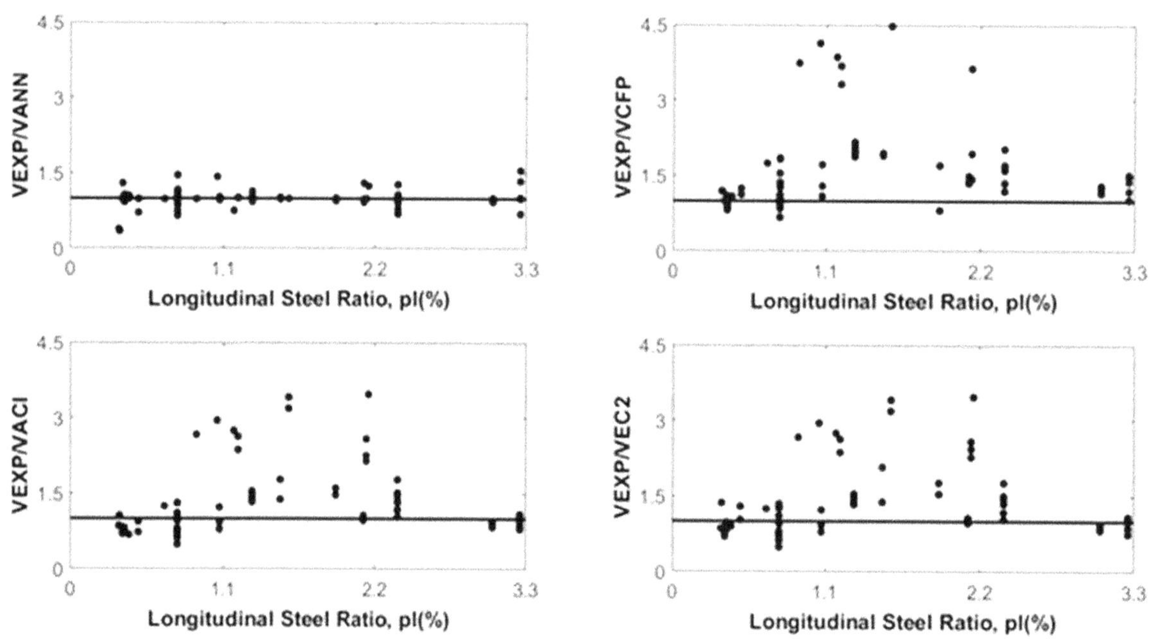

## 1.10 Conclusion

Following are the major conclusions made from current study:

- Through thorough literature review, a vast databank of 95 RCC shear wall samples have been compiled and nine (09) critical parameters controlling ultimate load carrying capacity of wall have been identified as, wall height (H), length of web portion of the wall ($L_w$), *wall* effective depth ($d$), thickness for wall boundary member ($b_w$), shear span ratio ($a_v/d$), concrete compressive strength ($f_c$), yield strength of horizontal and vertical steel ($f_y$), horizontal steel ratio ($\rho_h$) and vertical steel ratio ($\rho_v$).
- Four ANN models were developed, trained, validated and compared for determination of ultimate load carrying capacity (Vexp) of RCC shear wall and best model was proposed based on error metrices values (MSE, MAE and R values).
- The selected ANN model was compared with models based on compressive force design method and concrete design codes, i.e., ACI and EC2, to predict the experimental results. It was found that experimental values were predicted more closely by proposed the ANN model. This is in good agreement with the literature published by various researchers for different structural components. Therefore, soft computing techniques predict ultimate response of RC members of various geometry configurations and under various loading conditions better than the conventional methods.
- Effect of variation in wall depths, compressive strength, and shear span ratio, transverse and longitudinal steel ratio was also studied using conventional and non-conventional methods. Again, it was found that proposed ANN model incorporated the variation in properties well and predicted the experimental values better than its other counterparts.
- Once the ANN models are trained using certain dataset, they can accurately predict the response of RC members even with different geometric and material properties.

## REFERENCES

ACI. (2014). Building Code Requirements for Structural Concrete (ACI 318-14) and Commentary. American Concrete Institute.

Adeli, H. (2001). Neural Networks In Civil Engineering: 1989-2000. *Computer-Aided Civil and Infrastructure Engineering*, *16*(2), 126–142. doi:10.1111/0885-9507.00219

Ahmad, A., & Cotsovos, D. M. (2023). Reliability analysis of models for predicting T-beam response at ultimate limit response. *Proceedings of the Institution of Civil Engineers. Structures and Buildings*, *176*(1), 28–50. doi:10.1680/jstbu.20.00129

Ahmad, A., Cotsovos, D. M., & Lagaros, N. D. (2015). *Assessing The Reliability Of RC Code Predictions Through The Use Of Artificial Neural Networks.* The First International Conference on Structural Safety under Fire & Blast, Glasgow, UK, CONFAB.

Ahmad, A., Cotsovos, D. M., & Lagaros, N. D. (2020). Framework for the development of artificial neural networks for predicting the load carrying capacity of RC members. *SN Applied Sciences*, *2*(4), 1–21. doi:10.100742452-020-2353-8

Ahmad, A., Plevris, V., & Khan, Q.-Z. (2020). Prediction of Properties of FRP-Confined Concrete Cylinders Based on Artificial Neural Networks. *Crystals*, *10*(9), 811. doi:10.3390/cryst10090811

Alarcon, C., Hube, M., & De la Llera, J. (2014). Effect of axial loads in the seismic behavior of reinforced concrete walls with unconfined wall boundaries. *Engineering Structures*, *73*, 13–23. doi:10.1016/j.engstruct.2014.04.047

Altin, S., Kopraman, Y., & Baran, M. (2013). Strengthening of RC walls using externally bonding of steel strips. *Engineering Structures*, *49*, 686–695. doi:10.1016/j.engstruct.2012.12.022

Angus, J. E. (1991). *Criteria For Choosing The Best Neural Network-Part-I*. Naval Health Research Center.

Barkhordari, M. S., & Tehranizadeh, M. (2021). *Response estimation of reinforced concrete shear walls using artificial neural network and simulated annealing algorithm*. Structures. Elsevier.

Barkhordari, M. S., Tehranizadeh, M., & Scott, M. H. (2021). Numerical modelling strategy for predicting the response of reinforced concrete walls using Timoshenko theory. *Magazine of Concrete Research*, *73*(19), 988–1010. doi:10.1680/jmacr.19.00542

Beale, M. H., Hagan, M. T., & Demuth, H. B. (2015). Neural Network Toolbox™-User's Guide. The MathWorks, Inc.

Behnia, A., Chai, H. K., & Shiotani, T. (2014). Advanced structural health monitoring of concrete structures with the aid of acoustic emission. *Construction & Building Materials*, *65*, 282–302. doi:10.1016/j.conbuildmat.2014.04.103

Carmona, J. R., & Ruiz, G. (2014). Bond and size effects on the shear capacity of RC beams without stirrups. *Engineering Structures*, *66*, 45–56. doi:10.1016/j.engstruct.2014.01.054

Chakraborty, S., Kumar, P., & Chakraborty, S. K. (2012). Neural Network Approach to Response of Buildings Due to Earthquake Excitation. *International Journal of Geosciences*, *03*(03), 630–639. doi:10.4236/ijg.2012.33063

Christidis, K., Vougioukas, E., & Trezos, K. (2014). Deformation Capacity of Older RC Shear Walls: Experimental Assessment and Comparison with Eurocode 8–Part 3 Provisions. *Proceedings of the 2ECEES*. Research Gate.

Christidis, K., Vougioukas, E., & Trezos, K. G. (2013). Seismic assessment of existing RC shear walls non-compliant with current code provisions. *Magazine of Concrete Research*, *65*(17), 1059–1072. doi:10.1680/macr.13.00051

Cotsovos, D. M. & Lagaros, N. (2018). *Assessing the reliability of RC code predictions through the use of artificial neural networks*. Springer.

CSA (2004). *Design Of Concrete Structures. A23.3-04*. CSA.

Dackermann, U., Smith, W. A., & Randall, R. B. (2014). Damage identification based on response-only measurements using cepstrum analysis and artificial neural networks. *Structural Health Monitoring*, *13*(4), 430–444. doi:10.1177/1475921714542890

Das, S. K., Kumar, A., Das, B., & Burnwal, A. (2013). On soft computing techniques in various areas. *Computer Science & Information Technology (CS & IT)*, **3**, 59-68.

EC2. (2004). Eurocode 2: Design Of Concrete Structures - Part 1-1: General Rules And Rules For Buildings. EN 1992-1-1. *Management Centre: Avenue Marnix*, *17*, B-1000.

Greifenhagen, C., & Lestuzzi, P. (2005). Static cyclic tests on lightly reinforced concrete shear walls. *Engineering Structures*, *27*(11), 1703–1712. doi:10.1016/j.engstruct.2005.06.008

Günaydın, H. M., & Doğan, S. Z. (2004). A Neural Network Approach For Early Cost Estimation Of Structural Systems Of Buildings. *International Journal of Project Management*, *22*(7), 595–602. doi:10.1016/j.ijproman.2004.04.002

Hadi, M. N. S. (2003). Neural Networks Applications In Concrete Structures. *Computers & Structures*, *81*(6), 373–381. doi:10.1016/S0045-7949(02)00451-0

Karamodin, A., Kazemi, H. H., & Akbarzadeh, M. R. (2008). *Damage Index Monitoring of Structures Using Neural Networks.* The 14 World Conference on Earthquake Engineering, Beijing, China.

Kotsovos, G., Cotsovos, D., Kotsovos, M., & Kounadis, A. (2011). Seismic behaviour of RC walls: An attempt to reduce reinforcement congestion. *Magazine of Concrete Research*, *63*(4), 235–246. doi:10.1680/macr.10.00001

Kotsovos, M. D. (2012). Structural-concrete analysis and design: Need for a sound underlying theory. *Archive of Applied Mechanics*, *82*(10-11), 1439–1459. doi:10.100700419-012-0676-7

Krogh, A., & Vedelsby, J. (1995). Neural Network Ensembles, Cross Validation, and Active Learning. *Advances in Neural Information Processing Systems*, *7*, 21–238.

Lefas, I. D., & Kotsovos, M. D. (1990). Strength and deformation characteristics of reinforced concrete walls under load reversals. *Structural Journal*, *87*(6), 716–726.

Liao, F.-Y., Han, L.-H., & Tao, Z. (2009). Seismic behaviour of circular CFST columns and RC shear wall mixed structures: Experiments. *Journal of Constructional Steel Research*, *65*(8-9), 1582–1596. doi:10.1016/j.jcsr.2009.04.023

Malik, N. (2005). *Artificial neural networks and their applications*. arXiv preprint cs/0505019.

Mangalathu, S., Jang, H., Hwang, S.-H., & Jeon, J.-S. (2020). Data-driven machine-learning-based seismic failure mode identification of reinforced concrete shear walls. *Engineering Structures*, *208*, 110331. doi:10.1016/j.engstruct.2020.110331

Qian, Y., & Mita, A. (2007). Structural damage identification using Parzen-window approach and neural networks. *Structural Control and Health Monitoring*, *14*(4), 576–590. doi:10.1002tc.173

Raza, A., Khan, Q. Z., & Ahmad, A. (2020). Prediction of axial compressive strength for FRP-confined concrete compression members. *KSCE Journal of Civil Engineering*, *24*(7), 2099–2109. doi:10.100712205-020-1682-x

Sarle, W. S. (1995). Stopped Training and Other Remedies for Overfitting. Proceedings of the 27th Symposium, (pp. 1-10). Semantic Scholar.

Shuraim, A. B. (2014). A novel approach for evaluating the concrete shear strength in reinforced concrete beams. *Latin American Journal of Solids and Structures 11*, 093-112.

Tang, Y., & Zhang, J. (2011). Probabilistic seismic demand analysis of a slender RC shear wall considering soil–structure interaction effects. *Engineering Structures*, *33*(1), 218–229. doi:10.1016/j.engstruct.2010.10.011

Waris, M., Mir, J., Plevris, V., & Ahmad, A. (2020). Predicting compressive strength of CRM samples using Image processing and ANN. *IOP Conference Series: Materials Science and Engineering.* IOP Publishing. 10.1088/1757-899X/899/1/012014

Xu, B., Wu, Z., Chen, G., & Yokoyama, K. (2004). Direct identification of structural parameters from dynamic responses with neural networks. *Engineering Applications of Artificial Intelligence*, *17*(8), 931–943. doi:10.1016/j.engappai.2004.08.010

Yi, T. H., Lei, Y., Chen, H. P., Ye, X. W., Kang, F., & Wang, H. (2014). Technology developments in structural health monitoring and integrity maintenance. *TheScientificWorldJournal*, *2014*, 976927. doi:10.1155/2014/976927 PMID:25614901

Zenon Waszczyszyn, L. Z. (2001). Neural networks in mechanics of structures and materials-new results and prospects of applications. *Computers & Structures*, *79*(22-25), 2261–2276. doi:10.1016/S0045-7949(01)00083-9

Zhang, J., & Foschi, R. O. (2004). Performance-Based Design And Seismic Reliability Analysis Using Designed Experiments And Neural Networks. *Probabilistic Engineering Mechanics*, *19*(3), 259–267. doi:10.1016/j.probengmech.2004.02.009

# Chapter 10
# The Value Proposition of Machine Learning in Construction Management:
## Exploring the Trends in Construction 4.0 and Beyond

**Marina Marinelli**
https://orcid.org/0000-0002-6509-0612
*National Technical University of Athens, Greece*

**Mukund Janardhanan**
*University of Leicester, UK*

## ABSTRACT

*In the context of the fourth industrial revolution, machine –learning based applications, such as 3D Printing, computer vision and robotics, are envisaged to digitize construction. This chapter introduces the reader to the basics of AI and focuses on the role of machine learning in the development of technological applications aimed at the improvement of the construction industry's efficiency and sustainability. It reviews the potential benefits of these applications and identifies factors like e.g. high costs, skills gap, dynamic and risky working environment that typically keep them at a very low practical usage in construction. Furthermore, a bibliometric search reveals that, despite its key role in the digitization vision, machine learning is rarely acknowledged in the relevant 'Construction 4.0' literature. The chapter also identifies the cybersecurity issues emerging from machine learning applications' use and discusses mitigation strategies and technologies aimed at minimizing the impact of threats.*

DOI: 10.4018/978-1-6684-5643-9.ch010

## INTRODUCTION

The Fourth Industrial Revolution, also referred to as "Industry 4.0", is a recently introduced concept describing a disruptive innovation era, where the organizations and processes are digitally interconnected with the potential to reshape the value delivery mechanisms for services and products across the whole value chain (Bolpagni et al., 2022). There is no doubt that the current technological advancements have massively increased the potential quantity of information available, the resolution and frequency at which it is being captured, and the speed at which it can be processed, especially after the emergence of the Internet of Things (IoT), i.e., the radical evolution of internet into a network of interconnected objects with built-in sensors and computing capability (Konanahalli et al., 2022). As the volume and complexity of data available continuously grows, the learning abilities, the vast memory and the ever-increasing processing speed of computers, are more necessary than ever, to complement human intelligence. In this context, various subfields and applications of Artificial Intelligence (AI) such as machine learning, optimisation algorithms, natural language processing, robotics and computer vision can be applied to detect meaningful data patterns and support decision-making for complex real-world problems.

The disruptive innovation of Industry 4.0 has also brought an 'intelligent construction' era, widely reported as Construction 4.0, typified by technologies and applications which have been proven to effectively contribute to industrial operations' efficiency such as the IoT, sensors, cloud computing and big data analytics. Furthermore, well-established in the industrial practice, machine –learning based applications, such as Additive Manufacturing (widely known as 3D Printing) and robots are starting to pervade the construction industry and attract the intense interest of construction management researchers. In this context, the limits between different scientific fields become increasingly diffuse and transform construction into an interdisciplinary industry (You & Feng, 2020; Forcael et al., 2020).

However, the use of AI in construction management, is not new at all. In the past few decades, dozens of relevant articles have used machine learning and optimization algorithms to tackle a very diverse range of issues, indicatively including the management of health and safety risks, the prevention of disputes between the stakeholders, the optimization of the site layout, the management of decision processes with regards to bidding, maintenance and resource allocation, among others. In the above context, this chapter will explore the role of machine learning in Construction 4.0 applications and will attempt to quantitatively and qualitatively assess the impact of the advent of Construction 4.0 in the AI-related research trends.

Besides the above, the literature shows consensus on the fact that beyond the extensive bibliometric analyses on potential benefits and opportunities (e.g. Oesterreich & Teuteberg 2016; Boton et al., 2021; Forcael et al., 2020; Abioye et al., 2021), the actual adoption of the Construction 4.0 technologies by the construction professionals is very low. The reasons for the construction industry struggling so much with the beneficial adoption of AI and other digital technologies has been attributed to a variety of factors including cultural, cost, skill, technological and data security barriers (Abioye et al., 2021). Regarding the latter, machine learning algorithms are indeed vulnerable to attacks that can transform them into a liability for the system they are intended to aid (Joseph et al., 2019). However, at the same time, machine learning has been highlighted as a promising means for intrusion detection, especially over IoT data (Shahin & Sabri, 2022). In this context, this chapter will also discuss the cybersecurity issues emerging from the use of machine learning applications as well as mitigation strategies, which aim to select proper learning methods to minimize the impact of threats and prevent failures. This is a topic of critical importance for the construction field as such a failure can potentially result not only in cost and time implications, but also lead to injuries or loss of life (Abioye et al., 2021).

Furthermore, the construction management literature has shown over the past few years a constantly growing interest for the recently introduced technology of Blockchain, a decentralized and transparent mechanism able to safeguard AI applications from unauthorized manipulation and malicious attacks. As a considerable share of the vast relevant literature is devoted to the potential benefits of the Blockchain in the building sector - including construction, real estate and facilities management, this chapter will also discuss the value proposition of machine learning in its effectiveness of operation from a security defence perspective.

## BACKGROUND

The field of artificial intelligence, or AI, is concerned with building intelligent entities—machines that can compute how to act effectively and safely, in a wide variety of novel situations. According to the Turing test, proposed by Alan Turing in 1950, for a machine to be considered as 'intelligent', the following capabilities would be necessary: natural language processing to communicate successfully in a human language; knowledge representation to store what it knows or hears; problem solving and automated planning to answer questions and to draw new conclusions; machine learning to adapt to new circumstances and to detect and extrapolate patterns. On top of the above, later researchers expanded the test scope to include computer vision to perceive the world and robotics to manipulate objects and move about. These six disciplines compose most of AI (Russell & Norvig, 2022).

A brief presentation of the scope and interdependencies of the above AI subfields, with particular emphasis on Machine learning, is given below:

- **Natural language processing:** Concerns the development of language models i.e. models which get trained to predict what words are likely to come next in a text, and thereby suggest completions for an email or text message or suggest spelling or grammar corrections.
- **Knowledge representation**: Concerns how to represent diverse facts about the real world in a form that can be used to reason and solve problems. The organization of objects into categories and the development of taxonomies of hierarchy is a vital part of knowledge representation. Expert systems and case-based reasoning systems fall in this area of AI.
- **Problem solving and automated planning**: It is about how an agent can take advantage of the structure of a problem to efficiently construct complex plans of action. Planning combines search and logic to find a solution or prove the existence of a solution for Scheduling, temporal and resource constrained problems with millions of states and thousands of actions, among others. This AI branch is also known as Intelligence computing, representing the creation of a system with an independent thinking ability which has a wide application prospects in the optimization problems. In complex engineering problems, where the computation cost is too high or the development of an accurate model is not feasible, computer science proposes a range of intelligent algorithms with heuristic characteristics. Metaheuristics can be applied efficiently to large and non-uniform search spaces and can escape local optima, find the global optimum, and solve complex, non-linear, and discrete problems. Popular algorithms of this category include ant colony optimization (ACO), genetic algorithms (GA), greedy randomized adaptive search procedure (GRASP), iterated local search, particle swarm optimization (PSO), scatter search (SS), simulated annealing (SA), tabu search (TS) and variable neighbourhood search (VNS).

- **Machine learning**: Refers to the training of computers to simulate human learning and independently obtain knowledge with the aim to detect meaningful patterns in data. It encompasses a number of algorithms which learn from a dataset composed of examples or observations and are capable of making predictions. In this context, machine learning and data are complementary entities as machine learning can reveal the value of data and at the same time, machine learning models' performance depends on the volume of data used for their training. Machine learning shares common threads with the mathematical fields of statistics, information theory, game theory, and optimization; it is not trying to build automated intelligent behavior, but rather to use the strengths and special abilities of computers to complement human intelligence, often performing tasks that fall way beyond human capabilities (Shalev-Shwartz & Ben-David, 2014).

Machine learning models include linear regression, kernel (non-linear) regression, tree-based algorithms (decision trees, random forests, adaptively boosted trees), logistic regression, support vector machines, k-nearest neighbours, discriminant analysis, among others. Also machine learning includes Deep Learning, i.e. a broad family of techniques in which hypotheses take the form of complex algebraic circuits with tunable connection strengths. The word "deep" refers to the fact that the circuits are typically organized into many layers, which means that computation paths from inputs to outputs have many steps (Russell & Norvig, 2022). Generally, Deep Learning models contain three types of layers: input later (receives data), hidden layer (extracts patterns), and output layer (produces the results). The output of one layer is used as an input into the next one (Khallaf & Khallaf, 2021). The networks trained by deep learning methods are often called Artificial Neural Networks (ANN), to reflect the way that human brain works, even though the resemblance to real neural cells and structures is superficial (Russell & Norvig, 2022).

There are multiple architectures that can be implemented when it comes to ANNs. Each of these architectures has its uses and compatibilities with certain applications.

Commonly used types of networks are:

- The simple feed-forward networks where connections only have one direction,
- Recurrent neural networks which feed their intermediate or final outputs back into their own inputs Convolutional neural networks,
- Convolutional neural networks (CNNs) which contains spatially local connections, at least in the early layers, and has patterns of weights that are replicated across the units (Russell & Norvig, 2022)
- Encoder-decoder networks which involve an encoding and decoding process which forces the network to ignore the noisy part of the input and instead focus on encoding /representation of the more informative segments (Akinosho et al., 2020)

Machine learning encompasses different styles of learning: supervised, unsupervised, reinforcement and deep. The first relies on a set of procedures for function approximation based on training data (input) which has a clear label. The goal is to learn a mapping from inputs $x$ to outputs $y$, based on a labelled set of input – output pairs. In other words, the goal is to predict a response variable (or output) from the explanatory variables (or inputs). Supervised learning problems are described as classification or pattern-recognition when the response variables are categorical and regression when the outputs are numerical variables (Murphy, 2012). Common supervised learning methods include regression analysis, decision trees, support vector machines (SVMs), and deep learning (ANNs). Unsupervised learning does not include any response variable, and attempts to discover underlying relationships between the data. This knowledge discovery mission has no obvious error metric to use. The main tasks are: anomaly detection, dimension reduction, time-series modelling, and latent variable models (Calvet et al., 2017). Both supervised and unsupervised learning can be achieved using parametric and non-parametric models. The former are often easier to construct and implement but are constrained by the assumptions that they make about the data distribution. The latter are much more flexible but their complexity increases with the size of the dataset. Reinforcement learning, the least popular of the three categories, is used to acquire knowledge on how to act or behave (i.e. make decisions) when given occasional reward or punishment signals, i.e. according to the outcome of interactions with the environment (Murphy, 2012).

- **Computer vision**: A multidisciplinary field concerned with artificial simulation of the human sight. The two core processes of computer vision are reconstruction, where an image or a set of images is captured, processed and algorithmically analyzed with the aim to enable the agent to build a model of the world, and recognition, where visual and other information are used for distinctions to be drawn among the objects. As Pan and Zhang (2021) explain, deep learning has become the mainstream enabler of machine vision i.e. the tasks of image classification, object detection and semantic segmentation. To be more specific, image classification is to understand an entire image as a whole by assigning a certain label to it and is often conducted by the most popular architecture of neural networks i.e. the CNNs. Similarly, object detection uses the region-based convolutional neural network (R-CNN) to draw a bounding box around each object of interest in the image and give it a proper label (e.g. face). Finally, fully convolutional networks (FCNs) are used for semantic segmentation of the image which is to semantically understand the role of each pixel depending on certain visual properties, such as brightness, color, and texture.
- **Robots**: Highly automated devices equipped with effectors such as legs, wheels, joints, and grippers aimed at performing various tasks. Their operation is enabled by various sensors destined to measure the environment e.g. cameras, radars, lasers and their own state e.g. gyroscopes, strain and torque sensors, and accelerometers. Robots operate in real world's complexity and machine learning plays a pivotal role in their perception of the environment and the task they have to complete. Motion planning and trajectory optimization are typical problems emerging from the robots' nature as dynamical systems. The relevant literature is vast and reinforcement learning techniques are widely applied to give solutions.

The above demonstrate that Machine Learning has a pivotal role in the delivery of AI benefits as on the top of being an AI subfield on its own, it is also the AI-enabling factor of other AI subfields like computer vision and Robotics.

## THE ROLE OF MACHINE LEARNING IN CONSTRUCTION 4.0 TECHNOLOGIES AND APPLICATIONS

"Industrial revolutions" historically represent periods of disruptive technological advances that have led to paradigm changes such as the mechanization (the first industrial revolution), the intensive use of electricity (the second industrial revolution), and the wide use of IT –led automation (the third industrial revolution). The fourth Industrial revolution, also known as Industry 4.0, was publicly introduced in

2011 by the Director and CEO of the German Research Centre for Artificial Intelligence as an approach aimed to strengthen the competitiveness, safety and transparency of the German manufacturing industry. Industry 4.0 promotes the integration of information, objects, and people in cyber-physical scenarios which enable factories to transform their environment into one of intelligent manufacturing (Bolpagni et al., 2022).

Construction has long been adapting and incorporating knowledge, practices and tools from the manufacturing sector. Furthermore, construction is special in the uniqueness of its processes, products, and value chains, and works with diverse and semantically rich information (Turk et al., 2022). In this context, the disruptive innovation of Industry 4.0 has opened the discussion for the digital transformation of the building sector, widely reported as 'Construction 4.0', encompassing older and emerging technological applications such as the computer-aided design, the Internet of Things (IoT), cloud computing, 3D printing, and robotics. Those concepts are not only adding a new feature to the construction market but also transform the concept of construction itself. Although the industry has been traditionally labelled as unproductive, this trend might be changing and that change is expected to be stronger in the near future (Forcael et al., 2020).

## BIM and Digital Twin

In the field of computer-aided design, the popularization in the last decade of Building Information Modelling (BIM), i.e. the representation of 3-dimensional digital building components with computable graphic and data attributes and parametric rules (Sacks et al., 2018), has definitely paved the road for profound changes in the construction industry. Although BIM became popular in construction before Construction 4.0, achieving a high level of maturity in many countries, such as the UK, it continues to receive significant research attention in the context of the Construction 4.0 vision. The reason is that BIM can be reasonably considered as a digital backbone to work with AI (Pan & Zhang, 2021), as when combined with a network of wireless sensors, it establishes a real-time operation model (digital twin) which can become a data platform for the development of efficient decision support models based on machine learning mechanisms (Opoku et al., 2021; Zhao et al., 2022). As a result, BIM is amongst the most popular keywords in the Construction 4.0 era (Adekunle et al. 2021). Digital twins are generally understood as up-to-date digital representations of the physical and functional properties of a system, which may be a physical instrument (e.g. an aircraft engine), a social construct (e.g., a stock market), a biological system (e.g. a medical patient), or a composite system (such as a construction project, with aspects of physical products and social systems) (Sacks et al., 2020).

The digital twin is a well-known concept in the manufacturing industry, but only a nascent vision for construction. It envisages that the physical entity (building) and the digital twin model interact in real time, while the relevant data are analysed by machine learning applications (e.g. ANNs) to diagnose and predict the building status. Then, the digital twin processes the predictive information provided by the machine learning algorithm and makes intelligent decisions that are transmitted to physical entities (Zhao et al., 2022). While BIM is predominantly used in the design and construction phase, digital twins work on real-time and operational data. In other words, BIM models evolve into digital twins as we proceed towards operations and facilities management phase (Karmakar & Delhi, 2021).

The above paradigm takes advantage of the use of sensors and the IoT, where the latter describes a network of interconnected physical objects equipped with various sensors (e.g. RFID, video cameras, and laser scanners) which exchange and process data over the Internet or other communications networks. This data aims to support decisions related e.g. to resource optimization, risk monitoring, energy conservation, emission reductions, and project performance evaluation. However, IoT devices have a limited data-processing capability which calls for their integration with cloud technology, typified by nearly unlimited virtual computing resources (You & Feng 2020). Cloud computing is an emerging technology for transferring and storing data and perform calculations at third party data centres and is already widely used in the construction industry as a supporting technology for BIM which enables the collaboration of stakeholders on a common platform and promotes supply chain integration.

## Virtual/Augmented/Mixed Reality

Virtual reality (VR) aims to create artificial 3D environments that can replace a user's perception of the real world with a computer-generated artificial 3D environment which is not necessary based on a real one. The purpose of VR simulation is to enable users to experience unique insights into the way the world works. However, it has limitations in providing a high level of realism due to a lack of sensory feedback to accommodate perceptual and cognitive viewpoints.

Augmented Reality (AR) is also an emerging technology which can overlay computer-generated graphics onto the real world with the aim to create through some device such as a tablet or a smartphone, an augmented scene that enhances a person's perception of a virtual prototype within a real environment. AR technology can provide a more realistic experience and help users better understand and interact with virtual models like those generated by BIM software.

Mixed Reality (MR) is a technology that combines elements of both VR and AR to create a new type of immersive experience. In MR, users can interact with both virtual and real-world objects in a merged environment. Unlike AR, MR can use sensors to map and track the surrounding environment, allowing virtual objects to interact with real objects in a more realistic way.

A VR/AR system typically includes hardware components such as processors, displays, sensors, and input/output devices, as well as software and algorithms to achieve a realistic scenario. Input devices are used for user interaction with the virtual world, sending signals to the system for appropriate real-time output responses through output devices. The computer with VR/AR engine serves as an interface with the Input/Output (I/O) devices and handles the interaction with users (Li et al., 2018).

## 3D Printing

Three-dimensional printing, also known as additive manufacturing, is a manufacturing technology available since the '80s which aims to create physical objects through the deposition of layers of material by a robotic device (3D printer) which executes a digital plan. 3D Printing is particularly popular in the contemporary literature of construction, although its actual use is at experimental stage only. The employment of machine learning in 3D printing is particularly extensive and encompasses the stages of design, process optimization and quality control. As Goh et al., (2021) explain in detail, machine learning techniques can be used to make feature recommendations to existing CAD models, for feature recognition as well as for manufacturability analysis, including best printing direction. Data-driven models trained on FEA simulation results have also the potential to predict within milliseconds the final properties of the printed parts, such as the stress distribution, stiffness, toughness, and the elastoplastic behaviour. Then, the trained ANN models can potentially be incorporated into existing FEM frameworks to simulate the structural performance of larger parts of various scale and facilitate the development of new designs and materials. The search for new 3D printing materials with specific performance such as high compression and tensile properties, strong crack resistance and toughness, short setting time and high setting strength can be done by training the machine learning algorithms to detect features and patterns from a large database of available material properties. Similarly, machine learning algorithms can be used to determine the process–structure–properties relationships of Additive Manufacturing processes, and enable the proper selection of the parameters from relevant databases. he cloud platform has also a crucial impact on popularizing 3D printing as it enables the sharing of resources (e.g. 3D models) to a public repository and integrates them to form a comprehensive pool of resources.

## Machine Vision

Machine learning algorithms can also be used for quality monitoring, e.g. to detect defects by analyzing images obtained by cameras and processing failure measurements provided by suitably selected sensors. Anomaly detection relies heavily on machine vision which in turn most usually involves the use of CNNs trained to detect the difference between good printing samples and bad printing samples. Machine learning can also allow for greater efficiency in quantity surveying and cost control by making use of historic data to predict the amount of material needed and accurately estimate the budget (Goh et al., 2021).

## Robotics

Robotic applications form a long and well established reality in the Fourth Industrial Revolution, recognized for their precision, speed, and efficiency in the execution of various tasks. The collaborative work paradigm with human and robots sharing the same work space and working together towards a common goal, is very well researched in the industrial practice and has a strong AI dimension. The latter emerges from the need for robust machine learning -based controlling techniques to enable the robot to efficiently perform its tasks while respecting the human partner's safety and wellbeing. Gualtieri et al. (2021) and Proia et al. (2021) have presented relevant reviews of the available literature from the manufacturing sector.

In the field of construction though, the AI aspect of robotics has a much lower representation and the aspect of automation is much more prominent. Robots are perceived as an advanced form of operational mechanization which needs no human intervention and is mainly used with the aim to replace human operators in the execution of hazardous / strenuous tasks. This reflects the fact that the use of robots in construction sites is very difficult as a result of a variety of factors, including construction's unstructured organization and non-linear workflow, frequent spatial-temporal conflicts of resources and dynamic change of the plans with a high degree of uncertainty. In this context, although robotic technologies, including 3D Printing, have generated a strong literature trend structured around potential benefits and opportunities, their current level of actual maturity does not allow for their widespread adoption in construction practice. Gharbia et al. (2020) report that the great majority of robotic technologies belong to the categories of 'challenging' or 'achievable' and very few are mature enough to have been used in actual building projects. Liang et al. (2021) indeed confirm that in experimental level, there are robotic applications which have been developed and successfully trained through machine learning techniques in the execution of a variety of tasks, including brick or block laying, assembly of various structural elements such as timber structures or concrete elements, digging, concrete levelling, concrete finishing and painting, among others. They also support that Imitation learning methods present a promising opportunity to deploy robots on construction sites. This is the case of learning where skilled human workers continually train construction robots (learning from demonstration) and work with them to supervise their performance during the task execution.

## MACHINE LEARNING IN THE CONTEXT OF CONSTRUCTION 4.0 LITERATURE AND BEYOND

### Bibliometric Data

The analysis of the previous section clearly demonstrates that machine learning is of crucial importance in the implementation of the Construction 4.0 applications i.e. the BIM-based Digital Twin, 3D printing, Computer vision and Robotics. This section aims to investigate how accurately the literature presents this vital interrelation, taking into account the substantial literature trend that Construction 4.0 has created, since 2016, when it first emerged.

The literature search in Scopus returns at least 35 review articles (either generic or focused on a particular Construction 4.0 dimension) and another 257 journal and conference articles with 'Construction 4.0' as part of their title or selected keywords. Forcael et al. (2020) observe an exponential increase in the number of Construction 4.0 publications in the past 5 years and note that this is expected to dramatically increase in the years to come.

However, the investigation of the presence of machine learning in this literature, shows that only a 7% of those papers include machine learning -related words (e.g. deep learning, machine learning, reinforcement, ANNs etc) in the title or/and their keywords.

Examples of these very few papers are:

- 'Investigation for integration of deep learning and digital twins towards Construction 4.0' by Kor et al. (2022),
- 'development of regression-based deep neural networks for anatomical analysis of equipment in civil infrastructure projects' by Arashpour et al. (2022)
- 'Machine learning - based framework for construction delay mitigation' by Sanni-Anibire et al.'s (2021)
- 'Machine learning set to play a key role in the transition to Construction 4.by Venkatasubramanian (2020) but this however, is not adequately reflected in the Construction 4.0 literature.

Furthermore, a thorough scanning of the 35 review articles show that the references to the contribution of machine learning in Construction 4.0's mission are rare, extremely limited and unsystematic. The above show that the recently developed Construction 4.0 trend, systematically fails to acknowledge the inextricable role of machine learning in this new managerial paradigm.

The literature search beyond the Construction 4.0 concept, leads to congruent conclusions: Machine learning applications in construction represent a very well and long- established research theme with more than 9,000 papers published. This number though gets dramatically reduced to just 13 papers when the term 'Construction 4.0' is additionally requested as a keyword, or part of the title / abstract of these papers. Indicatively, the string for the aforementioned search is (TITLE (ANN) OR KEY (ANN) OR TITLE (Neural) OR KEY (Neural) AND KEY (Construction) OR SRCTITLE (Construction) OR

SRCTITLE (Building) OR SRCTITLE (Built) OR SRCTITLE (Structural) OR SRCTITLE (Civil)) AND TITLE-ABS-KEY (Construction 4.0)). Similarly,

- among the 967 papers contemplating SVM in construction, only 4 have also used the word 'Construction 4.0'
- among the 1,320 papers related to deep learning applications in construction, only 4 describe their mission as part of 'Construction 4.0'.
- for 3D Printing, the reviews by Perkins & Skitmore (2015), Labonnote et al. (2016) and Wu et al. (2016) clearly prove that the concept was well known even before the emergence of 'Construction 4.0'. The relevant research continues to evolve with 568 papers published to date, of which only 8 associate the method to the 'Construction 4.0' trend.
- As far as the use of robotics in construction is concerned, the relevant research theme also exists in the literature since the '90s (e.g. Kangari & Halpin, 1990; Skibniewski & Wooldridge, 1992; Rosenfeld et al., 1993) with more than 3,700 papers published to date. Only 18 of these papers though make the connection with Construction 4.0.

All the aforementioned quantitative metrics have been based on the search of the respective terms as keywords in construction related journals/ proceedings listed in Scopus.

The above clearly reveal a gap between the management side and the engineering side of construction research: the management – originated, emerging concept /vision of Construction 4.0 is not known (or of interest) to the vast majority of construction/structural/civil engineering researchers specialized in machine learning techniques, but very popular among construction management researchers. The latter typically emphasize on conceptual / organizational aspects of the new paradigm and don't seek to highlight the technical side of things. As a result, the machine learning aspect remains invisible in the Construction 4.0-specific literature. Furthermore, secondary weaknesses in technical accuracy can also be observed: for instance, in many cases the literature refers to AI as a standalone tool/application/technology and not as the resulting consequence of technological advancement; similarly 'big data', is commonly presented as 'technology' - which clearly is not. Also, the literature most usually fails to differentiate between the learning entities (ANNs and other machine learning algorithms) and the technologies that enable the collection of the data necessary for the learning aspect to materialise (IoT, sensors, cloud computing), as they are all interchangeably treated as Construction 4.0 'key concepts'.

## Expected Benefits for the Construction Industry

Regarding the gains that the literature envisages from the implementation of Construction 4.0, the most common expectations concern the attainment of higher productivity and efficiency, more accurate identification of risks, reduction of accidents, improvement of processes, higher quality and lower costs, which all translate into the achievement of competitive advantage for construction businesses. Regona et al. (2022) have presented a relevant extensive review of opportunities associated with the use of AI in construction. Furthermore, the literature contemplating the use of AI applications in construction, includes

- at least 409 papers (47% published on 2017 onwards) describing the use of machine learning techniques for cost estimation purposes in various kinds of construction projects,

- another at least 237 papers (51% published on 2017 onwards) devoted to machine learning- based scheduling.
- Furthermore, at least 284 papers (67% of which were published on 2017 onwards) are devoted to learning-based risk assessment and mitigation (e.g. structural performance risks, delay and scheduling risks, financial risks, fire risks, construction dispute risks),
- another at least 111 papers (38% of which were published on 2017 onwards) present the use of machine learning for the prediction/ improvement of productivity in construction,
- at least another 79 papers (71% of which were published on 2017 onwards) are contemplating the use of machine learning in problems related to accidents prevention and safety enhancement.
- Moreover, at least 98 machine learning-based papers (40% of which were published on 2017 onwards) are found in the field of procurement / bid decision making and tender evaluation
- another at least 47 papers have been published in the area of construction equipment management (70% of which on 2017 onwards).

It is to be noted that the publications' count is not exhaustive as specific keywords have been used at the Scopus database to show the publication trends and to give an idea of the importance of the various research areas. Indicatively, the search string used has the following structure, adapted according to the key area of interest: (KEY (Construction) OR SRCTITLE (Construction) OR SRCTITLE (Building) OR SRCTITLE (Built) OR SRCTITLE (Structural) OR SRCTITLE (Civil) AND TITLE (Neural) OR KEY (Neural) OR TITLE (ANN) OR KEY (ANN) OR TITLE (Learning) OR KEY (Learning) AND TITLE (schedul*) OR KEY (schedul*) AND NOT TITLE-ABS-KEY (student)). The exclusion of the term 'student' was deemed necessary as the inclusion of the term 'learning' had the potential to misdirect the search to educational papers.

Similarly, optimization algorithms and specifically metaheuristics, have also been long applied in construction due to their effectiveness in handling highly nonlinear, multi-modal, constrained, discontinuous, and non-differentiable optimization models, often encountered in construction project management. Evolutionary / genetic algorithms are the most popular metaheuristics, most frequently used for research problems of site layout selection, time-cost trade –off analysis for project activities, and resource selection (e.g. type and number of equipment) and allocation to activities (Liao et al., 2011).

The above show that the emergence of Construction 4.0 has not made any significant difference to the agenda of envisaged applications and benefits of technology in construction. Actually, the industry has been consistently turning to AI through time, with the aim to alleviate issues that traditionally plague the sector, i.e. cost and time overruns, accidents, increased risks, low productivity and procurement complexity. Nevertheless, despite the current research trend around AI concepts and applications, the pivotal role of machine learning is greatly understated, obscured by the technological buzzwords new to the field.

## Practical Application Examples and Limitations

As already mentioned in the previous section, the literature is full of references for numerous problems plaguing the construction industry including among others, safety issues, poor planning and scheduling, poor site management and supervision, lack of skilled labor, cost overruns, delays, excessive changes of design during construction, low productivity, absenteeism of labor, low plant utilization and availability, wastage and poor quality (e.g. Yap et al., 2019; Antoniou and Merkouri 2021; Antoniou 2021). With the use of advanced technology and AI tools, the construction industry aims to alleviate the above weaknesses and reap benefits of productivity, cost-efficiency, improved safety and sustainability, as widely reported for the manufacturing sector. Nevertheless, the construction industry faces distinct challenges when it comes to reaping benefits from technology, compared to the manufacturing sector. Firstly, construction takes place in an unpredictable environment, making it susceptible to high levels of variability and uncertainty, such as changes in weather conditions or customer requirements. Moreover, the construction process involves numerous interrelated steps and a large number of participants at different stages and locations, making it difficult to manage. Further challenges stem from the high degree of complexity involved in construction projects, which are often unique, time-limited, and require significant customization. Additionally, the temporary nature of construction projects means that complete and standardized specifications are often lacking. The construction industry also has a fragmented supply chain, made up of multiple SMEs, which are difficult to coordinate and have limited resources to invest in innovative technologies. In this context, only a small percentage of construction companies make full use of digital planning tools, despite agreement among decision-makers that digitization has the potential to positively impact every aspect of the industry. Given the above, a critical reflection on the actual potential of the various AI applications as levers of improvement, effectiveness and efficiency in construction, is necessary, for the reader to obtain a realistic as well as holistic understanding of the subject matter (Table 1).

*Table 1. Benefits, challenges and actual state of application in construction for various AI-enabled technologies*

| Technology/ Application | Benefits | Challenges / Disadvantages | State of application | Relevant sources |
|---|---|---|---|---|
| 3D printing | ▪ Labor cost 50–80% lower<br>▪ Construction waste 30–60% lower<br>▪ Faster construction<br>▪ Shorter supply chain<br>▪ Safer sites<br>▪ Eco-friendly materials<br>▪ Reduced formwork | ▪ Very strict material specifications for rheological and elastic properties and quick solidification<br>▪ The material flow must be continuous<br>▪ The printer size restricts the size of the projects.<br>▪ The load-bearing capacity is directional –dependent<br>▪ Reduced dimensional accuracy and limited geometrical possibilities<br>▪ High initial and set-up costs | ▪ Several 3D-printed building projects have been completed in the past few years, by various companies worldwide.<br>▪ The use is still very limited and experimental.<br>▪ Small-scale projects only are feasible. | El-Sayegh et al., (2020) |
| Technologies for tracking and locating items e.g. RFID, Global Positioning System (GPS), Ubiquitous Sensor Network (USN) etc<br>Potential applications of the same technologies include:<br>▪ Material tracking and handling onsite<br>▪ Smart scheduling based on real-time progress data<br>▪ Tracking workers movement and equipment<br>▪ Construction equipment management<br>▪ Quality control, inspection, maintenance, and operations | ▪ Error reduction<br>▪ Elimination of irregular and incidental processes<br>▪ Lower reliance on expensive material stocks<br>▪ Shorter task duration<br>▪ Reduction of labor costs<br>▪ Reduced tasks related to manual localization of materials<br>▪ Protection and quality assurance of materials<br>▪ Easier inter-organizational information in real-time<br>▪ Increased construction productivity<br>▪ Collection of real-time data to monitor processes<br>▪ Greater efficiency and safety | ▪ High cost<br>▪ Requires workers to be trained<br>▪ Potential resistance from workers fearing they could be replaced<br>▪ Regulations regarding the use of RFID and GPS antennas may restrict the use of such systems | The literature reports limited use in the construction industry. | Dardouri et al. (2022)<br>Dallasega et al. (2019)<br>Dallasega et al. (2018)<br>Perrier et al. (2020) |

*continues on following page*

*Table 1. Contiued*

| Technology/ Application | Benefits | Challenges / Disadvantages | State of application | Relevant sources |
|---|---|---|---|---|
| Deep learning | Increased accuracy in prediction of critical factors for project success including but not limited to<br>▪ Construction cost / duration<br>▪ Life cycle cost<br>▪ Accidents<br>▪ Resources Performance / Productivity<br>▪ Equipment maintenance needs<br>▪ Structural strength<br>▪ Ground condition<br>▪ Collusion in tendering<br>▪ Cyber threats<br>Tracking of objects/ workers/ equipment onsite | ▪ Criticized as a 'black-box' approach<br>▪ Each model is created for a specific purpose and is not necessarily transferable to other applications.<br>▪ Dataset size affects model performance.<br>▪ Large volumes of data are needed to train the models, which are not always available.<br>▪ Vision-based detection depends on camera angle and position, presence of obstructions, dense construction, and inconsistency in worker posture/ movement<br>▪ Deep learning algorithms are complex and require high technology and hardware. | Very widely used for research purposes. | Jacobsen et al. (2022)<br>Khallaf et al. (2021)<br>Aretoulis (2019)<br>Keshtegar et al., (2022)<br>Zhu et al., (2022) |
| VR/ AR for Information flow, accessibility, visualization | ▪ Increased customer's understanding on the final product (e.g. building)<br>▪ Avoidance of wasteful design changes<br>▪ Users are informed of the probability of an accident before it occurs.<br>▪ Accurate (post-) location of accidents and provision of emergency action guidance<br>▪ Can be used for health and safety training and to increase awareness of ergonomics amongst the workers<br>▪ Can be used to compare the actual project's progress to the plan<br>▪ Can be used for training purposes, i.e. to help workers through visualization understand their duties with the desired accuracy<br>▪ Can be used for quality inspection<br>▪ Can be used to familiarize engineering students with the actual construction project environment | ▪ All data (design, risks etc) must be identified and recorded in advance to make the most of such a mechanism.<br>▪ Increased Complexity of localization of items through RFID<br>▪ Interoperability issues between VR/ AR systems and other IT tools<br>▪ The user must constantly hold something with their hands. The alternative of glasses restricts the peripheral vision and the perception of the surrounding environment.<br>▪ Dizziness may occur as a result of the stereoscopic glasses<br>▪ Obtaining the required accurate human perception and cognition needs massive human experiments to be conducted<br>▪ The required devices i.e. smartphones, tablets, and AR glasses can be used by only one user, which seriously questions teamwork.<br>▪ A minimum distance between the device and the projection surface is needed to present information using this device.<br>▪ If users are exposed to direct projector light, it can cause damage to their eyes | The technology is still immature for wide use in practice, and needs further investigation | Kolaei et al. (2022)<br>Li et al. (2018) |
| Robots for<br>▪ Automatic installation / assembly of building elements<br>▪ Dangerous, strenuous, repetitive tasks<br>▪ Bricklaying<br>▪ Concrete spraying<br>▪ Foam concrete application | ▪ Improvement in productivity, quality, accuracy and safety<br>▪ Labor shortages mitigation | ▪ Dynamically changing, harsh and risky environment<br>▪ Workspace limitations<br>▪ Complexity in coordinating robot movement<br>▪ High cost | ▪ The great majority of robotic technologies remain at an experimental stage.<br>▪ The current level of robotic reasoning, perception and adaptability, is not sufficient for complex and dynamic construction environments | Kim et al. (2022)<br>Liang et al. (2021)<br>Gharbia et al. (2020) |

## MACHINE LEARNING AND CYBERSECURITY

### Understanding the Security Threats and Challenges

Handling data involves distinctive challenges resulting from the complexity of the data itself, the data processing, as well issues related to privacy and security risks, data governance and sharing practices. Cybersecurity and data privacy is a particularly critical issue as it involves huge implications in terms of compliance, data loss, exposure to non-legitimate users, and data quality. In this context, despite the undeniable usefulness of machine learning techniques, it cannot be overlooked that these methods are vulnerable to attacks. This could cause a notable decrease of their performance and eventually transform them into a liability for the system they are intended to aid. This results from the fact that traditional machine learning techniques were not designed originally to deal with clever and adaptive attacks (Dasgupta et al., 2022). According to Joseph et al. (2019), learning algorithms pose a risk to privacy because the behaviour of the learner reflects the data it was trained on and hence may reveal, if violated by an adversary, aggregate information on the patterns it gleans from that data.

- The security threats to machine learning systems can be divided into three categories based on the influence on classifiers, the security violation, and the knowledge perspectives. As Dasgupta et al. (2022) explain in detail, from the perspective of the influence on classifiers, the machine learning security threats can be classified as either causative or exploratory attacks.In a causative attack, the adversary alters the training data either directly or through injecting an adversarial training dataset. Similarly, it can distort the algorithm and model parameters. Attacks at the stage of training are also called poisoning attacks as the adversary's aim is to contaminate the training process with malicious activity (e.g. intentionally mislabelled datapoints) so that it can easily be accepted by the system in the future.
- On the other hand, in an exploratory attack, the adversary does not interfere with the training process but aims to cause misclassification at the testing stage. Attacks at the stage of testing are also called e*vasion* attacks.

From the perspective of security violation, the attacks can be distinguished in three categories, i.e. availability, integrity, or privacy violation attacks.

- Availability attacks attempt to compromise the normal system functionalities available to genuine users e.g. its quality, performance (e.g., speed), or access.
- Integrity attacks attempt to undermine the testing process, with the goal of inducing increased false negatives by the machine learning model when classifying adversarial samples.
- Privacy violation attacks attempt to extract confidential and private information from training data or the machine learning model by reverse-engineering.
- In addition, the attacks can be white-box, gray-box, or black-box attacks. In white-box (perfect-knowledge) attacks, the adversary is assumed to have knowledge of or access to the target model and its parameters, for example, full training data. As a result this may be a very powerful attack.
- In gray-box (limited-knowledge) attacks, the adversary only knows the feature representation and the learning algorithm type (e.g. SVM or ANN) but has no access in the training data or the algo-

rithm parameters. Still however, the adversary can attack the system through creating a surrogate dataset and model.
- In black-box (zero-knowledge) attacks, the adversary has no knowledge about the machine learning model except the output values and utilizes past inputs to infer the model's vulnerability (Dasgupta et al., 2022).

Sethi et al. (2018) distinguish two main schools of thought in the security of machine learning: the proactive one and the reactive one.

- Proactive security measures aim to prepare the system to anticipate possible attacks and make it more resistant to them. This practically translates into measures that increase the effort required for the attacker to evade the system.
- On the other hand, reactive security focuses on the system's performance, after it has been attacked, and emphasizes on how swiftly attacks can be detected and how soon the system's response can be restored.

## The Critical Role of Cybersecurity in Construction 4.0

The Construction 4.0 vision for use of Cyber-Physical Systems in construction, means that construction workers are equipped with devices such as tablets, for gathering and sharing project information to make their daily tasks easier and more efficient. Furthermore, many of the machinery and equipment used during construction (e.g. total stations, trucks, tower cranes) are no longer stand-alone pieces of equipment but are interconnected, and in some cases, operated or monitored remotely. The interconnectivity of the embedded Industrial Control Systems with the internet makes them a point of vulnerability that can be potentially exploited by malicious attackers (Turk et al., 2022). Also, digital twins, radio frequency sensors, automated earthmoving machinery, automated assembly machines for offsite construction, robotic concrete printers are some indicative examples of operational technology that needs to be effectively safeguarded so that digital applications do not evolve into a danger to projects, assets and people involved. For instance, having access to the digital twin of a building can provide significant information to malicious users for the building's operation such as how to disrupt power, where from to access it easily, how to hack its automation systems etc. Similarly, the safe use and lawful control of heavy machinery can be compromised in case of a cyber-attack. In the above context, due diligence in ensuring required security measures against cyber threats and raising employee cyber awareness are paramount for the construction sector, especially considering that a cybersecurity breach can lead to people injuries or even loss of lives (Sonkor and Garcia de Soto, 2021).

## Bibliometric Data

Cybersecurity is one more field where machine learning has a strong presence, with 165 articles associating machine learning with cybersecurity in their title, published from 2017 onwards. In the field of construction, the theme of cybersecurity has a comparatively limited presence but it certainly forms an emerging trend (e.g. Sonkor and Garcia de Soto, 2021; Mantha et al., 2021; Turk et al., 2022) which builds on the research momentum generated by 'smart' buildings / facilities / critical infrastructure (e.g. industrial plants, power grids, and water treatment plants) and their potential to leverage IoT-enabled big data towards operation and maintenance performance optimization (e.g. Grundy, 2017; Konanahalli et al., 2022).

## Adversarial Learning

Owing to the pressing need for secure machine learning systems, the models need to become more intelligent and proactive in anticipating an adversarial influence on the training data. In this context, there has been considerable recent attention towards improving the resilience of such systems with the emergence of adversarial Machine learning. Adversarial Machine learning studies malicious parties, i.e. inputs specifically crafted to deceive the model and cause misclassification. The aim of adversarial learning is to detect the prospective vulnerabilities of machine learning techniques both at the training and testing stages, create and launch attacks and estimate their impact on the machine learning algorithms. Then, devise counter-measures to proactively control the system's security by eliminating the adversary's ability to get past it (Dasgupta et al., 2022). Rahimi et al. (2020) suggest simulating attacks at testing time, evaluate their impact and improve the classifying algorithm accordingly, towards eliminating the impact observed. They are also in favour of introducing random data to the classification system in order to make it impossible for the adversary to completely understand its logic. However, this needs careful consideration of the trade-offs as excessive randomness can result in lower performance. Shaukat et al. (2020) add that machine learning techniques offer better results in cases that they are trained on diverse, massive, and real-time datasets while SVM is considered as the mostly used and successful technique for cyber security tasks.

## Federated Learning

Despite its large success, machine learning most usually is not used in real time due to data privacy and confidentiality concerns that would emerge from the use of a central server for the collection of data. This effectively means that a great amount of data and computing resources available at the network edge is not exploited effectively for learning improvement (Alazab et al., 2022). In this context, the concept of federated learning has been proposed for building decentralized and privacy-preserving learning systems. Federated learning promotes a collaborative, while distributed, approach that enables the development of learning models trained on datasets which are centrally coordinated but not actually shared. The federated learning process encompasses three major steps as follows: the Selection of the initial (global) model which is communicated to the local clients, the local training of the model for each local client separately and finally, the aggregation of updated parameters as fed back from the locally trained models to the central one, with the aim to update its parameters. The aggregator combines all local model updates and constructs a new improved global model. This updated global model is then shared with the clients to start a new learning iteration. As before, the local workers download again the global update and compute their next local update until the global training is complete. This way, the distributed computing power allows the aggregator to achieve higher quality training while minimizing cybersecurity risks.

With its innovative operational concept, federated learning can offer various important benefits for IoT applications as follows (Alazab et al., 2022; Nguyen et al., 2021):

- Minimization of privacy risks and enhanced data security can be achieved since the raw data and information is not communicated over the network, instead only updates are forwarded to the server.
- Confidentiality of data is also protected as local training of models in edge devices ensures that no data misuse by unauthorized access can occur.
- Availability of data is also ensured as local and global models are available in the edge device and the cloud, respectively
- Integrity of data is also safeguarded as the process is designed for privacy preservation and the sensitive data are not transferred out of the local environment
- Communication latency is minimized and network resources are saved as there is no requirement for transmitting IoT data to the server.
- Learning accuracy increases as a result of attracting diverse datasets compared to relying to centralized AI approaches with insufficient data and constrained computational capabilities
- The scalability of the network improves due to its distributed learning nature.

Yang et al. (2019) note that federated learning is not only an innovative modelling mechanism with promising applications in many industries, but also a business model providing a new paradigm for applications of big data. Specifically, when the isolated data occupied by each organization do not suffice to produce a robust model, the mechanism of federated learning makes it possible for multiple organizations to share a united model, without having to exchange any data, and form thus, a data alliance.

Federated learning – although recently introduced - is a concept with impressive presence in the literature. At least 4,300 papers, all published after 2017, can be found in Scopus, but only a handful of those are related to the construction industry.

## Blockchain

### How it Works

The concept of Blockchain was first introduced in 2008 as a system for electronic transactions which innovatively enables secure peer to-peer transactions for any amount of money, just like cash, but does not rely on a "trusted third party" (e.g. bank). Blockchain technology is based on a distributed ledger i.e. a database that exists across several locations or among multiple participants. When a transaction is registered on blockchain, it gets linked with other transactions in a block, and each block gets validated by a predefined consensus protocol which includes an index number, a unique hash code, timestamp data, the transaction data itself and a reference to the hash of the previous data block in the chain. This means that it is nearly impossible to alter the blocked data as the timestamp for transactions and proof-of-work are hashed into a sequential record as a chain that cannot be changed without redoing the proof-of-work on all the following blocks on a network (Tezel et al., 2021).

Blockchain serves as a foundation for Bitcoin and other cryptocurrencies and is considered as a potential disrupter of the status quo in many industries as it is applicable to any scenario involving validation, auditing, and sharing of data. The sectors of banking and finance, insurance, healthcare, higher education, energy, manufacturing, logistics, retail, food and beverage and construction are amongst the areas where Blockchain already holds or is expected to play a pivotal role in the future.

## Bibliometric Data

Though Construction 4.0 initiatives are launched with the goal of obtaining actionable data insights to tackle complex issues, cybersecurity concerns and other challenges drastically reduce the practical implementation of the envisaged opportunities. In this context, the innovative technology of Blockchain has also attracted significant research interest. Specifically, the construction literature shows an impressive 169 research papers published since 2017 with the aim to highlight the applications, advantages, limitations, challenges and opportunities of Blockchain in the industry and more widely, in the built environment.

## Expected Benefits for the Construction Industry

Tezel et al. (2021) note that Blockchain is one of the primary technologies facilitating the cyber-physical convergence in Construction 4.0 by enabling enhanced transparency in collaborative networks in a BIM environment. Plevris et al. (2022) have identified six important application areas in construction i.e. 1) Building information modelling (BIM) and Computer Aided Design (CAD); 2) Contract management and smart contracts; 3) Construction project management; 4) Smart buildings and smart cities; 5) Construction supply chain management; and 6) Property ownership, land titles, asset management and maintenance in real estate. They also note that Blockchain's potential impact on peoples' lives, societies and interactions is comparable to the impact of electricity, the computer and the internet. An alternative quantitative and conceptual analysis of the relevant literature published between 2017 and 2021 has also been provided by Scott et al. (2021).

The integration and synergies between Blockchain technology and machine learning has also attracted particular research interest. Shahin and Sabri (2022) present a system which applies various machine learning algorithms to filter data before being stored in the blockchain. Furthermore, according to Li et al. (2022), the integration of Blockchain with federated learning provides an opportunity to further improve the model's security and performance, besides increasing its scope of applications. Therefore, they present a comprehensive review of the literature contemplating the Blockchain-based federated learning framework. Moreover, according to Tanwar et al. (2020) the learning capabilities of machine learning can be applied to Blockchain- based applications to make them smarter. Machine learning may also be used to enhance the time taken to reach consensus by building better data sharing routes. Further, it creates an opportunity to build better models by taking advantage of the decentralized architecture of blockchain technology. This means that the data sets emerging from sensors, smart devices and IoT for use by machine learning models, could be stored on a blockchain network. This reduces errors in the data such as duplication, missing data value, errors, and noise. Along the same lines, Lu et al. (2020a) proposed a new architecture based on blockchain and deep reinforcement learning to improve the efficiency of data learning while ensuring the reliability of the data sharing process, minimizing privacy risks. Awan et al. (2019) also introduced a blockchain-based federated learning model for privacy protection and malicious actors' prevention. A similar architecture combining federated learning and blockchain technology was proposed by Lu et al. (2020b) with the aim to transform data-sharing problems into machine learning problems while ensuring an efficient as well as secure data sharing process. Kumar et al. (2022) proposed a model consisting of a blockchain module for data honesty control and a PCA-based module for privacy preservation. Furthermore, Vairagade and Brahmananda (2022) note that a blockchain may experience capacity issues if transactions exceed a certain level. They support that a machine-learning architecture can be used to control the creation of an additional, separate blockchain attached to the original one (sidechain), which can also be used to improve the overall security and quality of service.

## FUTURE RESEARCH DIRECTIONS

Although research studies have demonstrated the usefulness and relevance of different machine learning-based technological advancements in Construction, much of it remains at a stage of vision or experimental proof-of-concept. Especially as far as the robotic applications are concerned, although very well established in manufacturing environments, they are still amongst the least researched and least used areas of the ongoing technological transformation in the construction industry. Given that the use of robots in manufacturing environments has been widely associated to factors of vital importance for the construction industry, e.g. improved productivity, higher efficiency, quality and enhanced safety, it is imperative for AI research to prioritize the successful integration of robotic technologies in the construction sites' unique and unstandardized conditions. In this context, special attention needs to be given to collaborative scenarios between human and robotic partners, given that these are in the core of the emerging vision of Industry 5.0, which was recently endorsed by European Commission (Breque et al., 2021) and it is likely to bring a similar trend in the construction field in the near future. Therefore, particular research effort is needed towards the development of learning models for the effective training of smart robots, capable of safe, efficient and intuitive interactions with humans in construction environments. This further emphasizes the dire need for cybersecurity and data protection with the use of decentralized approaches such as the Blockchain and federated learning, whose actual application in the sector is currently very low.

## CONCLUSION

AI represents a long- and well- established research area in the field of construction management. It is typically considered as a key aspect for the improvement of construction effectiveness and efficiency, as well as a major ally in the industry's constant battle against cost and time overruns, accidents, increased risks and low productivity. Nevertheless, the unpredictable, dynamic, harsh and risky environment typifying construction, as well as the high cost of technological transformation combined with the related skills gap in the sector, pose significant challenges which must be overcome before the advanced technological applications of the $4^{th}$ Industrial Revolution can be widely employed in practice, as has long been achieved in the manufacturing sector.

There is no doubt that Machine Learning is the cornerstone of successful AI applications as besides being an AI subfield on its own, it is also the AI-enabling factor of other subfields like computer vision and Robotics. Nevertheless, in the context of the recently emerging 'Construction 4.0' paradigm, machine learning has not been given the due recognition it is entitled to, as the literature contemplating the construction digitization vision most usually makes no or very limited reference to the computational dimension of the new paradigm's key technologies and applications.

Moreover, despite its undeniable contribution in advancing construction engineering and management research, machine learning applications are vulnerable to attacks that can cause a notable decrease of their performance and reliability. In response to the need for higher data security and privacy protection, machine learning turns to adversarial and federated models of learning, capable for building decentralized privacy-preserving systems. Along with these, the innovative technology and applications of Blockchain are also rapidly gaining ground in the field of construction, providing a tamper-proof mechanism of transactions based on a distributed ledger. The integration of Blockchain with federated learning is also a promising path towards machine learning models of improved security and performance.

## REFERENCES

Abioye, S. O., Oyedele, L. O., Akanbi, L., Ajayi, A., Davila Delgado, J. M., Bilal, M., Akinade, O. O., & Ahmed, A. (2021). Artificial intelligence in the construction industry: A review of present status, opportunities and future challenges. *Journal of Building Engineering*, *44*, 103299. doi:10.1016/j.jobe.2021.103299

Adekunle, S. A., Aigbavboa, C., Ejohwomu, O., & Thwala, W. D. (2021). (in press). Digital Transformation in the Construction Industry: A Bibliometric Review. *Journal of Engineering. Design and Technology*. doi:10.1108/JEDT-08-2021-0442

Akinosho, T. D., Oyedele, L. O., Bilal, M., Ajayi, A. O., Delgado, M. D., Akinade, O. O., & Ahmed, A. A. (2020). Deep learning in the construction industry: A review of present status and future innovations. *Journal of Building Engineering*, *32*, 101827. doi:10.1016/j.jobe.2020.101827

Alazab, M., Rm, S. P., Parimala, M., Maddikunta, P. K. R., Gadekallu, T. R., & Pham, Q.-V. (2022). Federated Learning for Cybersecurity: Concepts, Challenges, and Future Directions. *IEEE Transactions on Industrial Informatics*, *18*(5), 3501–3509. doi:10.1109/TII.2021.3119038

Antoniou, F. (2021). Delay risk assessment models for road projects. *Systems*, *9* (3).

Antoniou, F., & Merkouri, M. (2021). Accident factors per construction type and stage: A synthesis of scientific research and professional experience. *International Journal of Injury Control and Safety Promotion*, *28*(4), 439–453. doi:10.1080/17457300.2021.1930061 PMID:34154518

Arashpour, M., Kamat, V., Heidarpour, A., Hosseini, M. R., & Gill, P. (2022). Computer vision for anatomical analysis of equipment in civil infrastructure projects: Theorizing the development of regression-based deep neural networks. *Automation in Construction*, *137*, 104193. doi:10.1016/j.autcon.2022.104193

Aretoulis, G. N. (2019). Neural network models for actual cost prediction in Greek public highway projects. *International Journal of Project Organisation and Management*, *11*(1), 41–64. doi:10.1504/IJPOM.2019.098712

Awan, S., Li, F., Luo, B., & Liu, M. (2019). Poster: A reliable and accountable privacy-preserving federated learning framework using the blockchain. In *Proceedings of the ACM Conference on Computer and Communications Security* (pp. 2561-2563). ACM. 10.1145/3319535.3363256

Bolpagni, M., Gavina, R., Ribeiro, D., & Arnal, I. P. (2022). Shaping the Future of Construction Professionals in Industry 4.0. *Structural Integrity*, *20*, 1–28. doi:10.1007/978-3-030-82430-3_1

Boton, C., Rivest, L., Ghnaya, O., & Chouchen, M. (2021). What is at the Root of Construction 4.0: A Systematic Review of the Recent Research Effort. *Archives of Computational Methods in Engineering*, *28*(4), 2331–2350. doi:10.100711831-020-09457-7

Breque, M., de Nul, L., & Petridis, A. (2021). *Industry 5.0: Towards a Sustainable, Human-Centric and Resilient European Industry*. European Commission, Directorate-General for Research and Innovation: Luxembourg https://op.europa.eu/en/publication-detail/-/publication/468a892a-5097-11eb-b59f-01aa75ed71a1/

Calvet, L., de Armas, J., Masip, D., & Juan, A. A. (2017). Learnheuristics: Hybridizing metaheuristics with machine learning for optimization with dynamic inputs. *Open Mathematics*, *15*(1), 261–280. doi:10.1515/math-2017-0029

Perrier, N. & Bourgault, M. (2020). Construction 4.0: A survey of research trends. *Journal of Information Technology in Construction,* 25, 416-437.

Dallasega, P., Rauch, E., & Linder, C. (2018). Industry 4.0 as an enabler of proximity for construction supply chains: A systematic literature review. *Computers in Industry*, *99*, 205–225. doi:10.1016/j.compind.2018.03.039

Dallasega, P., Rojas, R. A., Bruno, G., & Rauch, E. (2019). An agile scheduling and control approach in ETO construction supply chains. *Computers in Industry*, *112*, 103122. doi:10.1016/j.compind.2019.08.003

Dardouri, S., BuHamdan, S., Al Balkhy, W., Dakhli, Z., Danel, T., & Lafhaj, Z. (2022). RFID platform for construction materials management. *International Journal of Construction Management.*

Dasgupta, D., Akhtar, Z., & Sen, S. (2022). Machine learning in cybersecurity: A comprehensive survey. *Journal of Defense Modeling and Simulation*, *19*(1), 57–106. doi:10.1177/1548512920951275

Forcael, E., Ferrari, I., Opazo-Vega, A., & Pulido-Arcas, J. A. (2020). Construction 4.0: A literature review. *Sustainability (Switzerland)*, *12*(22), 1–28.

Gharbia, M., Chang-Richards, A., Lu, Y., Zhong, R. Y., & Li, H. (2020). Robotic technologies for on-site building construction: A systematic review. *Journal of Building Engineering*, *32*, 101584. doi:10.1016/j.jobe.2020.101584

Goh, G. D., Sing, S. L., & Yeong, W. Y. (2021). A review on machine learning in 3D printing: Applications, potential, and challenges. *Artificial Intelligence Review*, *54*(1), 63–94. doi:10.100710462-020-09876-9

Grundy, C. (2017). Cybersecurity in the built environment: Can your building be hacked? *Corporate Real Estate Journal*, *7*(1), 39–50.

Gualtieri, L., Rauch, E., & Vidoni, R. (2021). Emerging research fields in safety and ergonomics in industrial collaborative robotics: A systematic literature review. *Robotics and Computer-integrated Manufacturing*, *67*, 67. doi:10.1016/j.rcim.2020.101998

Joseph, A. D., Nelson, B., Rubinstein, B. I. P., & Tygar, J. D. (2019). *Adversarial Machine Learning*. Cambridge University Press. doi:10.1017/9781107338548

Kangari, R., & Halpin, D. W. (1990). Identification of factors influencing implementation of construction robotics. *Construction Management and Economics*, *8*(1), 89–104. doi:10.1080/01446199000000008

Karmakar, A., & Delhi, V. S. K. (2021). Construction 4.0: What we know and where we are headed? *Journal of Information Technology in Construction*, *26*, 526–545. doi:10.36680/j.itcon.2021.028

Keshtegar, B., Nehdi, M. L., Kolahchi, R., Trung, N.-T., & Bagheri, M. (2022). Novel hybrid machine leaning model for predicting shear strength of reinforced concrete shear walls. *Engineering with Computers*, *38*(S5), 3915–3926. doi:10.100700366-021-01302-0

Khallaf, R., & Khallaf, M. (2021). Classification and analysis of deep learning applications in construction: A systematic literature review. *Automation in Construction*, *129*, 103760. doi:10.1016/j.autcon.2021.103760

Kim, Y., Kim, H., Murphy, R., Lee, S., & Ahn, C. R. (2022). Delegation or Collaboration: Understanding Different Construction Stakeholders' Perceptions of Robotization. *Journal of Management Engineering*, *38*(1), 04021084. doi:10.1061/(ASCE)ME.1943-5479.0000994

Kolaei, A. Z., Hedayati, E., Khanzadi, M., & Amiri, G. G. (2022). Challenges and opportunities of augmented reality during the construction phase. *Automation in Construction*, *143*, 104586. doi:10.1016/j.autcon.2022.104586

Konanahalli, A., Marinelli, M., & Oyedele, L. (2022). Drivers and challenges associated with the implementation of Big Data within U.K. Facilities Management sector: An Exploratory Factor Analysis approach. *IEEE Transactions on Engineering Management*, *69*(4), 916–929. doi:10.1109/TEM.2019.2959914

Kor, M., Yitmen, I., & Alizadehsalehi, S. (2022). *An investigation for integration of deep learning and digital twins towards Construction 4.0*. Smart and Sustainable Built Environment.

Kumar, P. M., Rawal, B., & Gao, J. (2022). Blockchain-enabled Privacy Preserving of IoT Data for Sustainable Smart Cities using Machine Learning. In *14th International Conference on COMmunication Systems and NETworkS, COMSNETS 2022* (pp. 1-6). IEEE. 10.1109/COMSNETS53615.2022.9668530

Labonnote, N., Rønnquist, A., Manum, B., & Rüther, P. (2016). Additive construction: State- of-the-art, challenges and opportunities. *Automation in Construction*, *72*, 347–366. doi:10.1016/j.autcon.2016.08.026

Li, D., Han, D., Weng, T.-H., Zheng, Z., Li, H., Liu, H., Castiglione, A., & Li, K.-C. (2022). Blockchain for federated learning toward secure distributed machine learning systems: A systemic survey. *Soft Computing*, *26*(9), 4423–4440. doi:10.100700500-021-06496-5 PMID:34840525

Li, X., Yi, W., Chi, H.-L., Wang, X., & Chan, A. P. C. (2018). A critical review of virtual and augmented reality (VR/AR) applications in construction safety. *Automation in Construction*, *86*, 150–162. doi:10.1016/j.autcon.2017.11.003

Liang, C. J., Wang, X., Kamat, V. R., & Menassa, C. C. (2021). Human-Robot Collaboration in Construction: Classification and Research Trends. *Journal of Construction Engineering and Management*, *147*(10), 03121006. doi:10.1061/(ASCE)CO.1943-7862.0002154

Liao, T. W., Egbelu, P. J., Sarker, B. R., & Leu, S. S. (2011). Metaheuristics for project and construction management - A state-of-the-art review. *Automation in Construction*, *20*(5), 491–505. doi:10.1016/j.autcon.2010.12.006

Lu, Y., Huang, X., Dai, Y., Maharjan, S., & Zhang, Y. (2020a). Blockchain and Federated Learning for Privacy-Preserved Data Sharing in Industrial IoT. *IEEE Transactions on Industrial Informatics*, *16*(6), 4177–4186. doi:10.1109/TII.2019.2942190

Lu, Y., Huang, X., Zhang, K., Maharjan, S., & Zhang, Y. (2020b). Blockchain Empowered Asynchronous Federated Learning for Secure Data Sharing in Internet of Vehicles. *IEEE Transactions on Vehicular Technology*, *69*(4), 4298–4311. doi:10.1109/TVT.2020.2973651

Mantha, B. R. K., & García de Soto, B. (2021). Assessment of the cybersecurity vulnerability of construction networks. *Engineering, Construction, and Architectural Management*, *28*(10), 3078–3105. doi:10.1108/ECAM-06-2020-0400

Murphy, K. P. (2012). *Machine learning: A probabilistic perspective*. MIT Press.

Nguyen, D. C., Ding, M., Pathirana, P. N., Seneviratne, A., Li, J., & Vincent Poor, H. (2021). Federated Learning for Internet of Things: A Comprehensive Survey. *IEEE Communications Surveys and Tutorials*, *23*(3), 1622–1658. doi:10.1109/COMST.2021.3075439

Oesterreich, T. D., & Teuteberg, F. (2016). Understanding the implications of digitisation and automation in the context of Industry 4.0: A triangulation approach and elements of a research agenda for the construction industry. *Computers in Industry*, *83*, 121–139. doi:10.1016/j.compind.2016.09.006

Opoku, D.-G. J., Perera, S., Osei-Kyei, R., & Rashidi, M. (2021). Digital twin application in the construction industry: A literature review. *Journal of Building Engineering*, *40*, 102726. doi:10.1016/j.jobe.2021.102726

Pan, Y., & Zhang, L. (2021). Roles of artificial intelligence in construction engineering and management: A critical review and future trends. *Automation in Construction*, *122*, 1–21. doi:10.1016/j.autcon.2020.103517

Perkins, I., & Skitmore, M. (2015). Three-dimensional printing in the construction industry: A review. *International Journal of Construction Management*, *15*(1), 1–9. doi:10.1080/15623599.2015.1012136

PerrierN.BledA.BourgaultM.CousinN.DanjouC.PellerinR.RolandT. (2020).

Plevris, V., Lagaros, N. D., & Zeytinci, A. (2022). Blockchain in Civil Engineering, Architecture and Construction Industry: State of the Art, Evolution, Challenges and Opportunities. *Frontiers in Built Environment*, *8*, 840303. doi:10.3389/fbuil.2022.840303

Proia, S., Carli, R., Cavone, G., & Dotoli, M. (2021). Control Techniques for Safe, Ergonomic, and Efficient Human-Robot Collaboration in the Digital Industry: A Survey. *IEEE Transactions on Automation Science and Engineering*.

Rahimi, N., Maynor, J., & Gupta, B. (2020). Adversarial machine learning: Difficulties in applying machine learning existing cybersecurity systems. *EPiC Series in Computing*, *69*, 40–47. doi:10.29007/3xbb

Regona, M., Yigitcanlar, T., Xia, B., & Li, R. Y. M. (2022). Opportunities and Adoption Challenges of AI in the Construction Industry: A PRISMA Review. *Journal of Open Innovation*, *8*(1), 45. doi:10.3390/joitmc8010045

Rosenfeld, Y., Warszawski, A., & Zajicek, U. (1993). Full-scale building with interior finishing robot. *Automation in Construction*, *2*(3), 229–240. doi:10.1016/0926-5805(93)90043-W

Russell, S., & Norvig, P. (2022). *Artificial Intelligence - A Modern Approach*. Pearson.

Sacks, R., Brilakis, I., Pikas, E., Xie, H. S., & Girolami, M. (2020). Construction with digital twin information systems. *Data-Centric Engineering*, *1*(6), A32.

Sacks, R., Eastman, C., Lee, G., & Teicholz, P. (2018). BIM Handbook: A Guide to Building. John Wiley & Sons.

Sanni-Anibire, M. O., Zin, R. M., & Olatunji, S. O. (2021). Machine learning - Based framework for construction delay mitigation. *Journal of Information Technology in Construction*, *26*, 303–318. doi:10.36680/j.itcon.2021.017

Scott, D. J., Broyd, T., & Ma, L. (2021). Exploratory literature review of blockchain in the construction industry. *Automation in Construction*, *132*, 103914. doi:10.1016/j.autcon.2021.103914

Sethi, T. S., Kantardzic, M., Lyu, L., & Chen, J. (2018). A dynamic-adversarial mining approach to the security of machine learning. *Wiley Interdisciplinary Reviews. Data Mining and Knowledge Discovery*, *8*(3), 1245. doi:10.1002/widm.1245

Shahin, R., & Sabri, K. E. (2022). A Secure IoT Framework Based on Blockchain and Machine Learning. *International Journal of Computing and Digital Systems*, *11*(1), 671–683. doi:10.12785/ijcds/110154

Shalev-Shwartz, S., & Ben-David, S. (2014). *Understanding Machine Learning: From Theory to Algorithms*. Cambridge University Press. doi:10.1017/CBO9781107298019

Shaukat, K., Luo, S., Varadharajan, V., Hameed, I. A., & Xu, M. (2020). A Survey on Machine Learning Techniques for Cyber Security in the Last Decade. *IEEE Access : Practical Innovations, Open Solutions*, *8*, 222310–222354. doi:10.1109/ACCESS.2020.3041951

Skibniewski, M. J., & Wooldridge, S. C. (1992). Robotic materials handling for automated building construction technology. *Automation in Construction*, *1*(3), 251–266. doi:10.1016/0926-5805(92)90017-E

Sonkor, M. S., & García De Soto, B. (2021). Operational Technology on Construction Sites: A Review from the Cybersecurity Perspective. *Journal of Construction Engineering and Management*, *147*(12), 04021172. doi:10.1061/(ASCE)CO.1943-7862.0002193

Tanwar, S., Bhatia, Q., Patel, P., Kumari, A., Singh, P. K., & Hong, W.-C. (2020). Machine Learning Adoption in Blockchain-Based Smart Applications: The Challenges, and a Way Forward. *IEEE Access : Practical Innovations, Open Solutions*, *8*(8938741), 474–448. doi:10.1109/ACCESS.2019.2961372

Tezel, A., Papadonikolaki, E., Yitmen, I., & Bolpagni, M. (2022). Blockchain Opportunities and Issues in the Built Environment: Perspectives on Trust, Transparency and Cybersecurity. *Structural Integrity*, *20*, 569–588. doi:10.1007/978-3-030-82430-3_24

Turk, Ž., García de Soto, B., Mantha, B. R. K., Maciel, A., & Georgescu, A. (2022). A systemic framework for addressing cybersecurity in construction. *Automation in Construction*, *133*, 103988. doi:10.1016/j.autcon.2021.103988

Vairagade, R. S., & Brahmananda, S. H. (2022). Enabling machine learning-based side-chaining for improving QoS in blockchain-powered IoT networks. *Transactions on Emerging Telecommunications Technologies*, *33*(4). Advance online publication. doi:10.1002/ett.4433

Venkatasubramanian, K. (2020). Machine learning set to play a key role in the transition to Construction 4.0. *Proceedings of the Institution of Civil Engineers. Civil Engineering*, *174*(2), 54. doi:10.1680/jcien.2021.174.2.54

Wu, P., Wang, J., & Wang, X. (2016). A critical review of the use of 3-D printing in the construction industry. *Automation in Construction*, *68*, 21–31. doi:10.1016/j.autcon.2016.04.005

Yang, Q., Liu, Y., Chen, T., & Tong, Y. (2019). Federated machine learning: Concept and applications. *ACM Transactions on Intelligent Systems and Technology*, *10*(2), 12. doi:10.1145/3298981

Yap, J. B. H., Chow, I. N., & Shavarebi, K. (2019). Criticality of Construction Industry Problems in Developing Countries: Analyzing Malaysian Projects. *Journal of Management Engineering*, *35*(5), 04019020. doi:10.1061/(ASCE)ME.1943-5479.0000709

You, Z., & Feng, L. (2020). Integration of Industry 4.0 Related Technologies in Construction Industry: A Framework of Cyber-Physical System. *IEEE Access : Practical Innovations, Open Solutions*, *8*, 122908–122922. doi:10.1109/ACCESS.2020.3007206

Zhao, Y., Wang, N., Liu, Z., & Mu, E. (2022). Construction Theory for a Building Intelligent Operation and Maintenance System Based on Digital Twins and Machine Learning. *Buildings*, *12*(2), 87. doi:10.3390/buildings12020087

Zhu, H.-J., Wang, L.-M., Zhong, S., Li, Y., & Sheng, V. S. (2022). A Hybrid Deep Network Framework for Android Malware Detection. *IEEE Transactions on Knowledge and Data Engineering*, *34*(12), 5558–5570. doi:10.1109/TKDE.2021.3067658

## KEY TERMS AND DEFINITIONS

**Adversarial learning:** the field of machine learning which aims to study vulnerabilities of machine learning models and algorithms, and make them secure against adversarial manipulation.

**Artificial Neural Networks:** Machine learning models in which hypotheses take the form of complex algebraic circuits, typically organized into many layers, with tunable connection strengths.

**Computer vision:** AI sub-field concerned with artificial simulation of the human sight, enabled by deep learning techniques and encompassing image classification, object detection and image semantic segmentation.

**Construction 4.0:** Concept reflecting the vision for digital transformation of the construction industry, through the implementation of a wide range of emerging technologies and applications (e.g. cyber-physical systems, 3D Printing, robotics), in line with the forerunner paradigm of Industry 4.0 from the manufacturing sector.

**Cybersecurity:** Cybersecurity is the practice of protecting critical systems and sensitive information from digital attacks.

**Digital Twin:** A digital representation of the physical and functional properties of a physical object or process that serves as its real-time digital counterpart.

**Federated learning:** A machine learning approach where no data exchange is necessary, as it is based on local data samples, stored across multiple decentralized edge devices or servers.

**Robots:** Highly automated devices aimed at performing various tasks and capable of perceiving the environment through various sensors and machine learning -based control techniques.

# Chapter 11
# Explainable Safety Risk Management in Construction With Unsupervised Learning

**Fatemeh Mostofi**
https://orcid.org/0000-0003-0974-1270
*Department of Civil Engineering, Karadeniz Technical University, Turkey*

**Vedat Toğan**
https://orcid.org/0000-0001-8734-6300
*Department of Civil Engineering, Karadeniz Technical University, Turkey*

## ABSTRACT

*The success of Machine Learning (ML) approaches as promising solutions has encouraged their widespread implementation across different fields. Owing to the high accident rate, the construction industry embraced ML in the risk assessment procedure. What if the machine produces knowledge of the relationship between the risk features and accident outcomes contained in the safety dataset? What if machines can explain an accident dataset without human intervention? Unsupervised ML techniques offer several advantages over supervised approaches, including their explainability to analyze and understand complex datasets. This chapter demonstrates the practical implementation of the unsupervised learning method, clustering, and dimensionality reduction to explain the similarities, differences, variances, and patterns that exist between the feature spaces of an occupational safety risk dataset. Principal component analysis (PCA) and K-means clustering with silhouette analysis were selected as two unsupervised ML approaches to demonstrate their implementation in enhancing data-centric decision-making during the construction risk assessment procedure.*

DOI: 10.4018/978-1-6684-5643-9.ch011

## INTRODUCTION

The recent spread of Artificial Intelligence (AI) across industries has resulted in a significant increase in the productivity, accuracy, and quality of various processes. In particular, owing to its low productivity and high accident rate, the construction industry has widely adopted AI (Liu et al., 2022) within different processes throughout the project lifecycle. A variety of supervised, semi-supervised, and unsupervised Machine Learning (ML) algorithms have been used in civil engineering (Hacıefendioğlu et al., 2022a; Hacıefendioğlu et al., 2022b; Xu et al., 2021) and construction management (Al Jassmi et al., 2019; Fan, 2020; Karatas & Budak, 2022; Mostofi et al., 2022a). The implementation of these techniques demands considerable upfront data investment in terms of procuring datasets and manually labeling them. Unsupervised ML techniques discover hidden patterns underlying a dataset without human intervention or external guidance from the given labels. Explaining what is preserved by the machine provides interesting, useful, and unbiased insights into a dataset and its associated features. Existing state-of-the-art methods have proven the effectiveness of different unsupervised techniques in explaining the underlying patterns and variances among feature spaces. The unsupervised information gained provides managers with quick, meaningful, and precise insights that facilitate strategic and timely decision-making.

This section describes the practical implementation of unsupervised learning methods for construction risk management. This section addresses the common unsupervised learning methods of clustering and dimensionality reduction to explain the similarities, differences, variances, and patterns that exist between the feature spaces of a given dataset. To better illustrate how the dataset is explained, real-world construction safety datasets are processed using clustering and dimensionality reduction techniques, K-means clustering and Principal Component Analysis (PCA), respectively. Both methods have the potential for unsupervised explanation of features in a given dataset.

## BACKGROUND

Construction projects are massive and complex operations involving an enormous number of interlinked processes and activities. The occupational risk assessment of a construction site includes the identification, localization, and assessment of all safety risks involved in each of these activities. Therefore, even small construction sites are often exposed to a wide range of occupational risk. Most of the identified occupational risks are recognized as high-impact and severe risks; hence, they must be addressed using appropriate risk-response strategies. Avoidance, mitigation, transfer, and control are the most common risk-response strategies. Their selection and implementation demand an accurate assessment of each response method in terms of its budget and time requirements, as well as its level of control over a particular safety risk. Effective mitigation and control of construction risks require allocating the main portion of the risk budget to the area that has the most influence on controlling most of the identified risk items. The more information obtained about construction risks, the less exposure the project will have. Therefore, an important managerial challenge is to identify these risk areas and evaluate their contributions to reducing workplace risk. Unsupervised ML approaches offer several advantages over supervised techniques, including cost-effectiveness, dimensionality reduction, exploratory data analysis, clustering, handling of missing data, robustness to outliers, and anomaly detection. One of the benefits of unsupervised ML techniques is their ability to analyze and understand complex datasets. Techniques such as PCA and K-means clustering can provide a useful explanation of the patterns inside the dataset without the need for manual labeling of the dataset. PCA localizes these risk areas and their contribution to the overall reduction in severity of construction accidents. K-means clustering follows an unsupervised process for classifying an unlabeled dataset into clusters and identifying abnormal data points within a given dataset.

## OCCUPATIONAL RISK MANAGEMENT IN CONSTRUCTION

The uniqueness and complexity of construction projects pose considerable threats to human life and economy. Construction threats originate from various reasons, such as poor management practices, the use of faulty machinery, and the hazardous nature of different construction activities, i.e., working at height, using electrical equipment, and the existence of chemical substances in the workplace. Large budgets are allocated to increase the safety status of construction sites through different solutions. Conducting worker training sessions, improving safety audits, and/or enhancing existing safety inspections are just some of the examples. The practical implementation of any solution requires a factual assessment of existing site safety conditions. This allows for the effective implementation of improvement strategies, directed at locations with the highest influence on the overall site safety status. To improve the site safety status, the first step is to perform a risk assessment (RA), including the identification, analysis, and evaluation of construction risks: occurrence probability of an uncertain event. RA is the most substantial procedure to improve site safety status and needs to be implemented before taking any preventive action. The RA procedure is initiated with hazard identification, followed by analysis and evaluation of the registered hazard items. The core element of any RA procedure is the identification of all events, substances, and materials with a harmful potential. An effective RA must identify all construction hazards while offering a realistic prediction of the outcome risks. Past construction accident events were recorded and evaluated to realistically predict the site safety risk outcomes. Hence, evaluating and predicting construction risk levels utilizing collected accident records is essential for the effective mitigation and control of construction safety risks. Therefore, to achieve a comprehensive list of hazard elements and potential safety risk scenarios, in addition to the provisions of safety standards and guidelines, knowledge from past projects, either as experience or collected accident data, must be utilized.

The outcome of any RA procedure is used to plan the appropriate risk response strategies. Among the four response strategies, risk avoidance and transfer are the most preferred. However, construction safety risk cannot always be avoided. For example, rerouting from potential landslides, weak ground, and riverbanks is not always an option when dealing with road construction projects. Working with a high voltage is inseparable from the construction activity. If a construction risk cannot be avoided, the second-best option is to transfer it to other parties that are more capable of handling it, such as job delegation to subcontractors. However, risk transfer usually increases project costs. Construction professionals prefer not to accept risks unless they add them to contract prices. Consequently, several construction safety risks must be mitigated with appropriate control strategies or accepted by allocating an adequate contingency budget.

A risk registry contains a long list of items that must be addressed according to their estimated severity outcomes. However, different mitigation strategies exist for a single risk item. For example, a construction accident could be the result of an inexperienced worker incorrectly using unmaintained equipment. In addition, selecting the best mitigation strategy for every single task can be expensive; therefore, it is common practice to prioritize mitigation strategies that suit larger list items. PCA localizes the areas that require the most attention through an unbiased evaluation of the accident dataset. Therefore, PCA assists construction professionals in answering questions like: ''Which safety feature in the construction accident dataset has more impact in mitigating the construction risk?" On the other hand, K-means clustering can offer unbiased insight into the unlabeled construction accident dataset with a more cost-efficient approach than supervised ML methods. It aids construction practitioners to cluster the collected accident dataset, offering insights about its context and allowing the identification of the root causes of extreme accident records.

Supervised and unsupervised ML approaches, as a branch of AI, have proven to be successful solutions, aiding risk-based and risk-informed decision-making. Construction safety studies have investigated ML approaches as RA tools (Toğan et al., 2022) that support human experience-based judgment. ML publications have tripled since 2015 (Pan & Zhang, 2021) and have achieved state-of-the-art performance on a wide variety of RA methods. In the literature, ML approaches aid construction professionals in monitoring, identifying, evaluating, and prioritizing the risks associated with different construction activities (Hegde & Rokseth, 2020; Pan & Zhang, 2021).These approaches have been used to assess, identify, and prevent transportation accidents (Ding et al., 2016; Ma et al., 2009; M. Zhu et al., 2018). Likewise, the literature on building construction has widely benefited from ML approaches for a variety of risk-prevention tasks (ACI, 2018; Al-Ghamdi, 2002; Ayhan & Tokdemir, 2019a, 2019b, 2020; Goh & Chua, 2013; Koc et al., 2021, 2022; Mafi et al., 2018; Mostofi et al., 2022b; Sarkar et al., 2020; Tango & Botta, 2013; Toğan et al., 2022; Wang et al., 2019; R. Zhu et al., 2021). Unsupervised techniques, such as K-means and hierarchical clustering, DBSCAN, and Kohonen's self-organizing maps were used in different fields (AghaeiRad et al., 2017; Bao et al., 2019; Dabirian et al., 2020). For example, K-means clustering was used to process the traffic dataset and thus determine the signal timing plan (Ratrout, 2011). PCA has been utilized to determine the influence of air quality (Kumar & Goyal, 2013), house prices (Mostofi et al., 2022c), amount of pollutant elements (Azid et al., 2014), and identification of thematic competency areas within the field of project management (Dogbegah et al., 2011). Unsupervised techniques offer unbiased insights into a given dataset that allows understanding and exploring the context and structure of a dataset without any prior knowledge. They can divide a raw dataset into groups (clusters) based on the similarities between the data points while detecting errors, outliers, and anomalies in the dataset. The advantages of unsupervised techniques have inspired their use for different data-oriented purposes. Exploratory data analysis by unsupervised ML approaches provides an unbiased discovery of outliers, trends, and hidden patterns in a dataset. In addition, unsupervised ML approaches highlight the abnormal data points, label the dataset, and project the high-dimensional feature columns into a more efficient, lower-dimensional space. This makes them robust feature engineering solutions while their implementation prior to supervised ML techniques can improve the model performance. Moreover, they remove the need for laborious and manual labeling of the dataset, making them more cost-effective solutions than supervised ML approaches.

Unsupervised techniques can be utilized without any upfront investment. The unsupervised learning methods have been adopted for a variety of purposes within the field of safety management at construction sites. It was used to categorize sound pressures before predicting the associated noise level at the job site (Dabirian et al., 2020). In a recent study, PCA was used to select the most influential components for determining the severity of construction accidents (R. Zhu et al., 2021) and influential factors in determining house price values (Mostofi et al., 2021). Likewise, unsupervised ML approaches can benefit RA procedures at construction sites by providing unbiased, cost-efficient, and in-depth insights into the raw dataset and underlying hidden patterns. Integration of the unsupervised ML approaches with the supervised ML methods can synthesize hybrid ML approaches with enhanced prediction accuracy, reliability, and robustness of the supervised ML predictions. For example, unsupervised ML methods such as PCA allow the selection of the most influential safety features as the input of the severity predictor, which leads to the increased prediction accuracy of the adopted supervised ML method (Mostofi et al., 2021; Mostofi et al., 2022c). Furthermore, hybrid ML models comprising unsupervised methods can benefit from unbiased feature selection, which increases the reliability of the predictor for applications in different construction safety scenarios. The reliability of the adopted ML model plays a crucial role in construction safety. The reliability of an ML approach is not limited to its prediction accuracy, but also its ability to produce consistent results, the quality of the input dataset, the tendency of the model to produce biased predictions, and the ability of the model to provide insight about a particular prediction. The reliability of the ML-based RA solution, including the tendency of the model to bias, its robustness, quality, and explainability, is vital when it comes to the safety of construction personnel at the site.

This chapter provides practical guidelines for occupational risk management at construction sites, using unsupervised learning techniques. This chapter first discusses the state-of-the-art unsupervised ML methods for construction risk assessment. Then, the theoretical background of two unsupervised ML approaches, PCA and K-means clustering, is summarized. The practical implementation of these approaches is demonstrated while explaining the results within the context of safety risk management at the construction site. It should be noted that the scope of this chapter is limited to illustrating the application of PCA and K-means as two samples of explainable unsupervised ML approaches for RA in construction. Therefore, this chapter is not comprehensive in terms of the full range of benefits offered by ML methods and only emphasizes their transparency and explainability in facilitating informed decision-making for construction safety professionals. Correspondingly, this chapter does not focus on other approaches that can influence the outcomes of the adopted PCA and K-means clustering techniques.

## HOW PRINCIPAL COMPONENT ANALYSIS LOCALIZE THE INFLUENTIAL FEATURES?

One way to express the dataset is through visualization while plotting each feature column on a single axis. However, a safety dataset is commonly more complex than being presented with only the x, y, and z axes; thus, their direct visualization is impossible. Principal Component Analysis (PCA) is an unsupervised method that can drastically reduce the dimensionality of a high-dimensional dataset: a dataset in the number of features (columns) exceeds the number of observation records (rows). The PCA approach was first introduced by Karol Pearson (Pearson, 1901) and was later improved by Harold Hotelling (Hotelling, 1933). PCA projects a high-dimensional feature space onto the lower subspace. The projected features in the new space, principal components (PCs), provide insight into the contribution of the original features to explaining each PC and contribution of each feature to the primary PCs. For clarity, the column titles in the dataset were referred to as features, whereas the members in each column were referred to as attributes. For example, a sample accident dataset has a worker's experience as its feature, while 1-month, 3-months and 5-months are feature attributes.

The input dataset was arranged in form of $n \times d$ matrix, **X**, where $n$ and $d$ are the total number of rows and columns, respectively.

$$\mathbf{X} = \begin{bmatrix} x_{11} & \cdots & x_{1d} \\ \vdots & \ddots & \vdots \\ x_{n1} & \cdots & x_{nd} \end{bmatrix} \tag{1}$$

$x_{ij} \in R^{n \times d}, \forall i=1,2,\ldots,n \ \forall j=1,2,\ldots,d$

The unsupervised nature of the PCA algorithm requires removal of any output variable from the matrix **X**.

$$\mathbf{Y} = \begin{bmatrix} y_1 \\ \vdots \\ y_n \end{bmatrix} \tag{2}$$

$y_i \in R^n, \forall i=1,2,\ldots,n$

Where **Y** is the vector size $n$ that describes the risk level or severity outcome of each accident event.

PCA is initiated by centering the dataset around the intersection of the axes (origin) through a process called normalization. Normalization renders the feature columns unit-free without losing the covariance in the original dataset.

$$x_{ij}^{*} = \frac{x_{ij} - \mu_j}{\sigma_j} \tag{3}$$

Each matrix element, $x_{ij}$, is positioned in a unique row and column formatted as decimal points (float). The $\mu_j$ stands for the mean and $\sigma_j$ is the standard deviation of each feature column. In Eq. (3), the zero-division instances are replaced with one. Subsequently, the spread of each feature from the mean of the other features is measured, which forms a covariance matrix.

$$\mathbf{S} = \begin{bmatrix} s_{11} & \cdots & s_{1d} \\ \vdots & \ddots & \vdots \\ s_{d1} & \cdots & s_{dd} \end{bmatrix} \tag{4}$$

$s_{kr} \in R^{d \times d}, \forall k=1,2,\ldots,d \ \forall r=1,2,\ldots,d$

In Eq (4), **S** is $d \times d$ matrix in which each matrix element ($s_{kr}$) associated with a single correlation coefficient between the feature attributes (Eq. [5]).

$$s_{kr} = \frac{1}{n}\sum_{i=1}^{n} x_{ik}^{*} x_{ir}^{*} \tag{5}$$

The covariance matrix **S** decomposed into eigenvectors ($e_{ij}$) and eigenvalues ($\lambda$), representing the magnitude and importance, of the direction of the features in the newly projected space, respectively. So, the dataset dimension is reduced from $n\times d$ dimension into $n\times p$ dimension.

$$\mathbf{S}e_{ij} = \lambda e_{ij} \tag{6}$$

The eigenvalues associated with each feature column were used as the basis for sorting in the ascending order. At this stage, a threshold ($k$) was assigned to shortlist the most influential features in a given dataset.

$$\lambda_1 \geq \lambda_2 \geq \ldots \geq \lambda_k \tag{7}$$

$$X_k = \sum_{r=1}^{d} e_{rk} x_{ij}^*$$

$$\breve{\mathbf{X}} = \begin{bmatrix} \vdots & \vdots & & \vdots \\ X_1, & X_2, & \ldots, & X_k \\ \vdots & \vdots & & \vdots \end{bmatrix}$$

The shortlisted principal components ($X_k$) within the new dataset ( $\breve{\mathbf{X}}$ ) are multiplied by their normalized vector ( $x_{ij}^*$ ). The projected dataset contains the selected number of principal components ($X_k$), whereby selecting a $k$th number of principal components forms a new dataset ($\mathbf{X}_{\mathbf{PCA}}$). Here, the created dataset can be used as the input of ML, while the variance of each PC represents the amount of information it contains; thus, the higher the variance that the PC possesses, the more important it is in explaining the overall dataset (Eq. [8]).

$$\mathbf{X}_{\mathbf{PCA}} \subseteq \breve{\mathbf{X}} \tag{8}$$

$$X_{PCA} \in R^{n\times P}, \forall p<k$$

Therefore, the participation of each feature column in explaining the overall variance within the most influential feature columns defines its contribution in explaining the output variable (**Y**).

## HOW DOES K-MEANS CLUSTERING CATEGORIZE THE DATA POINTS?

K-means clustering is another well-known unsupervised technique that offers insight into important features and their influence within a given dataset. K-mean clusters the dataset into $k$-number of non-overlapping categories (Ahmad & Dey, 2007). K-means clustering was performed to find homogenous clusters within a given dataset. The number of clusters ($k$) are randomly selected which results in cluster centroids ($c_k$). Different accuracy measurements are used to determine the optimal number of clusters. In the chapter silhouette analysis (Rousseeuw, 1987) is also employed for analyzing the distance between the clusters and accordingly rating the success of the selected $k$ value (Eq. [9]).

$$c_1,c_2,\ldots,c_k \in R^m,\ \forall k=2,\ldots,k \tag{9}$$

The distance of each data point is measured from its nearest cluster centroid ($c_k$) (Chiang & Mirkin, 2007). This forms a distance matrix ($\boldsymbol{D}$) whereby K-mean clustering aimed at its minimization. In other words, the average distance between each data point and its closest cluster centroid ($c_k$) is to be minimized (Eq. [10]).

$$\boldsymbol{D} = \sum_{k=2}^{K} \sqrt{\sum_{j=1}^{m} x_j - c_k^{\,2}} \tag{10}$$

The K-mean algorithm finds the total distance between all data points and centroids within each cluster, referred to as clustering error ($D_k$), which is aimed at its minimization (Eq. [11]).

$$D= \{D_1,D_2,\ldots,D_k\} \tag{11}$$

This process is repeated for a predetermined number of iterations (attempts). In each iteration, the cluster centroids were updated.

## PRACTICAL IMPLEMENTATION OF UNSUPERVISED LEARNING FOR EXPLAINING THE CONSTRUCTION ACCIDENT DATASET

In this section, the application of the two unsupervised learning (PCA and K-means clustering) approaches is illustrated over the accident dataset that was used for predicting the severity of construction accidents using supervised ML techniques such as artificial neural networks (ANN) (Ayhan & Tokdemir, 2019a) and graph convolutional networks (GCN) (Mostofi et al., 2022b). The dataset contains information about 4479 construction accidents and was collected from 73 construction projects, with details about the hazard, possible risky behaviors, associated activity, and severity, as well as the worker's occupation, experience, and age. The dataset also contains an expert's evaluation of the shortcomings of human resources and workplace management that caused the recorded accident event. Unsupervised algorithms were used to evaluate the accident dataset using an unbiased machine judgment.

To illustrate the PCA implementation, the construction accident dataset must first be loaded using the script below. For clarity, the first five rows of the script outputs are listed in Table 1.

```
# Importing Pandas library for data analysis and manipulation
import pandas as pd
# Importing accident dataset
filename= 'saftey_data.csv'
df = pd.read_csv(filename)
df.head()
```

*Table 1. First row of the accident dataset*

| Hazard | Occupation | Activity | Risky behaviors | Human factors | Workplace factors | Experience | Age |
|---|---|---|---|---|---|---|---|
| Falling object | Engineer | Re-bar/formwork installation | Others | Others | Inadequate maintenance mechanism | 3-6 | 28 |
| Falling from hight | Engineer | Working at height | Inability to perceive external risks | Insufficient skill and perception | Method of statement problems | 1 | 22 |
| Sharp objects contraction | rough work crew | Repair/ Maintenance works | Tending to use a shortcut | Incorrect management system | Method of statement problems | 1-3 | 25 |
| Foreign body damaging the eye | finishing work crew | Finishing works | Violation of safe work policy | Physical disability | Inadequate communication | 3-6 | 21 |
| Foreign body damaging the eye | finishing work crew | Finishing works | Inability to perceive external risks | Incorrect management system | Inadequate communication | 3-6 | 21 |

The first dataset was controlled for any missing values within the feature columns using the script below. The output is shown in Figure 1.

```
df.isna().sum()
```

*Figure 1. Results of controlling for missing values*

```
Hazard               0
Occupation           0
Activity             0
Risky behaviors      0
Human factors        0
Workplace factors    0
Experience           0
Age                  0
dtype: int64
```

A copy of the unencoded dataset was used for subsequent analysis with the help of the script as follows:

```
DF = df.copy()
```

The dataset contains categorical (non-numerical) attributes, which cannot be understood by the algorithm. Converting categorical variables into a numerical format for subsequent processing using the algorithm is referred to as encoding. Before data processing, categorical attributes were converted into an integer format. For this, first, the numerical and categorical attributes are listed below.

```
categorical_variables =['Hazard',' Occupation', 'Experience',
                        'Human factors', 'Workplace factors',
                         'Activity', 'Risky behaviors']
numeric_variables = ['Age']
```

The Sklearn library (Buitinck et al., 2013; Pedregosa et al., 2011) offers various encoding methods for various purposes. Here, the Label Encoder is adapted to convert different attributes into integers, starting from 0, as shown in the following script. The resulting encoded datasets are listed in Table 2.

```
# Encoding the categorical variable
from sklearn.preprocessing import LabelEncoder
le = LabelEncoder()
# Label encoding
for column in categorical_variables:
  df[column] = le.fit_transform(df[column])
df.head()
```

*Table 2. Encoded dataset*

| Hazard | Occupation | Activity | Risky behaviors | Human factors | Workplace factors | Experience | Age |
|---|---|---|---|---|---|---|---|
| 7 | 1 | 19 | 0 | 3 | 8 | 0 | 22 |
| 18 | 6 | 13 | 4 | 2 | 8 | 2 | 25 |
| 10 | 3 | 6 | 5 | 4 | 1 | 3 | 21 |
| 10 | 3 | 6 | 0 | 2 | 1 | 3 | 21 |
| 19 | 6 | 0 | 0 | 2 | 1 | 3 | 39 |

To familiarize themselves with the dataset, a general statistical description was acquired using the method described in the script. The output is displayed in Table 3, considering the number of accident records (count), feature average (mean), standard deviation (std), minimum (min), maximum (max), and 25%, 50%, and 75% percentiles.

```
df.describe(include='all').round(2)
```

*Table 3. General description of the dataset*

| | Hazard | Occupation | Activity | Risky behaviors | Human factors | Workplace factors | Experience | Age |
|---|---|---|---|---|---|---|---|---|
| count | 4479 | 4479 | 4479 | 4479 | 4479 | 4479 | 4479 | 4479 |
| mean | 15.27 | 4.78 | 9.34 | 1.18 | 3.23 | 4.32 | 2.14 | 30.32 |
| std | 6.36 | 1.45 | 6.57 | 1.78 | 1.83 | 2.78 | 1.31 | 8.78 |
| min | 0 | 0 | 0 | 0 | 0 | 0 | 0 | 18 |
| 25% | 10 | 4 | 3 | 0 | 2 | 2 | 2 | 23 |
| 50% | 15 | 5 | 9 | 0 | 3 | 3 | 2 | 28 |
| 75% | 22 | 6 | 16 | 3 | 5 | 7 | 3 | 35 |
| max | 24 | 6 | 20 | 6 | 6 | 8 | 4 | 94 |

The statistical description (Table 3) represents the safety at different scales. The dataset must first be normalized based on the PCA procedure (Eq. [3]). Normalization also improves dataset integrity and computational accuracy. Normalization was performed using the following codes:

```
mu = np.mean(df, axis=0)          # Getting average of each column
std = np.std(df, axis=0)          # Getting standard deviation of each column
std[std==0] = 1                   # Solving zero division instances
df_norm = (df-mu)/std             # Normalizing the dataset
df_norm = df_norm.round(2)        # Rounding to 2 decimal points
df_norm.head()
```

*Table 4. Normalized dataset*

| Hazard | Occupation | Activity | Risky behaviors | Human factors | Workplace factors | Experience | Age |
|---|---|---|---|---|---|---|---|
| -1.3 | -2.6 | 1.47 | -0.66 | -0.13 | 1.32 | -1.63 | -0.95 |
| 0.43 | 0.84 | 0.56 | 1.58 | -0.67 | 1.32 | -0.1 | -0.61 |
| -0.83 | -1.22 | -0.51 | 2.14 | 0.42 | -1.2 | 0.66 | -1.06 |
| -0.83 | -1.22 | -0.51 | -0.66 | -0.67 | -1.2 | 0.66 | -1.06 |
| 0.59 | 0.84 | -1.42 | -0.66 | -0.67 | -1.2 | 0.66 | 0.99 |

The normalized dataset (Table 4) was used to better understand the feature behavior (Figure 2). The Seabron library was activated using the following script to obtain a pairwise plot between each feature column:

```
# importing the seaborn (a matplotlib-based data visualization library) li-
brary
import seaborn as sns
```

```
# Adjusting the plot theme, font type and size, and color
sns.set_theme(context='talk', style='white', palette='PiYG_r',
              font_scale=1.1, color_codes=True, rc=None)
# Plotting pairwise relationships between feature columns, using pairplot
sns.pairplot(df_norm, corner=True,  diag_kind='kde')
```

*Figure 2. Plotting pairwise relationships between features*

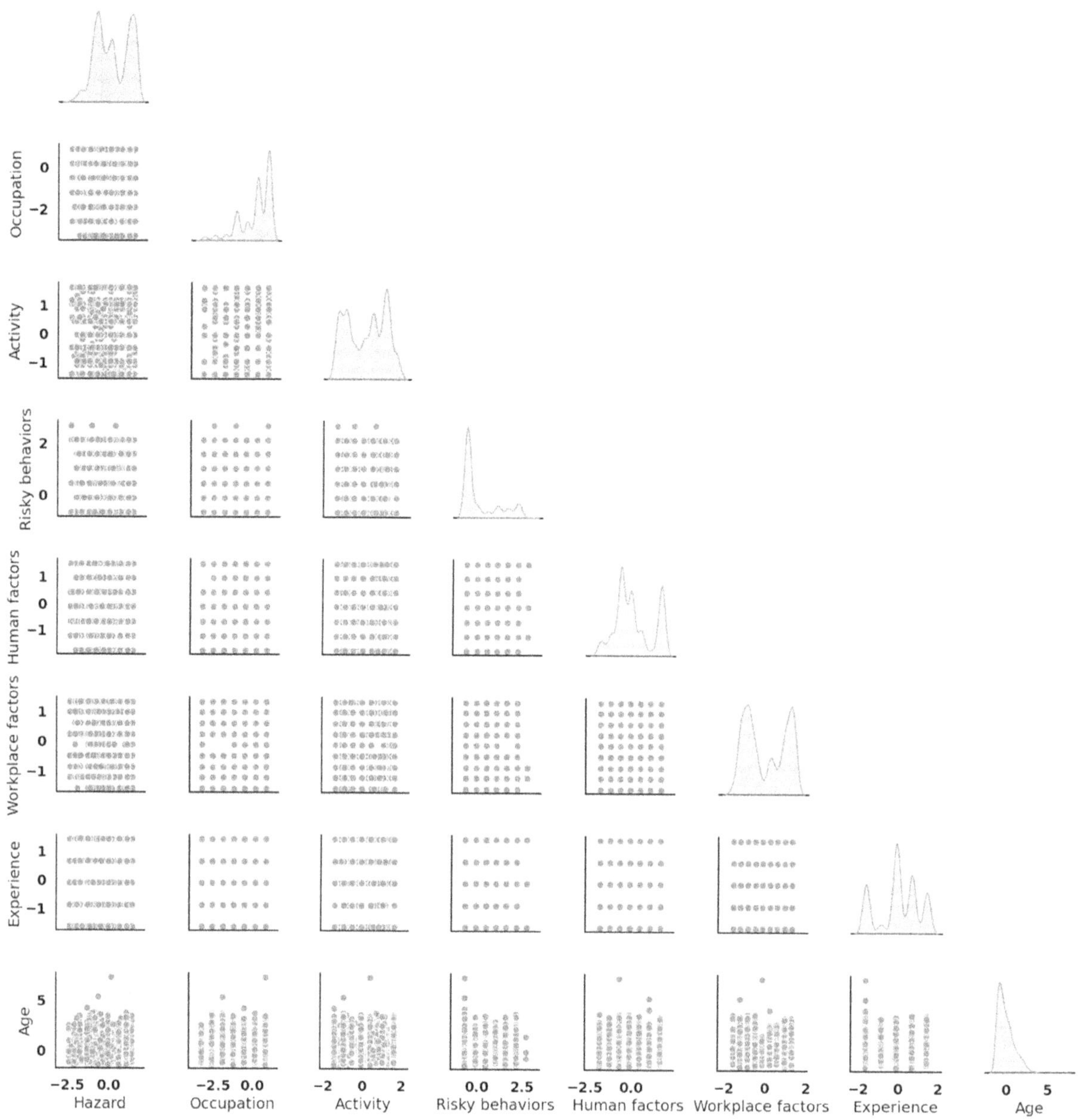

To demonstrate the PCA implementation, a PCA transformer from the Prince library (Max Halford, 2016) developed by MIT was used. For this purpose, the library must be installed using the following code:

```
!pip install prince
```

The PCA transformer was then employed using the following script: the first five rows are listed in Table 5. There were eight PCs, representing all eight feature columns.

```
import prince  # Improting PCA library from Prince
import pprint  #Library for printing datasets in a readable format
## Instantiate PCA object
pca = prince.PCA(n_components=8)
pca = pca.fit(df_norm)
df_PCA = pca.fit_transform(df_norm)
df_PCA.head()
```

*Table 5. Output from PCA transformation*

| **Principal component (PC) numbers** | | | | | | | |
|---|---|---|---|---|---|---|---|
| **0** | **1** | **2** | **3** | **4** | **5** | **6** | **7** |
| 1.359138 | 1.196515 | -0.58749 | -0.56948 | 1.554307 | 2.503239 | -1.7436 | -0.84354 |
| -1.24845 | -1.22552 | 0.239775 | 0.337343 | -0.72561 | 1.552671 | -0.03401 | -0.33915 |
| -0.07321 | -0.70432 | 1.290821 | -0.49037 | 1.967029 | 0.291242 | 1.649661 | -1.04756 |
| 1.175219 | -0.10157 | 0.867781 | -1.43267 | 0.940262 | -0.79002 | -0.28025 | -0.7318 |
| 0.432059 | -0.14899 | 0.87384 | 0.041887 | -1.03841 | -1.90572 | 0.457793 | 0.961764 |

The results of the PCA procedure transformed all eight feature columns from the normalized dataset into a new dimensional space. The fit-transform method generates a new matrix using the matrix elements from the original dataset. To obtain insight from the PCA procedure, the covariance matrix associated with the raw principal coordinates was computed via the transform method (see the script below). The corresponding computational results are listed in Table 6.

```
pca.transform(df_PCA).head()
```

Based on Eq. (6), the covariance matrix is decomposed by the code below to find the eigenvalues and eigenvectors associated with each safety column. The related outputs are shown in Figure 3.

```
pca.eigenvalues_
```

*Table 6. Raw principal coordinates*

| Principal component (PC) numbers | | | | | | | |
|---|---|---|---|---|---|---|---|
| **0** | **1** | **2** | **3** | **4** | **5** | **6** | **7** |
| -0.81043 | -0.13976 | -2.62167 | -0.69637 | -1.13461 | 1.833736 | 0.281991 | 1.911369 |
| 1.275005 | -0.74599 | -0.04888 | 0.264106 | 0.477914 | 1.605981 | -0.73962 | -0.82955 |
| -0.76405 | -0.54027 | -1.91038 | -0.24593 | 1.977653 | -0.71656 | -0.53412 | -0.93724 |
| -1.00635 | 1.132941 | -0.87339 | -1.11637 | 0.548124 | -0.82894 | -0.85299 | 0.543855 |
| -0.54316 | 0.853158 | 1.753919 | 0.560169 | 0.117035 | -1.37377 | -0.61829 | -0.32822 |

*Figure 3. The resulted eigenvalues*

```
array([1.26932804, 1.08265235, 1.05192452, 0.9939158 , 0.95724591,
       0.93089002, 0.86519681, 0.84884655])
```

In addition, the resulting eigenvalues were evaluated using the command line below based on the amount of variance preserved from the original dataset (Figure 4).

```
pca.explained_inertia_
```

*Figure 4. The variance is preserved for each eigenvalue*

```
array([0.158666  , 0.13533154, 0.13149056, 0.12423947, 0.11965574,
       0.11636125, 0.1081496 , 0.10610582])
```

The contribution of each accident record to explaining the variation in the original dataset was also observed. This allowed for the analysis of individual accident attributes (Table 7).

*Table 7. Contribution of each accident record towards explaining the variance in the original dataset*

| Principal component (PC) numbers | | | | | | | |
|---|---|---|---|---|---|---|---|
| **0** | **1** | **2** | **3** | **4** | **5** | **6** | **7** |
| 0.517434 | 0.018041 | 6.533865 | 0.487897 | 1.34483 | 3.612227 | 0.091909 | 4.303878 |
| 1.280708 | 0.514015 | 0.002271 | 0.070179 | 0.238603 | 2.770655 | 0.632278 | 0.810699 |
| 0.459902 | 0.269609 | 3.469421 | 0.060849 | 4.085796 | 0.551582 | 0.329736 | 1.034835 |
| 0.797848 | 1.185566 | 0.725151 | 1.2539 | 0.313859 | 0.73816 | 0.840964 | 0.348447 |
| 0.232425 | 0.672311 | 2.924383 | 0.31571 | 0.014309 | 2.027345 | 0.44185 | 0.12691 |

Considering the variance explained by each PC, unnecessary PCs must be removed for an efficient dimension reduction. In this example, 75% of the data variance is to be maintained, and thus the cumulative variance of each PC is to be obtained first using the Python script below, while the results are depicted in Figure 5.

```
pca_variance_explained = pca.explained_inertia_*100
pca_variance_explained = pca_variance_explained.round(2)
var_exp_sum = 0
for i in range(0, len(pca_variance_explained)):
    while (var_exp_sum < 75):
        var_exp_sum = var_exp_sum + pca_variance_explained[i]
        i += 1
        print(f"using on {i} vraibles results in exaplaing the {var_exp_sum}
of data")
        num_PCA = i
print ("\n", "Using ", num_PCA, 'of variables',  var_exp_sum, '% of data vari-
ance is kept using PCA')
pca_cumulative_variance_explained = np.cumsum(pca_variance_explained)
print("\n", pca_cumulative_variance_explained)
```

*Figure 5. Cumulative variance preserved within PCs*

```
using on 1 vraibles results in exaplaing the 15.87 of data
using on 2 vraibles results in exaplaing the 29.4 of data
using on 3 vraibles results in exaplaing the 42.55 of data
using on 4 vraibles results in exaplaing the 54.97 of data
using on 5 vraibles results in exaplaing the 66.94 of data
using on 6 vraibles results in exaplaing the 78.58 of data

 Using  6 of variables 78.58 % of data variance is kept using PCA

 [ 15.87  29.4   42.55  54.97  66.94  78.58  89.39 100.  ]
```

The cumulative variance demonstrates that six of the eight components can effectively represent 75% of the variance in the original accident dataset. The dataset was then projected from an eight-dimensional space into a new six-dimensional space (see the script below).

```
pca = prince.PCA(n_components=6,)
pca = pca.fit(df_norm)
```

The persevered variance of the dataset was drawn for each of the six PCs. Figures 6-8 show the projected accident datasets for all the six PCs using the scripts below.

```
plt.scatter(df_PCA_sel[:, 0], df_PCA_sel[:, 1], edgecolor='black', alpha=0.4)
#df_PCA_sel[:, 0] and df_PCA_sel[:, 1] select the 1st & 2nd PC
plt.xlabel(f'component 1 ({(pca_variance_explained[:1])[0].round(2)}%)') #
[:1] selects the first PC
plt.ylabel(f'component 2 ({(pca_variance_explained[1:2])[0].round(2)}%)') #
[1:2] selects the first PC
```

*Figure 6. Accident data project on first and second PCs*

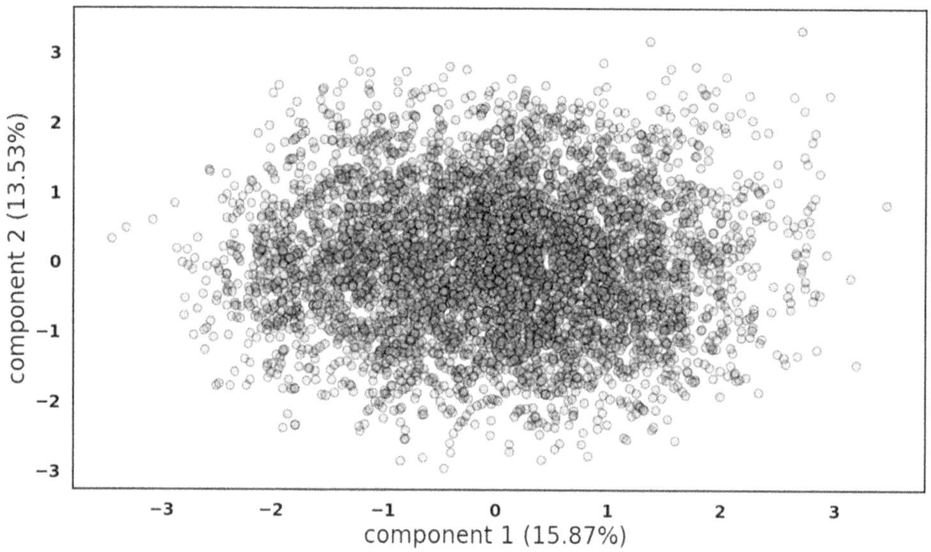


```
plt.scatter(df_PCA_sel[:, 3], df_PCA_sel[:, 4], edgecolor='black', alpha=0.4)
#df_PCA_sel[:, 0] and df_PCA_sel[:, 1] select the 1st & 2nd PC
plt.xlabel(f'component 1 ({(pca_variance_explained[2:3])[0].round(2)}%)') #
[:1] selects the first PC
plt.ylabel(f'component 2 ({(pca_variance_explained[3:4])[0].round(2)}%)') #
[1:2] selects the first PC
```

*Figure 7. Accident data project on third and fourth PCs*

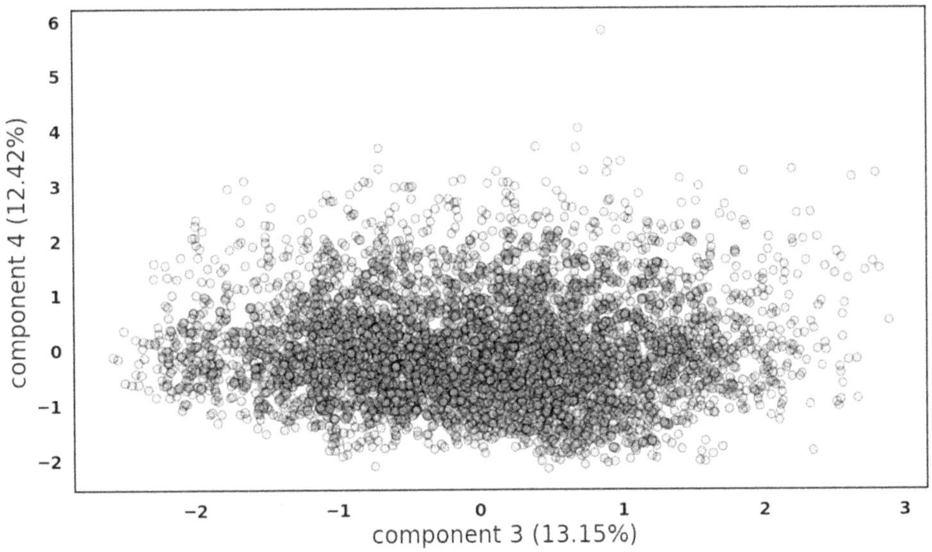


```
plt.scatter(df_PCA_sel[:, 5], df_PCA_sel[:, 6], edgecolor='black', alpha=0.4)
#df_PCA_sel[:, 0] and df_PCA_sel[:, 1] select the 1st & 2nd PC
plt.xlabel(f'component 1 ({(pca_variance_explained[4:5])[0].round(2)}%)') #
[:1] selects the first PC
plt.ylabel(f'component 2 ({(pca_variance_explained[5:6])[0].round(2)}%)') #
[1:2] selects the first PC
```

*Figure 8. Accident data project on fifth and sixth PCs*

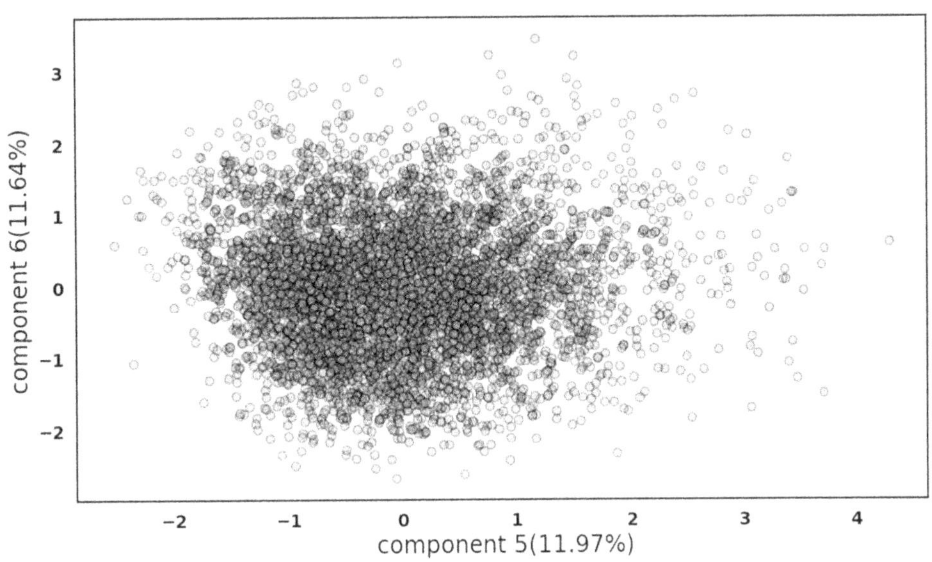

Similarly, the influence of accident features on each PC can be visualized via the following line of codes, resulting in the plot shown in Figure 9.

```
import numpy as np
from numpy import median
fig,ax= plt.subplots(2, 3, figsize=(18,9), sharex= True, sharey= True)
fig.tight_layout(pad=1.0)
for num, axes in enumerate(ax.flatten()):
  sns.barplot(x=PCA_Coef.index, y=PCA_Coef[num], data=PCA_Coef,
estimator=median, capsize=.2, palette="husl",  ax= axes)
  axes.set_xticklabels(axes.get_xticklabels(), rotation=90)
  axes.legend([f'Principal Component {num+1}'], loc='lower center')
  if num==0 or num==3: axes.set(ylabel='Influence factor')
  else: axes.set(ylabel='')
```

*Figure 9. Accident features influence PCs*

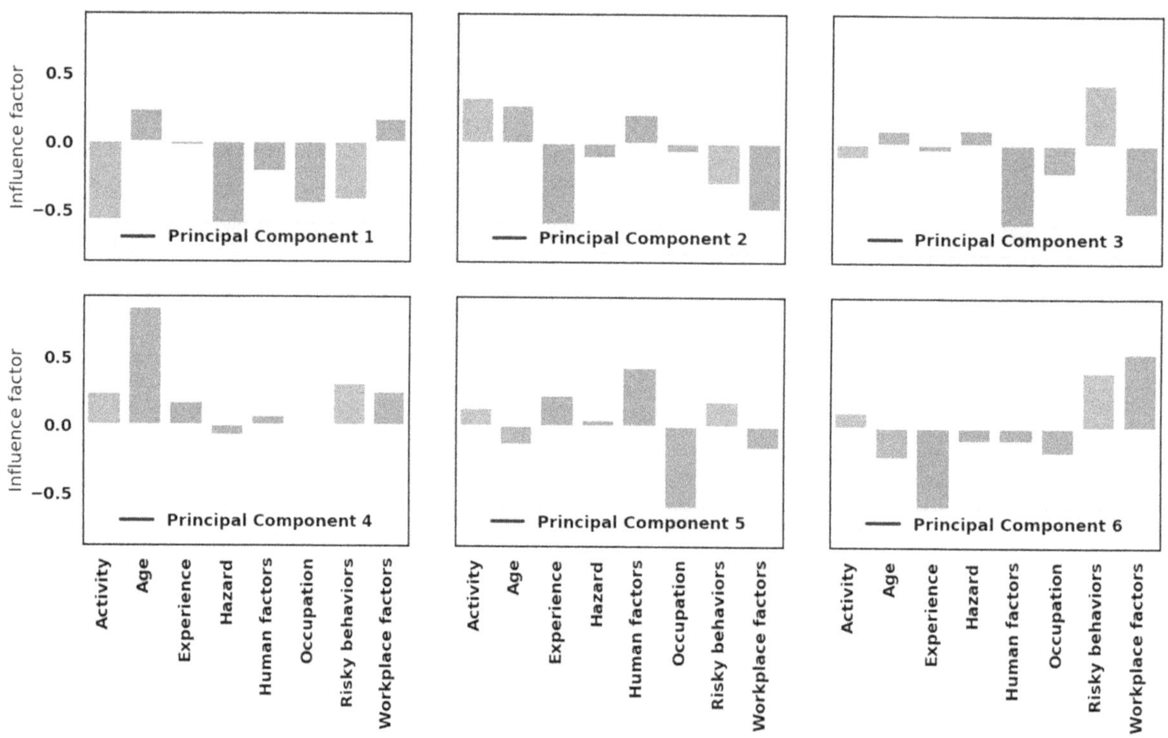


To obtain unbiased insight into PCA from the accident dataset, the column correlations (Table 8) were evaluated, as shown below.

```
PCA_Coef = pca.column_correlations(df_norm)
PCA_Coef
```

*Table 8. Feature scores are projected in a new dimensional space*

| Features | PC1 | PC2 | PC3 | PC4 | PC5 | PC6 |
|---|---|---|---|---|---|---|
| | **0** | **1** | **2** | **3** | **4** | **5** |
| **Activity** | -0.57 | 0.33 | -0.1 | 0.24 | 0.13 | 0.11 |
| **Age** | 0.24 | 0.27 | 0.1 | 0.86 | -0.13 | -0.22 |
| **Experience** | -0.03 | -0.71 | -0.05 | 0.17 | 0.23 | -0.58 |
| **Hazard** | -0.66 | -0.11 | 0.11 | -0.07 | 0.05 | -0.1 |
| **Human factors** | -0.22 | 0.21 | -0.72 | 0.07 | 0.43 | -0.1 |
| **Occupation** | -0.45 | -0.07 | -0.22 | 0.01 | -0.79 | -0.19 |
| **Risky behaviors** | -0.42 | -0.3 | 0.44 | 0.31 | 0.19 | 0.41 |
| **Workplace factors** | 0.17 | -0.49 | -0.51 | 0.25 | -0.16 | 0.55 |

As shown in Table 8, in the first PC, the hazard type (66%), followed by construction activity (57%), had the largest impact on the outcome of construction accidents. Hazard control is the first factor that requires the direction of accident prevention measures. Likewise, the type of construction activity has second importance in the formation of recorded accidents. Accordingly, examining the number of recorded hazard items provides insights into the most critical hazard elements. The script below shows the top ten hazard attributes with the highest frequency among the recorded accident data, and the results are presented in Figure 10.

*Figure 10. Most influential hazard types*

```
Vehicle                          728
Foreign body damaging the eye    597
Usage of hand tools              518
Pressing                         489
Falling object                   355
Moving object strike             314
Stationary object shock          300
Hazardous chemical exposure      195
Fire and explosion               180
Falling From Hight               176
Name: Hazard, dtype: int64
```

```
DF['Hazard'].value_counts()[:10]
```

Similarly, the related below command and Figure 11 indicate the most dangerous construction activities, which contributed to the majority of construction accidents.

```
DF['Activity'].value_counts()[:10]
```

*Figure 11. Most dangerous construction activities*

```
Assembly works                    726
Re-bar/formwork installation      695
Daily activities                  678
Usage of hand-tool/equipment      641
Welding/hot works                 415
Lifting operations                414
Finishing works                   312
Working with chemicals            246
Transportaion                     116
Concreting                         83
Name: Activity, dtype: int64
```

Thus, looking at the influential features of the first two PCs, construction professionals need to focus their attention on hazard prevention measures and training programs after performing construction activities. Improving vehicle passage, training drivers, and wearing safety goggles while working with construction equipment are examples of hazard prevention measures. Similarly, assembly works, rebar, and formwork installation need to be performed with better supervision, while the work statement of each activity needs to be clarified for the workers during the toolbox meetings. On the second PC, the experience of the worker is the most influential determinant of the resulting construction accidents. Therefore, to gain insight into the influential attributes associated with worker experience, the following code was used to obtain the outputs, as presented in Figure 12.

```
DF['Experience'].value_counts()
```

As expected, most construction accidents were associated with workers with less than six months of construction site experience. Interestingly, new workers experienced fewer accidents than workers with one–six months of working experience. This indicates the requirement for a continuous training program and tighter supervision throughout the first year of their construction experience.

The third and fourth PCA were responsible for the lower amount of variance in the original dataset. The third PC highlighted human factors (72%) and workplace factors (51%) as the most influential. The fourth PC identified the worker's age (86%) as the next risk-mitigation area.

*Figure 12. Frequency of accidents amongst construction workers with different experience levels*

```
1-3 month       1617
3-6 month       1052
1 month          887
6-12 month       750
12-24 month      173
Name: Experience, dtype: int64
```

The least importance was given to the features highlighted by the last PCs. The fifth PC prioritized construction occupations (79%). Risky behavior received 44% and 41% by the third and sixth PC, respectively, which had the least impact on the collected accident events.

Based on the obtained PCs and highlighted accident features, PCA suggests that construction professionals focus mostly on hazard prevention measures and improving safety procedures during the execution of construction activities. This was followed by additional safety training for the inexperienced workers. The other accident features received less importance according to the PC level and the amount of variance preserved. By examining the percentage of variance preserved in each PC, the contribution of safety features in the raw dataset can be determined. This information highlights the features that influence decision making based on the dataset. In addition, it can offer insight into the important features that lead to a specific prediction using the employed supervised ML predictors. For example, to evaluate the reliability of a severity prediction outcome, the entered hazard type and construction activity must be investigated among the entered attributes. Furthermore, integration of explainable unsupervised ML approaches such as PCA for feature selection with supervised ML predictors can enhance the prediction reliability from different perspectives, such as improved prediction accuracy and more transparency about the influential decision-making parameters of the adopted ML.

K-means clustering is an unsupervised learning technique that clusters a dataset near a selected number of cluster centers. When there is no reliable label for an accident dataset, the severity of construction accidents can be obtained using the K-means clustering approach. K-means clustering is appropriate for a small dataset, and to illustrate its application, 100 data samples were randomly selected using the following code:

```
df = df.sample(100)
```

Similar to the PCA approach, the dataset was encoded and normalized. Although K-means clustering does not necessitate data normalization, it improves the integrity of the dataset. To initiate the K-means clustering procedure, the tabular dataset is converted into a matrix form with the help of the code below.

```
X = df.to_numpy()
```

The K-means algorithm is then called from the Sklearn library (Buitinck et al., 2013; Pedregosa et al., 2011). The number of clusters was set to two and the number of iterations was set to 100. Besides, the initialization method is set as "K-means++" (Arthur & Vassilvitskii, 2007), which is a method that speeds up the K-mean procedure. K-means clustering was performed using the following script:

```
from sklearn.cluster import KMeans
kmeans = KMeans(n_clusters = 2, init='kmeans++', max_iter=100, n_init=10)
kmeans.fit(X)
kmeans.inertia_ # Getting the error
```

Employing two clusters resulted in 793.30 clustering errors, indicating that cluster numbers need to be increased. Three, four, and five clusters resulted in clustering errors of 728.10, 649.90, and 611.80 clustering errors, respectively. Hence, to determine the best number of clusters, the commonly used elbow method was applied by iterating through different numbers of clusters. Determining the optimum number of clusters is important to achieve efficient clustering. Alternatively, the silhouette method is adopted to determine the optimal number of clusters. The Silhouette approach was adopted using the script obtained from Sklearn (Buitinck et al., 2013; Pedregosa et al., 2011), while controlling the plot thickness of the clusters using a different number of clusters.

```
from sklearn.metrics import silhouette_samples, silhouette_score
import matplotlib.cm as cm
range_n_clusters = [2, 3, 4, 5, 6, 7, 8, 9]
for n_clusters in range_n_clusters:
    # Creating a subplot with 1 row and 2 columns
    fig, (ax1, ax2) = plt.subplots(1, 2)
    fig.set_size_inches(18, 6)
    fig.tight_layout(pad=3.0)
    # Initializing and number of clusters
    clusterer = KMeans(n_clusters=n_clusters, random_state=0)
    cluster_labels = clusterer.fit_predict(X)
    # The silhouette_score gives the average value for all the samples.
    silhouette_avg = silhouette_score(X, cluster_labels)
    print("For n_clusters =",n_clusters,"The average silhouette_score
is:",silhouette_avg,)
    # Computing the sample silhouette-scores
    sample_silhouette_values = silhouette_samples(X, cluster_labels)
    y_lower = 10
    for i in range(n_clusters):
        # Aggregating silhouette scores for i clusters and sorting them
        ith_cluster_silhouette_values = sample_silhouette_values[cluster_la-
bels == i]
        ith_cluster_silhouette_values.sort()
        size_cluster_i = ith_cluster_silhouette_values.shape[0]
        y_upper = y_lower + size_cluster_i
```

```
        color = cm.nipy_spectral(float(i) / n_clusters)
        ax1.fill_betweenx(np.arange(y_lower, y_upper), 0,ith_cluster_silhou-
ette_values, facecolor=color,edgecolor=color,alpha=0.7,)
        # Labelling the silhouette plots
        ax1.text(-0.05, y_lower + 0.5 * size_cluster_i, str(i))
        y_lower = y_upper + 10
    ax1.set_xlabel("The silhouette coefficient values")
    ax1.set_ylabel("Cluster label")
    # Highlighting the average silhouette-score
    ax1.axvline(x=silhouette_avg, color="red", linestyle="--")
    ax1.set_yticks([])
    ax1.set_xticks([-0.1, 0,0.1, 0.2, 0.3,  0.4,0.5, 0.6])
    # Plotting the created clusters
    colors = cm.nipy_spectral(cluster_labels.astype(float) / n_clusters)
    ax2.scatter(X[:, 0], X[:, 1], marker=".", s=1030, lw=0, alpha=1, c=colors,
edgecolor=None)
    # Labeling the clusters
    centers = clusterer.cluster_centers_
    ax2.scatter(centers[:, 0],centers[:, 1],marker="o",c="white",alpha=1,s=800
,edgecolor="k",)
    for i, c in enumerate(centers):
        ax2.scatter(c[0], c[1], marker="$%d$" % i, alpha=1, s=600,
edgecolor="k")
    ax2.set_xlabel("Space for the 1st feature")
    ax2.set_ylabel("\nSpace for the 2nd feature")
    plt.suptitle("\nSilhouette analysis for KMeans clustering on sample data
with n_clusters = %d"
        % n_clusters,fontsize=18,fontweight="normal")
plt.show()
```

The silhouette coefficient represents the fitness quality of clusters within an assigned cluster. The higher the silhouette coefficient, the better the dataset fits into the group. The two clusters resulted in a silhouette score of 0.103, and the resulting clusters had even thickness (Figure 13).

*Figure 13. Silhouette analysis of two clusters using K-means clustering*

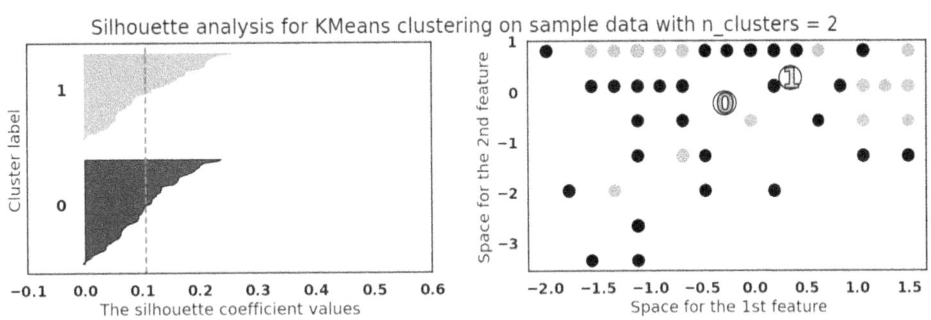

Increasing the number of clusters resulted in a silhouette score of 0.105 (Figure 14).

The scatter plots display clustering on the two feature spaces and do not fully visualize the clustering performed in the 8D space of the accident dataset. As illustrated in Figure 15, the silhouette score of the four clusters changed to 0.108.

The selection of five clusters improved the results while maintaining a similar clustering thickness, as shown in Figure 16.

*Figure 14. Silhouette analysis of three clusters using K-means clustering*

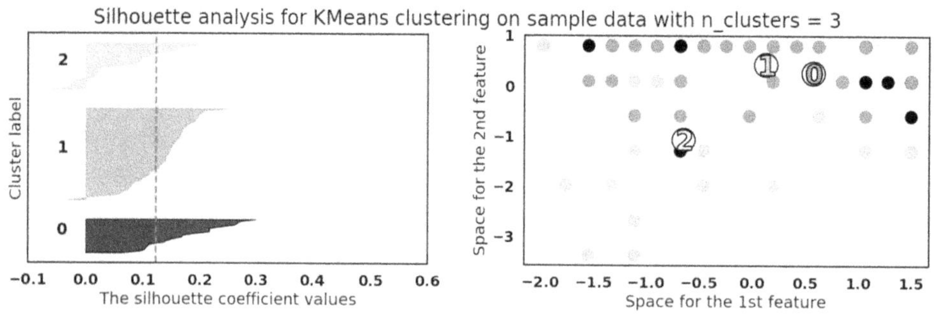


*Figure 15. Silhouette analysis of four clusters using K-means clustering*

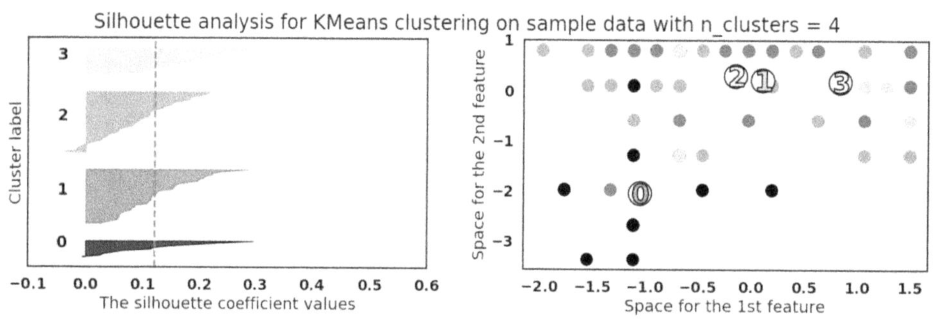


*Figure 16. Silhouette analysis of five clusters using K-means clustering*

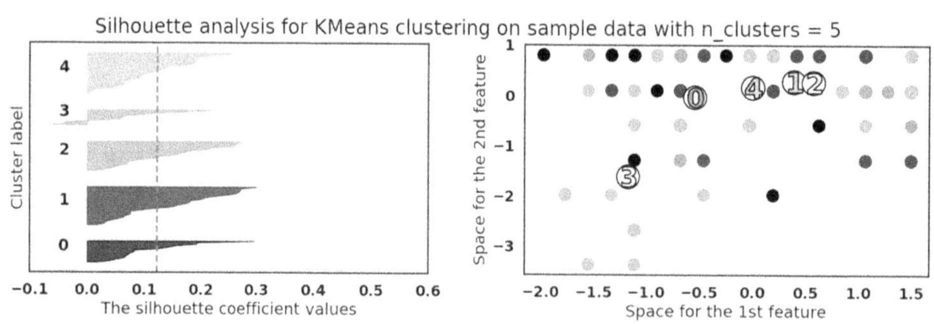

In the six clusters, the silhouette score increased from 0.119, and the evenness of the cluster thickness also improved (Figure 17). The procedure was continued for up to nine clusters, as described in the previous section.

Figure 18 shows the cluster thickness associated with the seven clusters, which resulted in a silhouette score of 0.124.

Better clustering performance was observed when grouping the dataset into eight clusters (Figure 19).

*Figure 17. Silhouette analysis of six clusters using K-means clustering*

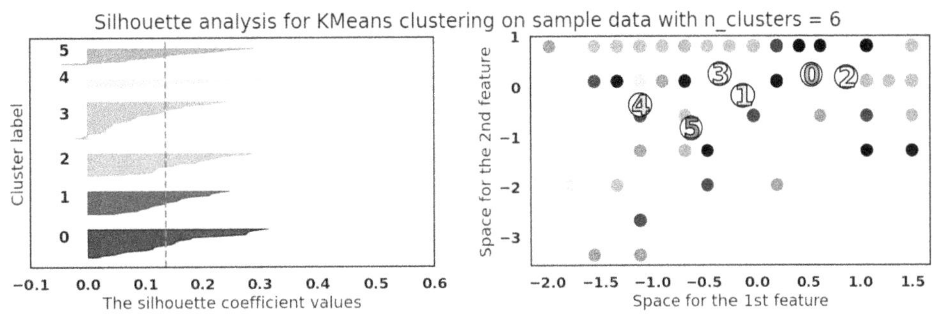


*Figure 18. Silhouette analysis of seven clusters using K-means clustering*

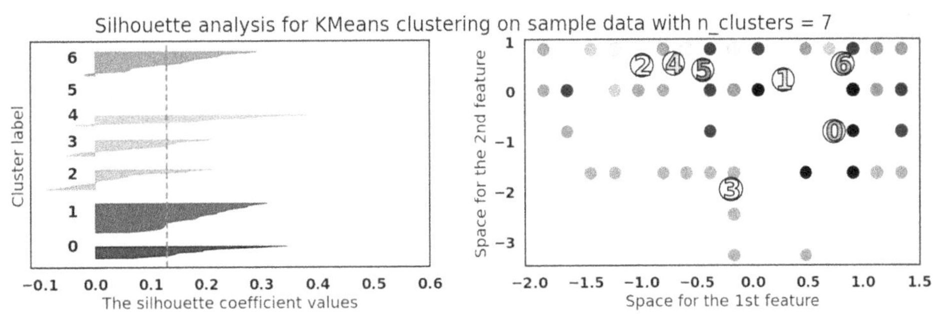


*Figure 19. Silhouette analysis of eight clusters using K-means clustering*

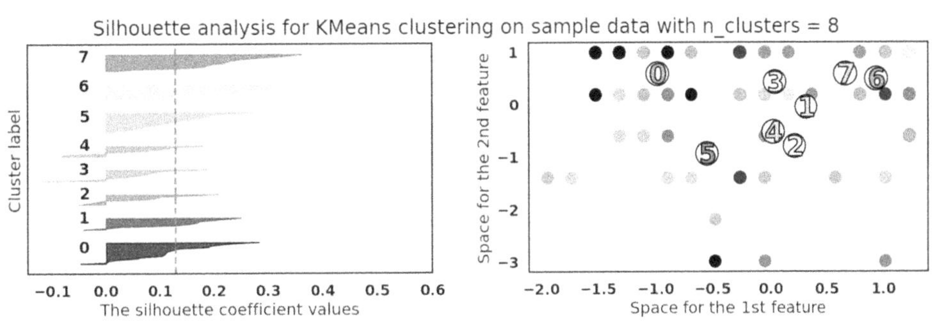

The obtained cluster thicknesses were improved using nine clusters. A better clustering performance was also indicated by a silhouette score of 0.129. Figure 20 shows the silhouette analysis results associated with eight clusters, in which the silhouette score decreased to 0.121.

*Figure 20. Silhouette analysis of nine clusters using K-means clustering*

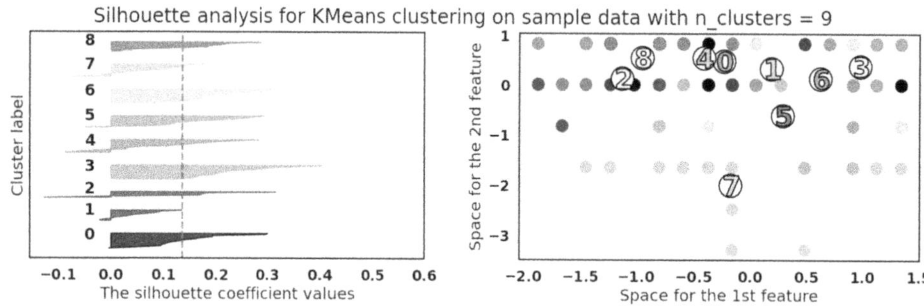


Therefore, the optimum number of clusters was selected to be eight. This resulted in all the accident datasets being placed into clusters with similar thicknesses, while they had a smaller distance to the cluster center compared to the other clustering values.

Above plots only displays the clustered dataset within the first and the second feature spaces. To explain dataset on the other feature spaces the silhouette coefficients obtained for the eight clusters resulted in the best clustering of the accident dataset. Accordingly, the coordinates of the eight cluster centers can be observed, where each coordinate represents an accident class (running the script below and obtaining the output in Figure 21).

```
kmeans = KMeans(n_clusters=8, random_state=0)   # 8 clusters
kmeans.fit(X)
kmeans.cluster_centers_
```

*Figure 21. Cluster centers of accident dataset*

```
array([[-1.30777778, -0.12222222,  1.09666667, -1.07666667, -0.63111111,
         0.95      ,  0.80555556, -0.10888889, -0.66444444],
       [-0.19315789, -0.17105263, -0.11736842, -0.20315789, -0.66210526,
        -0.46473684, -0.46526316,  0.75105263, -0.47578947],
       [ 0.90454545,  0.05454545,  0.83      ,  0.72727273, -0.64181818,
        -0.25545455, -0.58363636, -0.62454545,  0.84181818],
       [ 1.01538462,  0.27384615, -0.18538462,  1.01615385,  0.58923077,
         0.11461538,  1.01769231, -0.19615385, -0.51461538],
       [-0.10125   ,  0.146875  , -0.07      , -0.554375  , -0.28875   ,
        -0.31375   ,  1.111875  , -0.44125   ,  0.535     ],
       [-0.33428571,  0.40357143, -0.54142857,  0.56      , -0.30785714,
         1.34      , -0.84642857, -0.08571429, -0.27928571],
       [ 0.08857143,  0.04642857, -0.11285714, -0.65928571,  1.64785714,
        -0.78142857, -0.54285714,  0.53142857, -0.05214286],
       [-0.6625    , -2.1325    , -0.97      ,  0.6475    ,  0.9175    ,
        -0.28      , -0.925     , -0.7825    ,  2.125     ]])
```

Thus, unsupervised ML approaches, such as K-mean clustering, can be used to improve human performance for tasks such as labeling the accident dataset or grouping the accident records into the appropriate severity levels. In addition, looking at the accident dataset grouped within eight cluster groups, not only represents the K-mean clustering judgment about the eight-level severity of the recorded accident dataset but also highlights the abnormal data points.

*Figure 22. Grouping the accident dataset into eight clusters*

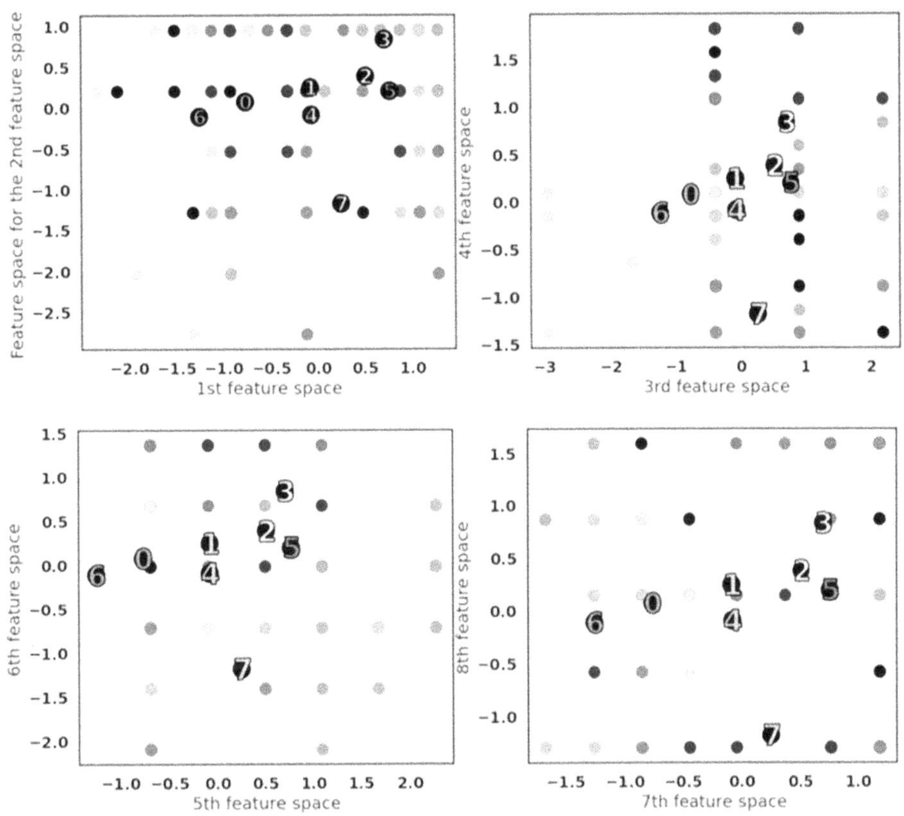


Figure 22 also demonstrates the K-means clustering ability for highlighting abnormal data points. Abnormal accident records blur the decision-making of both expert judgment and machines. Together, K-means clustering, and silhouette analysis can enhance the reliability of data-centric RA predictions. K-means clustering can categorize the unlabeled raw construction safety dataset in an unbiased and cost-efficient manner, while silhouette analysis reduces the uncertainty of the conventional K-means approach by selecting the best number of clusters. This approach offers insight into the structure of the raw dataset while highlighting outliers. The utilized K-means clustering allows decision makers to remove these outliers to improve the prediction performance of the adopted prediction model. In addition, they allow investigating the root causes of the highlighted outliers, aiding construction safety practitioners to focus on areas that cause extreme records. Alternatively, this can be used to improve the quality and accuracy of data collection, facilitating the detection of errors in recording the accident dataset, such as biased recorders, typos, or mistaken accident records. Hence, K-means clustering with silhouette analysis can increase the transparency of data-centric decision-making approaches, including supervised ML models, thereby increasing the reliability of the associated RA procedure and can in turn enhances the safety status of construction sites.

In brief, this section emphasizes unsupervised ML approaches as unbiased and cost-efficient solutions that can enhance the transparency and explainability of data-centric decision making and related RA procedures to improve safety conditions at construction sites. To achieve this, a real-world accident dataset was utilized to illustrate how unsupervised ML techniques can assist construction experts by improving their understanding of the collected accident dataset. Unsupervised ML approaches can increase the transparency of data-centric decision-making approaches such as supervised ML-based construction safety predictors. This also enhances the quality of the model input by selecting the most influential features and handling outliers using error removal or root cause analysis approaches.

## CONCLUSION

This chapter demonstrates how unsupervised learning algorithms can be leveraged in the Python environment for the increasing the transparency and explainability of ML. Data were first explored using representative scatter and bar plots. The features of the risk areas were evaluated and an explanation of the adopted PCA and K-means clustering analyses was provided. The details of the Python libraries, functions, and algorithms of each PCA and K-means clustering method in discovery of similarities and differences between feature spaces were provided.

However, most of the developed unsupervised ML methods are interpreted over publicly available datasets such as the IRIS dataset. The construction dataset is often described with a categorical attribute, which restricts the response of the collected data points to ML approaches. This section illustrates the real-world application of the construction dataset using two unsupervised techniques. PCA was used to project the dataset into a lower-dimensional space while ranking the importance of accident features in terms of their influence on the formation of the collected accident dataset. The accident data projected into the lower dimensional space displayed the percentage of influence for each PC associated with construction accident features, based on the amount of variance it preserved. In addition, the most influential PCs were analyzed for the localization of the important feature columns and their impact on explaining the collected accident records. By identifying the influential features in the raw dataset, PCA can explain the safety features that lead to a specific decision of an adopted prediction model. In addition, it improves the quality of the ML predictor input by selecting the most influential safety features.

In addition to PCA, the application of K-means clustering for the unsupervised categorization of accident records is illustrated. To determine the optimum number of clusters, a silhouette analysis was performed while scoring the density of each cluster. The thickness of the cluster plots and the silhouette score indicated the success of the K-mean clustering approach. Moreover, K-means clustering for the identification of abnormal data points within an accident dataset was discussed. The combination of K-means clustering with silhouette analysis results in a more reliable clustering of the construction accident dataset by detecting the outliers in the dataset. This offers an unbiased investigation of the collected accident dataset, which can enhance the reliability of data-centric decision-making and related RA procedures from different perspectives: (1) enhancing the quality and trustworthiness of the data collection system in the construction sites by detecting errors and biases in the dataset, (2) improving the accuracy of the data-centric prediction models by removing the outliers, and (3) root cause analysis of the identified extreme events to implement an appropriate accident prevention strategy based on the identified causes. Moreover, K-means clustering can be used as a cost-effective and unbiased approach for categorizing an unlabeled accident dataset, offering insight into the context of the raw construction accident dataset. This facilitates accident evaluation and accelerates the utilization of the collected dataset to assist in decision-making during the RA procedure. Overall, the utilization of unsupervised ML approaches can increase the reliability of the construction safety prediction model and associated RA procedure.

## REFERENCES

ACI. (2018). Predicting the Severity of Motor Vehicle Accident Injuries in Adana-Turkey Using Machine Learning Methods and Detailed Meteorological Data. *International Journal of Intelligent Systems and Applications in Engineering*, *1*(6), 72–79. doi:10.18201/ijisae.2018637934

AghaeiRad, A., Chen, N., & Ribeiro, B. (2017). Improve credit scoring using transfer of learned knowledge from self-organizing map. *Neural Computing & Applications*, *28*(6), 1329–1342. doi:10.100700521-016-2567-2

Ahmad, A., & Dey, L. (2007). A k-mean clustering algorithm for mixed numeric and categorical data. *Data & Knowledge Engineering*, *63*(2), 503–527. doi:10.1016/j.datak.2007.03.016

Al-Ghamdi, A. S. (2002). Using logistic regression to estimate the influence of accident factors on accident severity. *Accident; Analysis and Prevention*, *34*(6), 729–741. doi:10.1016/S0001-4575(01)00073-2 PMID:12371778

Al Jassmi, H., Ahmed, S., Philip, B., Al Mughairbi, F., & Al Ahmad, M. (2019). E-happiness physiological indicators of construction workers' productivity: A machine learning approach. *Journal of Asian Architecture and Building Engineering*, *18*(6), 517–526. doi:10.1080/13467581.2019.1687090

Arthur, D., & Vassilvitskii, S. (2007). K-means++: The advantages of careful seeding. *Proceedings of the Annual ACM-SIAM Symposium on Discrete Algorithms,* (pp. 1027–1035). ACM.

Ayhan, B. U., & Tokdemir, O. B. (2019a). Predicting the outcome of construction incidents. *Safety Science, 113*, 91–104. doi:10.1016/j.ssci.2018.11.001

Ayhan, B. U., & Tokdemir, O. B. (2019b). Safety assessment in megaprojects using artificial intelligence. *Safety Science*, *118*(March), 273–287. doi:10.1016/j.ssci.2019.05.027

Ayhan, B. U., & Tokdemir, O. B. (2020). Accident Analysis for Construction Safety Using Latent Class Clustering and Artificial Neural Networks. *Journal of Construction Engineering and Management*, *146*(3), 04019114. doi:10.1061/(ASCE)CO.1943-7862.0001762

Azid, A., Juahir, H., Toriman, M. E., Kamarudin, M. K. A., Saudi, A. S. M., Hasnam, C. N. C., Aziz, N. A. A., Azaman, F., Latif, M. T., Zainuddin, S. F. M., Osman, M. R., & Yamin, M. (2014). Prediction of the level of air pollution using principal component analysis and artificial neural network techniques: A case study in Malaysia. *Water, Air, and Soil Pollution*, *225*(8), 1–14. doi:10.100711270-014-2063-1

Bao, W., Lianju, N., & Yue, K. (2019). Integration of unsupervised and supervised machine learning algorithms for credit risk assessment. *Expert Systems with Applications*, *128*, 301–315. doi:10.1016/j.eswa.2019.02.033

Buitinck, L., Louppe, G., Blondel, M., Pedregosa, F., Mueller, A., Grisel, O., Niculae, V., Prettenhofer, P., Gramfort, A., Grobler, J., Layton, R., Vanderplas, J., Joly, A., Holt, B., & Varoquaux, G. (2013). *API design for machine learning software: experiences from the scikit-learn project.* https://doi.org//arxiv.1309.0238 doi:10.48550

Chiang, M. M. T., & Mirkin, B. (2007). Experiments for the number of clusters in K-means. In Lecture Notes in Computer Science (including subseries Lecture Notes in Artificial Intelligence and Lecture Notes in Bioinformatics): Vol. 4874 LNAI (pp. 395–405). Springer Berlin Heidelberg. doi:10.1007/978-3-540-77002-2_33

Dabirian, S., Han, S. H., & Lee, J. (2020). Stochastic-based noise exposure assessment in modular and off-site construction. *Journal of Cleaner Production*, *244*, 118758. doi:10.1016/j.jclepro.2019.118758

Ding, C., Wu, X., Yu, G., & Wang, Y. (2016). A gradient boosting logit model to investigate driver's stop-or-run behavior at signalized intersections using high-resolution traffic data. *Transportation Research Part C, Emerging Technologies*, *72*, 225–238. doi:10.1016/j.trc.2016.09.016

Dogbegah, R., Owusu-Manu, D., & Omoteso, K. (2011). A principal component analysis of project management competencies for the Ghanaian construction industry. *The Australasian Journal of Construction Economics and Building*, *11*(1), 26–40. doi:10.5130/AJCEB.v11i1.1680

Fan, C.-L. (2020). Defect Risk Assessment Using a Hybrid Machine Learning Method. *Journal of Construction Engineering and Management*, *146*(9), 04020102. doi:10.1061/(ASCE)CO.1943-7862.0001897

Goh, Y. M., & Chua, D. (2013). Neural network analysis of construction safety management systems: A case study in Singapore. *Construction Management and Economics*, *31*(5), 460–470. doi:10.1080/01446193.2013.797095

Hacıefendioğlu, K., Ayas, S., Başağa, H. B., Toğan, V., Mostofi, F., & Can, A. (2022a). Wood construction damage detection and localization using deep convolutional neural network with transfer learning. *Holz als Roh- und Werkstoff*, *80*(4), 791–804. doi:10.100700107-022-01815-5

Hacıefendioğlu, K., Mostofi, F., Toğan, V., & Başağa, H. B. (2022b). CAM-K: A novel framework for automated estimating pixel area using K-Means algorithm integrated with deep learning based-CAM visualization techniques. *Neural Computing & Applications*, *34*(20), 17741–17759. doi:10.100700521-022-07428-6

Hegde, J., & Rokseth, B. (2020). Applications of machine learning methods for engineering risk assessment – A review. *Safety Science, 122*(May 2019), 104492. doi:10.1016/j.ssci.2019.09.015

Hotelling, H. (1933). Analysis of a complex of statistical variables into principal components. *Journal of Educational Psychology*, *24*(6), 417–441. doi:10.1037/h0071325

Karatas, I., & Budak, A. (2022). Development and comparative of a new meta-ensemble machine learning model in predicting construction labor productivity. *Engineering, Construction and Architectural Management*. doi:10.1108/ECAM-08-2021-0692

Koc, K., Ekmekcioğlu, Ö., & Gurgun, A. P. (2021). Integrating feature engineering, genetic algorithm and tree-based machine learning methods to predict the post-accident disability status of construction workers. *Automation in Construction, 131*, 103896. doi:10.1016/j.autcon.2021.103896

Koc, K., Ekmekcioğlu, Ö., & Gurgun, A. P. (2022). Accident prediction in construction using hybrid wavelet-machine learning. *Automation in Construction, 133*, 103987. doi:10.1016/j.autcon.2021.103987

Kumar, A., & Goyal, P. (2013). Forecasting of Air Quality Index in Delhi Using Neural Network Based on Principal Component Analysis. *Pure and Applied Geophysics, 170*(4), 711–722. doi:10.100700024-012-0583-4

Liu, J., Luo, H., & Liu, H. (2022). Deep learning-based data analytics for safety in construction. *Automation in Construction, 140*, 104302. doi:10.1016/j.autcon.2022.104302

Ma, Y., Chowdhury, M., Sadek, A., & Jeihani, M. (2009). Real-time highway traffic condition assessment framework using vehicleInfrastructure integration (VII) with artificial intelligence (AI). *IEEE Transactions on Intelligent Transportation Systems, 10*(4), 615–627. doi:10.1109/TITS.2009.2026673

Mafi, S., AbdelRazig, Y., & Doczy, R. (2018). Machine Learning Methods to Analyze Injury Severity of Drivers from Different Age and Gender Groups. *Transportation Research Record: Journal of the Transportation Research Board, 2672*(38), 171–183. doi:10.1177/0361198118794292

Halford, M. (2016). *Prince Library*. MIT License. https://github.com/MaxHalford/prince/blob/master/LICENSE

Mostofi, F., Toğan, V., Ayözen, Y. E., & Tokdemir, O. B. (2022a). Predicting the Impact of Construction Rework Cost Using an Ensemble Classifier. *Sustainability (Basel), 14*(22), 14800. doi:10.3390u142214800

Mostofi, F., Toğan, V., Ayözen, Y. E., & Tokdemir, O. B. (2022b). Construction Safety Risk Model with Construction Accident Network: A Graph Convolutional Network Approach. *Sustainability (Basel), 14*(23), 15906. doi:10.3390u142315906

Mostofi, F., Toğan, V., & Başağa, H. B. (2021). House price prediction: A data-centric aspect approach on performance of combined principal component analysis with deep neural network model. *Journal of Construction Engineering. Management & Innovation, 4*(2), 106–116. doi:10.31462/jcemi.2021.02106116

Mostofi, F., Toğan, V., & Başağa, H. B. (2022c). Real-estate price prediction with deep neural network and principal component analysis. *Organization, Technology and Management in Construction. International Journal (Toronto, Ont.), 14*(1), 2741–2759. doi:10.2478/otmcj-2022-0016

Pan, Y., & Zhang, L. (2021). Roles of artificial intelligence in construction engineering and management: A critical review and future trends. *Automation in Construction, 122*(December 2020), 103517. doi:10.1016/j.autcon.2020.103517

Pearson, K. (1901). LIII. On lines and planes of closest fit to systems of points in space. *The London, Edinburgh and Dublin Philosophical Magazine and Journal of Science*, *2*(11), 559–572. doi:10.1080/14786440109462720

Pedregosa, F., Varoquaux, G., Gramfort, A., Michel, V., Thirion, B., Grisel, O., Blondel, M., Prettenhofer, P., Weiss, R., Dubourg, V., Vanderplas, J., Passos, A., Cournapeau, D., Brucher, M., Perrot, M., & Duchesnay, É. (2011). Scikit-learn: Machine learning in Python. *Journal of Machine Learning Research*, *12*(85), 2825–2830. https://dl.acm.org/doi/10.5555/1953048.2078195

Ratrout, N. T. (2011). Subtractive Clustering-Based K-means Technique for Determining Optimum Time-of-Day Breakpoints. *Journal of Computing in Civil Engineering*, *25*(5), 380–387. doi:10.1061/(ASCE)CP.1943-5487.0000099

Rousseeuw, P. J. (1987). Silhouettes: A graphical aid to the interpretation and validation of cluster analysis. *Journal of Computational and Applied Mathematics*, *20*(C), 53–65. doi:10.1016/0377-0427(87)90125-7

Sarkar, S., Pramanik, A., Maiti, J., & Reniers, G. (2020). Predicting and analyzing injury severity: A machine learning-based approach using class-imbalanced proactive and reactive data. *Safety Science*, *125*(February), 104616. doi:10.1016/j.ssci.2020.104616

Tango, F., & Botta, M. (2013). Real-time detection system of driver distraction using machine learning. *IEEE Transactions on Intelligent Transportation Systems*, *14*(2), 894–905. doi:10.1109/TITS.2013.2247760

Toğan, V., Mostofi, F., Ayözen, Y. E., & Behzat Tokdemir, O. (2022). Customized AutoML: An Automated Machine Learning System for Predicting Severity of Construction Accidents. *Buildings*, *12*(11), 1933. doi:10.3390/buildings12111933

Wang, S., Wei, C., Wei, Y., Wang, W., & Wu, F. (2019). A model of injury severity prediction in traffic accident based on GA-BP neural network. *CICTP 2019: Transportation in China - Connecting the World - Proceedings of the 19th COTA International Conference of Transportation Professionals*, (pp. 2470–2481). ACSE Library. 10.1061/9780784482292.215

Xu, Y., Zhou, Y., Sekula, P., & Ding, L. (2021). Machine learning in construction: From shallow to deep learning. *Developments in the Built Environment*, *6*, 100045. doi:10.1016/j.dibe.2021.100045

Zhu, M., Li, Y., & Wang, Y. (2018). Design and experiment verification of a novel analysis framework for recognition of driver injury patterns: From a multi-class classification perspective. *Accident; Analysis and Prevention*, *120*, 152–164. doi:10.1016/j.aap.2018.08.011 PMID:30138770

Zhu, R., Hu, X., Hou, J., & Li, X. (2021). Application of machine learning techniques for predicting the consequences of construction accidents in China. *Process Safety and Environmental Protection*, *145*(3), 293–302. doi:10.1016/j.psep.2020.08.006

## KEY TERMS AND DEFINITIONS

**Artificial Intelligence (AI):** Modelling the processes of human intelligence using a machine system.

**K-Means Clustering:** An unsupervised technique for classifying observations within a dataset into a smaller number of clusters.

**Machine Learning (ML):** A subset of AI that learns the input dataset using different algorithms for a variety of tasks.

**Principal Component Analysis (PCA):** An unsupervised method for identifying variables that persevere the maximum amount of variance in a dataset. PCA is useful for different applications such as dimensionality reduction, minimizing information loss, and increasing the explainability of high-dimensional datasets.

**Silhouette Analysis:** A visual method that evaluates the distance between clusters using the silhouette score.

**Supervised Machine Learning:** Training ML algorithms with labeled input data for delivery of a particular task.

**Unsupervised Machine Learning:** Training ML algorithms without labeled input data through an understanding of the patterns within the dataset.

# Chapter 12
# Enhanced Stochastic Paint Optimizer for Nonlinear Design of Fuzzy Logic Controllers in Steel Building Structures for the Near-Fault Earthquakes

**Mahdi Azizi**
*Near East University, Cyprus*

**Siamak Talatahari**
https://orcid.org/0000-0002-1211-7148
*Universtiy of Technology, Sydney, Australia*

## ABSTRACT

*Near-fault ground motion effects have been of great importance in recent decades due to the importance of these ground motions in vibration serviceability of structural systems. Fuzzy logic controller as one of the well-known control systems which is formulated based on the human expertise and knowledge is mostly incapable of providing proper control responses for structures under strong ground motions, and therefore the optimization of these controllers is concerned in recent years. In this chapter, optimal design of fuzzy controllers while considering the nonlinear characteristics of the structure is investigated. To fulfill this, an enhanced stochastic paint optimizer is proposed using the Levy flight. The proposed method is utilized for optimization of membership functions and rule base of fuzzy controllers implemented in the considered building structure. The overall performance of the proposed enhanced metaheuristic algorithm is compared with other methods. The obtained results demonstrate the capability of the new method in providing very competitive solutions.*

DOI: 10.4018/978-1-6684-5643-9.ch012

## 1. INTRODUCTION

Control systems are some sort of managing systems that directs or regulate the behavior of other systems and devices through control loops while different control devices and algorithms can be utilized for this purpose. Two of the well-known control categories are the "open loop" control procedure in which the control action is independent of the desired process variable while in the "close loop" control action, the control action is dependent on the considered process variable. Passive and active control systems alongside the semi-active and hybrid ones are the well-known control systems for vibration control of engineering systems which are utilized for performance improvement of different systems with different complexity. Regarding the advancement of technology in different areas including computer and material science, data processing, sensing technology, and stochastic procedures, active vibration control of large-scale building structures has been of great importance during the past decades. Based on the fact that active vibration control systems require multiple sensors and actuators for data acquisition from the considered building structure and providing the required control forces, a well-established control algorithm is also required for computing the amount of the control force which should be inserted to the building by actuators to complete the vibration control process. The Fuzzy Logic Controller (FLC) is one of the precisely developed control algorithms which have been used for active vibration control of building structures for several years. The previously developed FLCs are based on human expertise and knowledge in which the rule base and membership functions of these controllers are configured concerning human intelligence. It should be noted that human knowledge and expertise are not capable of providing the required optimal control responses in most cases (Al-Dawod et al. 2004). One of the most important disadvantages of the FLC is the knowledge-based configurations of the membership functions and rule base while the knowledge-based FLC is not the appropriate choice for complicated structural systems. However, most of the implemented FLCs in engineering applications demonstrate efficient and acceptable results (Edrees 2015).

Based on previous geological studies, most of the major cities in the world are located near or on active faults while the faults as one of the main causes of earthquakes have an important role in the characteristics determination of the earthquakes. After the Parkfield (California) earthquake in 1966 and the San Fernando (California) earthquake in 1971, the term "Near-Fault" was first introduced by Bolt (1975). However, in the following years, the effects of near-fault earthquakes were somewhat known, and the importance of this issue in the structural design was not well understood. After the Landers earthquake in 1992, Northridge in 1994, Kobe in 1995, and Chi-Chi in 1999, which resulted in severe damage to a wide range of structures, the importance of this issue had been revealed. Near-fault Earthquake (NFE) records have different characteristics than ordinary records. The NFE records have an impulsive motion with a short duration at the beginning which makes the structural system experience high input energy. The shear dislocation in the fault causes this pulse-type ground motion to be particularly oriented perpendicular to the fault. Therefore, the fault-normal component of the ground motion becomes more destructive than the fault-parallel one. In the NFE records, the normal component of the fault has higher peak ground acceleration than the parallel component of the fault at the same station (Somerville 2005). Due to the difference in vibration characteristics of NFE records, many research works, works and regulations have been conducted based on the far-field earthquakes while there is an urgent need to study the destructive effects of the NFE records and utilize some innovative systems for reducing these effects regarding the fact that many severe earthquakes happened in recent years in Asia (especially in the middle east) or even in USA and Canada. These earthquakes destroy many houses and endanger many lives so using the recent knowledge for enhancing the human lifestyle and securing their safety for a long time is one of the major concerns of structural engineering scientists.

Regarding the fact that the knowledge-based FLC's performance in vibration control purposes has been of great concern in recent decades due to the lack of reliability and also the necessity of keeping the occupants safe and also economic issues, the optimum design of FLCs for reducing the responses of the structures during earthquakes have been of great importance in recent years. In recent decades, metaheuristic optimization algorithms have been of use for optimal configuration of the FLC parameters to enhance its vibration control performance. Some of the most important research works in this field based on the utilized metaheuristic algorithms are provided. The main reason for providing such references is the fact that we need to have a perspective on the application of metaheuristic optimization algorithms in the optimum design of FLCs for vibration control purposes. Tairidis et al. (2016) discussed the FLC optimization implemented in the vibration suppression of smart structures utilizing the Genetic Algorithm (GA). Marinaki et al. (2011) proposed a developed version of the Particle Swarm Optimization (PSO) with a modified velocity equation, for optimal parameter configuration of the FLC in active control systems. Aliabadi and Taher (2017) presented the optimal tuning of the fuzzy Proportional-Integral-Derivative (PID) controller to utilize the control system in the automatic voltage regulator systems based on the Imperialist Competitive Algorithm (ICA). Marinaki et al. (2015) developed a new metaheuristic-based methodology utilizing the Differential Evolution (DE) algorithm for the parameter calculation of the FLC in active vibration control systems. Precup et al. (2016) utilized the Grey Wolf Optimization (GWO) algorithm for developing an innovatively fuzzy tuning approach for specific control systems regarding a decreased parametric sensitivity. Tiwari et al. (2016) discussed the applications of the Dragonfly Algorithm (DA) for temperature control of bioreactors with an optimized Fractional Order PID (FOPID) controller concerning the parameter evaluation of the controller. Hari and Patil (2017) discussed the applicability of the Jaya Algorithm (JA) for parameter tuning of the fuzzy controllers in multiple applications considering the system's nonlinearity.

In addition, many research works have been done in the field of earthquake engineering including the novel Type-2 fuzzy controller framework proposed by Rayegani and Nouri (2022). Meanwhile, Abdollahi et al. (2022) proposed a new paradigm for capturing ground motion uncertainty in seismic analysis and design of structures while instead of real ground motion records, the artificial intensifying acceleration is utilized. Haselton (2006) investigated the seismic collapse safety of reinforced concrete buildings. Amin et al. (2021) proposed the idea of evaluating the seismic collapse probability of an RC shaft-supported tank. Zheng et al. (2017) investigated the seismic collapse of a high-rise power transmission employing intelligent techniques.

Considering simple structural design models for control purposes (Abubaker et al. 2016), the main aim of this study is to utilize more accurate nonlinear structural models for control purposes. The Stochastic Paint Optimizer (SPO) (Kaveh et al. 2020) is utilized as the intelligent optimization technique for optimum control purposes while the Enhanced SPO (ESPO) is also proposed by the implementation of Levy flight into the standard algorithm for better performance. In the SPO, the beauty of colors and the art of painting in which the search space is determined as a painting canvas while a different combination of colors are utilized to represent the best configuration of colors. In this optimization algorithm, each of the solution candidates is considered as a color that is utilized for the production of new colors utilizing different color combination techniques while the quality of these colors develops iteratively by utilizing the combination of exploitation, exploration, and randomness of the solution results, as the optimization process continues. The SPO has lower computational complexity while this algorithm can reach optimal solutions in fewer computational efforts and it can converge to better solutions than other metaheuristic algorithms. It should be noted that SPO has been used for optimization purposes in different areas including engineering optimization by Chaotic SPO (CSPO) (Khodadadi et al. 2022) and the multi-objective version of SPO as MSPO (Khodadadi et al. 2022) in global engineering applications. Meanwhile, the SPO is selected among many other well-known metaheuristic algorithms including the Genetic Algorithm (GA) (Holland 1992), Particle Swarm Optimizer (PSO) (Kennedy and Eberhart 1995), Ant Colony Optimizer (ACO) (Dorigo et al. 2006), and Fire Fly Algorithm (FFA) (Yand and He 2013) as classic methods alongside the Fire Hawk Optimizer (FHO) (Azizi et al. 2023), jellyfish Optimizer (JO) (Chou et al. 2021), Geometric Octal Zones Distance Estimation" (GOZDE) (Kuyu & Vatansever 2022), Energy Valley Optimizer (EVO) (Azizi et al. 2023) and Atomic Orbital Search (AOS) (Azizi 2021).

The main motivation of this work is to assess the performance of a recently proposed metaheuristic algorithm in the optimum design of a fuzzy control system in such a way that a more controlled building structure will be achieved while destructive seismic inputs are considered for analytical purposes. Since the vibration control of structural systems in dealing with severe earthquakes has been of great concern in recent decades due to the advancement in vibration control technology and intelligent techniques, there is a great interest in applying these techniques to have well-behaved control platforms for vibration suppression of building structures.

*Figure 1. Rupture-directivity in different cases*
*(Naeim 1989)*

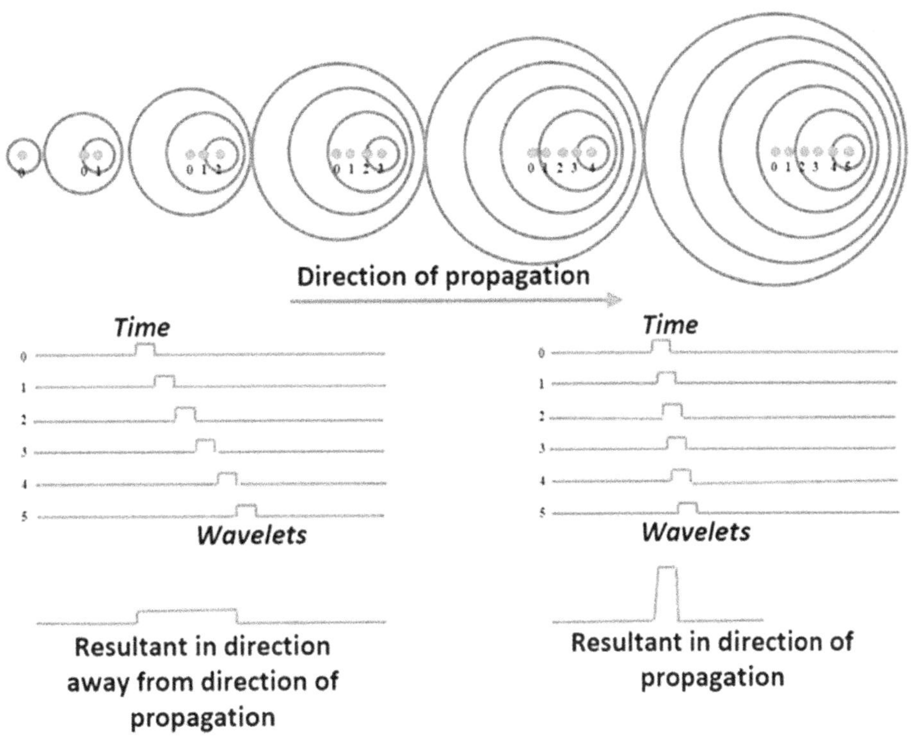

## 2. NEAR-FAULT GROUND MOTION

The approximate distance between 20 to 60 km from a ruptured fault is supposed to be a near-fault region in which rupture-directivity (Fig. 1) and fling step (Fig. 2) are two of the frequent pheromones.

In the near-fault region, some permanent ground displacements occur due to the static deformation field of the earthquake (called the "fling step"). Fling step displacements are not coupled with the rupture directivity displacement and mostly occur in the direction of the fault slip. In strike-slip faulting, the fling step occurs on the strike parallel component while the directivity pulse occurs on the strike-normal component. In dip-slip faulting, both the directivity pulse and the fling step occur on the component related to strike-normal (Stewart et al. 2002). These two aspects are illustrated in Fig. 2. The selected seismic records in this study are described in Table 1.

It should be noted that the selected earthquakes and the utilized seismic records of these earthquakes are considered based on the distances from the active faults so the behavior of the structures is evaluated using these records to demonstrate the capabilities of the controlled building structures in attenuating the seismic vibrations in the near-fault region.

*Table 1. Characteristics of the selected earthquake records*

| Abbr. | Earthquake - Date | $M_w$ | R (km) | Fault Mechanism | Station | Component | PGA (g) | PGV (cm/s) | PGA / PGV ($s^{-1}$) |
|---|---|---|---|---|---|---|---|---|---|
| **EQ1** | Tabas-1978 | 7.35 | 2.05 | Reverse | TABAS | TABT1 | 0.862 | 123.40 | 6.85 |
| | | | | | | TABL1 | 0.854 | 98.85 | 8.47 |
| **EQ2** | Imperial Valley-1979 | 6.53 | 2.66 | Strike Slip | Bonds Corner | BCR140 | 0.599 | 46.75 | 12.56 |
| | | | | | | BCR230 | 0.777 | 44.94 | 16.95 |
| **EQ3** | Loma Prieta-1989 | 6.93 | 3.85 | Reverse Oblique | Corralitos | CLS090 | 0.483 | 47.58 | 9.95 |
| | | | | | | CLS000 | 0.645 | 55.97 | 11.30 |
| **EQ4** | Landers-1992 | 7.28 | 2.19 | Strike Slip | Lucerne | LCN260 | 0.725 | 133.40 | 5.33 |
| | | | | | | LCN345 | 0.789 | 28.11 | 27.52 |
| **EQ5** | Northridge-1994 | 6.69 | 5.35 | Reverse | Sylmar | SCS052 | 0.623 | 116.25 | 5.26 |
| | | | | | | SCS142 | 0.923 | 88.53 | 10.22 |
| **EQ6** | Kobe-1995 | 6.9 | 0.96 | Strike Slip | KJMA | KJM090 | 0.630 | 76.11 | 8.12 |
| | | | | | | KJM000 | 0.834 | 91.11 | 8.98 |
| **EQ7** | Chi Chi-1999 | 7.62 | 3.12 | Reverse Oblique | CHY028 | CHY028E | 0.636 | 61.39 | 10.16 |
| | | | | | | CHY028N | 0.760 | 85.76 | 8.69 |

*Figure 2. The fling step phenomenon in different fault ruptures*
*(Stewart et al. 2002)*

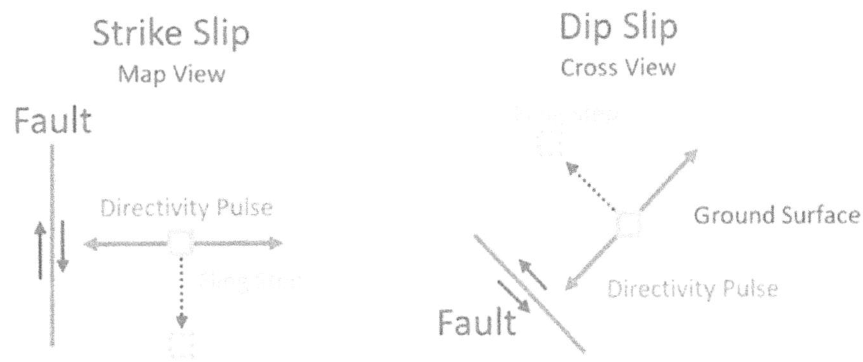


## 3. FUZZY LOGIC CONTROLLER

Nowadays, the importance of human expertise and knowledge are incredibly increasing based on the fact that human beings always move over the information age. Finding a proper mathematical model for modeling human intelligence and deduction has been of great interest over the past decades. Humankind always utilizes qualitative and quantitative information for data processing. The linguistic or verbal information (human language) is classified into qualitative information while the classical computer processors are not capable of analyzing this kind of information. In this regard, a strong hypothesis is required for introducing an expert model for the formulation of human knowledge or specifically human language in a systematized path to introduce a mathematical model for different systems with different applications. Fuzzy logic is one of these hypotheses.

In most of the conventional control algorithms and methods, a mathematical model is utilized for the configuration of the control systems and parameter identification of the control devices. In this regard, a specific mathematical model is utilized for the configuration of the previously proposed linear, nonlinear, and PID controllers. However, the utilized mathematical models for these controllers do not provide accurate and exact results in dealing with some specific applications. Based on the mentioned deficiencies, a fuzzy controller which is formulated based on the fuzzy sets and systems could provide an efficient control system for different applications through a specific model considering the human expertise and knowledge. In addition, fuzzy controllers could be helpful when a proper and exact mathematical model of the system problem is not generally available. One of the challenging applications of fuzzy controllers is for the closed-loop control of nonlinear systems in which the mathematical models of the systems are not accessible. A schematic view of a fuzzy logic controller implemented in a closed-loop control system is presented in Figure 3.

As presented in Fig. 5, a fuzzy control system has four main parts in which the control process for the considered system is conducted. In the first part (Fuzzification), the inputs of the control system are all fuzzified into linguistic terms. In the second one (Rule Base), the required control rules are all prepared through the classical modus-ponens rules described as the IF-THEN rules for control purposes. In the third part (Inference Mechanism), the fuzzy deduction is implemented in the control system which is utilized for extracting the outputs of the fuzzy rule base. In the last part (Defuzzification), the fuzzy outputs of the control system are interpreted into crisp data which is required for the control system.

*Figure 3. Fuzzy logic controller*

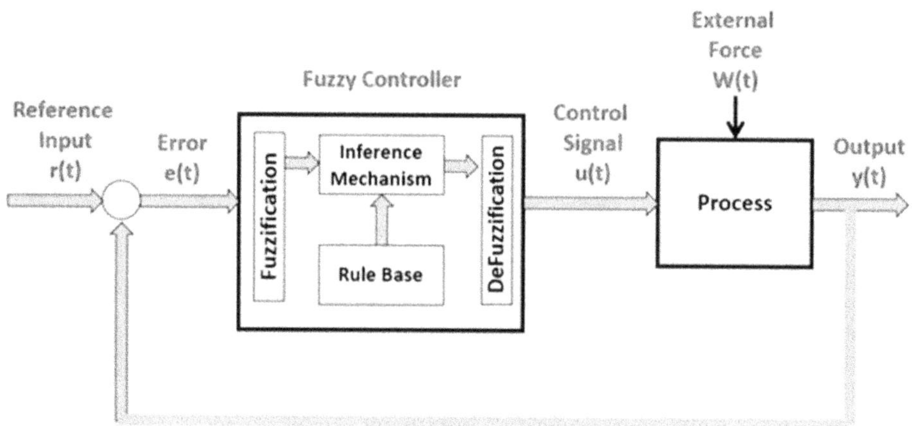


Based on the literature (Al-Dawod et al. 2004), the human expert and knowledge could not provide the required optimal control responses for specific applications. In other words, the provided configurations for membership functions and the rule base through human expertise could not lead the control responses into exact or even approximate solutions. By the way, a developed fuzzy system through human knowledge and expertise could be one of many feasible configurations of these systems in which the optimal responses are not guaranteed. For comparing purposes, a fuzzy system based on the human expert is provided in this section to verify the obtained results of the fuzzy optimization process. In this regard, 11 linguistic fuzzy variables are determined for the knowledge-based fuzzy system (Table 2).

A total of two inputs and one output are considered for the knowledge-based fuzzy system. As recommended by many fuzzy experts such as Al-Dawod et al. [1], the membership functions in the proposed fuzzy system is triangular-shaped. The membership functions for the fuzzy inputs and output of the knowledge-based fuzzy system are all depicted in Figure 4.

*Table 2. Fuzzy linguistic variables*

| *Variables* | *Definition* |
|---|---|
| *PVL* | *Positive and very Large* |
| *PL* | *Positive and Large* |
| *PM* | *Positive and Medium* |
| *PS* | *Positive and Small* |
| *PVS* | *Positive and very Small* |
| *ZR* | *Zero* |
| *NVS* | *Negative and very Small* |
| *NS* | *Negative and Small* |
| *NM* | *Negative and Medium* |
| *NL* | *Negative and Large* |
| *NVL* | *Negative and very Large* |

*Figure 4. Inputs (a) and the output (b) membership functions of the knowledge-based fuzzy system*

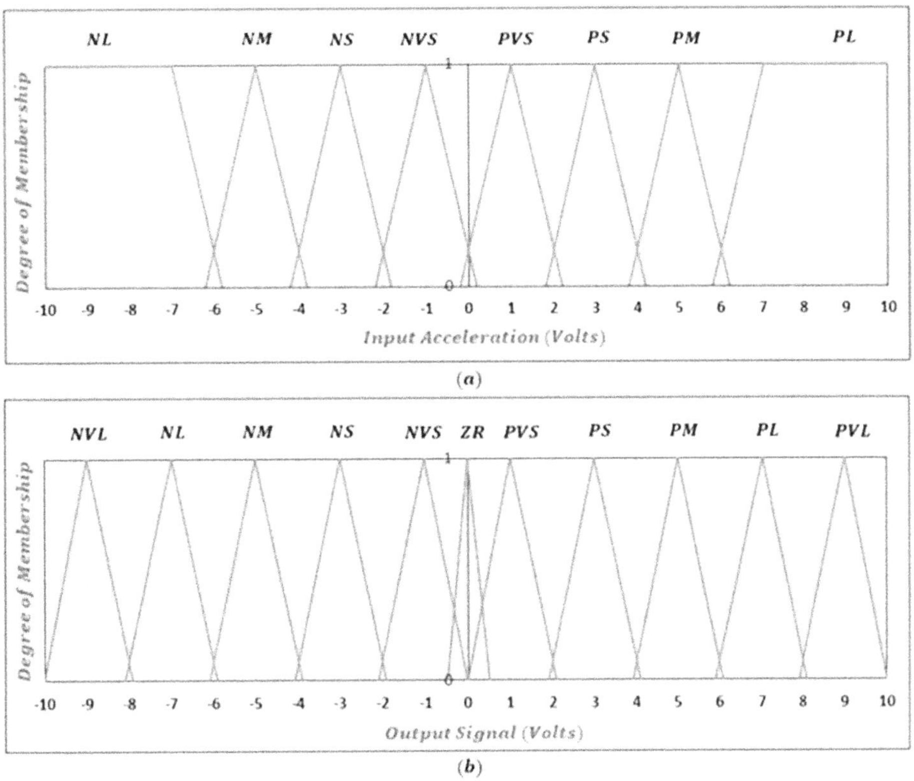


A total number of 64 fuzzy IF-THEN rules are determined in the fuzzy system as described in Table 3. Based on the fact that the presented fuzzy system is utilized for control purposes, it should be noted that the fuzzy membership functions are utilized in the "Fuzzification" part of the fuzzy controller, the provided rules are utilized in the "Rule Base" of the fuzzy controller and the linguistic variables are utilized in the "Inference Mechanism" part of the fuzzy controller. the "Centroid" method (Al-Dawod et al. 2004) is used for the "Defuzzification" purpose. Based on the provided information, the presented fuzzy system could be implemented in different systems for control purposes.

## 4. STOCHASTIC PAINT OPTIMIZER (SPO)

### 4.1. The Standard SPO

The core idea of the SPO is based on the beauty of colors and the art of painting in which the search space is determined as a painting canvas while a different combination of colors is utilized to represent the best configuration of colors. For many years, the color wheel (Figure 5) has been utilized as a key tool for artists and painters to present the basic principles of color mixtures. Based on Newton's color circle, red, yellow, blue, green, purple, orange and indigo are seven of the additive colors which have been of great importance in the color mixture for centuries. However, other beliefs in the determination of basic colors for a mixture have been proposed since now in which yellow, red, and blue or even the last two have been considered as basic colors for a mixture. Nowadays, the colors have been categorized into three different groups: primary, secondary, and tertiary colors.

*Table 3. The rule base of the knowledge-based fuzzy system*

| *Control Force* | | | | | | | | |
|---|---|---|---|---|---|---|---|---|
| | *First Input* | | | | | | | |
| *Second Input* | *NL* | *NM* | *NS* | *NVS* | *PVS* | *PS* | *PM* | *PL* |
| **NL** | *PVL* | *PL* | *PM* | *PS* | *PVS* | *ZR* | *NVS* | *NS* |
| **NM** | *PL* | *PM* | *PS* | *PS* | *PVS* | *ZR* | *NVS* | *NS* |
| **NS** | *PM* | *PS* | *PS* | *PVS* | *PVS* | *ZR* | *NVS* | *NS* |
| **NVS** | *PM* | *PS* | *PVS* | *PVS* | *ZR* | *NVS* | *NS* | *NM* |
| **PVS** | *PM* | *PS* | *PVS* | *ZR* | *NVS* | *NVS* | *NS* | *NM* |
| **PS** | *PS* | *PVS* | *ZR* | *NVS* | *NVS* | *NS* | *NS* | *NM* |
| **PM** | *PS* | *PVS* | *ZR* | *NVS* | *NS* | *NS* | *NM* | *NL* |
| **PL** | *PS* | *PVS* | *ZR* | *NVS* | *NS* | *NM* | *NL* | *NVL* |

*Figure 5. A schematic presentation of the color wheel*
*(Kaveh et al. 2020)*

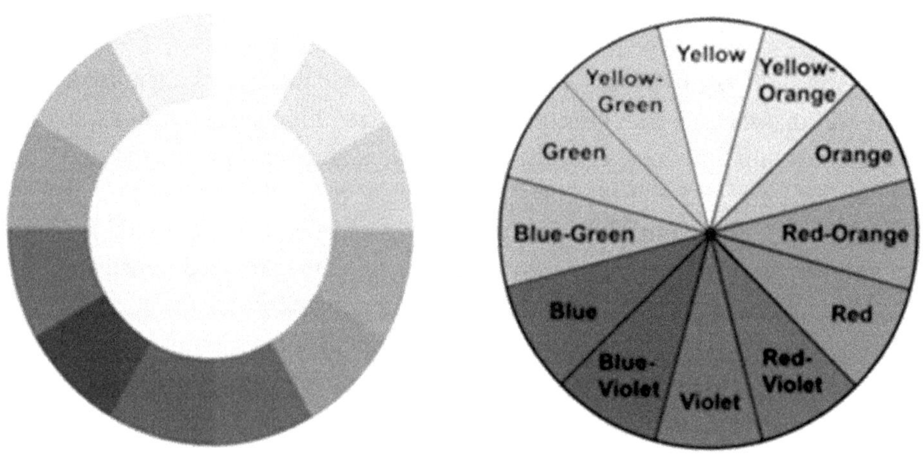


Red, blue, and yellow are determined as primary colors while the other colors are derived from these three colors so they are not mixed with other colors or even are not influenced by others. The secondary colors are orange, violet, and green which are achieved by mixing the primary colors. The tertiary colors (the rest of the colors in the wheel) are made of primary and secondary colors (Figure 6).

*Figure 6. A schematic presentation of the color classifications*
*(Kaveh et al. 2020)*

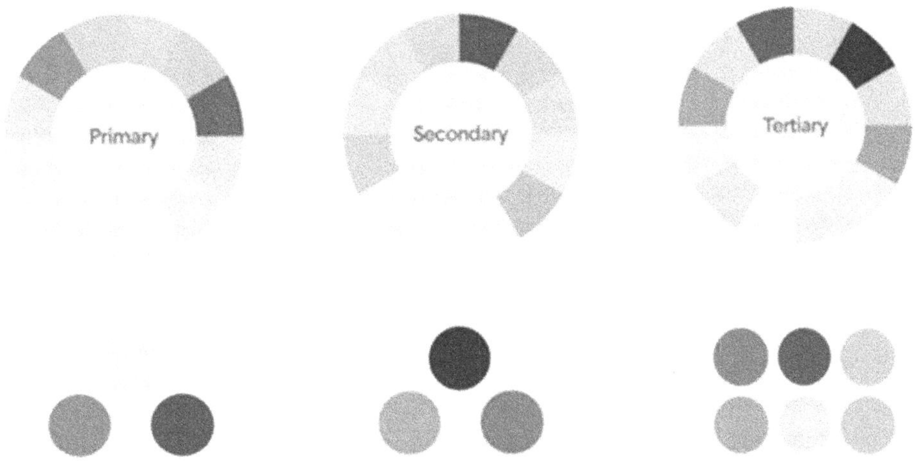


In the past decades, different techniques have been proposed for color combinations in which multiple factors including the visual experience balance and inner sense of order play important roles in this issue. The analogues color combination scheme denotes the mixture of the neighboring colors in the color wheel while one of the colors is determined as the dominant color while the others are considered as support and accent colors (Figure 7.a). In the complementary combination method for color mixture, the colors on the opposite side of the wheel which are determined as the complimentary colors are utilized for a mixture in which the high contrast of these colors is taken care of to be less distractive (Figure 7.b). In the Triadic color combination approach, three colors are utilized for this purpose which is equally apart from each other while a vivid unsaturated color will be created in this approach (Figure 7.c). The Tetradic combination method which is frequently utilized by painters represents the combination of four different colors in the color wheel which are grouped into two specific complimentary colors in a rectangular order (Figure 7.d).

*Figure 7. Different color combination techniques*
*(Kaveh et al. 2020)*

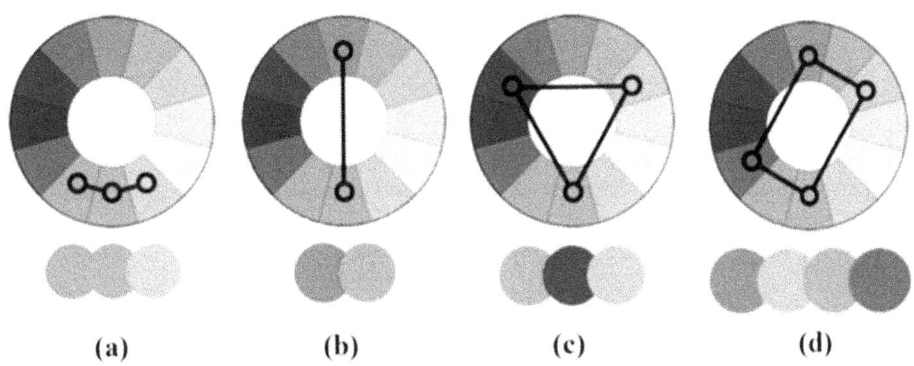

In the SPO, the initialization process is considered as follow:

$$C = \begin{bmatrix} C_1 \\ C_2 \\ \vdots \\ C_i \\ \vdots \\ C_n \end{bmatrix} = \begin{bmatrix} x_1^1 & x_1^2 & \cdots & x_1^j & \cdots & x_1^d \\ x_2^1 & x_2^2 & \cdots & x_2^j & \cdots & x_2^d \\ \vdots & \vdots & & \vdots & \ddots & \vdots \\ x_i^1 & x_i^2 & \cdots & x_i^j & \cdots & x_i^d \\ \vdots & \vdots & & \vdots & \ddots & \vdots \\ x_n^1 & x_n^2 & \cdots & x_n^j & \cdots & x_n^d \end{bmatrix}, \quad \begin{cases} i = 1,2,\ldots,n. \\ j = 1,2,\ldots,d. \end{cases} \tag{1}$$

where C is the vector for n solution candidates with d dimension.

Based on the objective function values of the solution candidates, the solution candidates are categorized into three groups as the primary colors regarding the best solutions ($C_P$), secondary colors considering the good solutions ($C_S$) and tertiary colors regarding the worst solutions ($C_T$). The position updating process for the initialized solution candidates is determined based on the previously mentioned techniques of color mixture. Regarding the analogous combination method, the position updating procedure is as follows:

$$C_{New,1} = C_i + rand.(C_{i+1} - C_{i-1}) \tag{2}$$

where $C_{New,1}$ is the new color created based on the analogues technique; $C_i$ is the *ith* color in the color wheel; $C_{i-1}$ and $C_{i+1}$ are the previous and next color to the ith one; *rand* is a randomly generated number in the range of (0, 1).

Based on the complementary combination method, the position updating procedure is as follows:

$$C_{New,2} = C_i + rand.(C_{Pi} - C_{Ti}) \tag{3}$$

where $C_{New,2}$ is the new color created based on the complementary technique; $C_i$ is the *ith* color in the color wheel; $C_{Pi}$ and $C_{Ti}$ are the randomly selected solutions from the primary and tertiary groups; *rand* is a randomly generated number in the range of (0, 1).

Considering the triadic combination method, the position updating procedure is as follows:

$$C_{New,3} = C_i + rand.\left(\frac{C_{Pi} + C_{Si} + C_{Ti}}{3}\right) \tag{4}$$

where $C_{New,3}$ is the new color created based on the triadic technique; $C_i$ is the *ith* color in the color wheel; $C_{Pi}$, $C_{Si}$ and $C_{Ti}$ are the randomly selected solutions from the primary, secondary, and tertiary groups; *rand* is a randomly generated number in the range of (0, 1).

Finally, the position updating procedure through the tetradic combination approach is as follows:

$$C_{New,4} = C_i + rand_1.C_{Pi} + rand_2.C_{Si} + rand_3.C_{Ti} + rand_4.C_{rand} \tag{5}$$

where $C_{New,4}$ is the new color created based on the tetradic technique; $C_i$ is the *ith* color in the color wheel; $C_{Pi}$, $C_{Si}$ and $C_{Ti}$ are the randomly selected solutions from the primary, secondary, and tertiary groups; $C_{rand}$ is a randomly selected color from the entire search space; $rand_1$, $rand_2$, $rand_3$ and $rand_4$ are randomly generated numbers in the range of (0, 1).

To completely formulate the mathematical model of the SPO, a boundary control is also implemented in the model in which the boundary violation by the decision variables is controlled. Besides, a termination criterion based on the maximum number of objective function evaluations or the maximum number of iterations is also considered in the model. The pseudo-code of the SPO is provided in Fig. 8.

*Figure 8. Pseudo code of the SPO*
*(Kaveh et al. 2020)*

```
procedure Stochastic Paint Optimizer (SPO)
        Inputs: The Number of Paints (Paint size)
        Outputs: The best Paint and its objective value
        Initialize the random paints
        Evaluate and sort the paints according the pre-defined objective function
        while The final criteria are not met
              Cluster the paints to three categories as Primary, Secondary and Tertiary
              for every paint
                  Create new paint (combination techniques) by Eqs.1-4
              end for
              Evaluate the fitness of all paints
              Update each paint if the new is better than the previous one
        end while
        Return the best solution.
end procedure
```

## 4.2. Enhanced Stochastic Paint Optimizer (ESPO)

Random movements are utilized for position updating of the search agents in most of the recent algorithms in which a stochastic process is performed for moving the solution candidates toward random trajectories. The first understanding of random movement in nature is about the well-known physical phenomenon including the sound, heat, and light transportations that have led to the birth of Brownian random motion. Random movements are some kind of moving process in which different series of consecutive random steps are included for different purposes. In the standard SPO, random numbers are frequently included in the position-updating process of the algorithm which alters the position of the considered solution candidate utilizing random steps (Eqs. 2 to 5). For enhancing the SPO, the well-known Levy distribution is used instead of random generations as follows:

$$Levy(s)=|s|^{-1-\beta} \quad 0<\beta\leq 2 \tag{6}$$

$$Levy(s,\gamma,\mu)=\begin{cases}\sqrt{\frac{\gamma}{2\pi}}exp\left[-\frac{\gamma}{2(s-\mu)}\right]\frac{1}{(s-\mu)^{\frac{3}{2}}} & 0<\mu<0\\ 0 & s\leq 0\end{cases} \tag{7}$$

where $\mu$, $\beta$, $s$, and $\gamma$ are the scaling parameter, shape factor, distribution factor, and position parameter respectively.

In the ESPO, the Levy flight is implemented in the mathematical model of the standard SPO in which the Levy flight is utilized instead of a random walk as presented in the following while the pseudo-code of ESPO is provided in Figure 9.

$$C_{New,1} = C_i + Levy(s).(C_{i+1} - C_{i-1}) \tag{8}$$

$$C_{New,2} = C_i + Levy(s).(C_{Pi} - C_{Ti}) \tag{9}$$

$$C_{New,3} = C_i + Levy(s).\left(\frac{C_{Pi} + C_{Si} + C_{Ti}}{3}\right) \tag{10}$$

$$C_{New,4} = C_i + Levy(s)_1.C_{Pi} + Levy(s)_2.C_{Si} + Levy(s)_3.C_{Ti} + Levy(s)_4.C_{Levy(s)} \tag{11}$$

where $C_{New,1}$, $C_{New,2}$, $C_{New,3}$ and $C_{New,4}$ are the new color created by utilizing Levy flight instead of random movement; $C_i$ is the *ith* color in the color wheel; $C_{i-1}$ and $C_{i+1}$ are the previous and next color to the *ith* one; $C_{Pi}$, $C_{Si}$ and $C_{Ti}$ are the randomly selected solutions from the primary, secondary, and tertiary groups; $C_{Levy(s)}$ is a Levy-based selected color from the entire search space.

*Figure 9. Pseudo code of the ESPO*

```
procedure Enhanced Stochastic Paint Optimizer (ESPO)
        Inputs: The Number of Paints (Paint size)
        Outputs: The best Paint and its objective value
        Initialize the random paints
        Evaluate and sort the paints according the pre-defined objective function
        while The final criteria are not met
            Cluster the paints to three categories as Primary, Secondary and Tertiary
            for every paint
                Perform the Levy flight
                Create new paint (combination techniques) by Eqs.8-11
            end for
            Evaluate the fitness of all paints
            Update each paint if the new is better than the previous one
        end while
        Return the best solution
end procedure
```

## 5. DESIGN EXAMPLE

### 5.1. Structural Details

A 20-story steel structure is considered for numerical investigations with complete details in (Ohtori et al. 2004). The steel moment-resisting frames are utilized as the load-bearing lateral system in this building. These frames are located in the perimeters of the plan in which the interior frames of the structure are composed of simple framing which has not to participate in the lateral load-resisting process. The high-strength steel materials are utilized for the beams and columns of the 20-story building in which the 248 Mega Pascal (MPa) steel material is utilized for the beams and the 345 MPa steel material is utilized for the columns. The structural design sections for the structural elements of this building are summarized in Table 4. The box-shaped columns are selected as design sections for the columns in the corner of the plan while the wide-flange sections are utilized for the beams and the columns in different positions of the plan. Bilinear material behavior is considered for the beams and columns of the 20-story steel structure with details in Table 4 and Figure 10. The structural analysis is conducted based on the Newmark-β method (Newmark 1959, Subbaraj & Dokainish 1989, Ohtori 1999). The fuzzy logic controller is implemented in the control process (Fig. 11)

*Figure 10. Hysteresis model*

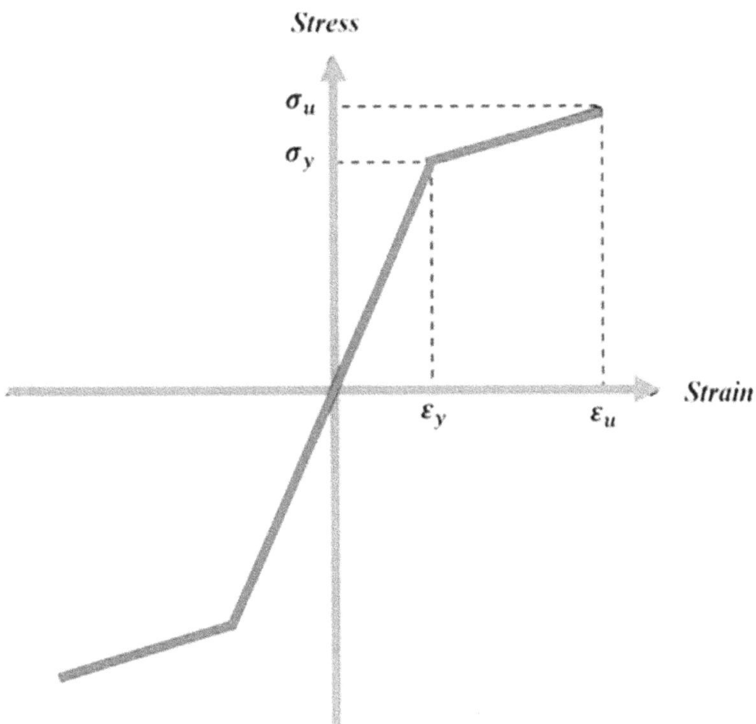

*Figure 11. Structural control simulation*

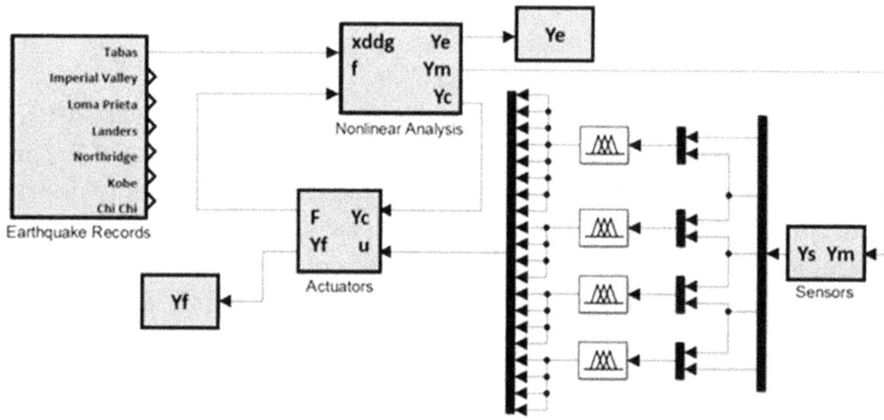


*Table 4. Characteristics of the nonlinear hysteresis model*

| Properties | Value | Units |
|---|---|---|
| ***Modulus of Elasticity (E)*** | *200,000* | *MPa* |
| ***Yield Strength ($\sigma_y$)*** | *345* | *MPa* |
| ***Tensile Strength ($\sigma_u$)*** | *450* | *MPa* |
| ***Yield Strain ($\varepsilon_y$)*** | *0.001725* | *mm/mm* |
| ***Tensile Stain ($\varepsilon_u$)*** | *0.18* | *mm/mm* |

## 5.2. Evaluation Criteria

The performance of the structural systems with implemented control systems should be considered separately in two different phases: the uncontrolled phase in which the structural system does not have any control devices or algorithms and the controlled phase in which a control system with control devices and algorithms have been provided in the structure. In this regard, the EC in this paper is considered as the response ratio of these two phases in which the response ratio of Controlled Structure (CS) to Uncontrolled Structure (US) is considered. Generally, these criteria are classified into three main categories as structural responses, structural damages, and the requirements of the control system (Table 5).

# 6. STATEMENT OF THE OPTIMIZATION PROBLEM

The fuzzy optimization is conducted for optimal tuning of the fuzzy membership function for two inputs and one output (Figure 12) and 64 rules (Table 6).

*Table 5. Details of the EC*

| Evaluation Criteria | Formulations |
|---|---|
| Inter-story drift ratio | $EC_1 = \max_{7EQs}\left\{\frac{\max_{t,i}\frac{\lvert d_i(t)\rvert}{h_i}}{\delta^{max}}\right\}$ |
| Level acceleration | $EC_2 = \max_{7EQs}\left\{\frac{\max_{t,i}\lvert \ddot{x}_{ai}(t)\rvert}{\ddot{x}_a^{max}}\right\}$ |
| Base shear | $EC_3 = \max_{7EQs}\left\{\frac{\max_{t,i}\left\lvert \sum_i m_i\ddot{x}_{ai}(t)\right\rvert}{F_b^{max}}\right\}$ |
| Ductility | $EC_3 = \max_{7EQs}\left\{\frac{\max_{t,i}\left\lvert \sum_i m_i\ddot{x}_{ai}(t)\right\rvert}{F_b^{max}}\right\}$ |
| Dissipated energy | $EC_3 = \max_{7EQs}\left\{\frac{\max_{t,i}\left\lvert \sum_i m_i\ddot{x}_{ai}(t)\right\rvert}{F_b^{max}}\right\}$ |
| Plastic connections | $EC_6 = \max_{7EQs}\left\{\frac{N_d^c}{N_d}\right\}$ |
| Control force | $EC_7 = \max_{7EQs}\left\{\frac{\max_{t,l}\lvert f_l(t)\rvert}{W}\right\}$ |
| Control device stroke | $EC_8 = \max_{7EQs}\left\{\frac{\max_{t,l}\lvert y_l^a(t)\rvert}{x^{max}}\right\}$ |
| Control power | $EC_9 = \max_{7EQs}\left\{\frac{\max_{t}\lvert \Sigma P_l(t)\rvert}{\dot{x}^{max}}\right\}$ |

*Figure 12. membership functions' variables inputs (a) and output (b)*

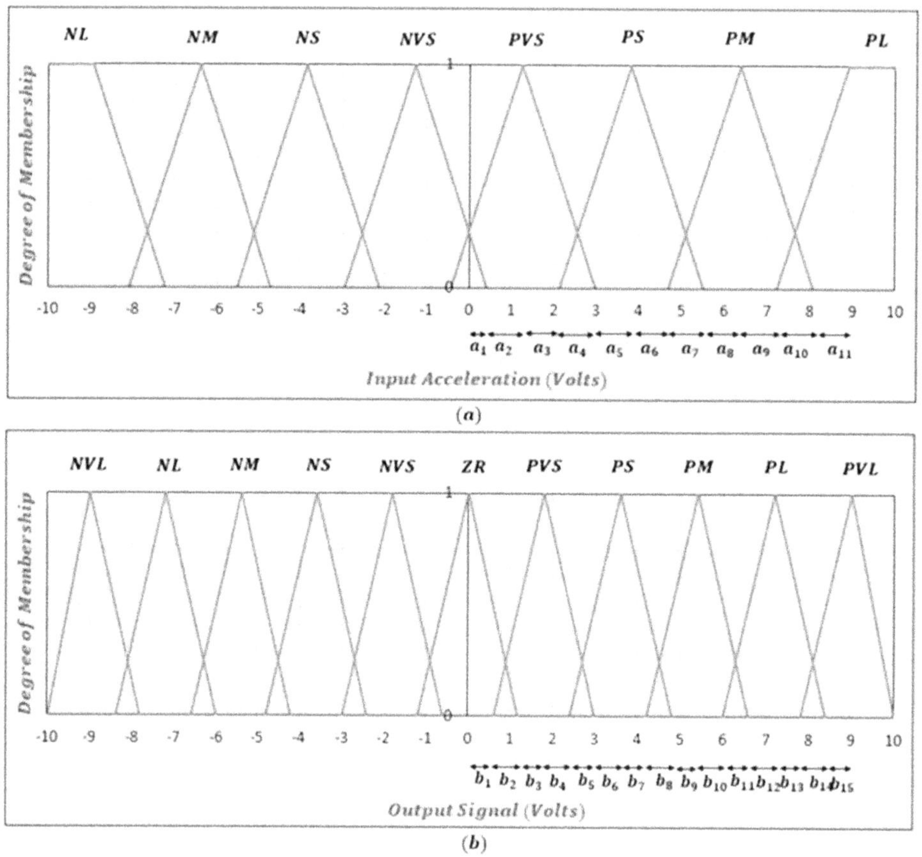


*Table 6. Optimization variables for the fuzzy rule base*

| Control Force | | | | | | | | |
|---|---|---|---|---|---|---|---|---|
| | First Input | | | | | | | |
| Second Input | NL | NM | NS | NVS | PVS | PS | PM | PL |
| **NL** | PVL/c1 | PL/c9 | PM/c17 | PS/c25 | PVS/c33 | ZR/c41 | NVS/c49 | NS/c57 |
| **NM** | PL/c2 | PM/c10 | PS/c18 | PS/c26 | PVS/c34 | ZR/c42 | NVS/c50 | NS/c58 |
| **NS** | PM/c3 | PS/c11 | PS/c19 | PVS/c27 | PVS/c35 | ZR/c43 | NVS/c51 | NS/c59 |
| **NVS** | PM/c4 | PS/c12 | PVS/c20 | PVS/c28 | ZR/c36 | NVS/c44 | NS/c52 | NM/c60 |
| **PVS** | PM/c5 | PS/c13 | PVS/c21 | ZR/c29 | NVS/c37 | NVS/c45 | NS/c53 | NM/c61 |
| **PS** | PS/c6 | PVS/c14 | ZR/c22 | NVS/c30 | NVS/c38 | NS/c46 | NS/c54 | NM/c62 |
| **PM** | PS/c7 | PVS/c15 | ZR/c23 | NVS/c31 | NS/c39 | NS/c47 | NM/c55 | NL/c63 |
| **PL** | PS/c8 | PVS/c16 | ZR/c24 | NVS/c32 | NS/c40 | NM/c48 | NL/c56 | NVL/c64 |

The objective function is formulated as follows;

$$Obj = \frac{\sum_{i=1}^{n} P_i \frac{CR_i}{UR_i}}{\sum_{i=1}^{n} P_i} \tag{21}$$

$$Obj = \frac{\begin{array}{c} 0.854\times(PC_4)_{Tabas} + 0.777\times(PC_4)_{Imperial\ Valley} + 0.654\times(PC_4)_{Loma\ Prieta} \\ +0.789\times(PC_4)_{Landers} + 0.923\times(PC_4)_{Northridge} + 0.834\times(PC_4)_{Kobe} + 0.760\times(PC_4)_{Chi\ Chi} \end{array}}{(0.854+0.777+0.654+0.789+0.923+0.834+0.760)} \tag{22}$$

where $UR_i$ and $CR_i$ denote the response of the structure in controlled and uncontrolled stages with $P_i$ as weight parameter.

## 7. NUMERICAL RESULTS

### 7.1. Mathematical Test Functions

To evaluate the overall performance of the ESPO, 4 unimodal (Table 7) and 4 multimodal (Table 8) are used as mathematical functions (Fig. 13).

*Table 7. Unimodal mathematical test functions*

| Function Name | Function Formula | Dim | Range | $f_{min}$ |
|---|---|---|---|---|
| Chung Reynolds Function | $F_1(x)=\left(\sum_{i=1}^{Dim} x_i^2\right)^2$ | 10 | [-100, 100] | 0 |
| Rosenbrock Function | $F_2(x)=\sum_{i=1}^{Dim}\left[100\left(x_{i+1}-x_i^2\right)^2+\left(x_i-1\right)^2\right]$ | 10 | [-30, 30] | 0 |
| Schwefel 1.2 Function | $F_3(x)=\sum_{i=1}^{Dim}\left(\sum_{j=1}^{i} x_j\right)^2$ | 10 | [-100, 100] | 0 |
| Step Function | $F_4(x)=\sum_{i=1}^{Dim}(\lvert x_i\rvert)$ | 10 | [-100, 100] | 0 |
| ***Dim:*** *Number of variables*<br>***Range:*** *Range of variables*<br>$f_{min}$: *Solution* | | | | |

*Table 8. Multimodal mathematical test functions*

| Function Name | Function Formula | Dim | Range | $f_{min}$ |
|---|---|---|---|---|
| Alpine 1 Function | $F_5(x)=\sum_{i=1}^{Dim}\lvert x_i \sin(x_i)+0.1x_i\rvert$ | 10 | [-10, 10] | 0 |
| Griewank Function | $F_6(x)=\sum_{i=1}^{Dim}\frac{x_i^2}{4000}-\prod_{i=1}^{Dim}\cos\left(\frac{x_i}{\sqrt{i}}\right)+1$ | 10 | [-100, 100] | 0 |
| Mishra 11 Function | $F_7(x)=\left[\frac{1}{Dim}\sum_{i=1}^{Dim}\lvert x_i\rvert-\left(\prod_{i=1}^{Dim}\lvert x_i\rvert\right)^{\frac{1}{Dim}}\right]^2$ | 10 | [-10, 10] | 0 |
| W/Wavy Function | $F_8(x)=1-\frac{1}{Dim}\sum_{i=1}^{Dim}\cos(kx_i)e^{\frac{-x_i^2}{2}}$ | 10 | [-π, π] | 0 |

***Dim:*** *Number of variables*
***Range:*** *Range of variables*
$f_{min}$*: Solution*

The convergence curves for the SPO and ESPO are depicted in Figures 14 and 15 in dealing with mathematical functions. Implementing the Levy flight had a positive impact on the performance of the ESPO.

## 7.2. Optimized FLC by ESPO and SPO

Considering the fuzzy optimization process, ESPO is capable of reaching *0.8846* for objective function while the SPO can calculate *0.9001* (Figure 16).

By evaluating the considered criteria (Table 11), the ESPO is capable of maximum reduction for $EC_1$ for the Kobe earthquake by up to 14% while for the $EC_2$, the response is decreased for five of the earthquakes with maximum and minimum reduction for Landers (8%) and Imperial Valley (3%) earthquakes. The response of the $EC_3$ is decreased for all of the earthquakes while the maximum and minimum reduction of $EC_4$ is 20% (Imperial Valley) and 4% (Landers) respectively. The energy dissipated ratio ($EC_5$) is reduced by up to 66%, 56%, 8%, 7%, 5%, and 2% for Imperial Valley, Loma Prieta, Tabas, Northridge, Kobe, Chi Chi, and Landers earthquakes respectively. The plastic hinges ratio ($EC_6$) is also reduced in most of the considered records. Based on the results of SPO (Table 12), it is obvious that the results of ESPO are better in most cases.

*Figure 13. 3-D graphs of the mathematical test functions*

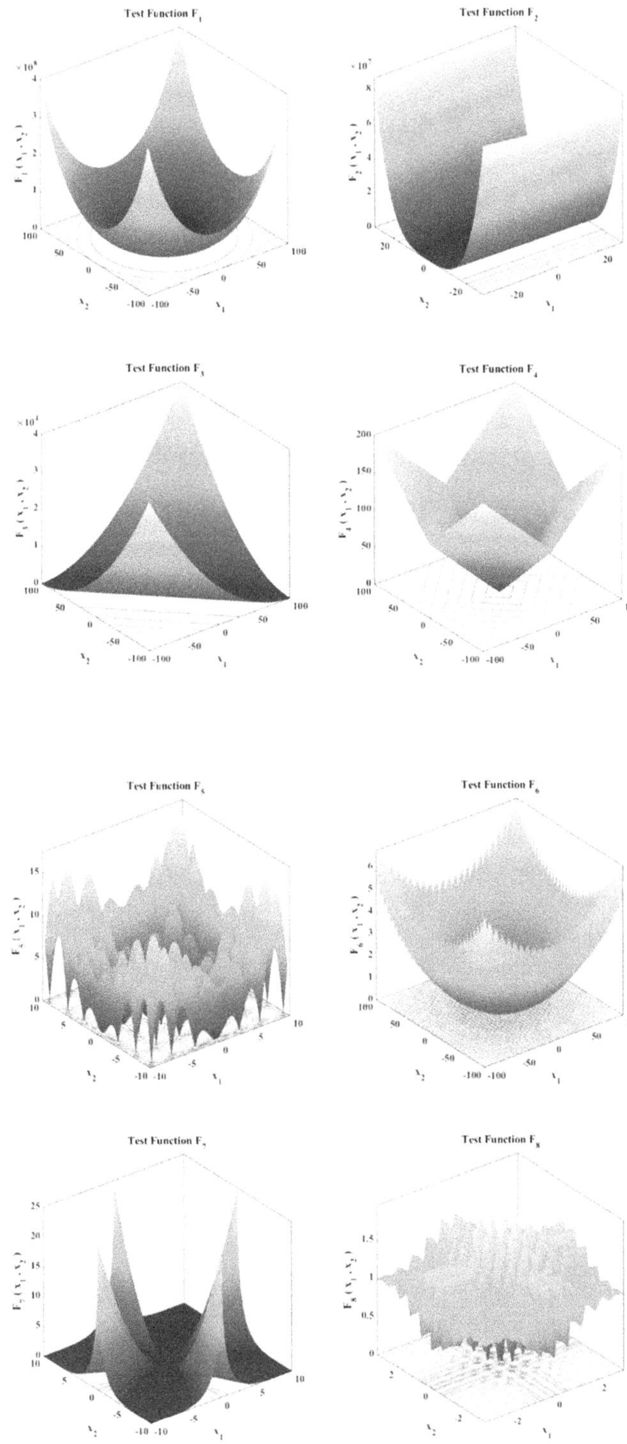

*Figure 14. Convergence history of the unimodal mathematical test functions*

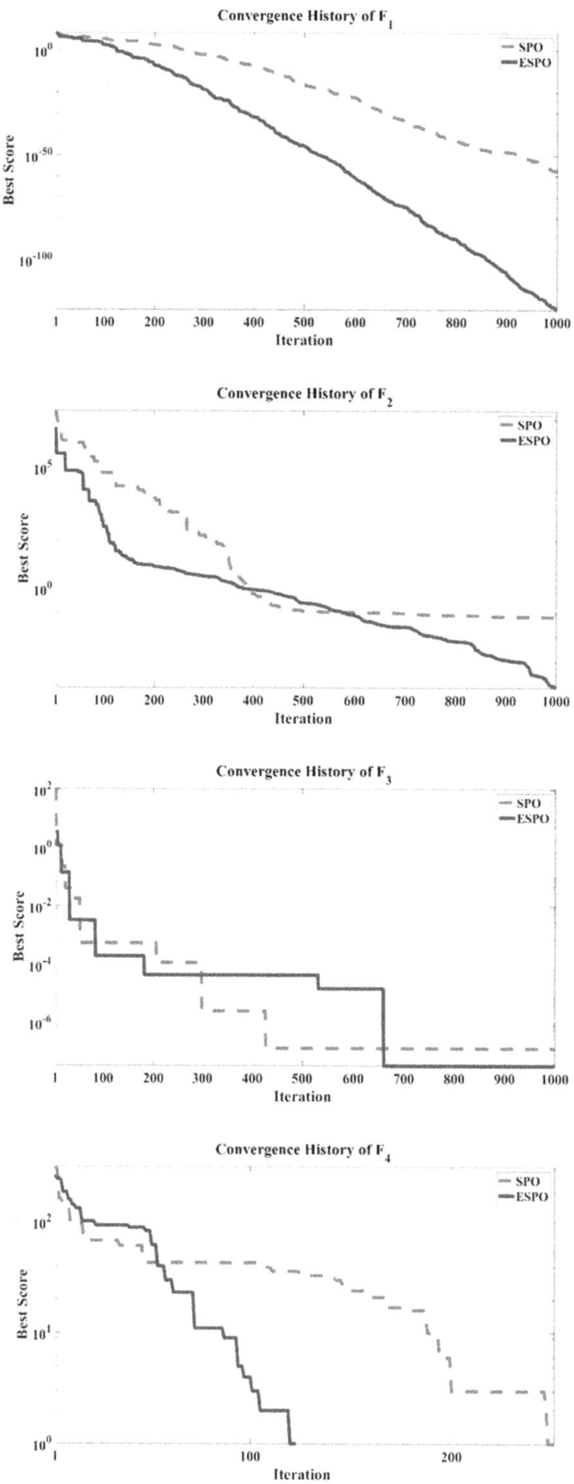

*Figure 15. Convergence history of the multimodal mathematical test functions*

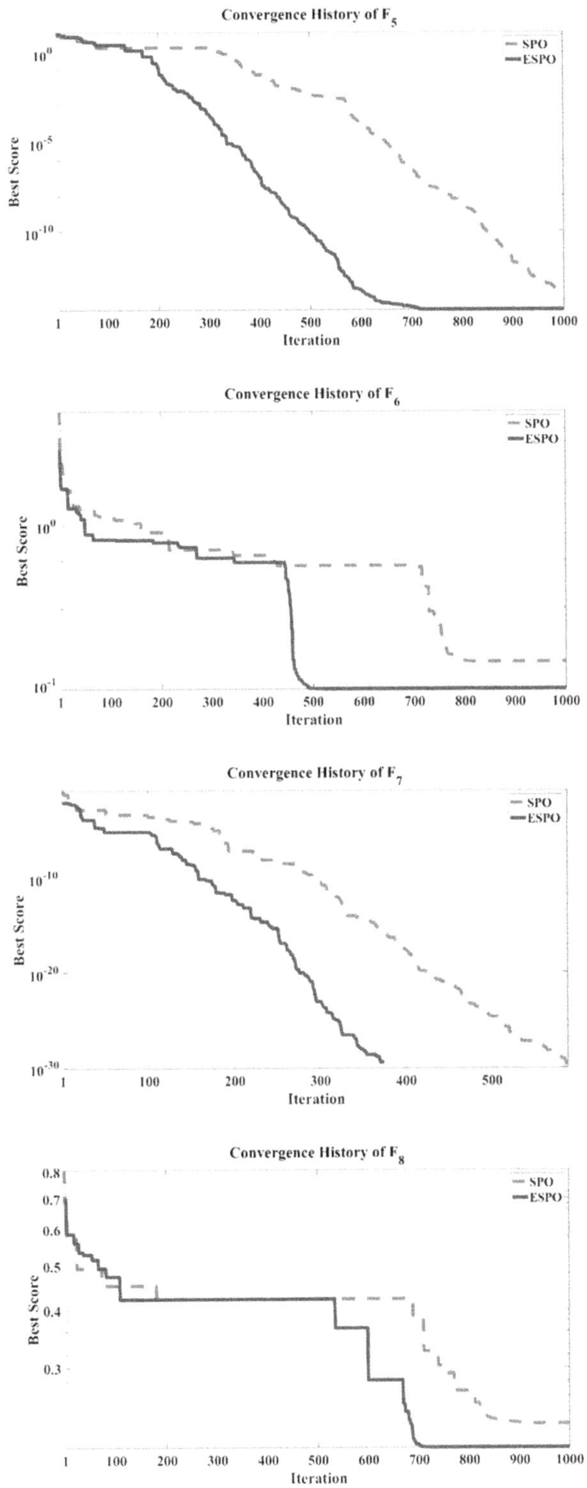

*Table 9. The best solutions obtained by ESPO and SPO*

| | Best Solutions | |
|---|---|---|
| Function | SPO | ESPO |
| F1 | 1.7560E-58 | 2.6229E-124 |
| F2 | 6.8478E-02 | 8.1978E-05 |
| F3 | 1.3645E-07 | 3.4659E-08 |
| F4 | 0.0000E+00 | 0.0000E+00 |
| F5 | 4.5932E-14 | 4.0612E-15 |
| F6 | 1.3797E-01 | 9.3496E-02 |
| F7 | 0.0000E+00 | 0.0000E+00 |
| F8 | 2.2677E-01 | 2.0106E-01 |

*Figure 16. The convergence history of Obj for the 20-story building by ESPO and SPO*

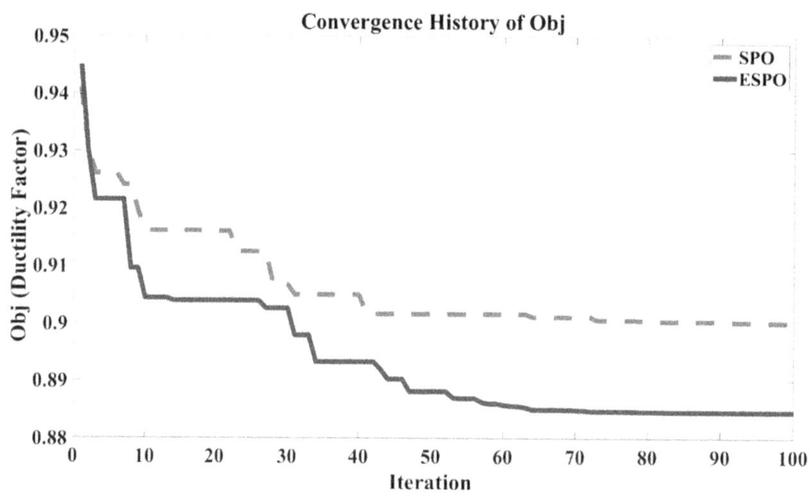


To perform a completely valid comparison between the results of the controlled structures based on the enhanced and the standard algorithms alongside the uncontrolled structure, some of the structural responses are evaluated separately. The $\phi_j/\phi_{yj}$ ratio for the ESPO- and SPO-based controlled structures alongside the uncontrolled structure is considered which denotes on the yield curvature of structural nodes. It should be mentioned that in order to suppress the damage distribution in the structural elements, there is an increasing interest in the reduction of this ratio due to the important effects of this ratio in the nonlinear behavior of the structural and control systems. For the controlled structure based on the ESPO, the calculated CR values are smaller than the ones for the uncontrolled structure and the controlled structure utilizing the SPO for all of the selected earthquakes (Figure 17).

*Table 11. Optimized EC utilizing ESPO*

| | | Earthquakes | | | | | | |
|---|---|---|---|---|---|---|---|---|
| Evaluation Criteria | | Tabas | Imperial Valley | Loma Prieta | Landers | Northridge | Kobe | Chi Chi |
| $EC_1$ | Drift Ratio | 0.9438 | 0.8905 | 0.9421 | 0.9755 | 0.9246 | 0.8693 | 0.9974 |
| $EC_2$ | Story Acceleration | 0.9264 | 0.9778 | 0.9484 | 0.9286 | 1.0114 | 0.9601 | 1.0385 |
| $EC_3$ | Bae Shear | 0.9815 | 0.8076 | 0.8075 | 0.9623 | 0.9304 | 0.8750 | 0.8115 |
| **$EC_4$** | **Ductility** | **0.9070** | **0.8094** | **0.9193** | **0.9686** | **0.8944** | **0.8138** | **0.8857** |
| $EC_5$ | Dissipated Energy | 0.9302 | 0.3415 | 0.4476 | 0.9906 | 0.9328 | 0.9504 | 0.9871 |
| $EC_6$ | Plastic Hinges | 0.9400 | 0.7500 | 0.5152 | 1.0000 | 0.9949 | 0.9881 | 0.9615 |
| $EC_7$ | Control Force | 0.0072 | 0.0072 | 0.0071 | 0.0048 | 0.0072 | 0.0072 | 0.0072 |
| $EC_8$ | Device Stroke | 0.0883 | 0.1796 | 0.1675 | 0.0928 | 0.1864 | 0.1364 | 0.1257 |
| $EC_9$ | Control Power | 0.0088 | 0.0117 | 0.0104 | 0.0045 | 0.0087 | 0.0090 | 0.0115 |

*Table 12. Optimized EC utilizing SPO*

| | | Earthquakes | | | | | | |
|---|---|---|---|---|---|---|---|---|
| Evaluation Criteria | | Tabas | Imperial Valley | Loma Prieta | Landers | Northridge | Kobe | Chi Chi |
| $EC_1$ | Drift Ratio | 0.9767 | 0.9348 | 0.9477 | 0.9720 | 0.9549 | 0.8791 | 1.1152 |
| $EC_2$ | Story Acceleration | 0.9617 | 1.0088 | 0.9620 | 0.9732 | 1.0099 | 0.9537 | 1.0407 |
| $EC_3$ | Bae Shear | 1.0318 | 0.8215 | 0.8839 | 0.9691 | 0.9447 | 0.8897 | 0.8638 |
| **$EC_4$** | **Ductility** | **0.9495** | **0.8103** | **0.9469** | **0.9665** | **0.9351** | **0.8231** | **0.8692** |
| $EC_5$ | Dissipated Energy | 0.9471 | 0.4986 | 0.4883 | 1.0183 | 0.9582 | 0.9575 | 1.0413 |
| $EC_6$ | Plastic Hinges | 0.9400 | 0.7917 | 0.5455 | 1.0000 | 1.0051 | 1.0060 | 0.9744 |
| $EC_7$ | Control Force | 0.0075 | 0.0075 | 0.0075 | 0.0055 | 0.0075 | 0.0075 | 0.0075 |
| $EC_8$ | Device Stroke | 0.0913 | 0.1886 | 0.1685 | 0.0937 | 0.1925 | 0.1379 | 0.1265 |
| $EC_9$ | Control Power | 0.0092 | 0.0131 | 0.0080 | 0.0052 | 0.0098 | 0.0093 | 0.0132 |

*Figure 17. Comparison of the curvature ratio for different earthquakes*

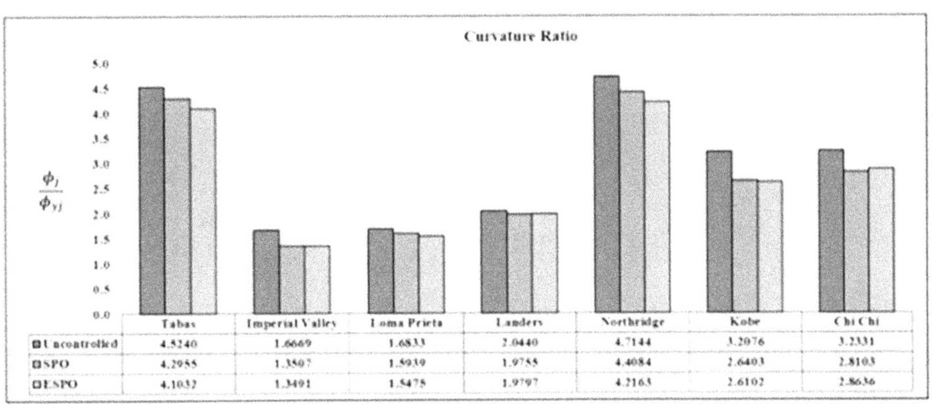

The Energy dissipation values for the controlled structure utilizing the ESPO are smaller than the uncontrolled structure (Figure 18). Based on the results, the ESPO-based controlled structure has smaller EDR values which makes this method capable of reducing the overall energy dissipation of the structural elements compared to the controlled structure with the SPO and the uncontrolled structure in dealing with some severe earthquakes.

*Figure 18. Comparison of the energy dissipation ratio for different earthquakes*

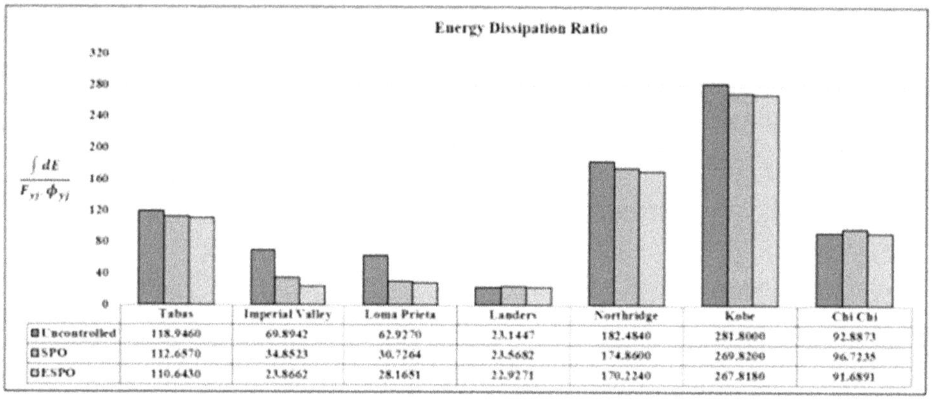


By evaluating the number of plastic hinges (Figure 19), the ESPO-based controlled structure has fewer plastic hinges than the uncontrolled and SPO-based controlled structures. The maximum difference between the ESPO-based controlled structure and the uncontrolled structure is about 49% (Loma Prieta) while this difference between the SPO-based controlled structure and the uncontrolled structure is for the same earthquake which is about 44%. It should be noted that for the Landers earthquake, this parameter is unchanged for the ESPO-based optimized approaches according to the uncontrolled structure while in the SPO-based controlled structure, this parameter is increased slightly for the Northridge and Kobe earthquakes.

*Figure 19. Number of plastic hinges for different earthquakes*

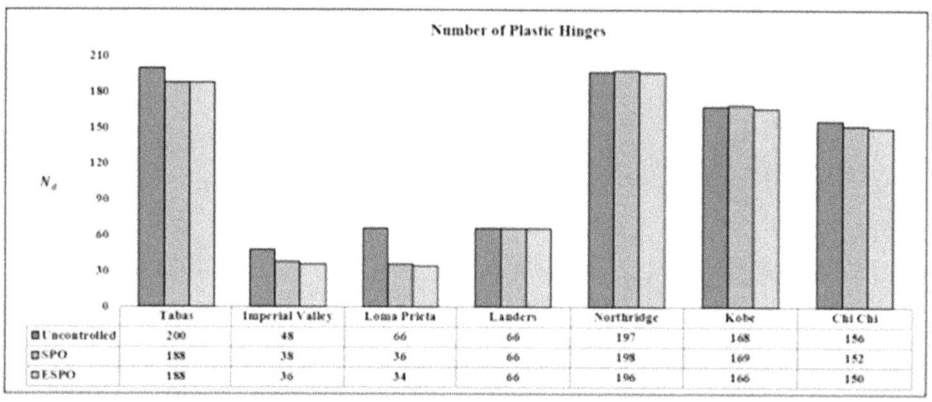

## 7.3. Comparing to Other Metaheuristics

For comparative purposes, the GA (Holland 1992), PSO (Kennedy and Eberhart 1995), ACO (Dorigo et al. 2006), and the Imperialist Competitive Algorithm (ICA) (Atashpaz-Gargari & Lucas 2007) are used for the optimum design of fuzzy controllers (Figure 20). ESPO algorithm can find *0.8846* for objective function which is the best one while the SPO with *0.9001,* PSO with *0.9004*, ICA with *0.9010*, GA with *0.9019*, and ACO with *0.9124*, have second to sixth places. The comparative data are described in Table 13.

*Table 13. Maximum EC of seven earthquake records obtained by different metaheuristics*

| *EC* | *Metaheuristics* | | | | | |
|---|---|---|---|---|---|---|
| | *GA* | *PSO* | *ACO* | *ICA* | *SPO* | *ESPO* |
| $EC_1$ | 1.0462 | 1.0644 | 1.0451 | 1.0394 | 1.1152 | **0.9974** |
| $EC_2$ | 1.0175 | **1.0103** | 1.0141 | 1.0194 | 1.0407 | 1.0385 |
| $EC_3$ | 1.0496 | 1.0394 | 1.0464 | 1.0444 | 1.0318 | **0.9815** |
| $EC_4$ | 0.9907 | 1.0024 | 0.9975 | 0.9976 | 0.9665 | **0.9686** |
| $EC_5$ | 1.0326 | 1.0689 | 1.0582 | 1.0756 | 1.0413 | **0.9906** |
| $EC_6$ | 1.0606 | 1.0303 | 1.0303 | 1.0305 | 1.0060 | **1.0000** |
| $EC_7$ | 0.0075 | 0.0073 | 0.0072 | 0.0078 | 0.0075 | **0.0072** |
| $EC_8$ | 0.1978 | 0.1966 | 0.2003 | 0.1982 | 0.1925 | **0.1864** |
| $EC_9$ | **0.0075** | 0.0127 | 0.0077 | 0.0078 | 0.0132 | 0.0117 |

*Figure 20. The convergence history of Obj for the 20-story building obtained by different metaheuristics*

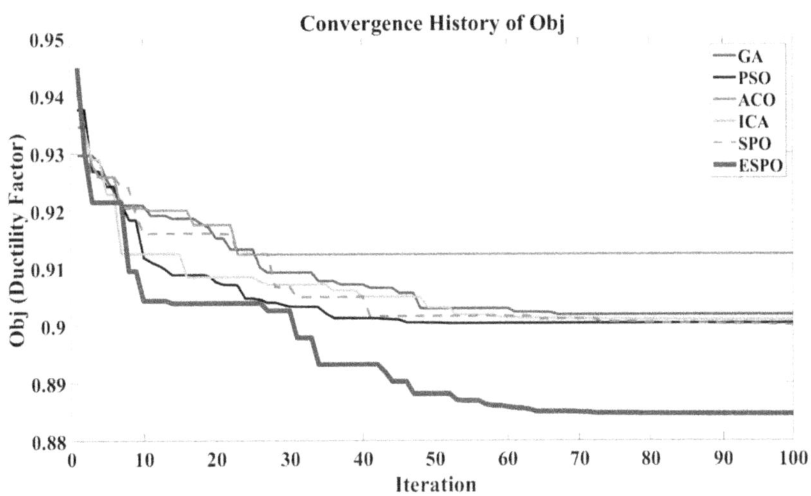

## 8. CONCLUSION

A fuzzy logic control is implemented in a 20-story structure for vibration control purposes. An enhanced version of the SPO algorithm is used for achieving better performance the random movements are replaced by the Levy flight which increases the capability of the enhanced method. The key findings of this work regarding the numerical investigations are as follows:

- The results of the mathematical function show that the ESPO is capable of converging to better results than SPO in most cases.
- ESPO is capable of calculating *0.8846* for the objective function of the fuzzy optimization problem which is better than the SPO one.
- ESPO is capable of reducing $EC_1$ ~ $EC_3$ by up to 14%, 8%, and 20% respectively. While SPO can reduce by up to 11%, 4%, and 17% respectively.
- For $EC_4$ ~ $EC_6$, ESPO can reduce by up to 20%, 66%, and 49% respectively which are smaller than the SPO results (17%, 51%, and 44%).
- ESPO performs better control action with less damage to structural elements.
- For $EC_7$ ~ $EC_9$, the ESPO requires a smaller amount of control force than the SPO.

In short, the practical configurations of the fuzzy logic controllers are in terms of human expertise and knowledge which are incapable of providing optimal responses for the system. The results of the proposed ESPO could improve the overall performance of these controllers in practice. For future works, the capability of the optimized FLC based on enhanced metaheuristics like ESPO would be investigated in practice and the results would be compared to the conventional fuzzy approaches. Meanwhile, the new practical applications of the fuzzy logic controllers should be formulated based on the algorithms like ESPO and the classical applications should be improved and modified accordingly.

## REFERENCES


Abdollahi, A., Amini, A., & Hariri-Ardebili, M. A. (2022). An uncertainty-aware dynamic shape optimization framework: Gravity dam design. *Reliability Engineering & System Safety*, *222*, 108402. doi:10.1016/j.ress.2022.108402

Abubaker, S., Nagan, S., & Nasar, T. (2016). Particle swarm optimized fuzzy control of structure with tuned liquid column damper. *Global Journal of Pure and Applied Mathematics*, *12*(1), 875–886.

Al-Dawod, M., Samali, B., Kwok, K., & Naghdy, F. (2004). Fuzzy controller for seismically excited nonlinear buildings. *Journal of Engineering Mechanics*, *130*(4), 407–415. doi:10.1061/(ASCE)0733-9399(2004)130:4(407)

Aliabadi, S. F., & Taher, S. A. (2017). Design of fuzzy-fopid controller optimized by ica for control of avr. *Majlesi Journal of Electrical Engineering*, *11*(4), 27–35.

Amin, J., Gondaliya, K., & Mulchandani, C. (2021). Assessment of seismic collapse probability of RC shaft supported tank. In *Structures* (Vol. 33, pp. 2639–2658). Elsevier.

Atashpaz-Gargari, E., & Lucas, C. (2007). Imperialist competitive algorithm: an algorithm for optimization inspired by imperialistic competition. In 2007 IEEE congress on evolutionary computation. IEEE.

Azizi, M. (2021). Atomic orbital search: A novel metaheuristic algorithm. *Applied Mathematical Modelling*, *93*, 657–683. doi:10.1016/j.apm.2020.12.021

Azizi, M., Aickelin, U., & Khorshidi, A., H., & Baghalzadeh Shishehgarkhaneh, M. (. (2023). Energy valley optimizer: A novel metaheuristic algorithm for global and engineering optimization. *Scientific Reports*, *13*(1), 226. doi:10.103841598-022-27344-y PMID:36604589

Azizi, M., Talatahari, S., & Gandomi, A. H. (2023). Fire Hawk Optimizer: A novel metaheuristic algorithm. *Artificial Intelligence Review*, *56*(1), 287–363. doi:10.100710462-022-10173-w

Bolt, B. A. (1975). *The San Fernando earthquake, 1971. Magnitudes, aftershocks, and fault dynamics. Bulletin 196*. Calif. Div. of Mines and Geology.

Chou, J. S., & Truong, D. N. (2021). A novel metaheuristic optimizer inspired by behavior of jellyfish in ocean. *Applied Mathematics and Computation*, *389*, 125535. doi:10.1016/j.amc.2020.125535

Dorigo, M., Birattari, M., & Stutzle, T. (2006). Ant colony optimization. *IEEE Computational Intelligence Magazine*, *1*(4), 28–39. doi:10.1109/MCI.2006.329691

Edrees, T. (2015). *Structural control and identification of civil engineering structures* [Doctoral dissertation, Luleå tekniska universitet].

Hari, M. J. V., & Patil, M. A. (2017). Optimization of FUZZY controller by JAYA algorithm. [IRJET]. *Int Res J Eng Technol*, *4*(7), 2182–2186.

Haselton, C. B. (2006). *Assessing seismic collapse safety of modern reinforced concrete moment frame buildings* [Doctoral dissertation, Stanford University].

Holland, J. H. (1992). Genetic algorithms. *Scientific American*, *267*(1), 66–73. doi:10.1038cientificamerican0792-66

Kaveh, A., Talatahari, S., & Khodadadi, N. (2020). Stochastic paint optimizer: Theory and application in civil engineering. *Engineering with Computers*, 1–32.

Kennedy, J., & Eberhart, R. (1995). *Particle swarm optimization. InProceedings of ICNN'95-international conference on neural networks* (Vol. 4). IEEE.

Khodadadi, N., Abualigah, L., & Mirjalili, S. (2022). Multi-objective stochastic paint optimizer (MOSPO). *Neural Computing & Applications*, *34*(20), 18035–18058. doi:10.100700521-022-07405-z

Khodadadi, N., Mirjalili, S. M., Mirjalili, S. Z., & Mirjalili, S. (2022, September). Chaotic Stochastic Paint Optimizer (CSPO). In *Proceedings of 7th International Conference on Harmony Search, Soft Computing and Applications: ICHSA 2022* (pp. 195-205). Springer Nature Singapore. 10.1007/978-981-19-2948-9_19

Kuyu, Y. Ç., & Vatansever, F. (2022). GOZDE: A novel metaheuristic algorithm for global optimization. *Future Generation Computer Systems*, *136*, 128–152. doi:10.1016/j.future.2022.05.022

Marinaki, M., Marinakis, Y., & Stavroulakis, G. E. (2011). Fuzzy control optimized by a multi-objective particle swarm optimization algorithm for vibration suppression of smart structures. *Structural and Multidisciplinary Optimization*, *43*(1), 29–42. doi:10.100700158-010-0552-4

Marinaki, M., Marinakis, Y., & Stavroulakis, G. E. (2015). Fuzzy control optimized by a multi-objective differential evolution algorithm for vibration suppression of smart structures. *Computers & Structures*, *147*, 126–137. doi:10.1016/j.compstruc.2014.09.018

Naeim, F. (Ed.). (1989). *The seismic design handbook.* Springer Science & Business Media. doi:10.1007/978-1-4615-9753-7

Newmark, N. M. (1959). A method of computation for structural dynamics. *Journal of the Engineering Mechanics Division*, *85*(3), 67–94. doi:10.1061/JMCEA3.0000098

Ohtori, Y., Christenson, R. E., Spencer, B. F. Jr, & Dyke, S. J. (2004). Benchmark control problems for seismically excited nonlinear buildings. *Journal of Engineering Mechanics*, *130*(4), 366–385. doi:10.1061/(ASCE)0733-9399(2004)130:4(366)

Ohtori, Y., & Spencer, B. F. Jr. (1999). A MATLAB-based tool for nonlinear structural analysis. In *Proc. of the 13th Engineering Mechanics Conf* (pp. 13-16). Scribd.

Precup, R. E., David, R. C., & Petriu, E. M. (2016). Grey wolf optimizer algorithm-based tuning of fuzzy control systems with reduced parametric sensitivity. *IEEE Transactions on Industrial Electronics*, *64*(1), 527–534. doi:10.1109/TIE.2016.2607698

Rayegani, A., & Nouri, G. (2022). Seismic collapse probability and life cycle cost assessment of isolated structures subjected to pounding with smart hybrid isolation system using a modified fuzzy based controller. In *Structures* (Vol. 44, pp. 30–41). Elsevier.

Somerville, P. G. (2005). Engineering characterization of near fault ground motions. In *Proc., NZSEE 2005 Conf.* New Zeland Society for Earthquake Engineering.

Stewart, J. P., Chiou, S. J., Bray, J. D., Graves, R. W., Somerville, P. G., & Abrahamson, N. A. (2002). Ground motion evaluation procedures for performance-based design. *Soil Dynamics and Earthquake Engineering*, *22*(9-12), 765–772. doi:10.1016/S0267-7261(02)00097-0

Subbaraj, K., & Dokainish, M. (1989). A survey of direct time-integration methods in computational structural dynamics—II. Implicit methods. *Computers & Structures*, *32*(6), 1387–1401. doi:10.1016/0045-7949(89)90315-5

Tairidis, G., Foutsitzi, G., Koutsianitis, P., & Stavroulakis, G. E. (2016). Fine tuning of a fuzzy controller for vibration suppression of smart plates using genetic algorithms. *Advances in Engineering Software, 101*, 123–135. doi:10.1016/j.advengsoft.2016.01.019

Tiwari, D., Pachauri, N., Rani, A., & Singh, V. (2016). Fractional order PID (FOPID) controller based temperature control of bioreactor. In 2016 international conference on electrical, electronics, and optimization techniques (ICEEOT) (pp. 2968-2973). IEEE.

Yang, X. S., & He, X. (2013). Firefly algorithm: recent advances and applications. *International journal of swarm intelligence, 1*(1), 36-50.

Zheng, H. D., Fan, J., & Long, X. H. (2017). Analysis of the seismic collapse of a high-rise power transmission tower structure. *Journal of Constructional Steel Research, 134*, 180–193. doi:10.1016/j.jcsr.2017.03.005

# Compilation of References


A.C. Institute. (2019). *Building Code Requirements for Structural Concrete (ACI 318-19): Commentary on Building Code Requirements for Structural Concrete (ACI 318R-19): an ACI Report*. American Concrete Institute.

Aalimahmoody, N., Bedon, C., Hasanzadeh-Inanlou, N., Hasanzade-Inallu, A., & Nikoo, M. (2021). BAT Algorithm-Based ANN to Predict the Compressive Strength of Concrete—A Comparative Study. *Infrastructures*, *6*(6), 80. doi:10.3390/infrastructures6060080

AASHTO. L. R. F. D. (2012). AASHTO LRFD bridge design specifications. American Association of State Highway and Transportation Officials: Washington, DC, USA.

Ababu, E. M., Markou, G., & Bakas, N. P. (2022). *Using Machine Learning and Finite Element Modelling to Develop a Formula to Determine the Deflection of Horizontally Curved Steel I-beams* [Paper presentation]. ICAART. doi:10.5220/0010982400003116

Abdel-Aal, S. I., Sabrah, R. E. A., Rabie, R. K., & Abdel Magid, H. M. (1997). Evaluation of groundwater quality for irrigation in Central Saudi Arabia. *Arab Gulf Journal of Scientific Research*, *15*(2), 361–377.

Abd-Elmaboud, M. E., Abdel-Gawad, H. A., El-Alfy, K. S., & Ezzeldin, M. M. (2021). Estimation of groundwater recharge using simulation-optimization model and cascade forward ANN at East Nile Delta aquifer, Egypt. *Journal of Hydrology. Regional Studies*, *34*, 100784. doi:10.1016/j.ejrh.2021.100784

Abdel-Qader, I., Abudayyeh, O., & Kelly, M. E. (2003). Analysis of Edge-Detection Techniques for Crack Identification in Bridges. *Journal of Computing in Civil Engineering*, *17*(4), 255–263. doi:10.1061/(ASCE)0887-3801(2003)17:4(255)

Abdollahi, A., Amini, A., & Hariri-Ardebili, M. A. (2022). An uncertainty-aware dynamic shape optimization framework: Gravity dam design. *Reliability Engineering & System Safety*, *222*, 108402. doi:10.1016/j.ress.2022.108402

Abed, F., El-Chabib, H., & AlHamaydeh, M. (2012). Shear characteristics of GFRP-reinforced concrete deep beams without web reinforcement. *Journal of Reinforced Plastics and Composites*, *31*(16), 1063–1073. doi:10.1177/0731684412450350

Abioye, S. O., Oyedele, L. O., Akanbi, L., Ajayi, A., Davila Delgado, J. M., Bilal, M., Akinade, O. O., & Ahmed, A. (2021). Artificial intelligence in the construction industry: A review of present status, opportunities and future challenges. *Journal of Building Engineering*, *44*, 103299. doi:10.1016/j.jobe.2021.103299

Abubaker, S., Nagan, S., & Nasar, T. (2016). Particle swarm optimized fuzzy control of structure with tuned liquid column damper. *Global Journal of Pure and Applied Mathematics*, *12*(1), 875–886.

ACI Committee 318 (2019). *Building code requirements for structural concrete (ACI 318-19) and commentary (ACI 318r-19)*. ACI.

ACI. (2014). Building Code Requirements for Structural Concrete (ACI 318-14) and Commentary. American Concrete Institute.

ACI. (2018). Predicting the Severity of Motor Vehicle Accident Injuries in Adana-Turkey Using Machine Learning Methods and Detailed Meteorological Data. *International Journal of Intelligent Systems and Applications in Engineering*, *1*(6), 72–79. doi:10.18201/ijisae.2018637934

Adekunle, S. A., Aigbavboa, C., Ejohwomu, O., & Thwala, W. D. (2021). (in press). Digital Transformation in the Construction Industry: A Bibliometric Review. *Journal of Engineering. Design and Technology*. doi:10.1108/JEDT-08-2021-0442

Adeli, H. (2001). Neural Networks In Civil Engineering: 1989-2000. *Computer-Aided Civil and Infrastructure Engineering*, *16*(2), 126–142. doi:10.1111/0885-9507.00219

Afzaal, H., Farooque, A. A., Abbas, F., Acharya, B., & Esau, T. (2020). *Groundwater estimation from major physical hydrology components using artificial neural networks and deep learning*. Water., doi:10.3390/w12010005

AGC Research Foundation. (2005). The Contractor's Guide to BIM (1st ed.). AGC Research Foundation.

AghaeiRad, A., Chen, N., & Ribeiro, B. (2017). Improve credit scoring using transfer of learned knowledge from self-organizing map. *Neural Computing & Applications*, *28*(6), 1329–1342. doi:10.100700521-016-2567-2

Ahmad, A., Cotsovos, D. M., & Lagaros, N. D. (2015). *Assessing The Reliability Of RC Code Predictions Through The Use Of Artificial Neural Networks.* The First International Conference on Structural Safety under Fire & Blast, Glasgow, UK, CONFAB.

Ahmad, A., & Cotsovos, D. M. (2023). Reliability analysis of models for predicting T-beam response at ultimate limit response. *Proceedings of the Institution of Civil Engineers. Structures and Buildings*, *176*(1), 28–50. doi:10.1680/jstbu.20.00129

Ahmad, A., Cotsovos, D. M., & Lagaros, N. D. (2020). Framework for the development of artificial neural networks for predicting the load carrying capacity of RC members. *SN Applied Sciences*, *2*(4), 1–21. doi:10.100742452-020-2353-8

Ahmad, A., & Dey, L. (2007). A k-mean clustering algorithm for mixed numeric and categorical data. *Data & Knowledge Engineering*, *63*(2), 503–527. doi:10.1016/j.datak.2007.03.016

Ahmad, A., Plevris, V., & Khan, Q.-u.-Z. (2020). Prediction of Properties of FRP-Confined Concrete Cylinders Based on Artificial Neural Networks. *Crystals*, *10*(9), 1-22. doi: 10.3390/cryst10090811

Ahmad, A., Plevris, V., & Khan, Q.-Z. (2020). Prediction of Properties of FRP-Confined Concrete Cylinders Based on Artificial Neural Networks. *Crystals*, *10*(9), 811. doi:10.3390/cryst10090811

Ahmadi, M., Kheyroddin, A., Dalvand, A., & Kioumarsi, M. (2020). New empirical approach for determining nominal shear capacity of steel fiber reinforced concrete beams. *Construction & Building Materials*, *234*, 117293. doi:10.1016/j.conbuildmat.2019.117293

Ahmadi, M., Kheyroddin, A., & Kioumarsi, M. (2021). Prediction models for bond strength of steel reinforcement with consideration of corrosion. *Materials Today: Proceedings*, *45*, 5829–5834. doi:10.1016/j.matpr.2021.03.263

Ahmed, M. (2020). Sustainable management scenarios for northern Africa's fossil aquifer systems. *Journal of Hydrology (Amsterdam)*, *589*, 125196. doi:10.1016/j.jhydrol.2020.125196

Akguzel, U., & Pampanin, S. (2010). Effects of variation of axial load and bidirectional loading on seismic performance of GFRP retrofitted reinforced concrete exterior beam-column joints. *Journal of Composites for Construction*, *14*(1), 94–104. doi:10.1061/(ASCE)1090-0268(2010)14:1(94)

Akguzel, U., & Pampanin, S. (2012). Assessment and design procedure for the seismic retrofit of reinforced concrete beam-column joints using FRP composite materials. *Journal of Composites for Construction*, *16*(1), 21–34. doi:10.1061/(ASCE)CC.1943-5614.0000242

Akinosho, T. D., Oyedele, L. O., Bilal, M., Ajayi, A. O., Delgado, M. D., Akinade, O. O., & Ahmed, A. A. (2020). Deep learning in the construction industry: A review of present status and future innovations. *Journal of Building Engineering*, *32*, 101827. doi:10.1016/j.jobe.2020.101827

Al Jassmi, H., Ahmed, S., Philip, B., Al Mughairbi, F., & Al Ahmad, M. (2019). E-happiness physiological indicators of construction workers' productivity: A machine learning approach. *Journal of Asian Architecture and Building Engineering*, *18*(6), 517–526. doi:10.1080/13467581.2019.1687090

Al-Adhaileh, M. H., & Alsaade, F. W. (2021). *Modelling and prediction of water quality by using artificial intelligence*. Sustainability., doi:10.3390u13084259

AlAjarmeh, O. S., Manalo, A. C., Benmokrane, B., Karunasena, K., Ferdous, W., & Mendis, P. (2020). Hollow concrete columns: Review of structural behavior and new designs using GFRP reinforcement. *Engineering Structures*, *203*, 109829. doi:10.1016/j.engstruct.2019.109829

Alamdari, M. M., Rakotoarivelo, T., & Khoa, N. L. D. (2017). A spectral-based clustering for structural health monitoring of the Sydney Harbour Bridge. *Mechanical Systems and Signal Processing*, *87*, 384–400. doi:10.1016/j.ymssp.2016.10.033

Alarcon, C., Hube, M., & De la Llera, J. (2014). Effect of axial loads in the seismic behavior of reinforced concrete walls with unconfined wall boundaries. *Engineering Structures*, *73*, 13–23. doi:10.1016/j.engstruct.2014.04.047

Alazab, M., Rm, S. P., Parimala, M., Maddikunta, P. K. R., Gadekallu, T. R., & Pham, Q.-V. (2022). Federated Learning for Cybersecurity: Concepts, Challenges, and Future Directions. *IEEE Transactions on Industrial Informatics*, *18*(5), 3501–3509. doi:10.1109/TII.2021.3119038

Al-Dawod, M., Samali, B., Kwok, K., & Naghdy, F. (2004). Fuzzy controller for seismically excited nonlinear buildings. *Journal of Engineering Mechanics*, *130*(4), 407–415. doi:10.1061/(ASCE)0733-9399(2004)130:4(407)

Al-Ghamdi, A. S. (2002). Using logistic regression to estimate the influence of accident factors on accident severity. *Accident; Analysis and Prevention*, *34*(6), 729–741. doi:10.1016/S0001-4575(01)00073-2 PMID:12371778

AlHamaydeh, M., Markou, G., Bakas, P. N., & Papadrakakis, M. (2022). AI-Based Shear Capacity of FRP-Reinforced Concrete Deep Beams without Stirrups. *Engineering Structures*, *264*, 264. doi:10.1016/j.engstruct.2022.114441

Aliabadi, S. F., & Taher, S. A. (2017). Design of fuzzy-fopid controller optimized by ica for control of avr. *Majlesi Journal of Electrical Engineering*, *11*(4), 27–35.

Al-Jerash, M. A. (1968). *Soils and agricultural development in the region of Al-Qassim*. Durham University.

Almuhaylan, M. R., Ghumman, A. R., Al-Salamah, I. S., Ahmad, A., Ghazaw, Y. M., Haider, H., & Shafiquzzaman, M. (2020). *Evaluating the impacts of pumping on aquifer depletion in arid regions using MODFLOW, ANFIS and ANN*. Water., doi:10.3390/w12082297

Al-Omran, A. M., Aly, A. A., Al-Wabel, M. I., Al-Shayaa, M. S., Sallam, A. S., & Nadeem, M. E. (2017). Geostatistical methods in evaluating spatial variability of groundwater quality in Al-Kharj Region, Saudi Arabia. *Applied Water Science*, *7*(7), 4013–4023. doi:10.100713201-017-0552-2

Al-Subhi, A. (2020). Parameters estimation of photovoltaic cells using simple and efficient mathematical models. *Solar Energy*, *209*, 245–257. doi:10.1016/j.solener.2020.08.079

Altin, S., Kopraman, Y., & Baran, M. (2013). Strengthening of RC walls using externally bonding of steel strips. *Engineering Structures*, *49*, 686–695. doi:10.1016/j.engstruct.2012.12.022

Amanat, K. M., & Hoque, E. (2006). A rationale for determining the natural period of RC building frames having infill. *Engineering Structures*, *28*(4), 495–502. doi:10.1016/j.engstruct.2005.09.004

American Concrete Institute. (1991). *ACI committee 352, Recommendations for design of beam-column joints in monolithic reinforced concrete structures (ACI 352R-91)*. ACI.

American Concrete Institute. (1998). *ACI committee 352, Recommendations for design of beam-column joints in monolithic reinforced concrete structures (ACI 352R-98)*. ACI.

American Concrete Institute. (2010). *ACI committee 352, Recommendations for design of beam-column joints in monolithic reinforced concrete structures (ACI 352R-10)*. ACI.

Amin, J., Gondaliya, K., & Mulchandani, C. (2021). Assessment of seismic collapse probability of RC shaft supported tank. In *Structures* (Vol. 33, pp. 2639–2658). Elsevier.

Anderson, M., Lehman, D., & Stanton, J. (2008). A cyclic shear stress-strain model for joints without transverse reinforcement. *Engineering Structures*, *30*(4), 941–954. doi:10.1016/j.engstruct.2007.02.005

Angelakos B. (1999). *The behavior of reinforced concrete roof corner joints under earthquake loads* [Ph.D dissertation]. University of Toronto.

Angus, J. E. (1991). *Criteria For Choosing The Best Neural Network-Part-I*. Naval Health Research Center.

Antoniou, F. (2021). Delay risk assessment models for road projects. *Systems*, *9* (3).

Antoniou, F., & Merkouri, M. (2021). Accident factors per construction type and stage: A synthesis of scientific research and professional experience. *International Journal of Injury Control and Safety Promotion*, *28*(4), 439–453. doi:10.1080/17457300.2021.1930061 PMID:34154518

Aranda-Mena, G., Crawford, J., Chevez, A., & Froese, T. (2009). Building information modelling demystified: Does it make business sense to adopt BIM? *International Journal of Managing Projects in Business*, *2*(3), 419–434. doi:10.1108/17538370910971063

Arashpour, M., Kamat, V., Heidarpour, A., Hosseini, M. R., & Gill, P. (2022). Computer vision for anatomical analysis of equipment in civil infrastructure projects: Theorizing the development of regression-based deep neural networks. *Automation in Construction*, *137*, 104193. doi:10.1016/j.autcon.2022.104193

Architectural Institute of Japan (AIJ). (1999). *Structural design guidelines for reinforced concrete buildings*. AIJ.

Aretoulis, G. N. (2019). Neural network models for actual cost prediction in Greek public highway projects. *International Journal of Project Organisation and Management*, *11*(1), 41–64. doi:10.1504/IJPOM.2019.098712

Arifin, F., Robbani, H., Annisa, T., & Ma'Arof, N. (2019). Variations in the Number of Layers and the Number of Neurons in Artificial Neural Networks: Case Study of Pattern Recognition. *Journal of Physics: Conference Series*, *1413*(1), 12016. doi:10.1088/1742-6596/1413/1/012016

Arthur, D., & Vassilvitskii, S. (2007). K-means++: The advantages of careful seeding. *Proceedings of the Annual ACM-SIAM Symposium on Discrete Algorithms*, (pp. 1027–1035). ACM.

ASCE-ACI Committee 445 on Shear and Torsion. (1998). Recent approaches to shear design of structural concrete. *Journal of Structural Engineering*, *124*(12), 1375–1417. doi:10.1061/(ASCE)0733-9445(1998)124:12(1375)

Ascheim, M., & Moehle, J. P. (1992). *Shear Strength and Deformability of RC Bridge Columns Subjected to Inelastic Cyclic Displacements.* Calif. Dept. Transp. Rep. No. UCBIEERC-92J04. Earthq. Eng. Res. Center, University of California, Berkeley.

Asteris, P. G., & Plevris, V. (2013, 12-14 June 2013). *Neural Network approximation of the masonry failure under biaxial compressive stress* 3rd South-East European Conference on Computational Mechanics (SEECCM III), Kos Island, Greece.

Asteris, P. G., & Plevris, V. (2013, 2013). Neural network approximation of the masonry failure under biaxial compressive stress. *ECCOMAS Special Interest Conference - SEECCM 2013: 3rd South-East European Conference on Computational Mechanics, Proceedings.* IACM. doi:10.7712/130113.4411.S2158

Asteris, P. G., & Plevris, V. (2017). Anisotropic masonry failure criterion using artificial neural networks. *Neural Computing & Applications, 28*(8), 2207–2229. doi:10.1007/s00521-016-2181-3 doi:10.100700521-016-2181-3

Asteris, P. G., Repapis, C. C., Cavaleri, L., Sarhosis, V., & Athanasopoulou, A. (2015). On the fundamental period of infilled RC frame buildings. *Structural Engineering and Mechanics, 54*(6), 1175–1200. doi:10.12989em.2015.54.6.1175

Atashpaz-Gargari, E., & Lucas, C. (2007). Imperialist competitive algorithm: an algorithm for optimization inspired by imperialistic competition. In 2007 IEEE congress on evolutionary computation. IEEE.

Athar, S. P. H., Taheri, M., Secrist, J., & Taheri, H. (2020). Neural network for structural health monitoring with combined direct and indirect methods. *Journal of Applied Remote Sensing, 14*(1), 1–13.

Avci, O., Abdeljaber, O., Kiranyaz, S., Hussein, M., Gabbouj, M., & Inman, D. J. (2021). A review of vibration-based damage detection in civil structures: From traditional methods to Machine Learning and Deep Learning applications. *Mechanical Systems and Signal Processing, 147*, 107077. doi:10.1016/j.ymssp.2020.107077

Awan, S., Li, F., Luo, B., & Liu, M. (2019). Poster: A reliable and accountable privacy-preserving federated learning framework using the blockchain. In *Proceedings of the ACM Conference on Computer and Communications Security* (pp. 2561-2563). ACM. 10.1145/3319535.3363256

Ayhan, B. U., & Tokdemir, O. B. (2019a). Predicting the outcome of construction incidents. *Safety Science, 113*, 91–104. doi:10.1016/j.ssci.2018.11.001

Ayhan, B. U., & Tokdemir, O. B. (2019b). Safety assessment in megaprojects using artificial intelligence. *Safety Science, 118*(March), 273–287. doi:10.1016/j.ssci.2019.05.027

Ayhan, B. U., & Tokdemir, O. B. (2020). Accident Analysis for Construction Safety Using Latent Class Clustering and Artificial Neural Networks. *Journal of Construction Engineering and Management, 146*(3), 04019114. doi:10.1061/(ASCE)CO.1943-7862.0001762

Azadi Kakavand, M. R., Sezen, H., & Taciroglu, E. (2019). *Rectangular column database*. designsafe-ci.

Azadi Kakavand, M. R., Sezen, H., & Taciroglu, E. (2021). Data-driven models for predicting the shear strength of rectangular and circular reinforced concrete columns. *Journal of Structural Engineering, 147*(1), 04020301. doi:10.1061/(ASCE)ST.1943-541X.0002875

Azhar, S., Nadeem, A., Mok, J. Y. N., & Leung, B. H. Y. (2008). Building Information Modeling (BIM): A new paradigm for visual interactive modeling and simulation for construction projects. *Proc., First International Conference on Construction in Developing Countries,* (*vol. 1*, 435–446). Scientific Research Publishing.

Azhar, S., Khalfan, M., & Maqsood, T. (2012). Building information modelling (BIM): Now and beyond. *Construction Economics and Building, 4*(12), 15–28. doi:10.5130/AJCEB.v12i4.3032

Azid, A., Juahir, H., Toriman, M. E., Kamarudin, M. K. A., Saudi, A. S. M., Hasnam, C. N. C., Aziz, N. A. A., Azaman, F., Latif, M. T., Zainuddin, S. F. M., Osman, M. R., & Yamin, M. (2014). Prediction of the level of air pollution using principal component analysis and artificial neural network techniques: A case study in Malaysia. *Water, Air, and Soil Pollution*, *225*(8), 1–14. doi:10.100711270-014-2063-1

Azimi, M., Eslamlou, A. D., & Pekcan, G. (2020). Data-driven structural health monitoring and damage detection through deep learning: State-ofthe- art review. In Sensors (Switzerland), 20(10).

Azimi, M., Eslamlou, A. D., & Pekcan, G. (2020). Data-driven structural health monitoring and damage detection through deep learning: State-of-the-art review. In Sensors, 20(10). doi:10.339020102778

Azizi, M. (2021). Atomic orbital search: A novel metaheuristic algorithm. *Applied Mathematical Modelling*, *93*, 657–683. doi:10.1016/j.apm.2020.12.021

Azizi, M., Aickelin, U., & Khorshidi, A., H., & Baghalzadeh Shishehgarkhaneh, M. (. (2023). Energy valley optimizer: A novel metaheuristic algorithm for global and engineering optimization. *Scientific Reports*, *13*(1), 226. doi:10.103841598-022-27344-y PMID:36604589

Azizi, M., Talatahari, S., & Gandomi, A. H. (2023). Fire Hawk Optimizer: A novel metaheuristic algorithm. *Artificial Intelligence Review*, *56*(1), 287–363. doi:10.100710462-022-10173-w

Babajanian Bisheh, H., Ghodrati Amiri, G., Nekooei, M., & Darvishan, E. (2019). Damage detection of a cable-stayed bridge using feature extraction and selection methods. *Structure and Infrastructure Engineering*, *15*(9), 1165–1177. do i:10.1080/15732479.2019.1599964

Baduge, S. K., Thilakarathna, S., Perera, J. S., Arashpour, M., Sharafi, P., Teodosio, B., Shringi, A., & Mendis, P. (2022). Artificial intelligence and smart vision for building and construction 4.0: Machine and deep learning methods and applications. *Automation in Construction*, *141*, 104440. doi:10.1016/j.autcon.2022.104440

Bakas, N.P., Langousis, A., Nicolaou, M. & Chatzichristofis, S.A. (2020), A Gradient Free Neural Network Framework Based on Universal Approximation Theorem. arXiv preprint arXiv:1909.13563

Bakas, N. P., Markou, G., Charmpis, D., & Hadjiyiannakou, K. (2021). *Performance and scalability of deep learning models trained on a hybrid supercomputer: Application in the prediction of the shear strength of reinforced concrete slender beams without stirrups* [Paper presentation]. COMPDYN.

Bakir, P. G., & Boduroğlu, H. M. (2002). A new design equation for predicting the joint shear strength of monotonically loaded exterior beam-column joints. *Engineering Structures*, *24*(8), 1105–1117. doi:10.1016/S0141-0296(02)00038-X

Banik, M. R., & Das, T. (2020). Application of neuro-GA hybrids in sensor optimization for structural health monitoring. *ACM International Conference Proceeding Series*, (pp. 6–12). ACM. 10.1145/3377049.3377131

Banzhaf, W., Nordin, P., Keller, R. E., & Francone, F. D. (1998). *Genetic programming: an introduction: on the automatic evolution of computer programs and its applications*. Morgan Kaufmann Publishers Inc.

Bao, W., Lianju, N., & Yue, K. (2019). Integration of unsupervised and supervised machine learning algorithms for credit risk assessment. *Expert Systems with Applications*, *128*, 301–315. doi:10.1016/j.eswa.2019.02.033

Bao, Y., Chen, Z., Wei, S., Xu, Y., Tang, Z., & Li, H. (2019). The State of the Art of Data Science and Engineering in Structural Health Monitoring. *Engineering (Beijing)*, *5*(2), 234–242. doi:10.1016/j.eng.2018.11.027

Bao, Y., Tang, Z., Li, H., & Zhang, Y. (2019). Computer vision and deep learning–based data anomaly detection method for structural health monitoring. *Structural Health Monitoring*, *18*(2), 401–421. doi:10.1177/1475921718757405

Bardak, S., Tiryaki, S., Nemli, G., & Aydın, A. (2016). Investigation and neural network prediction of wood bonding quality based on pressing conditions. *International Journal of Adhesion and Adhesives*, *68*, 115–123. doi:10.1016/j.ijadhadh.2016.02.010

Barkhordari, M. S., & Tehranizadeh, M. (2021). *Response estimation of reinforced concrete shear walls using artificial neural network and simulated annealing algorithm*. Structures. Elsevier.

Barkhordari, M. S., Tehranizadeh, M., & Scott, M. H. (2021). Numerical modelling strategy for predicting the response of reinforced concrete walls using Timoshenko theory. *Magazine of Concrete Research*, *73*(19), 988–1010. doi:10.1680/jmacr.19.00542

Barthel, R., Stangefelt, M., Giese, M., Nygren, M., Seftigen, K., & Chen, D. (2021). Current understanding of groundwater recharge and groundwater drought in Sweden compared to countries with similar geology and climate. *Geografiska Annaler. Series A. Physical Geography*, *0*(0), 1–23. doi:10.1080/04353676.2021.1969130

Barthorpe, R. J. (2010). *On Model and Data Based Approaches to Structural Health Monitoring*. White Rose University Consortium.

Bathe, K.-J. (1971). *Solution Methods for Large Generalized Eigenvalue Problems in Structural Engineering, Report UC SESM 71-20*. Civil Engineering Department, UC Berkeley.

Beale, M. H., Hagan, M. T., & Demuth, H. B. (2015). Neural Network Toolbox™-User's Guide. The MathWorks, Inc.

Beck, A., & Teboulle, M. (2009). Fast gradient-based algorithms for constrained total variation image denoising and deblurring problems. *IEEE Transactions on Image Processing*, *18*(11), 2419–2434. doi:10.1109/TIP.2009.2028250 PMID:19635705

Behera, A. K., Pradhan, R. M., Kumar, S., Chakrapani, G. J., & Kumar, P. (2022). Assessment of Groundwater Flow Dynamics Using MODFLOW in Shallow Aquifer System of Mahanadi Delta (East Coast), India. *Water (Basel)*, *14*(4), 611. doi:10.3390/w14040611

Behnia, A., Chai, H. K., & Shiotani, T. (2014). Advanced structural health monitoring of concrete structures with the aid of acoustic emission. *Construction & Building Materials*, *65*, 282–302. doi:10.1016/j.conbuildmat.2014.04.103

Belleri, A., Brunesi, E., Nascimbene, R., Pagani, M., & Riva, P. (2015). Seismic Performance of Precast Industrial Facilities Following Major Earthquakes in the Italian Territory. *Journal of Performance of Constructed Facilities*, *29*(5), 04014135. doi:10.1061/(ASCE)CF.1943-5509.0000617

Ben Seghier, M. E. A., Corriea, J. A. F. O., Jafari-Asl, J., Malekjafarian, A., Plevris, V., & Trung, N.-T. (2021). On the modeling of the annual corrosion rate in main cables of suspension bridges using combined soft computing model and a novel nature-inspired algorithm. *Neural Computing & Applications*, *33*(23), 15969–15985. doi:10.1007/s00521-021-06199-w doi:10.100700521-021-06199-w

Bentz, E. C., Vecchio, F. J., & Collins, M. P. (2006). Simplified modified compression field theory for calculating shear strength of reinforced concrete elements. *ACI Structural Journal*, *103*(4), 614.

Bianconi, F., Filippucci, M., & Buffi, A. (2019). Automated design and modeling for mass-customized housing. A web-based design space catalog for timber structures. *Automation in Construction*, *103*, 13–25. doi:10.1016/j.autcon.2019.03.002

Bigoni, C., Zhang, Z., & Hesthaven, J. S. (2020). Systematic sensor placement for structural anomaly detection in the absence of damaged states. In Computer Methods in Applied Mechanics and Engineering, 371. doi:10.1016/j.cma.2020.113315

Bishop, C. M. (2006). *Pattern Recognition and Machine Learning*. Springer.

Biskinis, D. E., Roupakias, G. K., & Fardis, M. N. (2004). Degradation of shear strength of reinforced concrete members with inelastic cyclic displacements. *Structural Journal, 101*(6), 773–783.

Bizhanimanzar, M., Leconte, R., & Nuth, M. (2019). Modelling of shallow water table dynamics using conceptual and physically based integrated surface-water-groundwater hydrologic models. *Hydrology and Earth System Sciences, 23*(5), 2245–2260. doi:10.5194/hess-23-2245-2019

Bizhanimanzar, M., Leconte, R., & Nuth, M. (2020). *Catchment-scale integrated surface water-groundwater hydrologic modelling using conceptual and physically based models: A model comparison study.* Water., doi:10.3390/w12020363

Blomfors, M., G. Berrocal, C., Lundgren, K., & Zandi, K. (2021). Incorporation of pre-existing cracks in finite element analyses of reinforced concrete beams without transverse reinforcement. *Engineering Structures, 229*, 111601. doi:10.1016/j.engstruct.2020.111601

Blomfors, M., Lundgren, K., & Zandi, K. (2020). Incorporation of pre-existing longitudinal cracks in finite element analyses of corroded reinforced concrete beams failing in anchorage. *Structure and Infrastructure Engineering, 0*(0), 1–17. doi:10.1080/15732479.2020.1782444

Blosch, M., & Fenn, J. (2018). *Understanding Gartner's hype cycles.* Gartner Research.

Bolpagni, M., Gavina, R., Ribeiro, D., & Arnal, I. P. (2022). Shaping the Future of Construction Professionals in Industry 4.0. *Structural Integrity, 20*, 1–28. doi:10.1007/978-3-030-82430-3_1

Bolt, B. A. (1975). *The San Fernando earthquake, 1971. Magnitudes, aftershocks, and fault dynamics. Bulletin 196.* Calif. Div. of Mines and Geology.

Borghini, A., Gusella, F., & Vignoli, A. (2016). Seismic vulnerability of existing R.C. buildings: A simplified numerical model to analyze the influence of the beam-column joints collapse. *Engineering Structures, 121*, 19–29. doi:10.1016/j.engstruct.2016.04.045

Bosche, F., Haas, C. T., & Akinci, B. (2009). Automated recognition of 3D CAD objects in site laser scans for project 3D status visualization and performance control. *Journal of Computing in Civil Engineering, 23*(6), 311–318. doi:10.1061/(ASCE)0887-3801(2009)23:6(311)

Boton, C., Rivest, L., Ghnaya, O., & Chouchen, M. (2021). What is at the Root of Construction 4.0: A Systematic Review of the Recent Research Effort. *Archives of Computational Methods in Engineering, 28*(4), 2331–2350. doi:10.100711831-020-09457-7

Bowden, G. J., Dandy, G. C., & Maier, H. R. (2005). Input determination for neural network models in water resources applications. Part 1—Background and methodology. *Journal of Hydrology (Amsterdam), 301*(1), 75–92. doi:10.1016/j.jhydrol.2004.06.021

Bozorg-Haddad, O., Baghban, S., & Loáiciga, H. A. (2021). Assessment of global hydro-social indicators in water resources management. *Scientific Reports, 11*(1), 17424. doi:10.103841598-021-96776-9 PMID:34465799

Breque, M., de Nul, L., & Petridis, A. (2021). *Industry 5.0: Towards a Sustainable, Human-Centric and Resilient European Industry.* European Commission, Directorate-General for Research and Innovation: Luxembourg https://op.europa.eu/en/publication-detail/-/publication/468a892a-5097-11eb-b59f-01aa75ed71a1/

Buitinck, L., Louppe, G., Blondel, M., Pedregosa, F., Mueller, A., Grisel, O., Niculae, V., Prettenhofer, P., Gramfort, A., Grobler, J., Layton, R., Vanderplas, J., Joly, A., Holt, B., & Varoquaux, G. (2013). *API design for machine learning software: experiences from the scikit-learn project.* https://doi.org//arxiv.1309.0238 doi:10.48550

Burrello, A., Brunelli, D., Malavisi, M., & Benini, L. (2020). Enhancing Structural Health Monitoring with Vehicle Identification and Tracking. *2020 IEEE International Instrumentation and Measurement Technology Conference (I2MTC)*, (pp. 1–6). IEEE. 10.1109/I2MTC43012.2020.9128641

Calvet, L., de Armas, J., Masip, D., & Juan, A. A. (2017). Learnheuristics: Hybridizing metaheuristics with machine learning for optimization with dynamic inputs. *Open Mathematics*, *15*(1), 261–280. doi:10.1515/math-2017-0029

Calvi, G. M., Pavese, A., Rasulo, A., & Bolognini, D. (2005). Experimental and numerical studies on the seismic response of RC hollow bridge piers. *Bulletin of Earthquake Engineering*, *3*(3), 267–297. doi:10.100710518-005-2240-0

Camelo, V., Beck, J. L., & Hall, J. F. (2001). *Dynamic Characteristics of Woodframe Buildings.*

Canadian Standards Association. (2004). *Design of Concrete Structures. A23*, 3–04.

Capellari, G., Chatzi, E., & Mariani, S. (2016). *Optimal Sensor Placement through Bayesian Experimental Design: Effect of Measurement Noise and Number of Sensors*. 41.

Carmona, J. R., & Ruiz, G. (2014). Bond and size effects on the shear capacity of RC beams without stirrups. *Engineering Structures*, *66*, 45–56. doi:10.1016/j.engstruct.2014.01.054

Caruana, R., & Niculescu-Mizil, A. (2006). An Empirical Comparison of Supervised Learning Algorithms. *Proceedings of the 23rd International Conference on Machine Learning*, (pp. 161–168). IEEE. 10.1145/1143844.1143865

Cassese, P., Ricci, P., & Verderame, G. M. (2017). Experimental study on the seismic performance of existing reinforced concrete bridge piers with hollow rectangular section. *Engineering Structures*, *144*, 88–106. doi:10.1016/j.engstruct.2017.04.047

Celik, O. C., & Ellingwood, B. R. (2010). Seismic fragilities for non-ductile reinforced concrete frames - Role of aleatoric and epistemic uncertainties. *Structural Safety*, *32*(1), 1–12. doi:10.1016/j.strusafe.2009.04.003

CEN. (2004). *Eurocode 8: Design of structures for earthquake resistance. Part 1: general rules, seismic actions and rules for buildings. European Standard EN 1998-1:2004.* Comite Europeen de Normalisation.

Chakraborty, S., Kumar, P., & Chakraborty, S. K. (2012). Neural Network Approach to Response of Buildings Due to Earthquake Excitation. *International Journal of Geosciences*, *03*(03), 630–639. doi:10.4236/ijg.2012.33063

Charef, R., Alaka, H., & Emmitt, S. (2018). Beyond the third dimension of BIM: A systematic review of literature and assessment of professional views. *Journal of Building Engineering*, *19*, 242–257. doi:10.1016/j.jobe.2018.04.028

Chen, C., He, W., Zhou, H., Xue, Y., & Zhu, M. (2020). A comparative study among machine learning and numerical models for simulating groundwater dynamics in the Heihe River Basin, northwestern China. *Scientific Reports*, *10*(1), 3904. doi:10.103841598-020-60698-9 PMID:32127583

Cheng, B., & Song, Y. (2010). Using Parametric Modeling to Improve the Equipment Systems Coordination Process for Buildings. *2010 International Conference on Management and Service Science*, (pp. 1–4). IEEE. 10.1109/ICMSS.2010.5577740

Chiang, M. M. T., & Mirkin, B. (2007). Experiments for the number of clusters in K-means. In Lecture Notes in Computer Science (including subseries Lecture Notes in Artificial Intelligence and Lecture Notes in Bioinformatics): Vol. 4874 LNAI (pp. 395–405). Springer Berlin Heidelberg. doi:10.1007/978-3-540-77002-2_33

Chong, J. W., Kim, Y., & Chon, K. H. (2014). Nonlinear multiclass support vector machine-based health monitoring system for buildings employing magnetorheological dampers. *Journal of Intelligent Material Systems and Structures*, *25*(12), 1456–1468. doi:10.1177/1045389X13507343

Chopra, A. K., & Goel, R. K. (2000). Building period formulas for estimating seismic displacements. *Earthquake Spectra*, *16*(2), 533–536. doi:10.1193/1.1586125

Chou, J. S., & Truong, D. N. (2021). A novel metaheuristic optimizer inspired by behavior of jellyfish in ocean. *Applied Mathematics and Computation*, *389*, 125535. doi:10.1016/j.amc.2020.125535

Christidis, K., Vougioukas, E., & Trezos, K. (2014). Deformation Capacity of Older RC Shear Walls: Experimental Assessment and Comparison with Eurocode 8–Part 3 Provisions. *Proceedings of the 2ECEES*. Research Gate.

Christidis, K., Vougioukas, E., & Trezos, K. G. (2013). Seismic assessment of existing RC shear walls non-compliant with current code provisions. *Magazine of Concrete Research*, *65*(17), 1059–1072. doi:10.1680/macr.13.00051

Cladera, A., & Marí, A. R. (2007). Shear strength in the new Eurocode 2. A step forward? *Structural Concrete*, *8*(2), 57–66. doi:10.1680tco.2007.8.2.57

Comité Euro-International du Béton. (1993). *CEB-FIP model code 1990: Design code*. Comité Euro-International du Béton.

Committee, A. C. I. (2015). *ACI 440.1 R-15: guide for the design and construction of structural concrete reinforced with FRP bars*. American Concrete Institute.

Cooper, H. H. Jr, & Jacob, C. E. (1946). A generalized graphical method for evaluating formation constants and summarizing well-field history. *Eos (Washington, D.C.)*, *27*(4), 526–534.

Cotsovos, D. M. & Lagaros, N. (2018). *Assessing the reliability of RC code predictions through the use of artificial neural networks*. Springer.

Coulibaly, P., Anctil, F., Aravena, R., & Bobée, B. (2001). Artificial neural network modeling of water table depth fluctuations. *Water Resources Research*, *37*(4), 885–896. doi:10.1029/2000WR900368

CSA (2004). *Design Of Concrete Structures. A23.3-04*. CSA.

Cury, A., Ribeiro, D., Ubertini, F., & Todd, M. D. (2022). *Structural Health Monitoring Based on Data Science Techniques, 21*. Research Gate.

Dabiri, H., Kioumarsi, M., Kheyroddin, A., Kandiri, A., & Sartipi, F. (2022). Compressive strength of concrete with recycled aggregate; a machine learning-based evaluation. *Cleaner Materials, 3*, 100044.

Dabirian, S., Han, S. H., & Lee, J. (2020). Stochastic-based noise exposure assessment in modular and off-site construction. *Journal of Cleaner Production*, *244*, 118758. doi:10.1016/j.jclepro.2019.118758

Dackermann, U., Smith, W. A., & Randall, R. B. (2014). Damage identification based on response-only measurements using cepstrum analysis and artificial neural networks. *Structural Health Monitoring*, *13*(4), 430–444. doi:10.1177/1475921714542890

Dahlberg, T. (2004). Procedure to calculate deflections of curved beams. *International Journal of Engineering Education*, *20*(3), 503–513.

Daliakopoulos, I. N., Coulibaly, P., & Tsanis, I. K. (2005). Groundwater level forecasting using artificial neural networks. *Journal of Hydrology (Amsterdam)*, *309*(1-4), 229–240. doi:10.1016/j.jhydrol.2004.12.001

Dallasega, P., Rauch, E., & Linder, C. (2018). Industry 4.0 as an enabler of proximity for construction supply chains: A systematic literature review. *Computers in Industry*, *99*, 205–225. doi:10.1016/j.compind.2018.03.039

Dallasega, P., Rojas, R. A., Bruno, G., & Rauch, E. (2019). An agile scheduling and control approach in ETO construction supply chains. *Computers in Industry*, *112*, 103122. doi:10.1016/j.compind.2019.08.003

Dardouri, S., BuHamdan, S., Al Balkhy, W., Dakhli, Z., Danel, T., & Lafhaj, Z. (2022). RFID platform for construction materials management. *International Journal of Construction Management.*

Das, S. K., Kumar, A., Das, B., & Burnwal, A. (2013). On soft computing techniques in various areas. *Computer Science & Information Technology (CS & IT), 3*, 59-68.

Dasgupta, D., Akhtar, Z., & Sen, S. (2022). Machine learning in cybersecurity: A comprehensive survey. *Journal of Defense Modeling and Simulation, 19*(1), 57–106. doi:10.1177/1548512920951275

Das, U. K., Roy, P., & Ghose, D. K. (2019). Modeling water table depth using adaptive Neuro-Fuzzy Inference System. *ISH Journal of Hydraulic Engineering, 25*(3), 291–297. doi:10.1080/09715010.2017.1420497

De Domenico, D., & Ricciardi, G. (2019). Shear strength of RC beams with stirrups using an improved Eurocode 2 truss model with two variable-inclination compression struts. *Engineering Structures, 198*, 109359. doi:10.1016/j.engstruct.2019.109359

Dehghani, H., & Bogdanovic, D. (2018). Copper price estimation using bat algorithm. *Resources Policy, 55*, 55–61. doi:10.1016/j.resourpol.2017.10.015

Del Vecchio, C., Del Zoppo, M., Di Ludovico, M., Verderame, G. M., & Prota, A. (2017). Comparison of available shear strength models for non-conforming reinforced concrete columns. *Engineering Structures, 148*, 312–327. doi:10.1016/j.engstruct.2017.06.045

Delgado, P. (2008). Avaliação da Segurança Estrutural em Pontes. [PhD thesis, FEUP, Porto]

Deng, H., Hong, H., Luo, D., Deng, Y., & Su, C. (2020). Automatic indoor construction process monitoring for tiles based on BIM and computer vision. *Journal of Construction Engineering and Management, 146*(1), 04019095. doi:10.1061/(ASCE)CO.1943-7862.0001744

Dhahir, M. K. (2017). Shear strength of FRP reinforced deep beams without web reinforcement. *Composite Structures, 165*, 223–232. doi:10.1016/j.compstruct.2017.01.039

Dhahir, M. K. (2018). Strut and tie modelling of deep beams shear strengthened with FRP laminates. *Composite Structures, 193*, 247–259. doi:10.1016/j.compstruct.2018.03.073

Dimopoulos, T., Tyralis, H., Bakas, N. P., & Hadjimitsis, D. (2018). Accuracy measurement of Random Forests and Linear Regression for mass appraisal models that estimate the prices of residential apartments in Nicosia, Cyprus. *Advances in Geosciences, 45*, 377–382. doi:10.5194/adgeo-45-377-2018

Ding, C., Wu, X., Yu, G., & Wang, Y. (2016). A gradient boosting logit model to investigate driver's stop-or-run behavior at signalized intersections using high-resolution traffic data. *Transportation Research Part C, Emerging Technologies, 72*, 225–238. doi:10.1016/j.trc.2016.09.016

Ding, J., Liu, Y., Zhang, L., Wang, J., & Liu, Y. (2016). An anomaly detection approach for multiple monitoring data series based on latent correlation probabilistic model. *Applied Intelligence, 44*(2), 340–361. doi:10.100710489-015-0713-7

Djurovic, N., Domazet, M., Stricevic, R., Pocuca, V., Spalevic, V., Pivic, R., Gregoric, E., & Domazet, U. (2015). Comparison of Groundwater Level Models Based on Artificial Neural Networks and ANFIS. *TheScientificWorldJournal, 2015*, 1–13. doi:10.1155/2015/742138 PMID:26759830

Dogbegah, R., Owusu-Manu, D., & Omoteso, K. (2011). A principal component analysis of project management competencies for the Ghanaian construction industry. *The Australasian Journal of Construction Economics and Building, 11*(1), 26–40. doi:10.5130/AJCEB.v11i1.1680

Dominguez Morales, M. (2000). *Fundamental period of vibration for reinforced concrete buildings*. University of Ottawa.

Dorigo, M., Birattari, M., & Stutzle, T. (2006). Ant colony optimization. *IEEE Computational Intelligence Magazine*, *1*(4), 28–39. doi:10.1109/MCI.2006.329691

Dubčáková, R. (2011). Eureqa: Software review. *Genetic Programming and Evolvable Machines*, *12*(2), 173–178. doi:10.100710710-010-9124-z

Eastman, C. M., Eastman, C., Teicholz, P., Sacks, R., & Liston, K. (2011). *BIM handbook: A guide to building information modeling for owners, managers, designers, engineers and contractors*. John Wiley & Sons.

EC2. (2004). Eurocode 2: Design Of Concrete Structures - Part 1-1: General Rules And Rules For Buildings. EN 1992-1-1. *Management Centre: Avenue Marnix*, *17*, B-1000.

Edelsbrunner, H., Kirkpatrick, D., & Seidel, R. (1983). On the shape of a set of points in the plane. *IEEE Transactions on Information Theory*, *29*(4), 551–559. doi:10.1109/TIT.1983.1056714

Edrees, T. (2015). *Structural control and identification of civil engineering structures* [Doctoral dissertation, Luleå tekniska universitet].

El-Sayed, A. K., El-Salakawy, E. F., & Benmokrane, B. (2005). Shear strength of concrete beams reinforced with FRP bars: design method. *7th International Symposium on Fiber-Reinforced (FRP) Polymer Reinforcement for Concrete Structures, 230*, (pp. 955-974). QuakeWrap.

El-Sayed, A. K., El-Salakawy, E. F., & Benmokrane, B. (2006). Shear capacity of high-strength concrete beams reinforced with FRP bars. *ACI Materials Journal*, *103*(3), 383.

Elvidge, S., Angling, M. J., & Nava, B. (2014). On the use of modified Taylor diagrams to compare ionospheric assimilation models. *Radio Science, 49*(9), 737-745. doi:10.1002/2014RS005435

Elwood, K., & Moehle, J.P. (2003). *Shake table tests and analytical studies on the gravity load collapse of reinforced concrete frames*. University of California, Berkeley. https://doi.org/ doi:10.16953/deusbed.74839

Elwood, K. J. (2004). Modelling failures in existing reinforced concrete columns. *Canadian Journal of Civil Engineering*, *31*(5), 846–859. doi:10.1139/l04-040

Elwood, K. J., & Moehle, J. P. (2005). Axial capacity model for shear-damaged columns. *ACI Structural Journal*, *102*(4), 578–587. doi:10.14359/14562

Emamgholizadeh, S., Moslemi, K., & Karami, G. (2014). Prediction the Groundwater Level of Bastam Plain (Iran) by Artificial Neural Network (ANN) and Adaptive Neuro-Fuzzy Inference System (ANFIS). *Water Resources Management*, *28*(15), 5433–5446. doi:10.100711269-014-0810-0

EN 1998-1, Eurocode 8. (2001). *Design of structures for earthquake resistance - Part 1: General rules, seismic actions and rules for buildings*. CENEuropean Committee for Standardization.

Ester, M., Kriegel, H.-P., Sander, J., & Xu, X. (1996). A density-based algorithm for discovering clusters in large spatial databases with noise. *KDD : Proceedings / International Conference on Knowledge Discovery & Data Mining. International Conference on Knowledge Discovery & Data Mining*, *96*(34), 226–231. doi:10.5120/739-1038

European Committee for Standardization (1991). *Eurocode 2: Design of concrete structures - Part 1-1: General rules and rules for buildings*. ECS.

European Committee for Standardization (2004). *Eurocode 2: Design of concrete structures – Part 1-1: General rules and rules for buildings*. ECS.

European Committee for Standardization (2005). *Eurocode 8: Design of structures for earthquake resistance – Part 3: Assessment and retrofitting of buildings*. ECS.

European Committee for Standardization. (2015). *Eurocode 2: Design of concrete structures: Part 1-1: General rules and rules for buildings*. European Committee for Standardization.

Fadipe, O. O., Abidoye, L. K., Adeosun, J. O., Oguntola, B. B., Adewusi, O., & Okeowo, T. E. (2021). Simulation of Groundwater Quality Characteristics using Artificial Neural Network. *Nigerian Journal of Technology*. doi:10.4314/njt.v40i1.21

Fallatah, O. A. (2020). Groundwater Quality Patterns and Spatiotemporal Change in Depletion in the Regions of the Arabian Shield and Arabian Shelf. *Arabian Journal for Science and Engineering*, *45*(1), 341–350. doi:10.100713369-019-04069-1

Fan, C.-L. (2020). Defect Risk Assessment Using a Hybrid Machine Learning Method. *Journal of Construction Engineering and Management*, *146*(9), 04020102. doi:10.1061/(ASCE)CO.1943-7862.0001897

Farrar, C. R., Doebling, S. W., & Nix, D. A. (2001). Vibration-based structural damage identification. *Philosophical Transactions of the Royal Society A: Mathematical, Physical and Engineering Sciences*, *359*(1778), 131–149.

Farrar, C. R., & Worden, K. (2013). *Structural health monitoring: a machine learning perspective*. John Wiley & Sons, Ltd.

Farrar, C., Worden, K., & Park, G. (2010). Complexity: A new axiom for Structural Health Monitoring. *Proceedings of the 5th European Workshop: Structural Health Monitoring*, 882–888.

Farsi, M. N., & Bard, P.-Y. (2004). Estimation des périodes propres de bâtiments et vulnérabilité du bâti existant dans l'agglomération de Grenoble. *Revue Française de Génie Civil*, *8*(2-3), 149–179. doi:10.1080/12795119.2004.9692601

Fathi, H., Nasir, V., & Kazemirad, S. (2020). Prediction of the mechanical properties of wood using guided wave propagation and machine learning. *Construction & Building Materials*, *262*, 120848. . doi:10.1016/j.conbuildmat.2020.120848

Favvata, M. J., Izzuddin, B. A., & Karayannis, C. G. (2008). Modelling exterior beam-column joints for seismic analysis of RC frame structures. *Earthquake Engineering & Structural Dynamics*, *37*(13), 1527–1548. doi:10.1002/eqe.826

Favvata, M. J., & Karayannisa, C. G. (2014). Influence of pinching effect of exterior joints on the seismic behavior of RC frames. *Earthquakes and Structures*, *6*(1), 89–110. doi:10.12989/eas.2014.6.1.089

Fédération Internationale du Béton (fib) (2013). *Model Code for concrete structures 2010*. FIB.

Feng, D. C., Wang, W. J., Mangalathu, S., Hu, G., & Wu, T. (2021). Implementing ensemble learning methods to predict the shear strength of RC deep beams with/without web reinforcements. *Engineering Structures*, *235*, 111979. doi:10.1016/j.engstruct.2021.111979

Figueiredo, E., & Cross, E. (2013). Linear approaches to modeling nonlinearities in long-term monitoring of bridges. *Journal of Civil Structural Health Monitoring*, *3*(3), 187–194. doi:10.100713349-013-0038-3

Figueiredo, E., Moldovan, I., Santos, A., Campos, P., & Costa, J. C. W. A. (2019). Finite Element–Based Machine-Learning Approach to Detect Damage in Bridges under Operational and Environmental Variations. *Journal of Bridge Engineering*, *24*(7), 04019061. doi:10.1061/(ASCE)BE.1943-5592.0001432

Figueiredo, E., Park, G., Farrar, C. R., Worden, K., & Figueiras, J. (2011). Machine learning algorithms for damage detection under operational and environmental variability. *Structural Health Monitoring*, *10*(6), 559–572. doi:10.1177/1475921710388971

Fiore, A., Quaranta, G., Marano, G. C., & Monti, G. (2016). Evolutionary polynomial regression–based statistical determination of the shear capacity equation for reinforced concrete beams without stirrups. *Journal of Computing in Civil Engineering*, *30*(1), 04014111. doi:10.1061/(ASCE)CP.1943-5487.0000450

Flah, M., Nunez, I., Ben Chaabene, W., & Nehdi, M. L. (2021). Machine Learning Algorithms in Civil Structural Health Monitoring: A Systematic Review. *Archives of Computational Methods in Engineering*, *28*(4), 2621–2643. doi:10.100711831-020-09471-9

Flynn, E. B., & Todd, M. D. (2010). A Bayesian approach to optimal sensor placement for structural health monitoring with application to active sensing. *Mechanical Systems and Signal Processing*, *24*(4), 891–903. doi:10.1016/j.ymssp.2009.09.003

Forcael, E., Ferrari, I., Opazo-Vega, A., & Pulido-Arcas, J. A. (2020). Construction 4.0: A literature review. *Sustainability (Switzerland)*, *12*(22), 1–28.

Fragiadakis, M., Lanese, I. A. P., & Papadrakakis, M. (2009). Experimental And Numerical Investigation of a Reinforced Concrete Building Designed For Gravity Loads Only. *2nd International Conference on Computational Methods in Structural Dynamics and Earthquake Engineering*. COMPDYN.

Friswell, M. I. (2008). Damage identification using inverse methods. *CISM International Centre for Mechanical Sciences, Courses and Lectures*, *499*(1993), 13–66.

Fu, J., Liu, J., Tian, H., Li, Y., Bao, Y., Fang, Z., & Lu, H. (2019). Dual attention network for scene segmentation. *Proceedings of the IEEE Computer Society Conference on Computer Vision and Pattern Recognition, 2019-June*, (pp. 3141–3149). IEEE. 10.1109/CVPR.2019.00326

García Cava, D., Avendaño-Valencia, L. D., Movsessian, A., Roberts, C., & Tcherniak, D. (2022). In A. Cury, D. Ribeiro, F. Ubertini, & M. D. Todd (Eds.), *On Explicit and Implicit Procedures to Mitigate Environmental and Operational Variabilities in Data-Driven Structural Health Monitoring BT - Structural Health Monitoring Based on Data Science Techniques* (pp. 309–330). Springer International Publishing.

Gatti, M. (2019). Structural health monitoring of an operational bridge: A case study. *Engineering Structures*, *195*(November 2018), 200–209. doi:10.1016/j.engstruct.2019.05.102

Ghannoum, W., Sivaramakrishnan, B., Pujol, S., Catlin, A. C., Fernando, S., Yoosuf, N., & Wang, Y. (2015). *NEES: ACI 369 rectangular column database*. National Science Foundation.

Gharbia, M., Chang-Richards, A., Lu, Y., Zhong, R. Y., & Li, H. (2020). Robotic technologies for on-site building construction: A systematic review. *Journal of Building Engineering*, *32*, 101584. doi:10.1016/j.jobe.2020.101584

Ghayeb, H. H., Razak, H. A., & Sulong, N. H. R. (2017). Development and testing of hybrid precast concrete beam-to-column connections under cyclic loading. *Construction & Building Materials*, *151*, 258–278. doi:10.1016/j.conbuildmat.2017.06.073

Gilles, D. C. (2011). *In situ dynamic characteristics of reinforced concrete shear wall buildings*. McGill University.

Gilles, D., McClure, G., & Chouinard, L. E. (2011). Uncertainty in fundamental period estimates leads to inaccurate design seismic loads. *Canadian Journal of Civil Engineering*, *38*(8), 870–880. doi:10.1139/l11-055

Goel Rakesh, K., & Chopra Anil, K. (1997). Period Formulas for Moment-Resisting Frame Buildings. *Journal of Structural Engineering*, *123*(11), 1454–1461. doi:10.1061/(ASCE)0733-9445(1997)123:11(1454)

Goel Rakesh, K., & Chopra Anil, K. (1998). Period Formulas for Concrete Shear Wall Buildings. *Journal of Structural Engineering*, *124*(4), 426–433. doi:10.1061/(ASCE)0733-9445(1998)124:4(426)

Goh, G. D., Sing, S. L., & Yeong, W. Y. (2021). A review on machine learning in 3D printing: Applications, potential, and challenges. *Artificial Intelligence Review*, *54*(1), 63–94. doi:10.100710462-020-09876-9

Goh, Y. M., & Chua, D. (2013). Neural network analysis of construction safety management systems: A case study in Singapore. *Construction Management and Economics*, *31*(5), 460–470. doi:10.1080/01446193.2013.797095

Golparvar-Fard, M., Savarese, S., & Peña-Mora, F. (2010). Automated model-based recognition of progress using daily construction photographs and IFC-based 4D models. *Construction Research Congress 2010: Innovation for Reshaping Construction Practice*, (pp. 51–60). ASCE Library. 10.1061/41109(373)6

Golparvar-Fard, M., Pena-Mora, F., & Savarese, S. (2011). Monitoring changes of 3D building elements from unordered photo collections. *2011 IEEE International Conference on Computer Vision Workshops (ICCV Workshops)*, (pp. 249–256). IEEE. 10.1109/ICCVW.2011.6130250

Golparvar-Fard, M., Pena-Mora, F., & Savarese, S. (2015). Automated progress monitoring using unordered daily construction photographs and IFC-based building information models. *Journal of Computing in Civil Engineering*, *29*(1), 04014025. doi:10.1061/(ASCE)CP.1943-5487.0000205

Gomes, G. F., de Almeida, F. A., da Silva Lopes Alexandrino, P., da Cunha, S. S. Jr, de Sousa, B. S., & Ancelotti, A. C. Jr. (2019). A multiobjective sensor placement optimization for SHM systems considering Fisher information matrix and mode shape interpolation. *Engineering with Computers*, *35*(2), 519–535. doi:10.100700366-018-0613-7

Gonand, F. (2016). *Impacts of Higher Energy Effciency on Growth and Welfare Across Generations in Saudi Arabia.* Kapsarc. https://www.kapsarc.org/wp-content/uploads/2016/10/KS-1657-DP051A-Impact-of-Higher-Energy-Efficiency-on-Growth-and-Welfare-Across-Generations-in-Saudi-Arabia.pdf

Gordan, M., Razak, H. A., Ismail, Z., Ghaedi, K., Tan, Z. X., & Ghayeb, H. H. (2020). A hybrid ANN-based imperial competitive algorithm methodology for structural damage identification of slab-on-girder bridge using data mining. *Applied Soft Computing*, *88*, 106013. doi:10.1016/j.asoc.2019.106013

Gordan, M., Sabbagh-Yazdi, S.-R., Ismail, Z., Ghaedi, K., Carroll, P., McCrum, D., & Samali, B. (2022). State-of-the-art review on advancements of data mining in structural health monitoring. *Measurement*, *193*, 110939. doi:10.1016/j.measurement.2022.110939

Gravett, D. Z., Mourlas, C., Markou, G. & Papadrakakis, M. (2019) *Numerical performance of a new algorithm for performing modal analysis of full-scale reinforced concrete structures that are discretized with the HYMOD approach [Paper presentation]*. COMPDYN, Athens.

Gravett, D. Z., Mourlas, C., Taljaard, V.-L., Bakas, N., Markou, G., & Papadrakakis, M. (2021). New fundamental period formulae for soil-reinforced concrete structures interaction using machine learning algorithms and ANNs. *Soil Dynamics and Earthquake Engineering*, *144*, 106656. doi:10.1016/j.soildyn.2021.106656

Greifenhagen, C., & Lestuzzi, P. (2005). Static cyclic tests on lightly reinforced concrete shear walls. *Engineering Structures*, *27*(11), 1703–1712. doi:10.1016/j.engstruct.2005.06.008

Grundy, C. (2017). Cybersecurity in the built environment: Can your building be hacked? *Corporate Real Estate Journal*, *7*(1), 39–50.

Gualtieri, L., Rauch, E., & Vidoni, R. (2021). Emerging research fields in safety and ergonomics in industrial collaborative robotics: A systematic literature review. *Robotics and Computer-integrated Manufacturing*, *67*, 67. doi:10.1016/j.rcim.2020.101998

Guldur, B., & Hajjar, J. (2014). *Laser-based structural sensing and surface damage detection.* Citeseer. doi:10.17760/d20009283

Günaydın, H. M., & Doğan, S. Z. (2004). A Neural Network Approach For Early Cost Estimation Of Structural Systems Of Buildings. *International Journal of Project Management*, *22*(7), 595–602. doi:10.1016/j.ijproman.2004.04.002

Guo, T., Wu, L., Wang, C., & Xu, Z. (2020). Damage detection in a novel deep-learning framework: A robust method for feature extraction. *Structural Health Monitoring*, *19*(2), 424–442. doi:10.1177/1475921719846051

Hacıefendioğlu, K., Ayas, S., Başağa, H. B., Toğan, V., Mostofi, F., & Can, A. (2022a). Wood construction damage detection and localization using deep convolutional neural network with transfer learning. *Holz als Roh- und Werkstoff*, *80*(4), 791–804. doi:10.100700107-022-01815-5

Hacıefendioğlu, K., Mostofi, F., Toğan, V., & Başağa, H. B. (2022b). CAM-K: A novel framework for automated estimating pixel area using K-Means algorithm integrated with deep learning based-CAM visualization techniques. *Neural Computing & Applications*, *34*(20), 17741–17759. doi:10.100700521-022-07428-6

Hadi, M. N. S. (2003). Neural Networks Applications In Concrete Structures. *Computers & Structures*, *81*(6), 373–381. doi:10.1016/S0045-7949(02)00451-0

Hafeez, G., Doudak, G., & McClure, G. (2018). Establishing the fundamental period of light-frame wood buildings on the basis of ambient vibration tests. *Canadian Journal of Civil Engineering*, *45*(9), 752–765. doi:10.1139/cjce-2017-0348

Hafeez, G., Doudak, G., & McClure, G. (2019). Dynamic characteristics of light-frame wood buildings. *Canadian Journal of Civil Engineering*, *46*(1), 1–12. doi:10.1139/cjce-2017-0266

Hafeez, G., Mustafa, A., Doudak, G., & McClure, G. (2014). Predicting the Fundamental Period of Light-Frame Wood Buildings. *Journal of Performance of Constructed Facilities*, *28*(6), A4014004. doi:10.1061/(ASCE)CF.1943-5509.0000519

Halford, M. (2016). *Prince Library.* MIT License. https://github.com/MaxHalford/prince/blob/master/LICENSE

Hallberg, D., & Tarandi, V. (2011). On the use of open bim and 4d visualisation in a predictive life cycle management system for construction works. [ITcon]. *Journal of Information Technology in Construction*, *16*, 445–466.

Han, J., & Moraga, C. (1995). The influence of the sigmoid function parameters on the speed of backpropagation learning. Lecture Notes in Computer Science (Including Subseries Lecture Notes in Artificial Intelligence and Lecture Notes in Bioinformatics), 930, 195–201. doi:10.1007/3-540-59497-3_175

Hari, M. J. V., & Patil, M. A. (2017). Optimization of FUZZY controller by JAYA algorithm. [IRJET]. *Int Res J Eng Technol*, *4*(7), 2182–2186.

Haselton, C. B. (2006). *Assessing seismic collapse safety of modern reinforced concrete moment frame buildings* [Doctoral dissertation, Stanford University].

Hatzigeorgiou, G. D., & Kanapitsas, G. (2013). Evaluation of fundamental period of low-rise and mid-rise reinforced concrete buildings. *Earthquake Engineering & Structural Dynamics*, *42*(11), 1599–1616. doi:10.1002/eqe.2289

Hebert, M., & Krotkov, E. (1992). 3D measurements from imaging laser radars: How good are they? *Image and Vision Computing*, *10*(3), 170–178. doi:10.1016/0262-8856(92)90068-E

Hegde, J., & Rokseth, B. (2020). Applications of machine learning methods for engineering risk assessment – A review. *Safety Science, 122*(May 2019), 104492. doi:10.1016/j.ssci.2019.09.015

Heinlein, S., Cawley, P., & Vogt, T. (2019). Validation of a procedure for the evaluation of the performance of an installed structural health monitoring system. *Structural Health Monitoring*, *18*(5–6), 1557–1568. doi:10.1177/1475921718798567

He, K., Zhang, X., Ren, S., & Sun, J. (2016). Deep residual learning for image recognition. *Proceedings of the IEEE Computer Society Conference on Computer Vision and Pattern Recognition, 2016-Decem*, (pp. 770–778). IEEE. 10.1109/CVPR.2016.90

Hill, M. (2009). *The business value of BIM: getting to the bottom line*. Opgehaald van BIM Construction. http://www.bim. construction. com/research

Ho, J. C. M., & Pam, H. J. (2003). Inelastic design of low-axially loaded high-strength reinforced concrete columns. *Engineering Structures*, *25*(8), 1083–1096. doi:10.1016/S0141-0296(03)00050-6

Holland, J. H. (1992). Genetic algorithms. *Scientific American*, *267*(1), 66–73. doi:10.1038cientificamerican0792-66

Hotelling, H. (1933). Analysis of a complex of statistical variables into principal components. *Journal of Educational Psychology*, *24*(6), 417–441. doi:10.1037/h0071325

Hou, L., Chen, H., Zhang, G., & Wang, X. (2021). Deep Learning-Based Applications for Safety Management in the AEC Industry: A Review. *Applied Sciences (Basel, Switzerland)*, *11*(2), 821. doi:10.3390/app11020821

Hsu, H.-C., Chang, S., Chen, C.-C., & Wu, I.-C. (2020). Knowledge-based system for resolving design clashes in building information models. *Automation in Construction*, *110*, 103001. doi:10.1016/j.autcon.2019.103001

Hsu, T. T. C. (1988). Softened truss model theory for shear and torsion. *ACI Structural Journal*, *85*(6), 624–635. doi:10.14359/2740

Huang, L., Duan, S. W., Zhu, K. M., & Chen, W. G. (2016). Study on Deformation Monitoring of Subway Station Deep Foundation Construction. *Applied Mechanics and Materials*, *847*, 425–430. . doi:10.4028/www.scientific.net/AMM.847.425

Huang, X., Wu, L., & Ye, Y. (2019). A Review on Dimensionality Reduction Techniques. *International Journal of Pattern Recognition and Artificial Intelligence*, *33*(10), 1950017. doi:10.1142/S0218001419500174

Hu, J., Shen, L., Albanie, S., Sun, G., & Wu, E. (2020). Squeeze-and-Excitation Networks. *IEEE Transactions on Pattern Analysis and Machine Intelligence*, *42*(8), 2011–2023. doi:10.1109/TPAMI.2019.2913372 PMID:31034408

Hutchinson, T. C., & Chen, Z. (2006). Improved Image Analysis for Evaluating Concrete Damage. *Journal of Computing in Civil Engineering*, *20*(3), 210–216. doi:10.1061/(ASCE)0887-3801(2006)20:3(210)

Hu, Y., & Castro-Lacouture, D. (2019). Clash relevance prediction based on machine learning. *Journal of Computing in Civil Engineering*, *33*(2), 4018060. doi:10.1061/(ASCE)CP.1943-5487.0000810

Huyan, J., Li, W., Tighe, S., Xu, Z., & Zhai, J. (2020). CrackU-net: A novel deep convolutional neural network for pixelwise pavement crack detection. *Structural Control and Health Monitoring*, *27*(8), 1–19. doi:10.1002tc.2551

Hwang, S. J., & Lee, H. J. (1999). Analytical model for predicting the shear strength of exterior reinforced concrete beam-column joints. *ACI Structural Journal*, *96*, 846–858.

IBM Cloud Education. (2020). *What is Artificial Intelligence (AI)?* IBM Cloud Learn Hub. https://www.ibm.com/cloud/learn/what-is-artificial-intelligence

Jabbari, E., Fathi, M., & Moradi, M. (2020). Modeling groundwater quality and quantity to manage water resources in the Arak aquifer, Iran. *Arabian Journal of Geosciences*, *13*(14), 663. doi:10.100712517-020-05681-4

Jacqueline, G. (2000). *The future for precast concrete in low-rise housing*. British Precast Concrete Federation.

Jafari, T., Kiem, A. S., Javadi, S., Nakamura, T., & Nishida, K. (2021). Fully integrated numerical simulation of surface water-groundwater interactions using SWAT-MODFLOW with an improved calibration tool. *Journal of Hydrology. Regional Studies*, *35*, 100822. doi:10.1016/j.ejrh.2021.100822

Jaillon, L., Poon, C. S., & Chiang, Y. H. (2009). Quantifying the waste reduction potential of using prefabrication in building construction in Hong Kong. *Waste Management (New York, N.Y.)*, *29*(1), 309–320. doi:10.1016/j.wasman.2008.02.015 PMID:18434128

Jáuregui, D. V., White, K. R., Woodward, C. B., & Leitch, K. R. (2003). Noncontact Photogrammetric Measurement of Vertical Bridge Deflection. *Journal of Bridge Engineering*, *8*(4), 212–222. doi:10.1061/(ASCE)1084-0702(2003)8:4(212)

Jeong, J., & Park, E. (2019). Comparative applications of data-driven models representing water table fluctuations. *Journal of Hydrology (Amsterdam)*, *572*, 261–273. doi:10.1016/j.jhydrol.2019.02.051

Jeon, J. S., Lowes, L. N., DesRoches, R., & Brilakis, I. (2015). Fragility curves for non-ductile reinforced concrete frames that exhibit different component response mechanisms. *Engineering Structures*, *85*, 127–143. doi:10.1016/j.engstruct.2014.12.009

Jiang, W., Hong, Y., Zhou, B., He, X., & Cheng, C. (2019). A GAN-Based Anomaly Detection Approach for Imbalanced Industrial Time Series. *IEEE Access : Practical Innovations, Open Solutions*, *7*, 143608–143619. doi:10.1109/ACCESS.2019.2944689

Joseph, A. D., Nelson, B., Rubinstein, B. I. P., & Tygar, J. D. (2019). *Adversarial Machine Learning*. Cambridge University Press. doi:10.1017/9781107338548

Kamali, A., & Niksokhan, M. H. (2017). Multi-objective optimization for sustainable groundwater management by developing of coupled quantity-quality simulation-optimization model. *Journal of Hydroinformatics*, *19*(6), 973–992. doi:10.2166/hydro.2017.007

Kandiri, A., Sartipi, F., & Kioumarsi, M. (2021). Predicting Compressive Strength of Concrete Containing Recycled Aggregate Using Modified ANN with Different Optimization Algorithms. *Applied Sciences (Basel, Switzerland)*, *11*(2), 485. doi:10.3390/app11020485

Kangari, R., & Halpin, D. W. (1990). Identification of factors influencing implementation of construction robotics. *Construction Management and Economics*, *8*(1), 89–104. doi:10.1080/01446199000000008

Karamodin, A., Kazemi, H. H., & Akbarzadeh, M. R. (2008). *Damage Index Monitoring of Structures Using Neural Networks*. The 14 World Conference on Earthquake Engineering, Beijing, China.

Karan, E., & Asadi, S. (2019). Intelligent designer: A computational approach to automating design of windows in buildings. *Automation in Construction*, *102*, 160–169. doi:10.1016/j.autcon.2019.02.019

Karatas, I., & Budak, A. (2022). Development and comparative of a new meta-ensemble machine learning model in predicting construction labor productivity. *Engineering, Construction and Architectural Management*. doi:10.1108/ECAM-08-2021-0692

Karmakar, A., & Delhi, V. S. K. (2021). Construction 4.0: What we know and where we are headed? *Journal of Information Technology in Construction*, *26*, 526–545. doi:10.36680/j.itcon.2021.028

Kaveh, A., Talatahari, S., & Khodadadi, N. (2020). Stochastic paint optimizer: Theory and application in civil engineering. *Engineering with Computers*, 1–32.

Kennedy, J., & Eberhart, R. (1995). Particle swarm optimization *IEEE International Conference on Neural Networks*, Piscataway, NJ, USA.

Kennedy, J., & Eberhart, R. (1995). *Particle swarm optimization. InProceedings of ICNN'95-international conference on neural networks* (Vol. 4). IEEE.

Kerschen, G., De Boe, P., Golinval, J. C., & Worden, K. (2005). Sensor validation using principal component analysis. *Smart Materials and Structures*, *14*(1), 36–42. doi:10.1088/0964-1726/14/1/004

Keshtegar, B., Nehdi, M. L., Kolahchi, R., Trung, N.-T., & Bagheri, M. (2022). Novel hybrid machine leaning model for predicting shear strength of reinforced concrete shear walls. *Engineering with Computers*, *38*(S5), 3915–3926. doi:10.100700366-021-01302-0

Khallaf, R., & Khallaf, M. (2021). Classification and analysis of deep learning applications in construction: A systematic literature review. *Automation in Construction*, *129*, 103760. doi:10.1016/j.autcon.2021.103760

Khedri, A., Kalantari, N., & Vadiati, M. (2020). Comparison study of artificial intelligence method for short term groundwater level prediction in the northeast Gachsaran unconfined aquifer. *Water Science and Technology: Water Supply*, *20*(3), 909–921. doi:10.2166/ws.2020.015

Khodadadi, N., Mirjalili, S. M., Mirjalili, S. Z., & Mirjalili, S. (2022, September). Chaotic Stochastic Paint Optimizer (CSPO). In *Proceedings of 7th International Conference on Harmony Search, Soft Computing and Applications: ICHSA 2022* (pp. 195-205). Springer Nature Singapore. 10.1007/978-981-19-2948-9_19

Khodadadi, N., Abualigah, L., & Mirjalili, S. (2022). Multi-objective stochastic paint optimizer (MOSPO). *Neural Computing & Applications*, *34*(20), 18035–18058. doi:10.100700521-022-07405-z

Kim, J., & LaFave, J.M. (2007). Key influence parameters for the joint shear behavior of reinforced concrete (RC) beam-column connections. *Eng Struct, 29*, 2523–39. https://doi.org/ruct.2006.12.012 doi:10.1016/j.engst

Kim, M. K., Thedja, J. P. P., & Wang, Q. (2020). Automated dimensional quality assessment for formwork and rebar of reinforced concrete components using 3D point cloud data. *Automation in Construction, 112*(August 2019). doi:10.1016/j.autcon.2020.103077

Kim, J., & LaFave, J. M. (2012). A simplified approach to joint shear behavior prediction of RC beam-column connections. *Earthquake Spectra*, *28*(3), 1071–1096. doi:10.1193/1.4000064

Kim, Y., Kim, H., Murphy, R., Lee, S., & Ahn, C. R. (2022). Delegation or Collaboration: Understanding Different Construction Stakeholders' Perceptions of Robotization. *Journal of Management Engineering*, *38*(1), 04021084. doi:10.1061/(ASCE)ME.1943-5479.0000994

Kiranyaz, S., Avci, O., Abdeljaber, O., Ince, T., Gabbouj, M., & Inman, D. J. (2021). 1D convolutional neural networks and applications: A survey. *Mechanical Systems and Signal Processing*, *151*, 107398. doi:10.1016/j.ymssp.2020.107398

Koc, K., Ekmekcioğlu, Ö., & Gurgun, A. P. (2021). Integrating feature engineering, genetic algorithm and tree-based machine learning methods to predict the post-accident disability status of construction workers. *Automation in Construction*, *131*, 103896. doi:10.1016/j.autcon.2021.103896

Koc, K., Ekmekcioğlu, Ö., & Gurgun, A. P. (2022). Accident prediction in construction using hybrid wavelet-machine learning. *Automation in Construction*, *133*, 103987. doi:10.1016/j.autcon.2021.103987

Kohiyama, M., Oka, K., & Yamashita, T. (2020). Detection method of unlearned pattern using support vector machine in damage classification based on deep neural network. *Structural Control and Health Monitoring*, *27*(8), e2552. doi:10.1002tc.2552

Kolaei, A. Z., Hedayati, E., Khanzadi, M., & Amiri, G. G. (2022). Challenges and opportunities of augmented reality during the construction phase. *Automation in Construction*, *143*, 104586. doi:10.1016/j.autcon.2022.104586

Konanahalli, A., Marinelli, M., & Oyedele, L. (2022). Drivers and challenges associated with the implementation of Big Data within U.K. Facilities Management sector: An Exploratory Factor Analysis approach. *IEEE Transactions on Engineering Management*, *69*(4), 916–929. doi:10.1109/TEM.2019.2959914

Kor, M., Yitmen, I., & Alizadehsalehi, S. (2022). *An investigation for integration of deep learning and digital twins towards Construction 4.0*. Smart and Sustainable Built Environment.

Kotsovos, G., Cotsovos, D., Kotsovos, M., & Kounadis, A. (2011). Seismic behaviour of RC walls: An attempt to reduce reinforcement congestion. *Magazine of Concrete Research*, *63*(4), 235–246. doi:10.1680/macr.10.00001

Kotsovos, M. D. (2012). Structural-concrete analysis and design: Need for a sound underlying theory. *Archive of Applied Mechanics*, *82*(10-11), 1439–1459. doi:10.100700419-012-0676-7

Kouhestani, S., & Nik-Bakht, M. (2020). IFC-based process mining for design authoring. *Automation in Construction*, *112*, 103069. doi:10.1016/j.autcon.2019.103069

Koza, J. R. (1994). Genetic programming as a means for programming computers by natural selection. *Statistics and Computing*, *4*(2), 87–112. doi:10.1007/BF00175355

Krishan, G. L. A. (2015). Groundwater Level Simulation Using Artificial Neural Network in Southeast, Punjab, India. *Journal of Geology & Geosciences*. doi:10.4172/2329-6755.1000206

Krogh, A., & Vedelsby, J. (1995). Neural Network Ensembles, Cross Validation, and Active Learning. *Advances in Neural Information Processing Systems*, *7*, 21–238.

Ku, K., & Mills, T. (2010). Research needs for building information modeling for construction safety. *International Proceedings of Associated Schools of Construction 45nd Annual Conference, Boston, MA*. Associated Schools of Construction.

Kulisz, M., Kujawska, J., Przysucha, B., & Cel, W. (2021). Forecasting water quality index in groundwater using artificial neural network. *Energies*, *14*(18), 5875. doi:10.3390/en14185875

Kumar, A., & Goyal, P. (2013). Forecasting of Air Quality Index in Delhi Using Neural Network Based on Principal Component Analysis. *Pure and Applied Geophysics*, *170*(4), 711–722. doi:10.100700024-012-0583-4

Kumar, P. M., Rawal, B., & Gao, J. (2022). Blockchain-enabled Privacy Preserving of IoT Data for Sustainable Smart Cities using Machine Learning. In *14th International Conference on COMmunication Systems and NETworkS, COMSNETS 2022* (pp. 1-6). IEEE. 10.1109/COMSNETS53615.2022.9668530

Kuyu, Y. Ç., & Vatansever, F. (2022). GOZDE: A novel metaheuristic algorithm for global optimization. *Future Generation Computer Systems*, *136*, 128–152. doi:10.1016/j.future.2022.05.022

Labonnote, N., Rønnquist, A., Manum, B., & Rüther, P. (2016). Additive construction: State- of-the-art, challenges and opportunities. *Automation in Construction*, *72*, 347–366. doi:10.1016/j.autcon.2016.08.026

Lagaros, N. D., & Papadrakakis, M. (2004). Learning improvement of neural networks used in structural optimization. *Advances in Engineering Software*, *35*(1), 9–25. doi:10.1016/S0965-9978(03)00112-1

Lagomarsino, S. (1993). Forecast models for damping and vibration periods of buildings. *Journal of Wind Engineering and Industrial Aerodynamics*, *48*(2), 221–239. doi:10.1016/0167-6105(93)90138-E

Langdon, W. B., Poli, R., McPhee, N. F., & Koza, J. R. (2008). Genetic programming: An introduction and tutorial, with a survey of techniques and applications. *Computational intelligence. Compendium (Newtown, Pa.)*, 927–1028.

Laory, I., Trinh, T. N., Smith, I. F. C., & Brownjohn, J. M. W. (2014). Methodologies for predicting natural frequency variation of a suspension bridge. *Engineering Structures*, *80*, 211–221. doi:10.1016/j.engstruct.2014.09.001

Laterza, M., D'Amato, M., & Gigliotti, R. (2017). Modeling of gravity-designed RC sub-assemblages subjected to lateral loads. *Eng Struct J, 130*, 242–60. doi:10.1016/j.engstruct.2016.10.044

LeCun, Y., Boser, B., Denker, J. S., Henderson, D., Howard, R. E., Hubbard, W., & Jackel, L. D. (1989). Backpropagation applied to digit recognition. *Neural Computation, 1*(4), 541–551. doi:10.1162/neco.1989.1.4.541

Lee, L.-H., Chang, K.-K., & Chun, Y.-S. (2000). Experimental formula for the fundamental period of RC buildings with shear-wall dominant systems. *The Structural Design of Tall Buildings, 9*(4), 295-307.

Lefas, I. D., & Kotsovos, M. D. (1990). Strength and deformation characteristics of reinforced concrete walls under load reversals. *Structural Journal, 87*(6), 716–726.

Liang, C. J., Wang, X., Kamat, V. R., & Menassa, C. C. (2021). Human-Robot Collaboration in Construction: Classification and Research Trends. *Journal of Construction Engineering and Management, 147*(10), 03121006. doi:10.1061/(ASCE)CO.1943-7862.0002154

Liao, F.-Y., Han, L.-H., & Tao, Z. (2009). Seismic behaviour of circular CFST columns and RC shear wall mixed structures: Experiments. *Journal of Constructional Steel Research, 65*(8-9), 1582–1596. doi:10.1016/j.jcsr.2009.04.023

Liao, T. W., Egbelu, P. J., Sarker, B. R., & Leu, S. S. (2011). Metaheuristics for project and construction management - A state-of-the-art review. *Automation in Construction, 20*(5), 491–505. doi:10.1016/j.autcon.2010.12.006

Li, D., Han, D., Weng, T.-H., Zheng, Z., Li, H., Liu, H., Castiglione, A., & Li, K.-C. (2022). Blockchain for federated learning toward secure distributed machine learning systems: A systemic survey. *Soft Computing, 26*(9), 4423–4440. doi:10.100700500-021-06496-5 PMID:34840525

Li, H., & Ou, J. (2016). The state of the art in structural health monitoring of cable-stayed bridges. *Journal of Civil Structural Health Monitoring, 6*(1), 43–67. doi:10.100713349-015-0115-x

Li, J., & Heap, A. D. (2008). *A Review of Spatial Interpolation Methods for Environmental Scientists.* G. Australia.

Li, J., Lu, W., & Fan, Y. (2021). Groundwater Pollution Sources Identification Based on Hybrid Homotopy-Genetic Algorithm and Simulation Optimization. *Environmental Engineering Science, 38*(8), 777–788. doi:10.1089/ees.2020.0117

Lin, F., Chen, X., & Yao, H. (2017). Evaluating the use of Nash-Sutcliffe efficiency coefficient in goodness-of-fit measures for daily runoff simulation with SWAT. *Journal of Hydrologic Engineering, 22*(11), 5017023. doi:10.1061/(ASCE)HE.1943-5584.0001580

Lin, T.-Y., Dollar, P., Girshick, R., He, K., Hariharan, B., & Belongie, S. (2017). Feature pyramid networks for object detection. *Proc. IEEE Conf. Comput. Vis. Pattern Recognit*, (pp. 936–944). IEEE. 10.1109/ISPA-BDCloud-SustainCom-SocialCom48970.2019.00217

Lin, W. Y., & Huang, Y.-H. (2019). Filtering of irrelevant clashes detected by BIM software using a hybrid method of rule-based reasoning and supervised machine learning. *Applied Sciences (Basel, Switzerland), 9*(24), 5324. doi:10.3390/app9245324

Lin, Y. Z., Nie, Z. H., & Ma, H. W. (2017). Structural Damage Detection with Automatic Feature-Extraction through Deep Learning. *Computer-Aided Civil and Infrastructure Engineering, 32*(12), 1025–1046. doi:10.1111/mice.12313

Li, S., Laima, S., & Li, H. (2018). Data-driven modeling of vortex-induced vibration of a long-span suspension bridge using decision tree learning and support vector regression. *Journal of Wind Engineering and Industrial Aerodynamics, 172*, 196–211. doi:10.1016/j.jweia.2017.10.022

Li, S., Zuo, X., Li, Z., & Wang, H. (2020). Applying Deep Learning to Continuous Bridge Deflection Detected by Fiber Optic Gyroscope for Damage Detection. *Sensors (Basel)*, *20*(3), 911. doi:10.339020030911 PMID:32046314

Li, T., Pan, Y., Tong, K., Ventura, C. E., & de Silva, C. W. (2022). Attention-Based Sequence-to-Sequence Learning for Online Structural Response Forecasting Under Seismic Excitation. *IEEE Transactions on Systems, Man, and Cybernetics. Systems*, *52*(4), 2184–2200. doi:10.1109/TSMC.2020.3048696

Liu, G., Niu, Y., Zhao, W., Duan, Y., & Shu, J. (2022). Data anomaly detection for structural health monitoring using a combination network of GANomaly and CNN. *Smart Structures and Systems*, *39*, 195–206. doi:10.12989ss.2022.29.1.053

Liu, J., Liu, P., Feng, L., Wu, W., Li, D., & Chen, F. (2020). Towards automated clash resolution of reinforcing steel design in reinforced concrete frames via Q-learning and building information modeling. *Automation in Construction*, *112*, 103062. doi:10.1016/j.autcon.2019.103062

Liu, J., Luo, H., & Liu, H. (2022). Deep learning-based data analytics for safety in construction. *Automation in Construction*, *140*, 104302. doi:10.1016/j.autcon.2022.104302

Liu, Y., Yao, J., Lu, X., Xie, R., & Li, L. (2019). DeepCrack: A deep hierarchical feature learning architecture for crack segmentation. *Neurocomputing*, *338*, 139–153. doi:10.1016/j.neucom.2019.01.036

Liu, Y., & Zhang, S. (2017). Probabilistic Baseline of Finite Element Model of Bridges under Environmental Temperature Changes. *Computer-Aided Civil and Infrastructure Engineering*, *32*(7), 581–598. doi:10.1111/mice.12268

Liu, Z., Cao, Y., Wang, Y., & Wang, W. (2019). Computer vision-based concrete crack detection using U-net fully convolutional networks. *Automation in Construction*, *104*(January), 129–139. doi:10.1016/j.autcon.2019.04.005

Li, X., Yi, W., Chi, H.-L., Wang, X., & Chan, A. P. C. (2018). A critical review of virtual and augmented reality (VR/AR) applications in construction safety. *Automation in Construction*, *86*, 150–162. doi:10.1016/j.autcon.2017.11.003

Lloyd, J. W. (2007). The difficulties of regional groundwater resources assessments in arid areas. *Arabian Journal for Science and Engineering*, *32*(1C), 35.

Lomio, F., Farinha, R., Laasonen, M., & Huttunen, H. (2018). Classification of building information model (BIM) structures with deep learning. *2018 7th European Workshop on Visual Information Processing (EUVIP)*, (pp. 1–6). Semantic Scholar.

Lowes, L. N., & Altoontash, A. (2003). Modeling reinforced-concrete beam-column joints subjected to cyclic loading. *Journal of Structural Engineering*, *129*(12), 1686–1697. doi:10.1061/(ASCE)0733-9445(2003)129:12(1686)

Lu, Q., Zhu, J., & Zhang, W. (2020). Quantification of Fatigue Damage for Structural Details in Slender Coastal Bridges Using Machine Learning-Based Methods. *Journal of Bridge Engineering*, *25*(7), 04020033. doi:10.1061/(ASCE)BE.1943-5592.0001571

Lu, R., & Brilakis, I. (2019). Digital twinning of existing reinforced concrete bridges from labelled point clusters. *Automation in Construction*, *105*(May), 102837. doi:10.1016/j.autcon.2019.102837

Lu, R., Brilakis, I., & Middleton, C. R. (2019). Detection of Structural Components in Point Clouds of Existing RC Bridges. *Computer-Aided Civil and Infrastructure Engineering*, *34*(3), 191–212. doi:10.1111/mice.12407

Lu, Y., Huang, X., Dai, Y., Maharjan, S., & Zhang, Y. (2020a). Blockchain and Federated Learning for Privacy-Preserved Data Sharing in Industrial IoT. *IEEE Transactions on Industrial Informatics*, *16*(6), 4177–4186. doi:10.1109/TII.2019.2942190

Lu, Y., Huang, X., Zhang, K., Maharjan, S., & Zhang, Y. (2020b). Blockchain Empowered Asynchronous Federated Learning for Secure Data Sharing in Internet of Vehicles. *IEEE Transactions on Vehicular Technology*, *69*(4), 4298–4311. doi:10.1109/TVT.2020.2973651

Lynn, A. C., Moehle, J. P., Mahin, S. A., & Holmes, W. T. (1996). Seismic evaluation of existing reinforced concrete building columns. *Earthquake Spectra*, *12*(4), 715–739. doi:10.1193/1.1585907

Lyons, R. T., Peralta, R. C., & Majumder, P. (2020). Comparing single-objective optimization protocols for calibrating the birds nest aquifer model—A problem having multiple local optima. *International Journal of Environmental Research and Public Health*, *17*(3), 853. doi:10.3390/ijerph17030853 PMID:32019060

Mafi, S., AbdelRazig, Y., & Doczy, R. (2018). Machine Learning Methods to Analyze Injury Severity of Drivers from Different Age and Gender Groups. *Transportation Research Record: Journal of the Transportation Research Board*, *2672*(38), 171–183. doi:10.1177/0361198118794292

Mai, S. H., Ben Seghier, M. E. A., Nguyen, P. L., Jafari-Asl, J., & Thai, D. K. (2022). A hybrid model for predicting the axial compression capacity of square concrete-filled steel tubular columns. *Engineering with Computers*, *38*(2), 1–18. doi:10.100700366-020-01104-w

Maisonobe, L. (2007). *Finding the circle that best fits a set of points*. SpaceRoots. https://www.spaceroots.org/documents/circle/circle-fitting.pdf

Majumder, P., & Eldho, T. I. (2020). Artificial Neural Network and Grey Wolf Optimizer Based Surrogate Simulation-Optimization Model for Groundwater Remediation. *Water Resources Management*, *34*(2), 763–783. doi:10.100711269-019-02472-9

Malekinezhad, H., & Banadkooki, F. B. (2018). Modeling impacts of climate change and human activities on groundwater resources using modflow. *Journal of Water and Climate Change*, *9*(1), 156–177. doi:10.2166/wcc.2017.147

Malekjafarian, A., Golpayegani, F., Moloney, C., & Clarke, S. (2019). A Machine Learning Approach to Bridge-Damage Detection Using Responses Measured on a Passing Vehicle. *Sensors (Basel)*, *19*(18), 4035. doi:10.339019184035 PMID:31546759

Malik, N. (2005). *Artificial neural networks and their applications*. arXiv preprint cs/0505019.

Mallardo, V., & Aliabadi, M. H. (2014). Optimal sensor placement for structural, damage and impact identification: A review. *SDHM Structural Durability and Health Monitoring*, *9*(4), 287–323. doi:10.32604dhm.2013.009.287

Mallick, J., Singh, C. K., Almesfer, M. K., Singh, V. P., & Alsubih, M. (2021). Groundwater quality studies in the kingdom of saudi arabia: Prevalent research and management dimensions. In Water (Switzerland). doi:10.3390/w13091266

Mamo, S., Birhanu, B., Ayenew, T., & Taye, G. (2021). Three-dimensional groundwater flow modeling to assess the impacts of the increase in abstraction and recharge reduction on the groundwater, groundwater availability and groundwater-surface waters interaction: A case of the rib catchment in the Lake Tana s. *Journal of Hydrology. Regional Studies*, *35*, 100831. doi:10.1016/j.ejrh.2021.100831

Mangalathu, S., Jang, H., Hwang, S.-H., & Jeon, J.-S. (2020). Data-driven machine-learning-based seismic failure mode identification of reinforced concrete shear walls. *Engineering Structures*, *208*, 110331. doi:10.1016/j.engstruct.2020.110331

Mansour, M. Y., Dicleli, M. U. R. A. T., Lee, J. Y., & Zhang, J. J. E. S. (2004). Predicting the shear strength of reinforced concrete beams using artificial neural networks. *Engineering Structures*, *26*(6), 781–799. doi:10.1016/j.engstruct.2004.01.011

Mantha, B. R. K., & García de Soto, B. (2021). Assessment of the cybersecurity vulnerability of construction networks. *Engineering, Construction, and Architectural Management, 28*(10), 3078–3105. doi:10.1108/ECAM-06-2020-0400

Marinaki, M., Marinakis, Y., & Stavroulakis, G. E. (2011). Fuzzy control optimized by a multi-objective particle swarm optimization algorithm for vibration suppression of smart structures. *Structural and Multidisciplinary Optimization, 43*(1), 29–42. doi:10.100700158-010-0552-4

Marinaki, M., Marinakis, Y., & Stavroulakis, G. E. (2015). Fuzzy control optimized by a multi-objective differential evolution algorithm for vibration suppression of smart structures. *Computers & Structures, 147*, 126–137. doi:10.1016/j.compstruc.2014.09.018

Mariniello, G., Pastore, T., Menna, C., Festa, P., & Asprone, D. (2021). Structural damage detection and localization using decision tree ensemble and vibration data. *Computer-Aided Civil and Infrastructure Engineering, 36*(9), 1129–1149. doi:10.1111/mice.12633

Markou, G., & AlHamaydeh, M. (2018). 3D finite element modelling of GFRP-reinforced concrete deep beams without shear reinforcement. *International Journal of Computational Methods, 15*(02), 1850001. doi:10.1142/S0219876218500019

Markou, G., & Bakas, N. P. (2021a). Prediction of the Shear Capacity of Reinforced Concrete Slender Beams without Stirrups by Applying Artificial Intelligence Algorithms in a Big Database of Beams Generated by 3D Nonlinear Finite Element Analysis. *Computers and Concrete, 28*(6), 433–447.

Markou, G., & Bakas, P. N. (2019). AI reinforced concrete: Algorithms predict concrete strength without training on experimental data. *Innovate (North Miami Beach, Fla.), 14*, 8–10.

Markou, G., & Bakas, P. N. (2021b). Developing reinforced concrete structures through AI algorithms and large-scale modelling. *Innovate (North Miami Beach, Fla.), 16*, 38–40.

Markou, G., Garcia, R., Mourlas, C., Guadagnini, M., Pilakoutas, K., & Papadrakakis, M. (2021). A new damage factor for seismic assessment of deficient bare and FRP-retrofitted RC structures. *Engineering Structures, 248*, 113152. doi:10.1016/j.engstruct.2021.113152

Markou, G., Mourlas, C., Bark, H., & Papadrakakis, M. (2018). Simplified HYMOD non-linear simulations of a full-scale multistory retrofitted RC structure that undergoes multiple cyclic excitations–An infill RC wall retrofitting study. *Engineering Structures, 176*, 892–916. doi:10.1016/j.engstruct.2018.08.002

Markou, G., Mourlas, C., & Papadrakakis, M. (2017). Cyclic nonlinear analysis of large-scale finite element meshes through the use of hybrid modelling (HYMOD). *International Journal of Mechanics, 11*, 218–225.

Markou, G., & Papadrakakis, M. (2013). Computationally efficient 3D finite element modelling of RC structures. *Computers and Concrete, 12*(4), 443–498. doi:10.12989/cac.2013.12.4.443

Markou, G., & Papadrakakis, M. (2015). A simplified and efficient hybrid finite element model (HYMOD) for non-linear 3D simulation of RC structures. *Engineering Computations, 32*(5), 1477–1524. doi:10.1108/EC-11-2013-0269

Mattas, C., Dimitraki, L., Georgiou, P., & Venetsanou, P. (2021). Use of factor analysis (Fa), artificial neural networks (anns) and multiple linear regression (mlr) for electrical conductivity prediction in aquifers in the gallikos river basin, Northern Greece. *Hydrology, 8*(3), 127. doi:10.3390/hydrology8030127

Ma, Y., Chowdhury, M., Sadek, A., & Jeihani, M. (2009). Real-time highway traffic condition assessment framework using vehicleInfrastructure integration (VII) with artificial intelligence (AI). *IEEE Transactions on Intelligent Transportation Systems, 10*(4), 615–627. doi:10.1109/TITS.2009.2026673

Mazhelis, O. (2006). One-class classifiers: A review and analysis of suitability in the context of mobile-masquerader detection. *South African Computer Journal = Suid-Afrikaanse Rekenaartydskrif, 2006*(36), 29–48. doi:10.46298/arima.1877

Mazzoni, S., Moehle, J. P., & Thewalt, C. R. (1991). *Cyclic response of RC beam-column roof corner joints test and retrofit.* Rep. No. UCB/EERC-91/14. Research Gate.

McCarthy, J. (2007). *What is artificial intelligence?* McKinsey & Company.

McConnell, S. W., & Wallace, J. W. (1995). *Behavior of Reinforced Concrete Beam-Column Roof corner-Joints Subjected to Reversed Cyclic Loading.* Rep. No. CU/CEE-95/07. NIH.

McCuen, R. H., Knight, Z., & Cutter, A. G. (2006). Evaluation of the Nash–Sutcliffe efficiency index. *Journal of Hydrologic Engineering, 11*(6), 597–602. doi:10.1061/(ASCE)1084-0699(2006)11:6(597)

Meggeta, L. M. (2003). The Seismic Design and Performance of Reinforced Concrete Beam-Column Roof corner Joints in Buildings. *Earthquake Spectra, 19*(4), 863–895. doi:10.1193/1.1623782

Megget, L. M. (1998). The seismic behavior of small reinforced concrete beam-column roof corner joints. *Bull New Zeal Natl Soc Earthq Eng, 31*(4), 215–245. doi:10.5459/bnzsee.31.4.215-245

Mezouary, L. El, & Mansouri, B. El. (2021). *Groundwater flow equation, overview, derivation, and solution. i*, 2–5.

Mishra, M. (2021). Machine learning techniques for structural health monitoring of heritage buildings: A state-of-the-art review and case studies. *Journal of Cultural Heritage, 47*, 227–245. doi:10.1016/j.culher.2020.09.005

Mitra, N., & Lowes, L. N. (2007). Evaluation, calibration, and verification of a reinforced concrete beam-column joint model. *Journal of Structural Engineering, 133*(1), 105–120. doi:10.1061/(ASCE)0733-9445(2007)133:1(105)

Mogili, S., & Kuang, J.S. (2019). Reversed cyclic performance of reinforced concrete roof corner joints under variable closing and opening stresses. *Eng Struct, 178,* 116–27. https://doi.org/ struct.2018.10.025 doi:10.1016/j.eng

Mogili, S., & Kuang, J. S. (2018). Cyclic Vulnerability of RC Roof corner Connections Under Relatively High Opening Shear Stresses. In *16th Eur Conf Earthq Eng.* NICEE.

Mogili, S., Kuang, J. S., & Hwang, S. J. (2020). Predicting Shear Strength of Reinforced Concrete Roof corner Joints in Closing and Opening Actions. *Journal of Structural Engineering, 146*(6), 1–11. doi:10.1061/(ASCE)ST.1943-541X.0002633

Mohanty, S., Jha, M. K., Kumar, A., & Panda, D. K. (2013). Comparative evaluation of numerical model and artificial neural network for simulating groundwater flow in Kathajodi-Surua Inter-basin of Odisha, India. *Journal of Hydrology (Amsterdam), 495*, 38–51. doi:10.1016/j.jhydrol.2013.04.041

Moretti, M. L., Tassios, T. P., & Vintzileou, E. (2014). Behavior and design of corner joints under opening bending moment. *ACI Structural Journal, 111*(1), 3–13. doi:10.14359/51686514

Moretti, M., & Tassios, T. P. (2007). Behavior of short columns subjected to cyclic shear displacements: Experimental results. *Engineering Structures, 29*(8), 2018–2029. doi:10.1016/j.engstruct.2006.11.001

Mostofi, F., Toğan, V., Ayözen, Y. E., & Tokdemir, O. B. (2022a). Predicting the Impact of Construction Rework Cost Using an Ensemble Classifier. *Sustainability (Basel), 14*(22), 14800. doi:10.3390u142214800

Mostofi, F., Toğan, V., Ayözen, Y. E., & Tokdemir, O. B. (2022b). Construction Safety Risk Model with Construction Accident Network: A Graph Convolutional Network Approach. *Sustainability (Basel), 14*(23), 15906. doi:10.3390u142315906

Mostofi, F., Toğan, V., & Başağa, H. B. (2021). House price prediction: A data-centric aspect approach on performance of combined principal component analysis with deep neural network model. *Journal of Construction Engineering. Management & Innovation*, *4*(2), 106–116. doi:10.31462/jcemi.2021.02106116

Mostofi, F., Toğan, V., & Başağa, H. B. (2022c). Real-estate price prediction with deep neural network and principal component analysis. *Organization, Technology and Management in Construction. International Journal (Toronto, Ont.)*, *14*(1), 2741–2759. doi:10.2478/otmcj-2022-0016

Mourlas, C., Gravett, D. Z., Markou, G., & Papadrakakis, M. (2019a). *Investigation of the soil structure interaction effect on the dynamic behavior of multistorey RC buildings. [Paper presentation].* VIII international conference on computational methods for coupled problems in science and engineering (coupled problems), Spain.

Mourlas, C., Khabele, N., Bark, H. A., Karamitros, D., Taddei, F., Markou, G., & Papadrakakis, M. (2020). Effect of Soil–Structure Interaction on Nonlinear Dynamic Response of Reinforced Concrete Structures. *International Journal of Structural Stability and Dynamics*, *20*(13), 2041013. doi:10.1142/S0219455420410138

Mourlas, C., Markou, G., & Papadrakakis, M. (2019b). Accurate and computationally efficient nonlinear static and dynamic analysis of reinforced concrete structures considering damage factors. *Engineering Structures*, *178*, 258–285. doi:10.1016/j.engstruct.2018.10.034

Mourlas, C., Markou, G., & Papadrakakis, M. (2019c). *3D Detailed Modeling of Reinforced Concrete Frames Considering accumulated damage during static cyclic and dynamic analysis–new validation case studies* [Paper presentation]. COMPDYN. doi:10.7712/120119.7207.19727

Mourlas, C., Papadrakakis, M., & Markou, G. (2017). A computationally efficient model for the cyclic behavior of reinforced concrete structural members. *Engineering Structures*, *141*, 97–125. doi:10.1016/j.engstruct.2017.03.012

Mousavi, M., & Gandomi, A. H. (2021a). Prediction error of Johansen cointegration residuals for structural health monitoring. *Mechanical Systems and Signal Processing*, *160*, 107847. doi:10.1016/j.ymssp.2021.107847

Mousavi, M., & Gandomi, A. H. (2021b). Deep learning for structural health monitoring under environmental and operational variations. *Proceedings of the Society for Photo-Instrumentation Engineers*, 11592. doi:10.1117/12.2582649

Mo, Y. L., & Nien, I. C. (2002). Seismic performance of hollow high-strength concrete bridge columns. *Journal of Bridge Engineering*, *7*(6), 338–349. doi:10.1061/(ASCE)1084-0702(2002)7:6(338)

Murphy, K. P. (2012). *Machine learning: A probabilistic perspective*. MIT Press.

Mutis, I., & Ramachandran, A. (2021). The Bimbot: Mediating technology for enacting coordination in teamwork collaboration. *J. Inf. Technol. Constr.*, *26*, 144–158. doi:10.36680/j.itcon.2021.009

Naderpour, H., & Mirrashid, M. (2020). Bio-inspired predictive models for shear strength of reinforced concrete beams having steel stirrups. *Soft Computing*, *24*(16), 12587–12597. doi:10.100700500-020-04698-x

Naeim, F. (Ed.). (1989). *The seismic design handbook*. Springer Science & Business Media. doi:10.1007/978-1-4615-9753-7

Namitha, M. R., Devi Krishna, J. S., Sreelekshmi, H., & Muhammed Ashik, P. (2019). Ground water flow modelling using visual modflow. *Journal of Pharmacognosy and Phytochemistry*, *8*(1), 2710–2714.

Navidi, W. (2019). *Statistics for Engineers and Scientists* (5th ed.). McGraw-Hill. doi:10.1017/CBO9781107415324.004

*New Greek Seismic Code*. (2000). NEAK.

Newmark, N. M. (1959). A method of computation for structural dynamics. *Journal of the Engineering Mechanics Division*, *85*(3), 67–94. doi:10.1061/JMCEA3.0000098

Ng, H. F. (2006). Automatic thresholding for defect detection. *Pattern Recognition Letters*, *27*(14), 1644–1649. doi:10.1016/j.patrec.2006.03.009

Nguyen, D. C., Ding, M., Pathirana, P. N., Seneviratne, A., Li, J., & Vincent Poor, H. (2021). Federated Learning for Internet of Things: A Comprehensive Survey. *IEEE Communications Surveys and Tutorials*, *23*(3), 1622–1658. doi:10.1109/COMST.2021.3075439

Ni, F. T., Zhang, J., & Chen, Z. Q. (2019). Pixel-level crack delineation in images with convolutional feature fusion. *Structural Control and Health Monitoring*, *26*(1), 1–18. doi:10.1002tc.2286

Ni, F. T., Zhang, J., & Noori, M. N. (2020). Deep learning for data anomaly detection and data compression of a long-span suspension bridge. *Computer-Aided Civil and Infrastructure Engineering*, *35*(7), 685–700. doi:10.1111/mice.12528

Niroomandi, A., Maheri, A., Maheri, M.R., & Mahini, S.S. (2010). Seismic performance of ordinary RC frames retrofitted at joints by FRP sheets. *Eng Struct, 32*, 2326–36. https://doi.org/.engstruct.2010.04.008 doi:10.1016/j

Niroomandi, A., Pampanin, S., & Dhakal, R. (2015). Comparison of alternative assessment procedures to predict seismic performance of RC columns. In *Tenth Pacific Conf. Earthq. Eng. (PCEE 2015),* (p. 1–10). AEES.

Niroomandi, A., Najafgholipour, M. A., & Ronagh, H. R. (2014). Numerical investigation of the affecting parameters on the shear failure of Nonductile RC exterior joints. *Engineering Failure Analysis*, *46*, 62–75. doi:10.1016/j.engfailanal.2014.08.003

Nordin, N. F. C., Mohd, N. S., Koting, S., Ismail, Z., Sherif, M., & El-Shafie, A. (2021). Groundwater quality forecasting modelling using artificial intelligence: A review. *Groundwater for Sustainable Development*, *14*, 100643. doi:10.1016/j.gsd.2021.100643

Nouali, A., & Matallah, M. (2017). A simplified approach to assess the size effect on the shear-flexure interaction in RC elements. *Engineering Structures*, *144*, 151–162. doi:10.1016/j.engstruct.2017.04.057

NRC/IRC. (2015). *National Building Code of Canada 2015*. National Research Council of Canada, Institute for Research in Construction.

NZS 3101, Code of practice for the design of concrete structure (NZS 3101-2006), Standard association Newzealands, 2006.

Oesterreich, T. D., & Teuteberg, F. (2016). Understanding the implications of digitisation and automation in the context of Industry 4.0: A triangulation approach and elements of a research agenda for the construction industry. *Computers in Industry*, *83*, 121–139. doi:10.1016/j.compind.2016.09.006

Ogunbo, J. N., Alagbe, O. A., Oladapo, M. I., & Shin, C. (2020). N-hidden layer artificial neural network architecture computer code: Geophysical application example. *Heliyon*, *6*(6), e04108. doi:10.1016/j.heliyon.2020.e04108 PMID:32566777

Oh, J., Hessel, M., Czarnecki, W. M., Xu, Z., van Hasselt, H., Singh, S., & Silver, D. (2020). Discovering reinforcement learning algorithms. Advances in Neural Information Processing Systems. NeurIPS.

Ohtori, Y., Christenson, R. E., Spencer, B. F. Jr, & Dyke, S. J. (2004). Benchmark control problems for seismically excited nonlinear buildings. *Journal of Engineering Mechanics*, *130*(4), 366–385. doi:10.1061/(ASCE)0733-9399(2004)130:4(366)

Ohtori, Y., & Spencer, B. F. Jr. (1999). A MATLAB-based tool for nonlinear structural analysis. In *Proc. of the 13th Engineering Mechanics Conf* (pp. 13-16). Scribd.

Omidi, M., & Behnamfar, F. (2015). A numerical model for simulation of RC beam-column connections. *Engineering Structures*, *88*, 51–73. doi:10.1016/j.engstruct.2015.01.025

Opoku, D.-G. J., Perera, S., Osei-Kyei, R., & Rashidi, M. (2021). Digital twin application in the construction industry: A literature review. *Journal of Building Engineering*, *40*, 102726. doi:10.1016/j.jobe.2021.102726

Oracle Cloud. (2022). *What is AI?* Oracle Cloud. https://www.oracle.com/artificial-intelligence/what-is-ai/

Osorio, E., Bairán, J. M., & Marí, A. R. (2017). Analytical modeling of reinforced concrete columns subjected to bidirectional shear. *Engineering Structures*, *138*, 458–472. doi:10.1016/j.engstruct.2017.02.029

Ousalem, H., Kabeyasawa, T., Tasai, A., & Ohsugi, Y. (2002). Experimental study on seismic behavior of reinforced concrete columns under constant and variable axial loadings. *Proc Japan Concr Inst*, *24*, 229–234.

Pal, J., & Chakrabarty, D. (2021). Effects of input/output parameters on artificial neural network model efficiency for breakthrough contaminant prediction. *Water Science and Technology: Water Supply*, *21*(7), 3614–3628. doi:10.2166/ws.2021.125

Pampanin, S., Calvi, G. M., & Moratti, M. (2002). Seismic behavior of RC beam-column joints designed for gravity loads. In *12th Eur Conf Earthq Eng.* (vol. 726, pp. 1–10). University of Canterbury.

Pampanin, S., Magenes, G., & Carr, A. (2003). Modelling of shear hinge mechanism in poorly detailed RC beam-column joints. In *Proc. Fib Symp. 2003 Concr. Struct. Seism. Reg.* (pp. 126–7). Research Gate.

Pang, X. B., & Hsu, T. T. C. (1996). Fixed angle softened truss model for reinforced concrete. *ACI Structural Journal*, *93*, 197–207.

Panthi, B., Dahal, P., Shrestha, P., & Thapa, K. B. (2021). Fundamental period of RC buildings with infill walls in Nepal. *Asian Journal of Civil Engineering*, *22*(5), 983–993. doi:10.100742107-021-00359-y

Pan, Y., & Zhang, L. (2020a). BIM log mining: Exploring design productivity characteristics. *Automation in Construction*, *109*, 102997. doi:10.1016/j.autcon.2019.102997

Pan, Y., & Zhang, L. (2020b). BIM log mining: Learning and predicting design commands. *Automation in Construction*, *112*, 103107. doi:10.1016/j.autcon.2020.103107

Pan, Y., & Zhang, L. (2021). Roles of artificial intelligence in construction engineering and management: A critical review and future trends. *Automation in Construction*, *122*, 1–21. doi:10.1016/j.autcon.2020.103517

Pan, Y., & Zhang, L. (2021a). A BIM-data mining integrated digital twin framework for advanced project management. *Automation in Construction*, *124*, 103564. doi:10.1016/j.autcon.2021.103564

Pan, Y., & Zhang, L. (2021b). Automated process discovery from event logs in BIM construction projects. *Automation in Construction*, *127*, 103713. doi:10.1016/j.autcon.2021.103713

Pan, Y., Zhang, L., & Li, Z. (2020). Mining event logs for knowledge discovery based on adaptive efficient fuzzy Kohonen clustering network. *Knowledge-Based Systems*, *209*, 106482. doi:10.1016/j.knosys.2020.106482

Paoletti, I. (2018). Mass customization in the era of industry 4.0: towards immaterial building technology. In *Informed Architecture* (pp. 77–87). Springer. doi:10.1007/978-3-319-53135-9_8

Papadrakakis, M., & Markou, G. (2022). Assessment of the Structural Response of Steel Reinforced and Steel-Fibre Reinforced Concrete Structures with 3D Detailed Modeling: Limitations and Remedies. In *Current Trends and Open Problems in Computational Mechanics* (pp. 395–405). Springer. doi:10.1007/978-3-030-87312-7_38

Park, H. G., Choi, K. K., & Wight, J. K. (2007). Strain-based shear strength model for slender beams without web reinforcement. *ACI Structural Journal*, *104*, 638–639.

Park, H. G., Kang, S., & Choi, K.-K. (2013). Analytical model for shear strength of ordinary and prestressed concrete beams. *Engineering Structures*, *46*, 94–103. doi:10.1016/j.engstruct.2012.07.015

Park, H. G., Yu, E. J., & Choi, K. K. (2012). Shear-strength degradation model for RC columns subjected to cyclic loading. *Engineering Structures*, *34*, 187–197. doi:10.1016/j.engstruct.2011.08.041

Park, J., Kim, K., & Cho, Y. K. (2017). Framework of Automated Construction-Safety Monitoring Using Cloud-Enabled BIM and BLE Mobile Tracking Sensors. *Journal of Construction Engineering and Management*, *143*(2), 05016019. doi:10.1061/(ASCE)CO.1943-7862.0001223

Park, S., Kim, S., & Choi, J. H. (2018). Gear fault diagnosis using transmission error and ensemble empirical mode decomposition. *Mechanical Systems and Signal Processing*, *108*, 58–72. doi:10.1016/j.ymssp.2018.02.028

Pauletta, M., Di Luca, D., & Russo, G. (2015). Exterior beam column joints - Shear strength model and design formula. *Engineering Structures*, *94*, 70–81. doi:10.1016/j.engstruct.2015.03.040

Paz, M., & Leigh, W. (2004). Uniform Building Code 1997: Equivalent Lateral Force Method. In M. Paz & W. Leigh (Eds.), *Structural Dynamics: Theory and Computation* (pp. 699–729). Springer US. doi:10.1007/978-1-4615-0481-8_23

Pearson, K. (1901). LIII. On lines and planes of closest fit to systems of points in space. *The London, Edinburgh and Dublin Philosophical Magazine and Journal of Science*, *2*(11), 559–572. doi:10.1080/14786440109462720

Pedregosa, F., Varoquaux, G., Gramfort, A., Michel, V., Thirion, B., Grisel, O., Blondel, M., Prettenhofer, P., Weiss, R., Dubourg, V., Vanderplas, J., Passos, A., Cournapeau, D., Brucher, M., Perrot, M., & Duchesnay, É. (2011). Scikit-learn: Machine learning in Python. *Journal of Machine Learning Research*, *12*(85), 2825–2830. https://dl.acm.org/doi/10.5555/1953048.2078195

Perkins, I., & Skitmore, M. (2015). Three-dimensional printing in the construction industry: A review. *International Journal of Construction Management*, *15*(1), 1–9. doi:10.1080/15623599.2015.1012136

Perrier, N. & Bourgault, M. (2020). Construction 4.0: A survey of research trends. *Journal of Information Technology in Construction,* 25, 416-437.

PerrierN.BledA.BourgaultM.CousinN.DanjouC.PellerinR.RolandT. (2020).

Peterson, L. E. (2009). K-nearest neighbor. *Scholarpedia Journal*, *4*(2), 1883. doi:10.4249cholarpedia.1883

Pham, D. T., & Liu, X. (1995). *Neural Networks for Identification, Prediction and Control*. Springer. doi:10.1007/978-1-4471-3244-8

Plevris, V., & Asteris, P. (2015). Anisotropic failure criterion for brittle materials using Artificial Neural Networks. 5th International Conference on Computational Methods in Structural Dynamics and Earthquake Engineering (COMPDYN 2015), Crete Island, Greece. doi:10.7712/120115.3537.3648

Plevris, V., & Asteris, P. G. (2014). Modeling of masonry compressive failure using Neural Networks OPT-i 2014. *1st International Conference on Engineering and Applied Sciences Optimization, Proceedings*. ASPETE.

Plevris, V., & Asteris, P. G. (2014). Modeling of Masonry Failure Surface under Biaxial Compressive Stress Using Neural Networks. *Construction & Building Materials*, *55*, 447–461. doi:10.1016/j.conbuildmat.2014.01.041

Plevris, V., & Solorzano, G. (2021, 28-30 June). *Prediction of The Eigenperiods of MDOF Shear Buildings using Neural Networks*. 8th ECCOMAS Thematic Conference on Computational Methods in Structural Dynamics and Earthquake Engineering (COMPDYN 2021), Athens, Greece. doi:10.7712/120121.8755.20415

Plevris, V., Solorzano, G., & Bakas, N. P. (2019, 24-26 June). *Literature review of historical masonry structures with machine learning.* 7th International Conference on Computational Methods in Structural Dynamics and Earthquake Engineering (COMPDYN 2019), Crete, Greece. doi:10.7712/120119.7018.21053

Plevris, V., Solorzano, G., Bakas, N. P., & Ben Seghier, M. E. A. (2022, 5-9 June). *Investigation of performance metrics in regression analysis and machine learning-based prediction models.* 8th European Congress on Computational Methods in Applied Sciences and Engineering (ECCOMAS 2022), Oslo, Norway. doi:10.23967/eccomas.2022.15510.23967/eccomas.2022.155

Plevris, V. (2009). *Innovative computational techniques for the optimum structural design considering uncertainties.* National Technical University of Athens.

Plevris, V., Bakas, N. P., Markeset, G., & Bellos, J. (2017, June 15-17). Literature review of masonry structures under earthquake excitation utilizing machine learning algorithms *6th ECCOMAS Thematic Conference on Computational Methods in Structural Dynamics and Earthquake Engineering (COMPDYN 2017)*, Rhodes, Greece. 10.7712/120117.5598.18688

Plevris, V., Lagaros, N. D., & Zeytinci, A. (2022). Blockchain in Civil Engineering, Architecture and Construction Industry: State of the Art, Evolution, Challenges and Opportunities. *Frontiers in Built Environment*, *8*, 840303. doi:10.3389/fbuil.2022.840303

Plevris, V., & Papadrakakis, M. (2011). A Hybrid Particle Swarm – Gradient Algorithm for Global Structural Optimization. *Computer-Aided Civil and Infrastructure Engineering*, *26*(1), 48–68. doi: 10.1111/j.1467-8667.2010.00664.x

Plevris, V., & Tsiatas, G. (2018). Computational Structural Engineering: Past Achievements and Future Challenges. *Frontiers in Built Environment*, *4*(21), 1–5. doi:10.3389/fbuil.2018.00021

Precup, R. E., David, R. C., & Petriu, E. M. (2016). Grey wolf optimizer algorithm-based tuning of fuzzy control systems with reduced parametric sensitivity. *IEEE Transactions on Industrial Electronics*, *64*(1), 527–534. doi:10.1109/TIE.2016.2607698

Priestley, M.J.N., Verma, R., & Xiao, Y. (1994). Seismic shear strength of reinforced concrete columns. *J Struct Eng*, *120*, 2310–29. https://doi.org/120:8(2310) doi:10.1061/(ASCE)0733-9445(1994)

Priestley, M. J. N., Seible, F., & Calvi, G. (1996). *Seismic Design and Retrofit of Bridge* (1st ed.). John Wiley & Sons, Inc. doi:10.1002/9780470172858

Proia, S., Carli, R., Cavone, G., & Dotoli, M. (2021). Control Techniques for Safe, Ergonomic, and Efficient Human-Robot Collaboration in the Digital Industry: A Survey. *IEEE Transactions on Automation Science and Engineering.*

Qian, Y., & Mita, A. (2007). Structural damage identification using Parzen-window approach and neural networks. *Structural Control and Health Monitoring*, *14*(4), 576–590. doi:10.1002tc.173

Quaranta, G., De Domenico, D., & Monti, G. (2022). Machine-learning-aided improvement of mechanics-based code-conforming shear capacity equation for RC elements with stirrups. *Engineering Structures*, *267*, 114665. doi:10.1016/j.engstruct.2022.114665

Quaranta, G., Lacarbonara, W., & Masri, S. F. (2020). A review on computational intelligence for identification of non-linear dynamical systems. *Nonlinear Dynamics*, *99*(2), 1709–1761. doi:10.100711071-019-05430-7

Qu, Z., Lin, L. D., Guo, Y., & Wang, N. (2015). An improved algorithm for image crack detection based on percolation model. *IEEJ Transactions on Electrical and Electronic Engineering*, *10*(2), 214–221. doi:10.1002/tee.22056

Radford, A., Metz, L., & Chintala, S. (2016). Unsupervised representation learning with deep convolutional generative adversarial networks. *4th International Conference on Learning Representations, ICLR 2016 - Conference Track Proceedings*, (pp. 1–16).

Rahimi, N., Maynor, J., & Gupta, B. (2020). Adversarial machine learning: Difficulties in applying machine learning existing cybersecurity systems. *EPiC Series in Computing*, *69*, 40–47. doi:10.29007/3xbb

Rainieri, C. (2022). In A. Cury, D. Ribeiro, F. Ubertini, & M. D. Todd (Eds.), *Environmental Influence on Modal Parameters: Linear and Nonlinear Methods for Its Compensation in the Context of Structural Health Monitoring BT - Structural Health Monitoring Based on Data Science Techniques* (pp. 269–288). Springer International Publishing.

Ramaswamy, S. (2017). How companies are already using AI. *Harvard Business Review*, *14*(April), 2017.

Ratrout, N. T. (2011). Subtractive Clustering-Based K-means Technique for Determining Optimum Time-of-Day Breakpoints. *Journal of Computing in Civil Engineering*, *25*(5), 380–387. doi:10.1061/(ASCE)CP.1943-5487.0000099

Rayegani, A., & Nouri, G. (2022). Seismic collapse probability and life cycle cost assessment of isolated structures subjected to pounding with smart hybrid isolation system using a modified fuzzy based controller. In *Structures* (Vol. 44, pp. 30–41). Elsevier.

Raza, A., Khan, Q. Z., & Ahmad, A. (2020). Prediction of axial compressive strength for FRP-confined concrete compression members. *KSCE Journal of Civil Engineering*, *24*(7), 2099–2109. doi:10.100712205-020-1682-x

Razaqpur, A. G., & Isgor, O. B. (2006). Proposed shear design method for FRP-reinforced concrete members without stirrups. *ACI Structural Journal*, *103*(1), 93.

Reconan, F. E. A. (2020). *v2.00, User's manual*. Research Gate. https://www.researchgate.net/publication/342361609_ReConAn_v200_Finite_Element_Analysis_Software_User's_Manual

Regona, M., Yigitcanlar, T., Xia, B., & Li, R. Y. M. (2022). Opportunities and Adoption Challenges of AI in the Construction Industry: A PRISMA Review. *Journal of Open Innovation*, *8*(1), 45. doi:10.3390/joitmc8010045

Reineck, K. H., & Dunkelberg, D. (Eds.). (2017). *ACI-DAfStb databases 2015 with shear tests for evaluating relationships for the shear design of structural concrete members without and with stirrups*. Beuth Verlag GmbH.

Reineck, K.-H., Bentz, E. C., Fitik, B., Kuchma, D. A., & Bayrak, O. (2013). ACI-DAFSTB Database of shear tests on slender reinforced concrete beams without stirrups. *ACI Structural Journal*, *110*(5). doi:10.14359/51686819

Ritto, T. G., & Rochinha, F. A. (2021). Digital twin, physics-based model, and machine learning applied to damage detection in structures. *Mechanical Systems and Signal Processing*, *155*, 107614. doi:10.1016/j.ymssp.2021.107614

Rodriguez, M. Z., Comin, C. H., Casanova, D., Bruno, O. M., Amancio, D. R., Costa, L. da F., & Rodrigues, F. A. (2019). Clustering algorithms: A comparative approach. *PLoS One*, *14*(1), e0210236. doi:10.1371/journal.pone.0210236 PMID:30645617

Ronneberger, O., Fischer, P., & Brox, T. (2015). U-net: Convolutional networks for biomedical image segmentation. Lecture Notes in Computer Science (Including Subseries Lecture Notes in Artificial Intelligence and Lecture Notes in Bioinformatics), 9351, 234–241. doi:10.1007/978-3-319-24574-4_28

Rosenfeld, Y., Warszawski, A., & Zajicek, U. (1993). Full-scale building with interior finishing robot. *Automation in Construction*, *2*(3), 229–240. doi:10.1016/0926-5805(93)90043-W

Roshni, T., Jha, M. K., & Drisya, J. (2020). Neural network modeling for groundwater-level forecasting in coastal aquifers. *Neural Computing & Applications*, *32*(16), 12737–12754. doi:10.100700521-020-04722-z

Rousseeuw, P. J. (1987). Silhouettes: A graphical aid to the interpretation and validation of cluster analysis. *Journal of Computational and Applied Mathematics*, *20*(C), 53–65. doi:10.1016/0377-0427(87)90125-7

Russell, S. J., & Norvig, P. (2010). *Artificial Intelligence: A Modern Approach*. Prentice Hall.

Russell, S., & Norvig, P. (2022). *Artificial Intelligence - A Modern Approach*. Pearson.

Rytter. (1993). *Aalborg Universitet Vibrational Based Inspection of Civil Engineering Structures*. Rytter, Anders.

Saatcioglu, M., Mitchell, D., Tinawi, R., Gardner, N. J., Gillies, A. G., Ghobarah, A., Anderson, D. L., & Lau, D. (2001). The August 17, 1999, Kocaeli (Turkey) earthquake—Damage to structures. *Canadian Journal of Civil Engineering*, *28*(4), 715–737. doi:10.1139/cjce-28-4-715

SABS Standards Division. (2009). Seismic actions and general requirements for building, South African National Standard (SANS), SABS Standards Division.

Sacks, R., Bloch, T., Katz, M., & Yosef, R. (2019). Automating design review with artificial intelligence and BIM: State of the art and research framework. *Computing in Civil Engineering 2019: Visualization, Information Modeling, and Simulation*, (pp. 353–360). Research Gate.

Sacks, R., Eastman, C., Lee, G., & Teicholz, P. (2018). BIM Handbook: A Guide to Building. John Wiley & Sons.

Sacks, R., & Barak, R. (2006). Quantitative assessment of the impact of 3D modelling of building structures on engineering productivity. *Joint International Conference on Computing and Decision Making in Civil and Building Engineering*, (pp. 14–16). IEEE.

Sacks, R., Brilakis, I., Pikas, E., Xie, H. S., & Girolami, M. (2020). Construction with digital twin information systems. *Data-Centric Engineering*, *1*(6), A32.

Sacks, R., Eastman, C., Lee, G., & Teicholz, P. (2018). *BIM handbook: A guide to building information modeling for owners, designers, engineers, contractors, and facility managers*. John Wiley & Sons. doi:10.1002/9781119287568

Sadowski, Ł., Nikoo, M., Shariq, M., Joker, E., & Czarnecki, S. (2019). The Nature-Inspired Metaheuristic Method for Predicting the Creep Strain of Green Concrete Containing Ground Granulated Blast Furnace Slag. *Materials (Basel)*, *12*(2), 293. doi:10.3390/ma12020293 PMID:30658508

Salehi, H., Biswas, S., & Burgueño, R. (2019). Data interpretation framework integrating machine learning and pattern recognition for self-powered data-driven damage identification with harvested energy variations. *Engineering Applications of Artificial Intelligence*, *86*, 136–153. doi:10.1016/j.engappai.2019.08.004

Salehi, H., Das, S., Chakrabartty, S., Biswas, S., & Burgueno, R. (2018). A methodology for structural health diagnosis and assessment using machine learning with noisy and incomplete data from self-powered wireless sensors. *Proceedings of the Society for Photo-Instrumentation Engineers*, 10598. doi:10.1117/12.2295990

Sanni-Anibire, M. O., Zin, R. M., & Olatunji, S. O. (2021). Machine learning - Based framework for construction delay mitigation. *Journal of Information Technology in Construction*, *26*, 303–318. doi:10.36680/j.itcon.2021.017

Santos, A., Figueiredo, E., Silva, M. F. M., Sales, C. S., & Costa, J. C. W. A. (2016). Machine learning algorithms for damage detection: Kernel-based approaches. *Journal of Sound and Vibration*, *363*, 584–599. doi:10.1016/j.jsv.2015.11.008

Sarkar, S., Pramanik, A., Maiti, J., & Reniers, G. (2020). Predicting and analyzing injury severity: A machine learning-based approach using class-imbalanced proactive and reactive data. *Safety Science*, *125*(February), 104616. doi:10.1016/j.ssci.2020.104616

Sarle, W. S. (1995). Stopped Training and Other Remedies for Overfitting. Proceedings of the 27th Symposium, (pp. 1-10). Semantic Scholar.

Schmidt, M., & Lipson, H. (2009). Distilling Free-Form Natural Laws from Experimental Data. *Science*, *324*(5923), 81–85. doi:10.1126/science.1165893 doi:10.1126cience.1165893 PMID:19342586

Schölkopf, B., Platt, J. C., Shawe-Taylor, J., Smola, A. J., & Williamson, R. C. (2001). Estimating the support of a high-dimensional distribution. *Neural Computation*, *13*(7), 1443–1471. doi:10.1162/089976601750264965 PMID:11440593

Scott, D. J., Broyd, T., & Ma, L. (2021). Exploratory literature review of blockchain in the construction industry. *Automation in Construction*, *132*, 103914. doi:10.1016/j.autcon.2021.103914

Seghier, M. E. A. B., Gao, X. Z., Jafari-Asl, J., Thai, D. K., Ohadi, S., & Trung, N. T. (2021a). Modeling the nonlinear behavior of ACC for SCFST columns using experimental-data and a novel evolutionary-algorithm. *Structures*, *30*, 692–709. doi:10.1016/j.istruc.2021.01.036

Seghier, M. E. A. B., Ouaer, H., Ghriga, M. A., Menad, N. A., & Thai, D. K. (2021b). Hybrid soft computational approaches for modeling the maximum ultimate bond strength between the corroded steel reinforcement and surrounding concrete. *Neural Computing & Applications*, *33*(12), 6905–6920. doi:10.100700521-020-05466-6

Sethi, T. S., Kantardzic, M., Lyu, L., & Chen, J. (2018). A dynamic-adversarial mining approach to the security of machine learning. *Wiley Interdisciplinary Reviews. Data Mining and Knowledge Discovery*, *8*(3), 1245. doi:10.1002/widm.1245

Sezen, H. (2002). *Seismic behavior and modeling of reinforced concrete building columns*. University of California at Berkeley.

Sezen, H., & Moehle, J. P. (2004). Shear strength model for lightly reinforced concrete columns. *Journal of Structural Engineering*, *130*(11), 1692–1703. doi:10.1061/(ASCE)0733-9445(2004)130:11(1692)

Shadbahr, E., Aminnejad, B., & Lork, A. (2021). Determining post-fire residual compressive strength of reinforced concrete shear walls using the BAT algorithm. *Structures, 32*, 651-661. doi: 10.1016/j.istruc.2021.03.002

Shafaei, J., & Nezami, S. A. (2019). Effect of different size of joint enlargement on seismic behavior of gravity load designed RC beam-column connections. *Struct Des Tall Spec Build, 28,* 1–27. doi:10.1002/tal.1653

Shafaei, J., Zareian, M. S., Hosseini, A., & Marefat, M. S. (2014). Effects of joint flexibility on lateral response of reinforced concrete frames. *Engineering Structures*, *81*, 412–431. doi:10.1016/j.engstruct.2014.09.046

Shahin, R., & Sabri, K. E. (2022). A Secure IoT Framework Based on Blockchain and Machine Learning. *International Journal of Computing and Digital Systems*, *11*(1), 671–683. doi:10.12785/ijcds/110154

Shalev-Shwartz, S., & Ben-David, S. (2014). *Understanding Machine Learning: From Theory to Algorithms*. Cambridge University Press. doi:10.1017/CBO9781107298019

Shang, Z., Sun, L., Xia, Y., & Zhang, W. (2021). Vibration-based damage detection for bridges by deep convolutional denoising autoencoder. *Structural Health Monitoring*, *20*(4), 1880–1903. doi:10.1177/1475921720942836

Shao, T., & Zhang, C. (2018). Architectural Design Model based on BIM Management System Model and Data Mining. *International Journal of Performability Engineering*, *14*(11), 2574. doi:10.23940/ijpe.18.11.p3.25742580

Sharaf, M. A., & Hussein, M. T. (1996). Groundwater quality in the Saq aquifer, Saudi Arabia. *Hydrological Sciences Journal*, *41*(5), 683–696. doi:10.1080/02626669609491539

Sharib, S., Ahmad, N., Plevris, V., & Ahmad, A. (2021, 28-30 June 2021). *Prediction Models for Load Carrying Capacity of RC Wall through Neural Network.* 14th ECCOMAS Thematic Conference on Evolutionary and Deterministic Methods for Design, Optimization and Control (EUROGEN 2021), Athens, Greece. doi:10.7712/140121.7956.18529

Sharma, A., Eligehausen, R., & Reddy, G. R. (2011). A new model to simulate joint shear behavior of poorly detailed beam-column connections in RC structures under seismic loads, Part I: Exterior joints. *Engineering Structures*, *33*(3), 1034–1051. doi:10.1016/j.engstruct.2010.12.026

Sharma, A., Saxena, A., Sethi, M., Shree, V., & Varun. (2011). Life cycle assessment of buildings: A review. *Renewable & Sustainable Energy Reviews*, *15*(1), 871–875. doi:10.1016/j.rser.2010.09.008

Shaukat, K., Luo, S., Varadharajan, V., Hameed, I. A., & Xu, M. (2020). A Survey on Machine Learning Techniques for Cyber Security in the Last Decade. *IEEE Access : Practical Innovations, Open Solutions*, *8*, 222310–222354. doi:10.1109/ACCESS.2020.3041951

Shayanfar, J., & Bengar, H.A. (2018). A practical model for simulating non-linear behavior of FRP strength ened RC beam-column joints. *Steel Compos Struct, 27,* 49–74. https://doi.org/.2018.27.1.049 doi:10.12989/scs

Shayanfar, J., Akbarzadeh Bengar, H., & Niroomandi, A. (2016). A proposed model for predicting non-linear behavior of RC joints under seismic loads. *Materials & Design*, *95*, 563–579. doi:10.1016/j.matdes.2016.01.098

Shayanfar, J., & Bengar, H. A. (2016). Numerical model to simulate shear behavior of RC joints and columns. *Computers and Concrete*, *18*(6), 877–901. doi:10.12989/cac.2016.18.6.877

Sheela, K. G., & Deepa, S. N. (2013). Review on methods to fix number of hidden neurons in neural networks. *Mathematical Problems in Engineering*, *2013*, 2013. doi:10.1155/2013/425740

Shelhamer, E., Long, J., & Darrell, T. (2017). Fully Convolutional Networks for Semantic Segmentation. *IEEE Transactions on Pattern Analysis and Machine Intelligence*, *39*(4), 640–651. doi:10.1109/TPAMI.2016.2572683 PMID:27244717

Shin, D. H., & Dunston, P. S. (2009). Evaluation of Augmented Reality in steel column inspection. *Automation in Construction*, *18*(2), 118–129. doi:10.1016/j.autcon.2008.05.007

Shin, M. H., Lee, H. K., & Kim, H. Y. (2018). Benefit-Cost analysis of Building Information Modeling (BIM) in a Railway Site. *Sustainability (Basel)*, *10*(11), 4303. Advance online publication. doi:10.3390u10114303

Shin, M., & Lafave, J. M. (2004). Modeling of cyclic joint shear deformation contributions in RC beam-column connections to overall frame behavior. *Structural Engineering and Mechanics*, *18*(5), 645–669. doi:10.12989em.2004.18.5.645

Shiri, N., Shiri, J., Yaseen, Z. M., Kim, S., Chung, I. M., Nourani, V., & Zounemat-Kermani, M. (2021). Development of artificial intelligence models for well groundwater quality simulation: Different modeling scenarios. *PLoS One*, *16*(5), e0251510. doi:10.1371/journal.pone.0251510 PMID:34043648

Shu, J., Li, J., Zhang, J., Zhao, W., Duan, Y., & Zhang, Z. (2022). An active learning method with difficulty learning mechanism for crack detection. *Smart Structures and Systems*, *29*(1), 195–206. doi:10.12989/SSS.2022.29.1.195

Shuler, C. K., & Mariner, K. E. (2020). Collaborative groundwater modeling: Open-source, cloud-based, applied science at a small-island water utility scale. *Environmental Modelling & Software*, *127*, 104693. doi:10.1016/j.envsoft.2020.104693

Shuraim, A. B. (2014). A novel approach for evaluating the concrete shear strength in reinforced concrete beams. *Latin American Journal of Solids and Structures 11*, 093-112.

Shyu, M., Chen, S., & Iyengar, S. S. (2020). a Survey on Deep Learning Techniques. *Strad Research, 7*(8).

Sigrist, V., Bentz, E., Ruiz, M. F., Foster, S., & Muttoni, A. (2013). Background to the fib Model Code 2010 shear provisions–part I: Beams and slabs. *Structural Concrete*, *14*(3), 195–203. doi:10.1002uco.201200066

Simonyan, K., & Zisserman, A. (2015). Very deep convolutional networks for large-scale image recognition. *3rd International Conference on Learning Representations, ICLR 2015 - Conference Track Proceedings*, (pp. 1–14).

Singha, S., Pasupuleti, S., Singha, S. S., Singh, R., & Kumar, S. (2021). Prediction of groundwater quality using efficient machine learning technique. *Chemosphere*, *276*, 130265. doi:10.1016/j.chemosphere.2021.130265 PMID:34088106

Skibniewski, M. J., & Wooldridge, S. C. (1992). Robotic materials handling for automated building construction technology. *Automation in Construction*, *1*(3), 251–266. doi:10.1016/0926-5805(92)90017-E

Solgi, R., Loáiciga, H. A., & Kram, M. (2021). Long short-term memory neural network (LSTM-NN) for aquifer level time series forecasting using in-situ piezometric observations. *Journal of Hydrology (Amsterdam)*, *601*, 126800. doi:10.1016/j.jhydrol.2021.126800

Solorzano, G., & Plevris, V. (2021). *Design of Reinforced Concrete Isolated Footings under Axial Loading with Artificial Neural networks.* 14th ECCOMAS Thematic Conference on Evolutionary and Deterministic Methods for Design, Optimization and Control (EUROGEN 2021), Streamed from Athens, Greece. doi:10.7712/140121.7955.18448

Somerville, P. G. (2005). Engineering characterization of near fault ground motions. In *Proc., NZSEE 2005 Conf.* New Zeland Society for Earthquake Engineering.

Song, J., Kim, J., & Lee, J.-K. (2018). NLP and deep learning-based analysis of building regulations to support automated rule checking system. *ISARC. Proceedings of the International Symposium on Automation and Robotics in Construction,* (*vol. 35*, pp. 1–7). IAARC.

Sonkor, M. S., & García De Soto, B. (2021). Operational Technology on Construction Sites: A Review from the Cybersecurity Perspective. *Journal of Construction Engineering and Management*, *147*(12), 04021172. doi:10.1061/(ASCE)CO.1943-7862.0002193

Sony, S., Dunphy, K., Sadhu, A., & Capretz, M. (2021). A systematic review of convolutional neural network-based structural condition assessment techniques. *Engineering Structures*, *226*, 111347. doi:10.1016/j.engstruct.2020.111347

Spijkerman, Z., Bakas, P. N., Markou, G., & Papadrakakis, M. (2021). Predicting the shear capacity of reinforced concrete slender beams without stirrups by applying artificial intelligence algorithms [Paper presentation]. COMPDYN 2021. Eccomas Proceedia. doi:10.7712/120121.8749.18602

Srivastava, S., & Sahana, S. K. (2019). Application of Bat Algorithm for Transport Network Design Problem. *Applied Computational Intelligence and Soft Computing*, *2019*, 9864090. doi:10.1155/2019/9864090

Standard, A. A. C. I. (2019). *An ACI Standard.* ACI.

Stewart, J. P., Chiou, S. J., Bray, J. D., Graves, R. W., Somerville, P. G., & Abrahamson, N. A. (2002). Ground motion evaluation procedures for performance-based design. *Soil Dynamics and Earthquake Engineering*, *22*(9-12), 765–772. doi:10.1016/S0267-7261(02)00097-0

Stylianoudaki, C., Trichakis, I., & Karatzas, G. P. (2022). Modeling Groundwater Nitrate Contamination Using Artificial Neural Networks. *Water (Basel)*, *14*(7), 1173. doi:10.3390/w14071173

Subbaraj, K., & Dokainish, M. (1989). A survey of direct time-integration methods in computational structural dynamics—II. Implicit methods. *Computers & Structures*, *32*(6), 1387–1401. doi:10.1016/0045-7949(89)90315-5

Suprayogi, I., Sandhyavitri, A., Anon, N., Anon, J., Anon, W., & Anon, A. (2020). Groundwater level forecasting model in tropical peatland using artificial neural network. *International journal of civil engineering and technology (IJCIET)*. doi:10.34218/IJCIET.11.2.2020.008

Su, Y., Guo, B., Zhou, Z., Zhong, Y., & Min, L. (2020). *Spatio-temporal variations in groundwater revealed by GRACE and its driving factors in the Huang-Huai-Hai Plain, China*. Sensors., doi:10.339020030922

Tabari, M. M. R., Azari, T., & Dehghan, V. (2021). A supervised committee neural network for the determination of aquifer parameters: A case study of Katasbes aquifer in Shiraz plain, Iran. *Soft Computing*, *25*(6), 4785–4798. doi:10.100700500-020-05487-2

Tairidis, G., Foutsitzi, G., Koutsianitis, P., & Stavroulakis, G. E. (2016). Fine tuning of a fuzzy controller for vibration suppression of smart plates using genetic algorithms. *Advances in Engineering Software*, *101*, 123–135. doi:10.1016/j.advengsoft.2016.01.019

Taljaard, V-L., Gravett, D. Z., Mourlas, C., Markou, G., Bakas, N., & Papadrakakis, M. (2021) *Development of a new fundamental period formula by considering soil-structure interaction with the use of machine learning algorithms [Paper presentation]*. COMPDYN 2021, Athens.

Tango, F., & Botta, M. (2013). Real-time detection system of driver distraction using machine learning. *IEEE Transactions on Intelligent Transportation Systems*, *14*(2), 894–905. doi:10.1109/TITS.2013.2247760

Tang, Y., & Zhang, J. (2011). Probabilistic seismic demand analysis of a slender RC shear wall considering soil–structure interaction effects. *Engineering Structures*, *33*(1), 218–229. doi:10.1016/j.engstruct.2010.10.011

Tanwar, S., Bhatia, Q., Patel, P., Kumari, A., Singh, P. K., & Hong, W.-C. (2020). Machine Learning Adoption in Blockchain-Based Smart Applications: The Challenges, and a Way Forward. *IEEE Access : Practical Innovations, Open Solutions*, *8*(8938741), 474–448. doi:10.1109/ACCESS.2019.2961372

Taylor, K. E. (2001). Summarizing multiple aspects of model performance in a single diagram. *Journal of Geophysical Research*, *106*(D7), 7183–7192. doi:10.1029/2000JD900719

Tezel, A., Papadonikolaki, E., Yitmen, I., & Bolpagni, M. (2022). Blockchain Opportunities and Issues in the Built Environment: Perspectives on Trust, Transparency and Cybersecurity. *Structural Integrity*, *20*, 569–588. doi:10.1007/978-3-030-82430-3_24

Thai, H-T. (2022). Machine learning for structural engineering: A state-of-the-art review. *Structures, 38*, 448-491.

The People's Republic of China National Standard. (2010). GB50010-2010: Code for Seismic Design of Buildings. Author.

Tiryaki, S., & Aydın, A. (2014). An artificial neural network model for predicting compression strength of heat treated woods and comparison with a multiple linear regression model. *Construction & Building Materials*, *62*, 102–108. doi:10.1016/j.conbuildmat.2014.03.041

Tiwari, D., Pachauri, N., Rani, A., & Singh, V. (2016). Fractional order PID (FOPID) controller based temperature control of bioreactor. In 2016 international conference on electrical, electronics, and optimization techniques (ICEEOT) (pp. 2968-2973). IEEE.

Toğan, V., Mostofi, F., Ayözen, Y. E., & Behzat Tokdemir, O. (2022). Customized AutoML: An Automated Machine Learning System for Predicting Severity of Construction Accidents. *Buildings*, *12*(11), 1933. doi:10.3390/buildings12111933

Tommaselli, A. M. G., & Reiss, M. L. L. (2005). A photogrammetric method for single image orientation and measurement. *Photogrammetric Engineering and Remote Sensing*, *71*(6), 727–732. doi:10.14358/PERS.71.6.727

Torres-Calderon, W., Chi, Y., Amer, F., & Golparvar-Fard, M. (2019). Automated mining of construction schedules for easy and quick assembly of 4D BIM simulations. *ASCE International Conference on Computing in Civil Engineering 2019: Visualization, Information Modeling, and Simulation, I3CE 2019*, (pp. 432–438). ASCE Library. 10.1061/9780784482421.055

Tureyen, A. K., & Frosch, R. J. (2002). Shear tests of FRP-reinforced concrete beams without stirrups. *ACI Structural Journal*, *99*(4), 427–434.

Turing, A. M. (1950). Computing Machinery and Intelligence. *Mind, New Series, 59*(236). Oxford University Press.

Turk, Ž., García de Soto, B., Mantha, B. R. K., Maciel, A., & Georgescu, A. (2022). A systemic framework for addressing cybersecurity in construction. *Automation in Construction*, *133*, 103988. doi:10.1016/j.autcon.2021.103988

Ustinovichius, L., Peckienė, A., & Popov, V. (2017). A model for spatial planning of site and building using BIM methodology. *Journal of Civil Engineering and Management*, *23*(2), 173–182. doi:10.3846/13923730.2016.1247748

Vairagade, R. S., & Brahmananda, S. H. (2022). Enabling machine learning-based side-chaining for improving QoS in blockchain-powered IoT networks. *Transactions on Emerging Telecommunications Technologies*, *33*(4). Advance online publication. doi:10.1002/ett.4433

van der Westhuizen, A. M., Markou, G., & Bakas, N. P. (2022). *Development of a New Fundamental Period Formula for Steel Structures Considering the Soil-structure Interaction with the Use of Machine Learning Algorithms* [Paper presentation]. ICAART. doi:10.5220/0010978400003116

Vecchio, F. J., & Collins, M. P. (1986). The modified compression-field theory for reinforced concrete elements subjected to shear. *ACI J.*, *83*(2), 219–231.

Venkatasubramanian, K. (2020). Machine learning set to play a key role in the transition to Construction 4.0. *Proceedings of the Institution of Civil Engineers. Civil Engineering*, *174*(2), 54. doi:10.1680/jcien.2021.174.2.54

Wallace, J. W., McConnell, S. W., Gupta, P., & Cote, P. A. (1998). Use of headed reinforcement in beam-column joints subjected to earthquake loads. *ACI Structural Journal*, *95*(5), 590–606. doi:10.14359/574 PMID:31500274

Walraven, J., Belletti, B., & Esposito, R. (2013). Shear capacity of normal, lightweight, and high-strength concrete beams according to Model Code 2010. I: Experimental results versus analytical model results. *Journal of Structural Engineering*, *139*(9), 1593–1599. doi:10.1061/(ASCE)ST.1943-541X.0000742

Walter, M. (2006). Return on Interoperability: the new ROL. UK: John Jageurs. *CAD User, March/April.*

Wang, S., Wei, C., Wei, Y., Wang, W., & Wu, F. (2019). A model of injury severity prediction in traffic accident based on GA-BP neural network. *CICTP 2019: Transportation in China - Connecting the World - Proceedings of the 19th COTA International Conference of Transportation Professionals*, (pp. 2470–2481). ACSE Library. 10.1061/9780784482292.215

Wang, Y., Zhao, Y., & Addepalli, S. (2020). Remaining useful life prediction using deep learning approaches: A review. *Procedia Manufacturing, 49*(2019), 81–88.

Wang, Y.-W., Ni, Y.-Q., & Wang, S.-M. (2022). Structural health monitoring of railway bridges using innovative sensing technologies and machine learning algorithms: a concise review. *Intelligent Transportation Infrastructure, 1*, liac009.

Wang, G. L., Dai, J. G., & Teng, J. G. (2012). Shear strength model for RC beam-column joints under seismic loading. *Engineering Structures*, *40*, 350–360. doi:10.1016/j.engstruct.2012.02.038

Wang, Q., Cheng, J. C. P., & Sohn, H. (2017). Automated Estimation of Reinforced Precast Concrete Rebar Positions Using Colored Laser Scan Data. *Computer-Aided Civil and Infrastructure Engineering*, *32*(9), 787–802. doi:10.1111/mice.12293

Wang, Q., Sohn, H., & Cheng, J. C. P. (2019). Development of high-accuracy edge line estimation algorithms using terrestrial laser scanning. *Automation in Construction*, *101*, 59–71. doi:10.1016/j.autcon.2019.01.009

Wang, Y., Wang, X., Wang, J., Yung, P., & Jun, G. (2013). Engagement of facilities management in design stage through BIM: Framework and a case study. *Advances in Civil Engineering*, *2013*, 2013. doi:10.1155/2013/189105

Waris, M., Mir, J., Plevris, V., & Ahmad, A. (2020). Predicting compressive strength of CRM samples using Image processing and ANN. *IOP Conference Series: Materials Science and Engineering*. IOP Publishing. 10.1088/1757-899X/899/1/012014

Widyaningrum, E., Peters, R. Y., & Lindenbergh, R. C. (2020a). Building outline extraction from als point clouds using medial axis transform descriptors. *Pattern Recognition*, *106*, 107447. doi:10.1016/j.patcog.2020.107447

Wong, S.H.F., & Kuang, J.S. (2011). Predicting shear strength of RC exterior beam-column joints by modified rotating-angle softened-truss model. *Comput Concr*, *8*, 59–70. https://doi.org/ac.2011.8.1.059 doi:10.12989/c

Wong, H. F. (2005). *Shear strength and seismic performance of non-seismically designed reinforced concrete beam-column joints*. Hong Kong Univ Sci Technol. doi:10.14711/thesis-b914043

Worden, K., Farrar, C. R., Manson, G., & Park, G. (2007). The fundamental axioms of structural health monitoring. *Proceedings - Royal Society. Mathematical, Physical and Engineering Sciences*, *463*(2082), 1639–1664. doi:10.1098/rspa.2007.1834

Wu, H.-N., Shen, S.-L., Chen, R.-P., & Zhou, A. (2020). Three-dimensional numerical modelling on localised leakage in segmental lining of shield tunnels. *Computers and Geotechnics*, *122*, 103549. doi:10.1016/j.compgeo.2020.103549

Wunsch, A., Liesch, T., & Broda, S. (2021). Groundwater level forecasting with artificial neural networks: A comparison of long short-term memory (LSTM), convolutional neural networks (CNNs), and non-linear autoregressive networks with exogenous input (NARX). *Hydrology and Earth System Sciences*, *25*(3), 1671–1687. doi:10.5194/hess-25-1671-2021

Wu, P., Wang, J., & Wang, X. (2016). A critical review of the use of 3-D printing in the construction industry. *Automation in Construction*, *68*, 21–31. doi:10.1016/j.autcon.2016.04.005

Xie, S., Girshick, R., Dollár, P., Tu, Z., & He, K. (2017). Aggregated residual transformations for deep neural networks. *Proceedings - 30th IEEE Conference on Computer Vision and Pattern Recognition, CVPR 2017, 2017-Janua*, (pp. 5987–5995). IEEE. 10.1109/CVPR.2017.634

Xu, B., Wu, Z., Chen, G., & Yokoyama, K. (2004). Direct identification of structural parameters from dynamic responses with neural networks. *Engineering Applications of Artificial Intelligence*, *17*(8), 931–943. doi:10.1016/j.engappai.2004.08.010

Xu, L., Pan, J., & Cai, J. (2019). Seismic performance of precast RC and RC/ECC composite columns with grouted sleeve connections. *Engineering Structures*, *188*(March), 104–110. doi:10.1016/j.engstruct.2019.03.022

Xu, Y. L. (2018). Making good use of structural health monitoring systems of long-span cable-supported bridges. *Journal of Civil Structural Health Monitoring*, *8*(3), 477–497. doi:10.100713349-018-0279-2

Xu, Y., Zhou, Y., Sekula, P., & Ding, L. (2021). Machine learning in construction: From shallow to deep learning. *Developments in the Built Environment*, *6*, 100045. doi:10.1016/j.dibe.2021.100045

Yan, H., & Demian, P. (2008). *Benefits and barriers of building information modelling*. Loughborough University.

Yang, X. S., & He, X. (2013). Firefly algorithm: recent advances and applications. *International journal of swarm intelligence, 1*(1), 36-50.

Yang, K.-H., Chung, H.-S., Lee, E.-T., & Eun, H.-C. (2003). Shear characteristics of high-strength concrete deep beams without shear reinforcements. *Engineering Structures, 25*(10), 1343–1352. doi:10.1016/S0141-0296(03)00110-X

Yang, Q., Liu, Y., Chen, T., & Tong, Y. (2019). Federated machine learning: Concept and applications. *ACM Transactions on Intelligent Systems and Technology, 10*(2), 12. doi:10.1145/3298981

Yang, Y., & Nagarajaiah, S. (2014). Data Compression of Structural Seismic Responses via Principled Independent Component Analysis. *Journal of Structural Engineering, 140*(7), 04014032. doi:10.1061/(ASCE)ST.1943-541X.0000946

Yan, Y., & Hajjar, J. F. (2021). Automated extraction of structural elements in steel girder bridges from laser point clouds. *Automation in Construction, 125*, 103582. doi:10.1016/j.autcon.2021.103582

Yap, J. B. H., Chow, I. N., & Shavarebi, K. (2019). Criticality of Construction Industry Problems in Developing Countries: Analyzing Malaysian Projects. *Journal of Management Engineering, 35*(5), 04019020. doi:10.1061/(ASCE)ME.1943-5479.0000709

Ye, X. W., Jin, T., & Yun, C. B. (2019). A review on deep learning-based structural health monitoring of civil infrastructures. *Smart Structures and Systems, 24*(5), 567–585. doi:10.12989ss.2019.24.5.567

Yin, H., & Gai, K. (2015). An empirical study on preprocessing high-dimensional class-imbalanced data for classification. *Proceedings - 2015 IEEE 17th International Conference on High Performance Computing and Communications, 2015 IEEE 7th International Symposium on Cyberspace Safety and Security and 2015 IEEE 12th International Conference on Embedded Software and Systems, H*, (pp. 1314–1319). IEEE. 10.1109/HPCC-CSS-ICESS.2015.205

Yi, T. H., Lei, Y., Chen, H. P., Ye, X. W., Kang, F., & Wang, H. (2014). Technology developments in structural health monitoring and integrity maintenance. *TheScientificWorldJournal, 2014*, 976927. doi:10.1155/2014/976927 PMID:25614901

Yi, T. H., Li, H. N., Song, G., & Guo, Q. (2016). Detection of Shifts in GPS Measurements for a Long-Span Bridge Using CUSUM Chart. *International Journal of Structural Stability and Dynamics, 16*(4), 1–21. doi:10.1142/S0219455416400241

Yoon, H., Jun, S. C., Hyun, Y., Bae, G. O., & Lee, K. K. (2011). A comparative study of artificial neural networks and support vector machines for predicting groundwater levels in a coastal aquifer. *Journal of Hydrology (Amsterdam), 396*(1-2), 128–138. doi:10.1016/j.jhydrol.2010.11.002

Younes, A., Fahs, M., & Ackerer, P. (2021). *Modeling of Flow and Transport in Saturated and Unsaturated Porous Media*. Multidisciplinary Digital Publishing Institute. doi:10.3390/w13081088

Young, K. C. (2011). *An Investigation of the Fundamental Period of Vibration of Irregular Steel Structures* [Unpublished Thesis, The Ohio State University, Columbus].

You, Z., & Feng, L. (2020). Integration of Industry 4.0 Related Technologies in Construction Industry: A Framework of Cyber-Physical System. *IEEE Access : Practical Innovations, Open Solutions, 8*, 122908–122922. doi:10.1109/ACCESS.2020.3007206

Yu, F., & Koltun, V. (2016). Multi-scale context aggregation by dilated convolutions. *4th International Conference on Learning Representations, ICLR 2016 - Conference Track Proceedings*. Research Gate.

Yuen, K. V., & Mu, H. Q. (2012). A novel probabilistic method for robust parametric identification and outlier detection. *Probabilistic Engineering Mechanics, 30*, 48–59. doi:10.1016/j.probengmech.2012.06.002

Yu, K., Zhang, C., Shooshtarian, M., Zhao, W., & Shu, J. (2021). Automated finite element modeling and analysis of cracked reinforced concrete beams from three dimensional point cloud. *Structural Concrete*, *22*(6), 3213–3227. doi:10.1002uco.202100194

Zabin, A., González, V. A., Zou, Y., & Amor, R. (2022). Applications of machine learning to BIM: A systematic literature review. *Advanced Engineering Informatics*, *51*, 101474. Advance online publication. doi:10.1016/j.aei.2021.101474

Zandi, K., Ransom, E. H., Topac, T., Chen, R., Beniwal, S., Blomfors, M., Shu, J., & Chang, F. (2019, November 15). A Framework for Digital Twin of Civil Infrastructure—Challenges & Opportunities. *Structural Health Monitoring, 2019*. Advance online publication. doi:10.12783hm2019/32288

Zenon Waszczyszyn, L. Z. (2001). Neural networks in mechanics of structures and materials-new results and prospects of applications. *Computers & Structures*, *79*(22-25), 2261–2276. doi:10.1016/S0045-7949(01)00083-9

Zhang, N. (2017). Seismic Performance and Shear Strength of Reinforced Concrete Beam-column Roof corner Joints [Ph.D. Dissertation]. The Hong Kong University of Science and Technology.

Zhang, N., Kuang, J. S., & Mogili, S. (2017). Cyclic Behavior of Reinforced Concrete Beam-Column Roof corner Joints. In *16th World Conf. Earthq. Eng. 16WCEE*. Semantic Scholar.

Zhang, C., Shu, J., Shao, Y., & Zhao, W. (2022). Automated generation of FE models of cracked RC beams based on 3D point clouds and 2D images. *Journal of Civil Structural Health Monitoring*, *12*(1), 29–46. doi:10.100713349-021-00525-5

Zhang, J., & Foschi, R. O. (2004). Performance-Based Design And Seismic Reliability Analysis Using Designed Experiments And Neural Networks. *Probabilistic Engineering Mechanics*, *19*(3), 259–267. doi:10.1016/j.probengmech.2004.02.009

Zhang, R., Chen, Z., Chen, S., Zheng, J., Büyüköztürk, O., & Sun, H. (2019). Deep long short-term memory networks for nonlinear structural seismic response prediction. *Computers & Structures*, *220*, 55–68. doi:10.1016/j.compstruc.2019.05.006

Zhang, R., Liu, Y., & Sun, H. (2020a). Physics-guided convolutional neural network (PhyCNN) for data-driven seismic response modeling. *Engineering Structures*, *215*, 110704. doi:10.1016/j.engstruct.2020.110704

Zhang, R., Liu, Y., & Sun, H. (2020b). Physics-informed multi-LSTM networks for metamodeling of nonlinear structures. *Computer Methods in Applied Mechanics and Engineering*, *369*, 113226. doi:10.1016/j.cma.2020.113226

Zhang, T., Visintin, P., & Oehlers, D. J. (2016). Shear strength of RC beams with steel stirrups. *Journal of Structural Engineering*, *142*(2), 04015135. doi:10.1061/(ASCE)ST.1943-541X.0001404

Zhang, Z., & Sun, C. (2021). Structural damage identification via physics-guided machine learning: A methodology integrating pattern recognition with finite element model updating. *Structural Health Monitoring*, *20*(4), 1675–1688. doi:10.1177/1475921720927488

Zhao, W., Jiang, Y., Liu, Y., & Shu, J. (2022). Automated recognition and measurement based on three-dimensional point clouds to connect precast concrete components. *Automation in Construction*, *133*, 104000. doi:10.1016/j.autcon.2021.104000

Zhao, W., Liu, Y., Zhang, J., Shao, Y., & Shu, J. (2022). Automatic pixel-level crack detection and evaluation of concrete structures using deep learning. *Structural Control and Health Monitoring*, *e2981*(8). doi:10.1002tc.2981

Zhao, Y., Wang, N., Liu, Z., & Mu, E. (2022). Construction Theory for a Building Intelligent Operation and Maintenance System Based on Digital Twins and Machine Learning. *Buildings*, *12*(2), 87. doi:10.3390/buildings12020087

Zheng, H. D., Fan, J., & Long, X. H. (2017). Analysis of the seismic collapse of a high-rise power transmission tower structure. *Journal of Constructional Steel Research*, *134*, 180–193. doi:10.1016/j.jcsr.2017.03.005

Zhou, K., Wu, Z. Y., Yi, X. H., Zhu, D. P., Narayan, R., & Zhao, J. (2017). Generic Framework of Sensor Placement Optimization for Structural Health Modeling. *Journal of Computing in Civil Engineering*, *31*(4), 04017018. doi:10.1061/(ASCE)CP.1943-5487.0000662

Zhou, Q., Zhou, H., Zhou, Q., Yang, F., & Luo, L. (2014). Structure damage detection based on random forest recursive feature elimination. *Mechanical Systems and Signal Processing*, *46*(1), 82–90. doi:10.1016/j.ymssp.2013.12.013

Zhu, H.-J., Wang, L.-M., Zhong, S., Li, Y., & Sheng, V. S. (2022). A Hybrid Deep Network Framework for Android Malware Detection. *IEEE Transactions on Knowledge and Data Engineering*, *34*(12), 5558–5570. doi:10.1109/TKDE.2021.3067658

Zhu, M., Li, Y., & Wang, Y. (2018). Design and experiment verification of a novel analysis framework for recognition of driver injury patterns: From a multi-class classification perspective. *Accident; Analysis and Prevention*, *120*, 152–164. doi:10.1016/j.aap.2018.08.011 PMID:30138770

Zhu, R., Hu, X., Hou, J., & Li, X. (2021). Application of machine learning techniques for predicting the consequences of construction accidents in China. *Process Safety and Environmental Protection*, *145*(3), 293–302. doi:10.1016/j.psep.2020.08.006

Zimina, D., Ballard, G., & Pasquire, C. (2012). Target value design: Using collaboration and a lean approach to reduce construction cost. *Construction Management and Economics*, *30*(5), 383–398. doi:10.1080/01446193.2012.676658

Zollinger, W. R. III, Sutton, D. L., Montler, G., & Seifried, M. M. (2010). *BIM: Sharing project data reduces conflict, AACE International Transactions*. BIM.

# About the Contributors

**Vagelis Plevris** is an Associate Professor at the Department of Civil & Architectural Engineering of Qatar University in Doha, Qatar. He serves as Chief Editor for "Computational Methods in Structural Engineering", a section of the journal Frontiers in Built Environment, by Frontiers in Switzerland. He holds a Diploma in Civil Engineering (5-year studies) from the National Technical University of Athens (NTUA) with specialization in Structural Engineering. He also holds an MSc from NTUA on "Structural Design and Analysis of Structures", a Master in Business Administration (MBA) from Athens University of Economics and Business (AUEB) and a PhD from the School of Civil Engineering of NTUA. The title of his Doctoral Thesis was "Innovative Computational Techniques for the Optimum Structural Design Considering Uncertainties". His research interests include (i) Finite Element Method (FEM); (ii) Static and Dynamic Analysis of Structures with FEM; (iii) Earthquake Engineering; (iv) Optimum Design of Structures; (v) Reliability and Probabilistic Analysis of Structures; and (vi) Neural Networks and their Applications in Engineering. His published research work includes more than 100 contributions in journal papers, peer reviewed conference papers and book chapters.

**Afaq Ahmad** is a highly ambitious, innovative, and hardworking professional engineer with over 13 years of combined experience in design and research. Currently, he is working as Associate Professor at the University of Engineering & Technology Taxila. Founder of Scotant Engineer Consultant and member of Opti-Structure International. Currently, his research group is working on structural health monitoring through CNN and Image processing, prediction of reinforced concrete and composite material through Artificial Neural Networks. His published research work contains 45 peer-reviewed journal papers and 13 papers in international and national conferences, with a cumulative impact factor of 69+ and a ResearchGate score of 24. His research group is working with international research groups (Prof. Vagelis Plevris, Qatar University, Prof. Mohamed Elchalakani, University of Western Australia, Prof. Dr. Farid Abed, American University of Sharjah and Dr. Omar Alajarmeh, University of Southern Queensland).

**Nikos Lagaros** is the Dean of the School of Civil Engineering at NTUA and Professor of Structural Optimization at the School of Civil Engineering. Prior moving to NTUA, he was Assistant Professor at the Civil Engineering Department of the University of Thessaly, Greece. In the past he also served as Visiting Professor at Department of Biological Engineering, Laboratory for Computational Biology & Biophysics, Massachusetts Institute of Technology, Boston, USA and the Department of Mechanical Engineering, Faculty of Engineering, McGill University, Montreal, Canada. Dr. Lagaros obtained his BEng and PhD, both in Civil Engineering, from NTUA and he teaches classes on Structural Analysis, Optimization and Computer Programming. Dr. Lagaros is an active member of the computational mechanics research community. Among others, his publication track record includes more than 150 peer reviewed journal paper, 10 books and 25 book chapters while he has presented his work in numerous international venues. His h-index is 45 according to Google Scholar, with more than 6500 citations on his work.

***

**Zeinep Achmet** is a Ph.D student at the Laboratory for Earthquake Engineering of NTUA. The subject of her research is the application of machine learning techniques to structural health monitoring systems against natural hazards. She received a diploma in Civil Engineering from NTUA (2017) and she holds an MSc degree in Analysis and Design of Earthquake Resistant Structures (2017-2019).

**Afzal Ahmed** has specialization in Water Resources & Irrigation Engineering. Ahmed is serving at the Civil Engineering Department, UET Taxila Pakistan, and is active in the areas of research of Experimental Hydraulics, Water Resources Management, Eco-Hydraulics, Sediment Transport and Environmantal Hydrology.

**Shiraz Ahmed** has 10 years' experience of teaching and research in the field of Civil Engineering. He is a graduate from UET Taxila with a bachelor's degree obtained in 2011. He has worked as Research Assistant at TITE UET Taxila from August 2011 to February 2013. He had a position of Lecturer at Department of Civil Engineering, UCET UOS for 3 years. In 2016, he was admitted in UHasselt Belgium for pursuing his PhD in Transportation engineering. In June 2020, he successfully defended his PhD thesis by publishing research work in A1 journals with Cumulative Impact Factor of 13.5. His main interests are in Transport planning, Mobility/Construction Management, Travel Behavior Prediction and Modelling. He is currently working as an Assistant Professor at GIK Institute of Engineering, Sciences and Technology, Swabi.

**Mahdi Azizi** published research papers in the fields of structural engineering, structural optimization, and structural vibration control, while his main purpose has been on developing metaheuristic algorithms for engineering applications. He teaches basic and advanced courses of civil engineering.

**Nikolaos Bakas** is a Senior Data Scientist at the National Infrastructures for Research and Technology, GRNET. He has been a Lecturer at Neapolis University Pafos and an Associate Research Scientist at the Computation-based Science and Technology Research Center, at The Cyprus Institute. His research interests comprise the fundamental mathematical modelling of machine learning algorithms, as well as applications in a variety of thematic areas. Apart from research, he has been a PI for multiple industrial applications of AI.

**Dario De Domenico** is Assistant Professor of Structural Engineering at the Department of Engineering, University of Messina, Italy, and was awarded the National Scientific Habilitation for the qualification of Associate Professor in 2022. His research activity has led to over 100 papers published in peer-reviewed ISI journals in the following areas: (i) earthquake engineering and seismic behavior of new and existing structures; (ii) performance-based design and optimization of passive vibration control devices (isolators, dampers, TMDs and inerter-based control systems) for wind- and earthquake-resistant structures;

(iii) limit-state behavior of reinforced concrete structures through experimental tests and development of analytical, numerical and code-oriented procedures; (iv) diagnostics, monitoring and safety assessment of existing bridges. He serves as Associate Editor of 6 referred journals, Editorial Board Member of over 10 referred journals in the area of Structural Engineering, and regular invited Reviewer of over 40 international journals with impact factor.

**Ghasan Doudak**, Ph.D, P.Eng. is a Professor of structural engineering at the University of Ottawa. His area of expertise includes multi-scale understanding of how complete structural systems function, encompassing issues like how complete buildings respond to effects of wind storms, ground shaking during earthquakes, or other actions like impacts and blasts. He is a member of the National Building Code Standing Committee on Structural Design, CSA O86 Technical Committee on Engineering Design in Wood, CSA s850 Technical Committee on Blast Resistant Buildings, and board member and treasurer of Canadian Association for Earthquake Engineering (CAEE). Dr. Doudak's research contribution, through collaboration with Canadian as well as international researchers, is in establishing solutions to meet specific performance requirements for mid-rise wood and hybrid buildings. Dr. Doudak has established strong affiliations to complementary research groups at national and international levels. National collaborations include projects with researchers from the University of New Brunswick, Concordia, and McGill. International collaborations include projects with the Technical University of Denmark and National Research Council of Italy (IVALSA). Prior to joining the Engineering Faculty, Dr. Doudak held the position of Manager, Wood and Structural Standards at the Canadian Wood Council, where he was actively involved in work related to the development of building Codes and standards, developing tools for designers, and communicating to key audiences including government agencies, building designers and researchers. Dr. Doudak received his Master of Science degree from the Technical University of Denmark (DTU). Following his graduation, he worked as a structural engineer where he designed commercial and residential buildings. In 2001, Dr. Doudak was awarded a fellowship from the Danish Research Agency for a 3-year research program in Civil Engineering at McGill University. His PhD research was aimed at determining the load paths in wood light frame buildings under various stages of construction using a holistic design approach. Based on his valuable contribution to the field of wood engineering, Dr. Doudak was awarded the Forest Products Laboratory Young Engineer Award, 2008. This award is intended to encourage young (40 years of age or younger) engineers, researchers, or scientists with "promising potential for significant contribution to the field".

**Michalis Fragiadakis** is the Director of the Laboratory for Earthquake Engineering and Associate Professor at the School of Civil Engineering of the National Technical University of Athens (NTUA), Greece. He is also Visiting Professor at EMGCU/IFSTTAR laboratory of Gustave Eiffel University. He received his PhD from NTUA and holds MSc degrees form NTUA and Imperial College, London. He has worked as adjunct lecturer at the University of Thessaly and as a Postdoctoral researcher at the University of Pavia and the University of Cyprus. His research activity is focused on structural engineering and on the seismic risk assessment of structures and lifelines. His interests also extend to computational methods in structural analysis and on the application of novel soft-computing methodologies for structural analysis and design. Dr Fragiadakis has been involved in many national and European research projects. He has been co-chairman of 6 international conferences (COMPDYN series) and has co- organised several mini-symposia and special sessions in international conferences. He has authored and co-authored more than 200 publications and his work is extensively disseminated.

**Abdul Razzaq Ghumman** works in the Department of Civil Engineering, College of Engineering, Qassim University, Saudi Arabia.

**Ghazanfarah Hafeez** is an Assistant Professor at Concordia University in Building Science and Environmental Engineering Department (BCEE). She obtained her Master of engineering sciences degree from Western University and her Ph.D. from the University of Ottawa. Her research focuses on Structural Health Monitoring (SHM) of Wood Buildings, Lateral Load Resisting Systems, Sustainable Design, Artificial Neural Networks and Optimization Algorithms in Timber Structures and Materials.

**Husnain Haider** is presently responsible for teaching and research in the areas of water resources and environmental engineering at Qassim University, Saudi Arabia. He completed his first PhD from the University of Engineering and Technology (UET) Lahore Pakistan on surface water quality management and second PhD from University of British Columbia (UBC) on infrastructure asset management. He worked on sustainability assessment of infrastructure and industrial projects during his Postdoctoral at UBC, Canada. Dr. Haider is author of more than 100 publications, including peer-reviewed journal articles, book chapters, and international conference papers. His research interests are: sustainable infrastructure, natural resource conservation, life-cycle thinking, environmental modeling and assessment, risk-based decision-making under uncertainty, and customer-focused infrastructure management.

**Fahad Iqbal** is a senior-year student of Civil Engineering at Ghulam Ishaq Khan Institute of Engineering Sciences and Technology, Pakistan. He has been honored with the prestigious KP Merit Scholarship and has consistently achieved a place on the Dean's Honor Roll, recognizing his outstanding academic performance. His research interests primarily revolve around Building Information Modeling, Artificial Intelligence, Virtual Reality, Construction Management, and Entrepreneurship.

**Mukund Nilakantan Janardhanan** received his Ph.D. degree in Manufacturing Engineering from Monash University, in 2015. From 2015 to 2018, he was a Postdoctoral Fellow with Aalborg University, Denmark. Since March 2018, he has been a Lecturer in Engineering Management with the University of Leicester, U.K. He has published over 50 articles in reputed journals and conferences. His research interests include manufacturing systems optimization, production planning and metaheuristic algorithms.

**Marina Marinelli** received her PhD degree in Construction Engineering and Management from the National Technical University of Athens (NTUA) in 2012. She is an Assistant Professor in Engineering Project Planning and Management at the School of Civil Engineering, NTUA, and an Honorary Lecturer with the School of Engineering, University of Leicester, U.K. She has published over 30 articles in reputed journals and conferences in various areas including machine learning and big data applications in construction, cost management and sustainability strategies.

**George Markou** received his Ph.D. degree at the National Technical University of Athens, Greece. His specialty involves numerous scientific fields related to modeling of structures through the use of the finite element method. His recent research work addresses the modeling of composite structures under seismic loading conditions, soil-structure interaction problems, assessment of retrofitted structures, smart monitoring of structures, and the use of machine learning algorithms for solving engineering problems. Prof Markou is also using high-performance computing in the development of predictive models through machine learning algorithms. He is currently an Associate Professor at the University of Pretoria in the Department of Civil Engineering.

**Giorgio Monti** has graduated in Civil Engineering at the University of Roma La Sapienza, then he has obtained a Master of Science at the University of California at Berkeley, and a PhD at the University of Roma La Sapienza in Structural Engineering. Since 2001 he has been Full Professor at the Sapienza University of Roma, where he is Director of the Master Course on "Advanced Structural Design according to Eurocodes". He is Expert Member of the Ministry of Infrastructures and Transportations of Italy and serves in several EU pre-normative Committees. Since 2005 he has been National Coordinator of a Civil Protection program in Italy on "Assessment and Risk Reduction of Reinforced Concrete Buildings". His scientific activity addresses the topics of: modelling and analysis of reinforced concrete structures under seismic excitation, risk assessment of existing structures in reinforced concrete and masonry, reliability analysis of structures and infrastructures in seismic zones and code calibration, use of sustainable materials and dry prefabrication in constructions, strategies for seismic protection of buildings, including base isolation, strategies for the preservation of cultural heritage, advanced solutions and techniques for structural strengthening. In these fields, he has so far produced more than 400 publications, of which more than 110 in peer-reviewed international journals, and 6 books.

**Fatemeh Mostofi** is a Ph.D. student at Karadeniz Technical University. Her research areas include the application of machine learning approaches in construction management, specifically in risk, quality, productivity, and cost management. Currently, she is focused on graph neural network applications in construction management. Fatemeh holds a bachelor's degree in civil engineering from Heriot-Watt University, where she graduated with first-class honors in 2013. Fatemeh also worked as a teaching assistant and part-time lecturer at Heriot-Watt University from 2014 to 2016.

**Mehdi Nikoo** is pursuing his Ph.D. at the Department of Building, Civil and Environmental Engineering, Concordia University in Montréal, Canada. He holds a Bachelor of Civil Engineering from Islamic Azad University, Ahvaz Branch and an MSc from Islamic Azad University, Dezful Branch in Iran. His research focuses on Wood-Based Sustainable Design implementing Artificial Neural Networks and Optimization Algorithms.

**Muhammad Noman** has a Doctorate in Structural Engineering with major interests in Machine Learning, Structural Fire Engineering, BIM, and concrete Structures.

**Ghufran Ahmed Pasha** is currently working as an Assistant Professor in the Civil Engineering Department, University of Engineering & Technology (UET) Taxila. He obtained his Ph.D. from the Saitama University, Japan specializing in Hydraulic Engineering. His main research interests are Flow vegetation interaction, Eco System based Disaster Risk reduction, Sediment Transport, Hydraulic Scour, and Flood Risk Management. Overall, he has more than 12 years of rich teaching, research, and professional industry experience with more than 50 International and National publications. In addition, he is a member of the editorial board of more than 10 International journals. He is also working in an International Researcher Collaboration with Saitama University, Japan.

**Giuseppe Quaranta** is Associate Professor in Structural Engineering at Sapienza University of Rome (Italy) since 2019, and holds the National Scientific Habilitation for the position of Full Professor in Structural Engineering since 2020. His main research fields are the following: i) structural monitoring of civil structures and infrastructures (including design of traditional and smart sensor networks, dynamic identification as well as structural diagnostic); ii) optimum design of seismic protection devices (including seismic isolators, tuned mass dampers, and viscous dampers); iii) analysis and design of concrete structures, concrete-filled steel tubes and truss-reinforced composite steel-concrete beams.

**Jiangpeng Shu** is a 'Hundred Talent' Professor at Zhejiang University. He obtains Ph.D. from Chalmers University of Technology in Structural Engineering. Before, joint Zhejiang University, he has been working as an Assistant Professor at University of Southern Denmark and Visiting Professor at Stanford University. His research focus on intelligent inspection, monitoring and assessment of existing structures, especially with artificial intelligence and digital twin. He has made outstanding achievements in scientific research, with more than 60 high-quality refereed journal articles. He has earned international visibility as peer reviewer and member of the editorial boards of scientific journals. He is involved in several national and international scientific societies (including ACI/ASCE/fib) and he has given significant contributions to many technical committees, which brought him to establish fruitful collaborations with renowned international institutes.

**Siamak Talatahari** is a researcher in data science (DS), machine learning (ML), and artificial intelligence (AI) and their application to engineering. Thanks to introducing, improving, hybridizing, and applying DS/AI/ML methods, he has published over 200 refereed international journal articles, three edited books in Elsevier, and several chapters in international books. He is recognized as a "Top One Percent Scientist of the World" in the field of "Engineering" and "Computer Sciences" for several years. He served as the lead or guest editor of some special issues in different journals.

**Vedat Toğan** received the BSc and MSc degrees in 2000 and 2004 from the Department of Civil Engineering at Karadeniz Technical University, Turkey respectively. He received his Ph.D. degree from the same university in 2009. He is working as a Professor at the Civil Engineering Department of Karadeniz Technical University. His research interests are focused on meta-heuristic algorithms, optimization, construction management, and the application of artificial intelligence in construction management. He has published extensively in these areas, including numerous journal articles, conference papers, and book chapters.

**Muhammad Uzzaman** has been conducting research on Water and wastewater treatment and recycling, Water quality management, Water resources management, Ground water pollution and remediation, Solid waste management. Also working on several projects on low cost wastewater recycling technique.

**Ashley van der Westhuizen** is currently enrolled as a Masters student at the University of Pretoria specializing in Structural Engineering. The focus of her dissertation is on seismic design specifically looking at the fundamental period of steel structures with the use of artificial intelligence and machine learning algorithms. She completed her undergraduate Civil Engineering degree at the University of Pretoria. Following her undergraduate degree, she completed an Honours degree at the University of Pretoria where she expanded her understanding of structures and the use of machine learning to determine the dynamic behaviour of steel structures. Her work was published and presented at the 14th International Conference on Agents and Artificial Intelligence. While undertaking her Masters she assisted in work involving the use of machine learning in the evaluation of shear strength of existing columns.

**Qingcong Zeng** is a PhD candidate in Zhejiang University.

# Index

www.ingramcontent.com/pod-product-compliance
Lightning Source LLC
LaVergne TN
LVHW060405300926
849162LV00006B/1349